Fundamentals of TOXICOLOGY

Fundamentals of TOXICOLOGY

Dr K Pandey

Professor, Head of the Department of Zoology
Dean, Faculty of Science, DDU—Gorakhpur University
Gorakhpur [UP], India

Dr J P Shukla

MSc [Zoology], PhD, FAEB, FASI

Reader, PG Department of Zoology
S Kisan PG College, Basti 272001 [UP], India

Dr S P Trivedi

MSc [Zoology], PhD, CES [IGNOU], FAScAW, FZS [Cal], FAEB

Reader, Environmental Toxicology Laboratory
Department of Zoology, University of Lucknow
Lucknow 226007 [UP], India

New Central Book Agency (P) Ltd

LONDON

HYDERABAD ERNAKULAM BHUBANESWAR

NEW DELHI KOLKATA PUNE GUWAHATI

NCBA

REGD OFFICE
8/1 Chintamoni Das Lane, Kolkata 700 009, India
email: ncbapvtltd@eth.net

OVERSEAS
NCBA (UK) Ltd, 149 Park Avenue North, Northampton, NN3 2HY, UK
email: ncbauk@yahoo.com

EXPORT
NCBA Exports Pvt. Ltd, 212 Shahpur Jat, New Delhi 110 049
email: ncbaexports@ncbapvtltd.com

BRANCHES

2/27 Ansari Road, 1st Floor
Daryaganj, **New Delhi** 110 002
email: ncbadel@ncbapvtltd.com

Shop Nos. 3 & 4, Vinayak Towers
681/B Budhwar Peth, Appa Balwant Chowk
Pune 411 002
email: ncbapune@ncbapvtltd.com

Shop No. 15, Bharati Towers, Ground Floor
Forest Park
Bhubaneswar 751 001
email: ncbabub@ncbapvtltd.com

House No. 3-1-315, 1st Floor
Nimboliadda, Kachiguda, **Hyderabad** 500 027
email: ncbahydb@ncbapvtltd.com

GSS Shopping Complex, 1st Floor
Opposite College Post Office, Convent Road
Ernakulam 682 035
0484-4021054 / 53

Radhamayee Complex [Second Floor]
Opp. Axis Bank, Panbazar, Jaswanta Road
Guwahati 781 001
email: ncbaguwahati@ncbapvtltd.com

FUNDAMENTALS OF TOXICOLOGY ● Dr K Pandey, Dr J P Shukla and Dr S P Trivedi

First Published: February 2005
Reprinted: January 2006, 2009, 2011, July 2012, August 2013,
June 2019

PUBLISHER AND TYPESETTER
New Central Book Agency (P) Ltd
8/1 Chintamoni Das Lane, Kolkata 700 009

PRINTER
New Central Book Agency (P) Ltd
Web-Offset Division, Dhulagarh, Sankrail, Howrah

GRAPHICS: Sukumar Patra

PROOF CHECKING: Subas Maitra

PROJECT TEAM
Sukumar Patra, Prabhat Jas, Soumen Paul, Kushaldev Bhattacharya, Ananta Bhowmik and Arun Kabri

ISBN: 81 7381 476 7
978 81 7381 476 1

Price ₹ 485.00

DEDICATED

TO

THE HAPLESS VICTIMS

OF

TOXICOLOGICAL IGNORANCES

Preface

Toxicology, as it is commonly viewed, is not a newly emerging field of scienҫe; yet it drew attention for its systematic study and teaching at U. G. and P. G. level only a few decades ago as a result of the environmental abuses and contamination and destruction of valuable and fast-depleting natural resources.

When the authors undertook the preparation of this monograph what provided the principal impetus was a complete absence of a single elaborate conceptual presentation of information on Toxicology. It remained diffused and confined to research papers, scientific reports and, maximally, to a few petty books dealing with superficial information regarding the elementary principles of toxicology. Additionally, our interaction with the problems of fish toxicology and ecotoxicology for more than 35 years provided considerable factual strength and conceptual background to our united endeavour.

This book is intended for a wide audience. Yet there were two particular readerships in our mind — the U. G. and P. G. students in the universities and colleges, together with those interested in the study of toxicology but not undertaking formal academic studies. And, secondly, those associated with the research, development, and ultimate use of various chemicals in multifarious forms. The book is exhaustive, yet the students of all Indian universities and colleges will find an integrated and comprehensive account on various aspects of toxicology along with useful data.

Our readership-ambition is high. It is further hoped that besides the students, teachers, researchers and administrators, our far-seeing political leaders and industrialists would also find valuable ready references herein on toxicological problems which would be invaluable to them in their macro-decisions on large financial outlayings for social or business purposes.

The detailed data and inferences presented in the book have come from a large variety of sources and those not our own are gratefully acknowledged. We still feel that there may be some deficiencies in this book as it is difficult to keep pace with the fast rate of research in this magnificently progressive field.

One of us (Shukla) expresses his heartfelt obligation towards Dr. B. S. Chandel, Dr. A. K. Karnataka, Dr. R. C. Gupta, Dr. M. Mishra, Dr. I. M. Tripathi, Dr. V. K. Mahendroo, Dr. A. K. Srivastava, Dr. A. Capoor, Dr. A. K. Tripathi, (Dr.) Miss Nirupama Tewary, Dr. P. L. Shukla (Principal), Sri R. P. Goswami, Sri H. P. Tewary, Sri S. K. Shukla (IPS) and Sri D. C. Mishra (IPS) for their sincere affection and encouragement. He extends his deep affection to his wife, sister Suman, nephew Suryansh, and sons Anjaney and Maruti, for their cooperations in many invisible, yet highly valuable, ways.

The authors are greatly indebted to Shri Amitabha Sen, Director, New Central Book Agency (P) Ltd. and its production personnel and offer their heartfelt thanks for their sincere endeavour in the production of the book.

Dr. Kamleshwar Pandey

Dr. J. P. Shukla

Dr. S. P. Trivedi

Contents

1

Concept and Scope of Toxicology

SYNOPSIS

Introduction, General account, Definition of toxicology and ecotoxicology, History of toxicology, Disciplines of toxicology (Environmental, Economic, Forensic, Clinical, Mechanistic, Biochemical, Regulatory, Industrial, Genetic, Preventive, Behavioural, Systemic, Comparative toxicology, Toxinology and Wildlife Toxicology, Ethnic or Geographical Toxicology), Survey and classification of toxicants, Definition of toxicity and scope and importance of toxicology.

Introduction

Of all the branches of life science, the study of toxicology has generated greater excitement and became a subject of general interest in the present century. Today toxicology is not merely a study of poisons — as many conceive falsely, based on the definitions given in dictionaries. It clearly indicates that toxicology has left the lexicographers way behind. Toxicology, therefore, may be defined as the study of interaction of materials (drugs, chemicals, drinks, foods, polymers, pesticides, etc.) with a biological system and depiction of adverse responses. The word toxicology is derived from two Greek words : *toxikon* = poison + *logos* = discourse or study. However, in the present scenario, the word 'poison' has lost its original relevance. The 'Poison Act' was legislated in 1918 and since then much water has flown over it.

Centuries ago, Paracelsus (1493-1541) — the *father of toxicology* — categorically said that all substances are poisons. It is the right dose that differentiates a poison from a remedy. An old adage emphasises that even nectar (*amrit* or that which stops death) is also a poison, if consumed in excess. It has been proven more than correct in recent times. Actually, response is related to the dose as the same chemical that acts effectively like a drug — in low doses — could be a poison in higher doses. This statement has evolved a *dose-response relationship* in toxicology. Arsenic and mercury were remarkable remedies for many illness in the early days of allopathy; at present no country will ever permit a new drug containing arsenic or mercury to be introduced without in-depth toxicological evaluation.

Apart from the normal responses, many xenobiotics could produce adverse reactions on widespread use. An interesting example relates to hexachlorophene, widely used since 1945 as mild bactericide in talcum powders and soaps. In 1970s, hundreds of children were severely affected and over 30 expired after applying a talcum powder containing 5% hexachlorophene in France.

In fact, the chemical had been cleared after essential sub-acute study by administration to rats for 30 days. Further, on a re-evaluation of the toxicology of the drug, it was found to react with white matter in the brain and spinal cord; and interfered with the mitochondrial metabolism leading to myelin (covering of nerves) degeneration producing spasms, convulsions, coma and consequent death. The chemical was subsequently banned for cosmetic use in various countries of the world. The role of the toxicologist may well be judged by this episode and we can thus relate toxicology as the science that defines *the safety threshold of chemicals.* However, in the broad sense, toxicology may be referred to as the basic science of poisons. A poison or toxicant may be any chemical substance which, on entering the body of an animal in insignificant amounts, cause malfunctioning of vital activities and may lead to impaired health. Practically, there is no chemical which may be completely safe under all the exposure conditions. Contrarily, it is also true that there is no chemical which cannot be used safely by limiting the dose or the exposure.

Poison may be either a natural product of animal or plant origin or it may be a synthetic one. The poisons that are synthesized within the organism itself are referred to as *endogenous poisons* or *toxins* and those which enter an organism from the outside (environment) are referred to as *exogenous poisons* or *toxins.* Exogenous chemicals are also referred to as xenobiotics.

The basic science of toxicology encompasses biochemistry, pharmacology, pathology, and, to a great measure, chemistry. Statistical probability being an important ingredient of prediction, one cannot ignore the role of biostatistics in toxicology.

Toxicology is essentially a predictive or speculative discipline of science, as humans cannot be used in experimenting with unknown chemicals. Almost all toxicological studies start with rodents, that is, rats and mice, which are referred to as target or test animals. This may also be carried on to larger animals like dogs, monkeys and rabbits, if experiment demands. The results of the animal studies have to be extrapolated to humans and actually this is the most difficult exercise, considering the wide difference in the organ systems of human and test animals.

In a nutshell, toxicology provides a sound basis for formulating measures to protect the health of workers against occupational toxicants in factories, mines and other occupational environments in particular and other population in general. It is also valuable in the protection of public health against hazards associated with toxic substances in food, air and water. Undoubtedly, toxicology has played — and will continue to play — a significant role for human health and welfare in the world.

The term Environmental Toxicology is relatively new and its exact usage is not established yet fully. The term ought to be reserved for all applications of toxicology to organisms other than man and his domestic animals. Defined in this way, Environmental Toxicology is the study of toxins in the restricted sense (poisons produced by living organisms including microbes).

Ecological Toxicology is the study of all toxicants produced by living organisms and their effects on overall ecological relationship. This discipline is much broader than toxicology.

Wildlife Toxicology involves study of the effects of toxicants of any origin on wildlife, in exactly the same sense as the veterinary toxicology is related to domestic animals.

General Account

Toxicology is becoming an increasingly important subject of discussion of the modern society because of the adverse effects of the growing use of chemicals and high-tech radioactive appliances. It is a multidisciplinary subject with a vast and diffused literature.

Most of the living beings are exposed to a great variety of natural as well as man-made chemical substances. Under certain conditions, such exposures cause health hazards, ranging in severity from death to subtle biological alterations. Society's ever-increasing desire to identify and prevent these effects has resulted in the dramatic evolution of toxicology from only a study of poisons to the present-day complex science.

Having realized the importance of toxicology, WHO (World Health Organization) in 1982 organized a Toxicology Training Course (TTC) in China as a part of the on-going China-WHO collaborative programme on medical sciences. At the beginning of the 21st century 'Toxicology' has emerged as an important branch of pharmacology. It provides a sound basis for formulating measures to protect the health of factory workers, farmers, miners and others exposed to different toxicants. It is also valuable in the protection of public health against hazards associated with toxic substances in food, air and water. In fact, toxicology has played — and will continue to play — a significant role in the health and welfare of the world.

Definition of Toxicology

Toxicology is a Greek word—*toxin* or *toxicum* or *toxicon* = poison + *logus* = discourse or knowledge, i.e., "the science of poison". In other words "the branch of pharmacology which deals with the various aspects of poisons and poisoning is known as toxicology". Poison may be defined as a substance which, even in small dose, produces adverse effect in the metabolism of an organism and, consequently, may cause death.

Presently, toxicology is not restricted to the study of poisons only. Rather, it is the study of the nature and mechanism of toxic effects of substances on living organisms and other biological systems. Toxicology also deals with the quantitative assessment of the severity and frequency of these effects in relation to the exposure of the organisms. Along with other sciences, toxicology contributes to the development of safer chemicals used as drugs, food additives, pesticides etc.

Du Bois and Geiling (1959) defined toxicology as a branch of medical science that deals with the nature, properties, effects and the detection of poisons. It is, therefore, the science of poisons. In this definition are included studies on the metabolism and excretion of poisons, on the action of poisons, and on the treatment of poisoning, as well as systematic chemical and physical analyses and diagnoses.

The first research institute of toxicology in India is ITRC (Industrial Toxicology Research Centre) established in 1969 in Lucknow (U.P.)

Ecotoxicology

The term ecotoxicology was described by Truhaut in 1974. It is the study of the adverse effects of the environmental pollutants on an ecosystem — studied under natural conditions. At present, ecotoxicology is considered in a wider sense to include — in addition to toxic effects of chemicals (toxicants) on organisms — also their processes to environmental concern.

History of Toxicology

Toxicology predates human beings and in many ways is a part of their history, which began as an interest in poisons — from plants and animals.

The Early Era : Hippocrates was the first to introduce primitive principles of toxicology around 400 BC. He gave quite a few tips to prevent the absorption of toxins.

The earliest man was well aware of toxic effects of a number of substances, such as snake venom, the poisonous plants, hemlock and aconite and the toxic mineral substances such as arsenic (As), lead (Pb), antimony (Sb) etc. Some of these were actually used intentionally for their toxic effects to commit homicide and suicide.

For centuries homicides (toxic substances) were not uncommon in Europe. To protect poisoning, there were continual efforts directed towards the discovery and development of preventive and antidotal measures. However, a more critical evaluation of these measures was only taken up by Maimonides (1135-1204) with his famous book entitled "*Poisons and their Antidotes*" published in 1198.

The Middle Ages (1300-1600) : More significant contributions to the evaluation of toxicology were made in the 16th century and Paracelsus (1493-1541) (*father of toxicology*) stated : 'No substance is a poison by itself', 'It is the dose that makes a substance poison' and, 'The right dose differentiates a poison and a remedy'. These statements laid the foundation of the concept of 'dose-response-relationship' and 'therapeutic index' developed later. Further, Bergsucht (1533-1534) described in his book the clinical manifestations of chronic arsenic and mercury poisoning as well as miner's disease. He may, therefore, be considered the *forefather of occupational toxicology*.

Later on Orfila, M.J.B. (1787-1853), a Spanish physician, also known as the *father of modern toxicology*, wrote an important treatise (1814-1815) describing a systematic correlation between the chemical and biological information on certain poisons. He also devised methods for chemical analysis and detecting poisons for legal proof of lethal intoxication. The introduction of this approach developed a new discipline of modern toxicology, namely, *forensic toxicology*. Orfila became the Professor of Jurisprudence at the University of Paris and pioneered Forensic Toxicology.

Another renowned toxicologist of this era was Francois Mogendie (1783-1855), a physiologist, who studied the mechanism of action of emetine and strychnine. Claude Bernard (1813-1878) used poisons as tools in analyzing physiological aspects of organic systems.

Initially there was a tendency to treat toxicology as a part of pharmacology; but, with all the developments and catastrophies, toxicology could not catch public attention until Alexandre Dumas devoted a separate chapter to Toxicology in his world-famous novel 'Count of Monte Cristo'.

Recent Developments in Toxicology

With increasing population, the modern society demands improvement of the health and living conditions, including nutrition, clothing, dwelling and transportation. In order to fulfil these growing needs, a large variety of chemicals, many of them in large quantities, must be manufactured and used. It may be emphasized that tens of thousands of various chemicals are in commercial production in industrialized countries. In one way or the other, these chemical compounds come in contact with various segments of the population, e.g., people engaged in their manufacture, handling, use, consumption, or even misuse. Furthermore, people may also be exposed to the more persistent chemicals via various environmental media as well.

It is, therefore, essential to assess the toxicity of a vast number of chemicals to make them more manageable. As an attempt to fulfil this need, criteria have been proposed and adopted for the selection of chemicals to be tested according to their priority.

In modern disciplines a variety of toxicological testings have been used viz., carcinogenicity, mutagenicity, immunotoxicity etc.

Advances made in biochemical and chemobiokinetic studies — as well as those in the genetic toxicology, immunotoxicology and morphological studies at subcellular level — have contributed to a better understanding of the nature, site and mechanism of action of toxicants. Recent studies have shown that the responses to toxicants are better correlated with the effective dose, i.e., the concentration of the toxicant at the site of action, rather than the administered dose. Furthermore, where the effect results mainly or entirely from an active metabolite, the concentration of the metabolite is more important rather than that of the parent chemical.

To translate scientific knowledge into basic tools for health profession, the toxicologists are more involved in the determination of safe exposure limits or assessment of the risks. The former includes the acceptable daily intake (WHO, 1962) and the threshold limit values (Federal Register, 1971), whereas, the latter is applied to such substances which are believed to have no threshold or their threshold cannot be determined.

These determinations include comprehensive studies of the toxic properties, demonstration of dosages that produce no remarkable adverse effects, establishment of dose-effect and dose-response relationship, chemobiokinetic and biotransformational studies.

Disciplines of Toxicology

Modern toxicology is regarded as a multidisciplinary science. It is subdivided into four principal disciplines — Environmental, Economic, Clinical, and Forensic toxicology. However, Loomis (1968) refers only three basic subdivisions of toxicology, i.e., Environmental, Economic, and Forensic toxicology. He refers clinical and medicolegal aspects of toxicology under the head 'Forensic Toxicology'. Further, Klaasen (1986) incorporated three more categories of toxicology — Descriptive, Mechanistic and Regulatory toxicology. Few other branches of toxicology — which could not be covered by the aforenamed authors — are Analytical toxicology, Behavioural toxicology, Aquatic toxicology, Industrial toxicology, Comparative toxicology, Speculative toxicology, Ethnic or Geographical toxicology, Immunotoxicology, Wildlife toxicology and Genetic toxicology.

The principal divisions and branches of toxicology are :

1. Environmental Toxicology

The environmental degradation due to the presence of various pollutants is referred as environmental toxicology. Thus environmental toxicology is the study of causes, conditions and effects of such chemicals on the living system and biosphere.

Environmental toxicology incorporates pollution studies, residue analysis, industrial hygiene and occupational health. In general, pollution can be defined as 'matter in the wrong place' or any substance released into the environment which degrades it. According to Edward (1972) 'Pollution is the release of substances or energy into the environment by mankind in quantities that damage either his health or resources'. The individual's health status vis-a-vis his work in environment — termed 'occupational health' — and the efforts to keep the work place clean — called 'industrial hygiene' — go hand in hand. Occupational health can be defined as the science and art devoted to the recognition, evaluation and control of those environmental factors or stresses arising in or from the work place which may cause sickness, impaired health and well-being or significant discomfort and inefficiency among workers. In simple words, the health status of individuals in relation to the environment of the working place is referred to as *occupational health.*

If the same definition is extended to include the environment outside the work place, and illness or death among the citizens of a community, then it is called environmental health. Thus, occupational health relates to work place pollution while environmental health relates to environmental pollutions. However, the methodology for both studies is the same. The stresses at work place can be caused by noise, heat, chemical substances, gases etc., while, in the general environment, it can be the pollution of air, water and food.

Therefore, environmental toxicology may be regarded as a multi-disciplinary science involving chemistry, life sciences, agriculture, medical science, public health, sanitary engineering etc. In simpler terms, it is the science of chemical phenomenon in the environment. In this way, environmental toxicology may be treated as a branch of Environmental Chemistry which deals with the nature and action of toxicants on the living organisms.

In view of their nature and mode of action, the environmental components may be grouped into three terms :

A. **Environmental Pollutant** : A substance present in nature, in greater than natural abundance due to human activity, which ultimately has a detrimental effect on the environment and, therefrom, on living organisms and human beings. Examples — Arsenic, Lead, Mercury, Cadmium, SO_2, CO_2, CO etc.

Definition of Pollutant

1. A pollutant is anything including a chemical, geochemical substance, or a biological product that deteriorates the natural environment.
2. According to Holister and Porteus, a pollutant is a substance which adversely changes the natural environment by changing the growth rate of species, interferes with the food chain, is deleterious, or interferes with health, comfort, amenities, or property values of people.
3. According to Odum : "Pollutants are the residues of the things we make use, and throw away".
4. According to Holdgate (1971), a pollutant may be defined as 'something that is present in the wrong place, at the wrong time and in the wrong quantity'.
5. In the Environmental Protection Act, 1986, of India, 'Environmental Pollutant' means any solid, liquid or gaseous component present in such concentration as may be, or tend to be, injurious to environment, and 'Environmental Pollution' means the presence of any environmental pollutant in the environment.

B. **Environmental Contaminant**: A component or material which does not occur in nature, but is introduced by human activity into the environment, affecting its composition. A contaminant is classified as pollutant when it exerts a detrimental effect, e.g.

1. MIC (Methyl Isocyanide) tragedy of Bhopal (M.P.) in 1984 Dec. 3rd;
2. Chlorine gas escaped from a derailed railway tanker near Youngstown, Florida, in 1978 and killed eight persons driving on a nearby road.

   Both the gases, i.e., MIC and Cl_2, do not occur in the atmosphere, so both are, undoubtedly, contaminants. Because of their deleterious effects, both are also pollutants.

C. **Environmental Toxicant** : Any substance in the environment which exerts toxic effect over the organisms is referred to as environmental toxicant and/or their metabolites adversely affecting certain structures of the organism. The structure may be nonspecific such as any tissue in direct contact with corrosive chemicals. More often, it is specific, involving a particular subcellular structure. A variety of structures, including the 'receptors', may be affected.

The term environmental toxicology is relatively new and its appropriate use is not yet established. The term ought to be reserved for all applications of toxicology to organisms other than man and his domestic animals.

To sum up, environmental toxicology is the qualitative and quantitative study of injurious effects of chemical and physical agents.

2. Forensic Toxicology

Detection of causes of mortality due to any toxicant (poison) via medical examination is an allied aspect of toxicology and referred as *Forensic Toxicology*. Therefore, forensic toxicology deals with the poison, its type, symptoms and possible treatment. In precise form, forensic toxicology is concerned with the medicolegal aspects of deleterious effects of chemicals on human and other animals.

Definition of Poison

It is difficult to find a short but comprehensive definition of poison. Broadly speaking, a poison may be defined as a substance which, when absorbed into the living organisms, or when acting locally on its tissues, injures the health or destroys life.

Classification of Poisons : For practical purposes, in toxicology, 'Poisons' may be divided into four classes :

1. Corrosives
2. Irritants
3. Neurotics (Nerve Poisons)
4. Cardiac (heart) Poisons.

1. **Corrosives** : Corrosive poisons destroy by chemical actions all tissues with which they come in contact. Strong minerals, inorganic acids such as H_2SO_4, HNO_3, HCl, organic acids such as acetic acid (CH_3COOH), oxalic acid and also caustic alkalis — belong to the class of corrosive poisons.

Symptoms : The symptoms commence at once. Pain is immediately felt in the parts which have come in contact with the poison. When swallowed, there is a violent burning pain from the mouth to the alimentary tract, followed by immediate and uncontrolled vomiting. Symptoms of collapse quickly set in, the skin becomes cold and covered with sweat and the pulse becomes feeble. There is difficulty in breathing and intense thirst. Swallowing becomes difficult, rather, impossible. The mind remains clear till the last.

Treatment : There are three ways of counteracting the action of the corrosives :

(a) Neutralization through an antidote. In case of acids, neutralization is done by giving mild alkalis and chalk, egg shells; while, in case of alkalis, by dilute acids viz., vinegar, lime juice etc.

(b) Dilution of the poison by giving plenty of water to drink.

(c) Administration of white of egg, butter, rice water, barley water etc. which provides soothing effect.

2. **Irritants** : Irritant poisons produce inflammation of the part with which they come in contact, but do not, as a rule, destroy the tissues. Vegetable acids, juices of nuts, mader, lalchita etc. are local irritants.

Many other important poisons of this class, e.g., arsenic, mercury, antimony, besides causing local inflammation, produce certain specific symptoms when absorbed into the system.

Symptoms : When swallowed, the symptoms do not usually appear immediately, but there is usually an interval of about half an hour or so between the swallowing of the poison and the commencement of the symptoms. Pain is felt from the mouth to the stomach. Violent vomiting and purging set in. Thirst is always intense. Gradually, symptoms of collapse develop.

Treatment : As in case of corrosives, acids should be neutralised by alkalis, and vice versa. Demulcent drinks should be freely given. Specific treatment only by a physician.

3. **Neurotics** : Neurotic poisons — after absorption into the system — act on the nerve centres, brain and spinal cord. They may be divided into four groups, each exhibiting specific symptoms :

A. Narcotics (Opium — Morphin).
B. Deliriants (Datura - thorn-apple, its fruit or flower. Belladonna).
C. Excitants (Alcohol, Hemp, Cocaine).
D. Spasmodics (Strychin, Nux vomica).

Symptoms of Narcotics : Narcotics produce drowsiness, passing on to deep sleep and coma.

Symptoms of Deliriants : Deliriants produce active delirium, sudden fits of laughter, catching at imaginary objects, staggering and becoming insensate etc.

Symptoms of Excitants : Excitants produce excitement, reduced control over movements, deep sleep passing on to coma.

Symptoms of Spasmodics : Spasmodics produce muscular spasms and fixation of chest. Death results from suffocation or exhaustion.

Treatment : Neurotically affected person should be kept under close examination and soon be treated by a competent physician. However, an attempt to induce vomiting should always be made as quickly as possible.

4. **Cardiac Poisons** : Cardiac poisons, prussic acid, aconite, tobacco etc., attack the heart and circulation.

Symptoms of Cardiac Poisoning : The ordinary symptoms are giddiness, fainting, slow pulse and respiratory failure. Death results usually from cardiac failure.

Treatment : Immediate intake of excess water followed by hospitalization.

Special Poisons

Poisoning may be homicidal, suicidal or accidental. The poisons chiefly used in India are opium, arsenic, datura, aconite, mercury, copper salts, nux vomica, methylated spirit etc. For homicidal purposes, arsenic is most commonly used, and for suicidal purposes opium; opium is also used as infanticide.

Two frequently used poisons are arsenic and opium from which lives may be saved by timely treatment. Description of their usual symptoms and treatment is given below :

1. **Arsenic** : Killers select this poison due to its easy availability from the market. It is tasteless, colourless and odourless and can be mixed with food, only a small quantity is required to destroy life, and the symptoms can be mistaken for that of cholera.

Symptoms : The symptoms usually commence about half an hour after the poison is administered. An irritant poison, causes burning pain in the throat, stomach and abdominal pain, thirst, vomiting, purging and straining, also presence of blood in the vomitus and stools, scantiness or suppression of urine, cramps, feeble pulse, and, in some cases, headache, delirium etc. Death usually takes place within twenty-four hours.

Distinction between Cholera and Arsenical Poisoning : Very often the symptoms of acute arsenical poisoning are mistaken for that of cholera. In arsenical poisoning the stool usually contains blood and there is burning in the throat before vomiting. In cholera, the stool from beginning is like 'rice water', and there is no pain or irritation of the throat before vomiting. In arsenical poisoning the vomiting usually starts first, followed by purging, while, in cholera, it is just the reverse.

Treatment : Vomiting should be encouraged and the patient should be given mild warm water. Freshly prepared hydrated ferric oxide is used as an antidote of Arsenic. It should be given in tablespoonful dose every 5 or 10 minutes, until the symptoms are relieved. Doctor should be consulted at once.

2. **Opium** : Solid opium, often mixed with mustard oil, is often swallowed to commit suicide.

Symptoms : The symptoms commence in about half an hour. There is slight excitement lasting for a short time, followed by giddiness and drowsiness, followed by deep sleep, ending in coma and unconsciousness. The odour of the opium may be detected in the breath. The breathing becomes shallow and slow, the pulse is weak and irregular, the skin gets bathed in sweat and is cold, and the pupil gets contracted. Death usually occurs in about 12 hours.

Treatment : Emetics should be induced as quickly as possible by using common salt and hot water. Tickling the throat excites vomiting. Patient should be kept awake by sprinkling of cold water on the face and head, and wet towel should be kept on the head. Patient should be given hot coffee to drink if he is conscious. Artificial respiration may be given. Patient should be kept under observation of physician.

3. Economic Toxicology

Subjects classified under this head — food, drinks, drugs, etc. — are directly and indirectly involved in the economics of the country, e.g., pure food contributes to good health and is of real value in the commercial export market. In other words, economic toxicology is directly or indirectly related to the economics of the nation. It deals with the harmful effects of chemicals which are intentionally administered for the purpose of achieving a specific effect in the biological system.

Two aspects are closely related : (i) Food toxicology, and (ii) Drug toxicology.

(i) **Food Toxicology** : The contamination of the food could be accidental, unintentional, or intentional adulteration for higher profits. In freshly cooked food, many of the contaminants would be eliminated at a high temperature. An interesting example is the poisoning due to cyanogenic glycosides from eating uncooked raw cassava (tapioca) in Africa. In India, people eat cooked cassava and this problem does not exist. The possibilities of contamination are more in the processed foods. Food adulteration is an art in itself and not much can scientifically be said except about the common adulterants.

The Beer Epidemic of 1990 : Large scale poisonings used to occur in the early part of the 20th century and an interesting episode was the Beer Epidemic in U.K. In June 1990, a few people —

alleged to be beer-drinkers — died in the north of U.K., around Liverpool. More deaths occurred in the following months at Manchester, Salsford, Lancashire and adjacent areas. Initially, it was attributed to alcoholism, peripheral neuritis or multiple neuritis. But when the numbers affected reached a few thousands, the government appointed a commission of enquiry with Lord Kelvin as chairman. Analysis of the beer showed 1.45 - 2.6% Arsenic. Close investigation proved that the beer was from the brewery of Bostock at Garston near Liverpool.

Poisonings are not limited to the olden days. Cooking oil contaminated with Triorthocresol Phosphate (TOCP) killed 300 people in Spain in 1981 and left 20,000 with neurotoxic lesions. In 1960, a large number of people died in Kerala after using sugar contaminated with an insecticide, Folidol. In 1971, 113 babies (out of 12,000 victims) died from dried milk powder contaminated with arsenic.

The use of Kesari dal (a banned item) for extended period of time produces lathyrism due to the presence of Oxalydiaminopropionic Acid (ODPA) and β-aminopropionitrile.

The Minamata disease due to consumption of fish contaminated with methyl mercury dumped as industrial waste into the Minamata Bay of Japan killed large number of people.

In 1972, in Iraq, about 6,000 people were affected seriously by consuming bread made from cereals contaminated with alkyl-mercury fungicides. Itai-Itai disease in Japan, which produced severe bone and joint pain like arthritis, was due to higher cadmium content in rice receiving effluent from a zinc cadmium mine upstream.

Major Contaminants in food : The major contaminants of food are microbial organisms, viz., small round virus (SRV) which causes gastro-enteritis, environmental contaminants, viz., *Pb, Hg,* chromate, arsenic etc., natural toxicants, viz., mycotoxins, pesticide residues and food additives (for detail see Chapter 8).

(ii) **Drug Toxicology** : Actually the evolution of toxicology began with the work on poisons used as drugs. Paracelsus, *father of toxicology,* reported that it is the dose which determines the toxicity and not the drug by itself. The chemical would become a drug only if the side-actions are not significant in comparison to the major desired effect. Morphine, which was being used as an analgesic, earlier produces respiratory depression and also addiction. So the use of morphine as an analgesic today is highly restricted. Cortisone and penicillin can produce immunological reactions, but their good effects outweigh the adverse ones and so are yet in use.

In Japan (1960) a number of people were affected by Sub-acute Myelo-Optico-Neuropathy (SMON). It caused abdominal pain and persistent diarrhea followed by a kind of paralysis of the extremities of the legs. In severe cases, blue-red visions, blindness and sensory disturbances preceded death. Investigations traced that SMON was due to intake of clioquinol, which is actually iodochloro-8 hydroxy quinoline, marketed as Mexaform and Entero Vioform in India.

It was previously thought to be a neurotoxic response of clioquinol, but toxicological investigations proved this to be due to the amine-rich diet of the Japanese, mainly fish, interacting with the clioquinol, which was found to be a mono-amino oxidase inhibitor — causing disturbances of amine metabolism and accumulation. This is an excellent example of geographical toxicology where drugs and food habits are contradictory to each other. SMON may not be produced in India where dietary intake of amines (fish products) are far less.

Drugs like antipyrine and theophylline are metabolised at a faster rate by high protein-low carbohydrate intake. Quite a lot of information has been collected on drug toxicity on newborn and immature rats, and higher toxicity has been observed.

Antihistamines, chlorophenilamine maleate and diphenyl hydramine show high toxicity in very young rats due to rapid absorption from the site of ingestion. These drugs are very poorly tolerated by young children too.

Causes of higher toxicity of drugs in infants : The following are the possible causes of higher toxicity of drugs in infants :

1. The larger volume of extracellular fluid in infants (almost thrice that of adults) which increases drug retention and slows down its excretion.
2. The immaturity of drug metabolising enzymes. Chloromycetin (chloramphenicol) can cause cardiovascular collapse due to this particular reason.

3. Slower excretory mechanism can also be a reason, for example, Niliclixic acid is retained in blood for 15 days in newborn calves, while its half-life is one and a half hour in seven month old calves.

The margin of safety (therapeutic index) is the guiding factor in developing a chemical as a drug. Actually *margin of safety* is the difference between LD_{50} and ED_{50}; the wider the gap between the two values, better the chemical as a drug.

4. Clinical Toxicology

It deals with diagnosis and treatment of the effects of harmful diseases caused by toxic substances of exogenous origin i.e., xenobiotics. These include accidental and intentional abuse of chemical substances, including therapeutic agents; undesirable excessive and non-therapeutic drug effects; injurious interaction of xenobiotics; accidental exposure to toxic substances in house and industry, intentional and inadvertent food additives; environmental contaminants, and naturally-occurring toxic hazards. Clinical toxicology is closely related to forensic toxicology. The application of antidotal substances to prevent the toxic actions of poisons are also important aspects of clinical toxicology.

5. Mechanistic Toxicology

It is a branch of toxicology which is associated with the elucidation of mechanisms by which xenobiotics exert their deleterious effects on living beings.

6. Biochemical Toxicology

The study of changes at the cellular and subcellular levels as a result of toxic action of chemicals is designated as biochemical toxicology.

7. Regulatory Toxicology

It is a branch of toxicology which deals with the imposition of certain restrictions on the entry of chemicals in the environment with the help of law. For this purpose, legislations are framed and passed by the government (State and Central) and regulatory agencies enforce these legislations.

8. Industrial Toxicology

It deals with the study of the nature and types of chemical substances emitted by the industries in ambient air and water, and possible effects of such contaminants on the living biota including human beings. It is much close to environmental toxicology. It also considers the suggestions regarding proper measures to treat hazardous substances released from the industries and, thus, to minimize their release into the environment.

9. Genetic Toxicology

It deals with the study of deleterious effects of xenobiotics, if any, on chromosomes and, more specifically, upon the genetic constitution of individuals. It is closely related to the mechanistic as well as biochemical toxicology.

10. Preventive Toxicology

This branch deals with the evaluation and recommendation of suitable measures for the protection of environment and its living resources.

11. Behavioural Toxicology

It is chiefly concerned with the study of alterations in the behaviour of organisms following exposure to specific toxicant.

12. Systemic Toxicology

This branch deals with the adverse effects of chemicals on a particular system of organisms, viz., respiratory system, immune system, nervous system, cardiovascular system etc.

13. Comparative Toxicology

This branch deals with the study of relative-toxicities of xenobiotics to various organisms.

14. Toxinology

Toxinology is the study of toxins in the restricted sense (poisons produced by micro-organisms, viz., bacteria and viruses etc.).

15. Wildlife Toxicology

It involves the study of the effects of toxicants of any origin on wildlife, in exactly the same way as the veterinary toxicology is related to domestic animals.

16. Ethnic or Geographical Toxicology

It is a recent area concerned with the puzzling phenomenon that certain toxic drug reaction seems to be more frequent in particular countries or in one continent. It analyses various factors (climate, nutritional practices, genetic factors and other environmental influences) which might account for these differences in responses of the human body to a xenobiotic.

Toxicants and their Classification

Definition of Toxicant

In simple words, a toxicant may be defined as an agent that causes an adverse effect or response in a biological system, seriously damaging its structure or function or producing death. The adverse effect or response may be defined in terms of a measurement that is outside the normal range for healthy organisms.

In view of the National Institute for Occupational Safety and Health a toxicant may be defined as "a substance which demonstrates the potential to induce cancer, produces long term disease or bodily injury; affects health adversely; produces acute discomfort; or endangers the life of man or animals through exposure via the respiratory tract, skin, eyes, mouth, or the other routes."

Introduction of Toxicants into the Ecosystem

A toxicant or foreign substance may be introduced deliberately or accidently into the ecosystem, impairing the quality of environment and making it unfavorable for organisms.

Toxicants find their way in the ecosystem from :

- Nonpoint sources such as agricultural runoff from land, contaminated ground water and bottom sediments, urban run-off, dredged sediment disposal, and atmospheric fall out, and
- Point sources such as discharges (effluents) from manufacturing plants, hazardous waste disposal sites, and municipal waste treatment plants.

Survey and Classification of Toxicants

There are a number of chemical substances in the environment — some of these are toxic and rest non-toxic. The toxicants are released from various sources into air, water and soil. They get into the human food chain from the environment. Once the toxicants enter our biological system, they interfere with the biochemical processes and may lead, in some, cases to fatal results.

According to the International Register of Potentially Toxic Chemicals of the **UNEP (United Nations Environment Programme)**, there are four million known chemicals in the world today and another 30,000 new chemical compounds are added to this list every year. Among these, 60,000 - 70,000 chemicals are very commonly used. Apart from their benefit in increasing production, living standards and health, many of them are potentially toxic.

Survey of toxicants in various sections of environment may precisely by explained in the following points :

I. Toxicants in air
II Toxicants in water
III Toxicants contaminating food
IV Others viz., noise and radiation pollution.

I. Toxicants in Air

Quite a large number of chemicals presumably pose serious problems of health when present in air. It is, therefore, necessary to exercise strict control during their manufacture and handling.

Environmental Protection Agency (EPA), Duluth, U.S.A. and Occupational Safety and Health Administration and Consumer Product Safety Commission (1978), listed 24 extremely hazardous substances in the atmosphere which are :

1. Arsenic
2. Acrylonitrite
3. Asbestos
4. Berilium
5. Benzene
6. Cadmium
7. Chlorinated solvents
8. Chromates
9. Chloroflurocarbons (CFCs)
10. Coke oven emission
11. Diethylstilbestrol
12. Dibromochloropropane
13. Ethylene dibromide
14. Ethylene oxide
15. Lead
16. Mercury
17. Ozone
18. Nitrosoamines
19. PCBs
20. PBBs
21. Sulphur dioxide
22. Vinyl chloride
23. Radiations
24. Toxic waste disposal emissions and leachates.

The common toxicants present in air are :

- Carbon monoxide
- Nitrogen gases, viz., NO, NO_2, NH_3, and N_2O
- Sulphur gases, viz., SO_2 and H_2S
- Hydrocarbons viz., CH_4
- Photochemical oxidants viz., O_3, and other photochemical products like benzopyrene, peroxybenzoyl nitrate (PB_zN) and peroxyacetyl nitrate (PAN).
- Lead from automobile emission (for detailed account refer *Chapter 10)*

2. Toxicants in Water

A large number of toxicants — intentionally or otherwise — reach in the aquatic environment. The principal sources of toxicants in aquatic environment are

- Sewage and other wastes of domestic nature
- Agricultural discharges
- Industrial effluents, and
- Wastes from thermal and nuclear power plants.

Toxicants in aquatic environment may, however, be grouped into two principal categories :

1. Toxic trace elements and heavy metals found in natural and waste waters
2. Pesticides.

1. **Toxic trace elements and heavy metals found in natural and waste waters** : These toxicants are listed in Table 1.1. Some of these are essential at low level, serving as nutrients for animal and plant life, but certainly deleterious at higher levels. Heavy metals as toxicants are described in *Chapter 13.*

2. **Pesticides in water** : The aquatic bodies contain a large number of pesticides, primarily from the drainage of agricultural field. Detailed account on pesticides is described in *Chapter 3.*

Table 1.1 : Toxic trace and heavy metals found in natural and waste waters and their effects

Sl. No.	*Name of Toxicants*	*Symbol*	*Sources*	*Effects and Significance*
1.	Arsenic	As	Mining by-products, Pesticides, Chemical wastes, Fertilizer factory effluent (Pandey and Shukla, 1982)	Toxic, possibly carcinogenic
2.	Beryllium	Be	Coal, nuclear power and space industries	Acute chronic toxicity, possibly carcinogenic
3.	Boron	B	Coal, detergent, industrial wastes	Toxic to plants
4.	Cadmium	Cd	Industrial discharge, mining waste, metal plating, water pipes	Replaces zinc biochemically, causes high B.P., kidney damage, deformities of testicular tissue and RBC toxicity to aquatic biota
5.	Copper	Cu	Metal plating, industrial and domestic waste, mining, mineral leaching	Essential trace elements, not very much toxic to animals, toxic to plants and algae at moderate levels
6.	Chromium	Cr	Metal plating, cooling tower water additive	Essential trace elements, possibly carcinogenic as Cr (VI)
7.	Fluorine	F	Natural geological sources, industrial waste, water additive	Prevents tooth decay at about 1 ppm, causes mottled teeth and bone damage at about 5 ppm
8.	Lead	Pb	Industry, mining, plumbing, coal, gasoline, automobile exhausts	Causes anemia, kidney disease, nervous disorders
9.	Mercury	Hg	Industrial wastes, mining, pesticides	Highly toxic

Contd...

Sl. No.	Name of Toxicants	Symbol	Sources	Effects and Significance
10.	Manganese	Mn	Mining, industrial waste, acid mine drainage	Produces sterility
11.	Molybdenum	Mb	Industrial wastes, Natural sources	Possibly toxic to animals and essential for plants
12.	Nickel	Ni	Mining	Carcinogenic
13.	Selenium	Se	Natural, Geological sources	Essential at low level but toxic at higher level
14.	Zinc	Zn	Industrial waste, metal plating, plumbing	Essential for many metabolic enzymes but, at high level, toxic to animals

3. Toxicants Contaminating Food

A food additive may be defined as "a substance or mixture of substances other than a basic food, which is present in food as a result of any aspect of production, processing, storage or packaging". In other words, food additive is any substance not normally present in the food in question and which is added, either deliberately or incidentally, in order that some quality of the food may be improved.

Food additives may be of two types, namely, *Intentional additives*, for example, sodium benzoate, sorbic acid, lecithin, monosodium glutamate, saccharin etc., and *Incidental additives,* which are mixed with food during processing or packaging etc.

Food additives are used to provide better colour, flavour and texture and as a preservative. Several food additives have been found associated with the toxic hazards and health risks. Many of the additives like cyclomates, safrole and diethyl stilbestrol (DES) have been banned. Several others like saccharin, monosodium glutamate, nitrite, nitrate, etc. are under close investigation and are likely to be banned. For detailed account on food additives, refer *Chapter 8*.

4. Others (Noise and Raditation Pollution)

I. Noise

Noise is an environmental pollutant, but it differs from air and water pollutants in that it disappears fast and thus does not remain in environment for long. Presence of unwanted sound in the atmosphere that produces adverse effect in human beings is termed as 'Noise Pollution'.

Vehicles do not just cause air pollution, they are also a major cause of noise pollution, exposure to which over an extended period of time can even lead to hearing loss.

Actually, sound, as such, is not considered as a pollutant. Many sounds are agreeable and pleasing to human ear. But the constant presence of unavoidable and irritating sound, for periods longer than the tolerance limit of human ear, is now considered as one of the serious air pollutants.

Noise can interfere with all our activities, such as work, rest, recreation, and sleep. It is not only a nuisance but also a threat to our health and, thus deserves the attention of all concerned.

Sources of noise are many. The main sources of man-made noise in developed urban areas are mechanized vehicles such as trucks, buses, cars, scooters, fire engines, tractors, generators etc. Industrial machinery, trains, aeroplanes, and accessory noise producers such as horns, sirens, loud-speakers etc. also contribute to noise pollution. Musical instruments, TV, radio, washing machines, vacuum cleaners, food mixers, air-conditioners, shouting, social gatherings, etc. are some other sources of noise pollution. The main menace of noise, however, comes from transport. It mainly includes noise from road traffic, rail traffic, aircraft, in addition to vehicle horns. The number of road vehicles, particularly the diesel engine vehicles, has increased enormously in recent years. A survey conducted in metropolitan cities has shown that the average noise level in Delhi, Mumbai, and Kolkata is as high as 90 dB. Inhabitants of the major cities, subjected continuously to this level of noise, run the risk of impairing and gradually losing their hearing ability.

Noise is transmitted as sound energy through any elastic medium like air, water or steel by the vibrations of the molecules in the medium. When there are no molecules, as in a vacuum, sound cannot travel. Sound travels faster in solids than in air. The speed of sound in air, at 20°C, is nearly 344 m/s, while the speed of sound in wood is 3,962 m/s and in steel 5,029 m/s.

Sound is usually measured either by (i) Sound pressure, or (ii) Sound intensity.

There are two important parameters for describing sound :

1. **The sound wave frequency**: As sound waves travel through air, the air undergoes periodic increase and decrease in density (compression and rarefaction). The number of compressions and rarefactions per unit time is known as the sound wave frequency. It determines the pitch of the sound. Its unit is Hertz (Hz) or cycles per second. The human auditory system is normally able to discriminate pitches in the range of frequencies between 16 and 20,000 Hz — called the audio-frequency range.

2. **The sound wave amplitude or magnitude**: This is a measure of loudness. Since audible sounds may vary in intensity by many levels of magnitude, a logarithmic scale is used to measure sound intensity. This unit is decibel (dB), named after Alexander Graham Bell. The threshold for detection of sound is defined as 0 dB. One dB is equal to the faintest sound a human ear can hear. Normal conversation generates about 50 to 60 dB, a city street has a sound level around 75 dB, and an auto horn at close range is about 90 dB. Sounds become painful and potentially damaging to auditory receptors at 130 dB. The most immediate and acute effect of noise pollution is impairment of hearing. Long exposure to loud noise can cause some hearing loss that may become permanent. Generally, hearing loss is a result of neural damage involving injury to the hair cells. Two theories are offered to explain noise-induced injury. The first is that excessive hearing forces mechanically damage the hair cells. The second is that intense noise stimulation forces the hair cells into high metabolic activity that overdrives them to the point of metabolic failure and consequent cell death. Once destroyed, hair cells are not capable of regeneration. It has been established that a person exposed to a noise level of 90 dB for more than eight hours per day is liable to severe hearing problems after 20 years. This is also true in case of traffic police personnel on duty and also for shopkeepers and residents near noise-polluted intersections. In addition to hearing damage, continuous noise can induce non-auditory physiological effects. Noise pollution can interfere with speech communication, sleep, acoustic privacy and cause annoyance — thus affecting human health, comfort and efficiency. Noise pollution is also known to increase the rate of heart beat. It causes constriction of blood vessels, makes the skin pale, muscles to contract leading to nervous breakdown and tension. It increases the digestive spasms and causes dilatation of the pupils of the eye. Fluctuations in arterial blood pressure, decrease in heart output, impairment of night vision, and decrease in the rate of colour perception have also been reported. Noise also interferes in inter-personal communication. It causes headache, irritability (annoyance) and extreme emotional disturbances. It aggravates existing diseases by disturbing the peace of mind and sleep. Effects of noise on the fetus are not fully known. Medical scientists have found that an unborn child moves and kicks about when there is loud noise. It also responds with a sudden increase in the heart beat as though it were disturbed or frightened. However, more research is required to establish a noise index consistent with human comfort and well-being. Overall effects of high intensity noise on human health is shown in Table 1.2 (for detailed account refer *Chapter 9*) :

Table 1.2 : Effects of high intensity noise on human health

Noise (dB)	*Effects observed*
0	Threshold of hearing
110	Stimulation of receptors in skin
120	Pain threshold
130	Nausea, vomiting, dizziness, interference with all activities
140	Pain in ear, extreme limit of noise tolerance
150	Significant change in pulse rate, prolonged exposure causing burning of skin, extreme emotional disturbances
160	Minor permanent damage to hearing on prolonged exposure
190	Major permanent damage to hearing in a short time

II. Radiation Pollution

The atoms of certain elements are known to undergo spontaneous disintegration with the emission of atomic particles, i.e. radiation. Such elements or substances that emit radiations due to spontaneous nuclear disintegration are termed radioactive substances. The well-known radioactive elements are : Carbon^{14}, Thorium^{234}, Uranium^{238}, Phosphorus^{32}, Potassium^{40}, Cesium^{137}, Radon^{222}, Radium^{226} etc.

The radioactive rays and subatomic particles enter the environment mainly from two sources :

I. Natural sources
II. Anthropogenic (man-made) sources.

I. Natural sources

The non-ionizing ultraviolet rays from the sun, cosmic rays and natural decomposition of radioactive substances emit radiation and sub-atomic particles.

II. Anthropogenic (man-made) radiation

1. X-rays used in medical practice
2. Radiation therapy in treatment of cancer
3. Application of radio-isotopes in research work, medical practice and archeological dating
4. Atomic power plant
5. Nuclear generator
6. Radioactive fall-out resulting from nuclear weapons testings.

Contamination of the environment with the radioactive waves may be by :

(i) **Accidental spilling** : Example—Chernobyl (Russia), Three Miles Island (U.S.A.) Nuclear disasters.

(ii) **Deliberate action** : Example — Atomic explosion as in Hiroshima, Nagasaki (Japan) and explosives in wars.

(iii) **Radioactive substances** : Example — Cesium, Uranium, Zirconium, Tritium, Strontium, Radium, Plutonium etc.

(iv) **Radio-Isotopes** : Example — Isotopes of Uranium^{238}, Carbon^{14}, Cobalt^{60}, Manganese^{54} and Strontium^{90}.

(v) **Ignorance** : Example—Disposal of nuclear wastes indiscriminately and cutting of forests.

Types of Radiation

Radiations are chiefly of two types :

1. Non-ionizing UV
2. Ionizing (X-rays, alpha, beta, neutrons and gamma radiations).

Ionizing radiations are of two types :

(i) **Electromagnetic Radiation** : It is in the form of short wave, high energy radiation similar to X-rays. Example, gamma radiation.

(ii) **Corpuscular or Particulate Radiation** : It consists of subatomic particles which transfer their energy to whatever object they strike. Example, alpha and beta radiations.

Radiation effects

(i) **Non-Ionizing Radiation** : The radiations which do not possess ionizing effect are called non-ionizing radiation :

UV rays : UV rays emitted by sun may penetrate living tissues up to 0.1 mm thickness, depending upon the length of exposure and are specifically absorbed by nucleic acids and generate pyrimidine dimers. These suppress separation of DNA strand during replication. These may produce mutation.

(ii) **Ionizing Radiations** : Very high-energy radiations that are able to remove electrons from atoms and attach them to other atoms, thereby producing +ve and –ve ion pairs, are known as ionizing radiations. Isotopes of elements that emit ionizing radiations are called radionuclides or radioisotopes :

(a) **X-rays** : The X-rays possess high speed almost equal to the speed of light — and have ionizing power but do not penetrate deep in the living tissues.

(b) **Gamma rays** : The gamma rays possess enormous penetrating power and travel with a speed that equals speed of light. It may penetrate a lead sheet of about 15 cm (6 inch) thickness. It also damages living tissues.

For further detail on radiative pollution, refer *Chapter 7*.

Classification of toxicants on the basis of organ/system affected

From the human health view-point and the organ/system affected, the toxicants may be of the following types :

1. Environmental Carcinogens

In various countries, cancer is one of the leading cause of death. It is supposed to be induced by a variety of agents and also certain life styles, e.g., smoking cigarettes, drinking alcohol, chewing tobacco etc. However, toxicants in the community and at work places, besides certain therapeutic agents, also play certain role in inducing cancer. Attempts at reducing cancer incidence are, therefore, aimed at minimizing these hazardous life styles along with identifying these chemical carcinogens, in the hope that human exposure to them can be eliminated or minimized. Chemical carcinogens may thus be defined as any chemical which initiates carcinogenicity. In Table 1.3 a list of toxicants that are considered as definite or probable human carcinogens has been identified and reported by IARC (1979) :

2. Cardiotoxicants

Heart may not be a common target organ, yet it is damaged by a variety of chemical or drugs referred to as cardiotoxicants. They act either directly on the myocardium or indirectly through the nervous system or blood vessels. A list of such chemicals as compiled by Van Stea (1980) is reproduced in Table 1.4 :

Table 1.4 : List of Cardiotoxicants

1. Phenytoin
2. Clotibrate
3. Lithium
4. Ethanol
5. Cobalt
6. Allylamine
7. Phenothiazines
8. Carbromide
9. Carbromal
10. Brominated vegetable oils

Table 1.3 : Environmental carcinogens, their applications and hazards

Sl. No.	Carcinogen	Uses	Hazards
1.	α-naphthylamine	Antioxidant, dye manufacture, colour film manufacture	It may cause bladder cancer
2.	β-naphthylamine	Dye manufacture, reagent	Bladder cancer
3.	4-nitrobiphenyl	Chemical analysis	It also may cause bladder cancer
4.	Methylchloride	Ion exchange, resin manufacture	Carcinogenic
5.	3,3'-dichlorobenzidine	Dye manufacture	Carcinogenic
6.	Bis (chloromethyl) ether	Ion exchange, resin manufacture	Lung cancer
7.	Benzidine	Manufacture of dyes, rubber, plastics, printing ink	Bladder cancer
8.	Ethylynamine	Paper and textile processing	Carcinogenic
9.	β-propiolactone	Plastic manufacture	Potential carcinogen
10.	Vinylchloride	PVC plastics	Hepatic carcinogen
11.	Ethylene dichloride	Industrial solvent, gasoline, additive for lead scavenging, grain fumigant	Suspected carcinogen
12.	N-ethyl-N-nitroso-N-butylamine	Gasoline and lubricant additive, insecticide	Carcinogenic
13.	Benz (α-) Pyrene, Polycyclic aromatic hydrocarbons (PAH)	Coal furnace, also occurs in organic particulate matter adsorbed on soot in urban atmosphere up to 20/μg/m^3	Carcinogenic

3. Immunotoxicants

Actually, the immune system serves an important role in the body's defense against infection by microbes viz., viruses, bacteria, fungi and unicellular and multicellular parasites as well as against neoplastic cells.

It has now become an established fact that a variety of toxicants (chemicals) can impair the function of the system. A variety of chemicals may induce hypersensitivity reactions. The site of such reactions are mainly confined to the respiratory tract and the skin. Dean et al (1979) reported that a variety of substances may adversely affect the immune system.

These chemicals (toxicants) are listed in Table 1.5 :

Table 1.5 : List of Immunotoxicants

A. Heavy Metals	1. Lead, 2. Mercury, 3. Cadmium, 4. Nickel, 5. Chromium, 6. Methyl mercury, 7. As (III) oxide, 8. Sodium arsenate.
B. Pesticides	1. DDT, 2. Dieldrin, 3. Carbaryl, 4. Carbafuran, 5. Methylparathion, 6. Hexachlorobenzene (HCB).
C. Halogenated Hydrocarbons	1. Polychlorinated Bi-phenyls, 2. PBB, 3. TCDD, 4. Trichloroethylene, 5. Chloroform, 6. Pentachlorophenol.
D. Miscellaneous compounds	1. Benz pyrene, 2. Benzene, 3. Methylchalanthrene.

4. Teratogens

Chemical agents causing teratogenesis are called teratogens. Teratogenesis is the formation of congenital defects. Various teratogens in animal models are compiled in Table 1.6 :

Table 1.6 : List of Teratogenic Agents

A. Natural toxins	1. Nicotine
B. Solvents	1. Benzene, 2. CCl_4, 3. Di-methyl sulfoxide, 4. Propylene, 5. Xylene, 6. Glycol.
C. Insecticides	
D. Herbicides	
E. Fungicides	
F. Azodyes	1. Trypon blue, 2. Evans blue
G. Antibiotics	1. Penicillin, 2. Streptomycin, 3. Tetracyclines
H. Drugs and chemicals	1. Caffeine, 2. Carbutamide, 3. Chloropromazine, 4. Hydroxy urea, 5. Nitrosamines, 6. Quinine, 7. Rauwolffia, 8. Alkaloids.

5. Occupationally Inhaled Toxicants

Following industrialization, since the Industrial Revolution begun in 1970, the human respiratory system is increasingly exposed to airborne toxicants. Airborne substances exist in the form of gases, vapours, liquid droplets and solid particulate matters of different sizes. The uptake and effects of inhaled toxicants, therefore, depend not only on their toxic nature but also on their physical characteristics. Various adverse effects may result from inhalation exposure to toxicants (Table 1.7).

6. Hepatotoxicants

Some of the hepatic (liver) injuries are produced by hepatotoxicants like CCl_4; Halothane; α-naphthylisocyanate (ANIT); Chloroform, Ethanol; Bromobenzene; Bromsulphathalein (BSP); Aflatoxins and Phosphorus. These hepatotoxicants induce several types of injuries like steatosis, necrosis, cirrhosis, neoplasia etc. These injuries are relatively easily produced in experimental animals.

7. Mutagenic agents

Mutagenesis can occur as a result of interaction between mutagenic agents (like As, Cd, Co, Hg, Ni, Pb and Pt) and the genetic material, i.e., DNA of organisms. While spontaneous mutation and natural selection are the major means of evolution, a number of toxicants in recent decades have been reported to be mutagenic.

Table 1.7 : List of Occupationally Inhaled Toxicants, their site of action and effects

Sl. No.	*Name of toxicants*	*Common name of diseases*	*Site of action*	*Acute effects*	*Chronic effects*
1.	Aluminium	Aluminosis	Upper airways, alveolar interstitium	Cough, shortness of breath	Interstitial fibrosis
2.	Ammonia		Upper airways	Immediate upper and lower respiratory tract irritation, edema	Chronic bronchitis
3.	Arsenic		Upper airways	Bronchitis, headache	Lung cancer, bronchitis, Laryngitis
4.	Asbestos	Asbestosis	Parenchyma		Lung cancer, Pulmonary fibrosis, pleural calcification
5.	Beryllium	Berylliosis	Alveoli	Severe pulmonary edema, pneumonia	Pulmonary fibrosis; Progressive dyspnea, Interstitial granulomatosis
6.	Cadmium oxide		Alveoli	Cough, pneumonia	Emphysema
7.	Carbides of tungsten	Hard metal disease	Upper and Lower airways	Hyperplasia and metaplasia of bronchial epithelium	Fibrosis, Peribronchial and perivascular fibrosis
8.	Chlorine		Upper airways	Cough, dyspnea, bronchopneumonia	
9.	Chromium		Nasopharynx, upper airways	Nasal Irritation, bronchitis	Lung tumors and cancer
10.	Coal dust	Pneumoconiosis	Lung parenchyma, lymph nodes		Pulmonary fibrosis
11.	Cotton dust	Byssinosis	Upper airways	Tightness in chest, wheezing	Reduced pulmonary function, chronic bronchitis
12.	Manganese	Manganese pneumonia	Lower airways and alveoli	Acute pneumonia — often fatal	Recurrent pneumonia
13.	Nickel		Parenchyma nasal mucosa, bronchi	Pulmonary edema delayed by 2 days	Squamous cell carcinoma of nasal cavity and lung
14.	Oxides of nitrogen		Terminal respiratory bronchi and alveoli	Pulmonary edema	Emphysema
15.	Silica	Silicosis, pneumoconiosis	Lung parenchyma, lymph nodes		Pulmonary fibrosis
16.	SO_2		Upper airways	Bronchoconstriction, cough, tightness in chest	
17.	Toluene		Upper airways	Acute bronchitis, bronchospasm, pulmonary edema	
18.	Vanadium		Upper and Lower airways	Upper airway irritation and mucus production	Chronic bronchitis
19.	Xylene		Lower airways	Pulmonary edema	

Toxicity

Actually, toxicity is a relative term generally applied in comparing one chemical with another. Toxicity of a chemical may be defined as the capability to cause injury in a living organism. A highly toxic substance is that which causes damage to an organism if administered in a very small amount. But a substance of low toxicity will not produce an effect unless the amount is very large. In fact, toxicity may not be defined without reference to the quantity of a substance administered or absorbed (dose), the way in which this quantity is administered (e.g., inhalation, ingestion, injection), and distributed in time (e.g., single dose, repeated doses), the type and severity of injury and the time required to produce that injury. Thus, toxicity is a relative property of a chemical which refers to its potential to have a deleterious effect on a living organism. Also, toxicity is a function of the concentration of the chemical and the duration of exposure. Toxicity data are generally used in comparing toxic potentialities of chemicals.

The nature and extent of toxicity varies depending upon the origin of toxicants. They may be natural or synthetic in origin.

Naturally occurring toxicants may originate from animal, plant or mineral sources. Animal toxins as venom belong to this category. Many plants contain toxic agents and many of these are developed for pharmacological applications. Some among these are atropine, opium, reserpine, quinine, picrotoxin etc. A number of antibiotics come from microbes, including mycotoxins. The majority of toxicants stem from industrial or commercial pool of synthetic chemicals.

The toxic effects of the chemical may be physiological, biochemical and pathological in nature. The changes produced by the toxicants may be complex, damaging one or different organs/tissues/cells. These, in action, may be corrosives, irritants, narcotics etc.

Categories of Toxic effects

Toxic effects may be grouped into :

- Local and systemic effect
- Reversible and irreversible effect
- Immediate and delayed effect
- Morphological effect
- Functional effect
- Biochemical effect.

All these effects are described in proceeding *chapter*. But it has become customary to classify toxicity as acute, subacute and chronic toxic effects or toxicity as described in *Chapter 15*.

Principal aspects of Toxicology

The following are the principal aspects of toxicology :

1. **Toxicometrics** : It is the study of measurement of poisons.
2. **Toxicodynamics** : It deals with the biochemical and physiological effects of xenobiotics and mechanisms of their action.
3. **Toxicokinetics** : It deals with the absorption, distribution, biotransformation and excretion of xenobiotics.

Scope and Importance of Toxicology

Toxicology incorporates a very broad scope. It is primarily concerned with the identification of deleterious agents and, secondly, with the treatment of these agents for detoxification, i.e., removal of the toxin. Apart from this a toxicologist is also concerned with the character, properties, signs and symptoms and the chemical detection of the toxin, where offensive intent is suspected.

Actually, toxicology has its impact on human life from the day of conception in the womb till the last moments of life. Nevertheless, certain toxicants are even deleterious for the sperms and the ova.

Toxicology deals with the toxicity studies of chemicals used :

1. In medicine for diagnostic, preventive and therapeutic purposes.
2. In food industry as direct and indirect additives.
3. In agriculture as pesticides, growth regulators, artificial pollinators, and animal food additives.
4. In chemical industry as solvent, components and intermediates of plastics and many other types of chemicals.
5. Toxicology is also concerned with the health, effects of metals (as in mines and smelters), petroleum products, paper and pulp, toxic plants and animals toxins.

The study of toxicology may be system-wise and also agent-wise. In this way, we may have systemic toxicology involving toxicology of central nervous system, hepatic system, respiratory system, ophthalmic system etc. The study would include defences of each of these systems against assault from foreign chemicals, responses and reactions as also injuries caused through excesses.

An agent-wise study would take up drugs, pesticides, food additives, chemicals, heavy metals, trace elements, polymers, radiation, chemical carcinogens, teratogens. There is hardly much of generalization which can be made and many of the agents have to be studied individually in different species before being precisely conclusive.

Application of toxicology for practical purposes can be broadly divided into three areas which are, however, flexible and overlapping :

1. Forensic toxicology where medical science has a role.
2. Economic or industrial toxicology which includes drugs, pesticides, food additives, containers (like cans and packing materials) etc.
3. Environmental toxicology which includes pollutional studies, residue analysis, industrial hygiene and occupational health.

Besides, many other divisions and names are also given by different scientists.

When standard procedures may not be able to predict the toxicity in human, then the predictions have to be based on chemical and physical properties, molecular structure, biological activity relations, pharmacokinetics etc. This is termed as *Speculative Toxicology* and is based upon certain speculative studies.

Usefulness of the study of toxicology to human beings may be described in the following points :

1. **Toxicology may be helpful in the development of :**

(i) Suitable and safer food additives,

(ii) Suitable and safer pesticides,

(iii) Suitable drugs against any specific disease.

2. **The data on acute toxicity tests for various xenobiotics against different fauna may be valuable in following ways** :

(i) Provides an idea of toxic dose of a specific toxicant for specific animal.

(ii) Makes certainty of sublethal doses of the toxicants for specific animal.

(iii) Provides precise idea about maximum permissible limits for the pollutants in the ambient air or the drinking water.

(iv) Assists in evaluation of maximum acceptable daily intake of specific chemical.

(v) Aids in determination of sensitive species.

(vi) Aids in ascertaining sublethal dose of any chemical for long-term toxicity tests.

3. **The data on long-term toxicity tests provides reliability for :**

(i) In evaluation of safer levels of xenobiotics.

(ii) In determining the most sensitive stage of a particular target animal against any toxicant.

(iii) In elucidation of mode of toxicity of chemicals.

(iv) In rational therapy of intoxication.

(v) In establishment of sensitive predictive tests useful in achieving information for risk assessment.

4. Toxicology may be helpful in the antidotal therapy.

5. It may be helpful in the monitoring of environmental pollution.

6. Toxicology may be helpful in monitoring of risk assessment.

7. An understanding of the mechanism of toxic action also contributes to the basic knowledge of pharmacology, physiology, biochemistry and cytology.

8. Analytical toxicology provides suitable procedures to evaluate the presence or absence of different types of substances and their levels in the environment.

9. It also contributes an idea about the permissible limit of any toxicant in any segment of the environment.

10. Comparative toxicology is concerned with the study of different species of animals exposed to chemical stresses to evaluate how close these are to human systems.

11. It gives an idea about the factors affecting the toxicity of any toxicant.

Revision Questions

A. Long-Answer Questions

1. Write an essay on toxicology and its scope.
2. Describe in detail about forensic toxicology.
3. Define toxicology and illustrate its various branches.
4. Give an account of the survey and classification of toxicants.
5. Write an essay on the scope and importance of toxicology.

B. Short-Answer Questions

1. Write short notes on :

(i) Environmental toxicology

(ii) Economic toxicology

(iii) Forensic toxicology.

2. Explain in brief :

(i) Early era of toxicology

(ii) Recent developments in toxicology

(iii) Clinical toxicology

(iv) Toxic trace and heavy metals.

3. Describe briefly :

(i) Noise pollution

(ii) Radioactive contaminants

(iii) Important environmental carcinogens

(iv) Occupationally inhaled toxicants

(v) Environmental carcinogens.

2

Environmental Pollution and Public Health

SYNOPSIS

Introduction, Definitions of Pollution and Pollutants, Classification of Pollutants, Causes of Environmental Pollution, Types of Pollution, Pollution in relation to Public Health.

Introduction

A human being has brain capacity of around 1400 cc and that has empowered him to alter his environment according to his needs and comfort, which the ancestral apes were devoid of as their brain capacity was up to 400 cc only.

Environmental pollution is one of the most dreadful ecological crises to which we are subjected today. Ancient India sages quoted that in the current age of *Kaliyug*, the sins of mankind are being so bad that the earth and mankind would be destroyed in the near future. Today India, which occupies 7th place among the industrialized developing countries of the world, is provided with a good industrial infrastructure in several industries like chemicals, power, nuclear energy, petroleum, plastics, food, pesticides etc. A rapid progress in atomic and nuclear energy has added a huge amount of radioactive substances in the atmosphere. Various industrial effluents and emissions — especially deleterious gases — are released into the air daily. Thus the environment is deteriorated to such an extent that it has crossed the critical limit and has become lethal to most of the organisms including human. The consequences of the industrial developments are that we find warnings everywhere as "Air unfit for breathing", "Water unfit for drinking", "Vegetable unfit for eating" and so on!

In fact, environmental pollution is the necessary evil of all man-made progress. It is not only the industrialized countries which are threatened with environmental pollution, but the menace is growing in India too. The developed countries dump lots of effluents into the environment, polluting the entire globe. Needless to mention here that pollution has been exported to developing countries by the developed countries of the world. Undoubtedly, US is the most polluted country of the world today.

Pollution is not just the emission of deleterious gases, CFCs, radioactive wastes etc. but it is also the rubbish we dump on our roadsides, the plastic bags or covers we throw away, the empty packs and cans we toss away. Actually the list is endless.

Continued economic growths, mismanagement of resources and population explosion have an explosive impact on the environment. At present our earth has become a sick planet and urgently requires a cure.

At UN Conference on Environment at Stockholm (Sweden) on June 14, 1972, Indira Gandhi, the late P.M. of India, had said : "Modern man must reestablish his unbroken link with nature and with life". He must again learn to invoke the energy and recognize, as did the ancients in India millennia ago, that we can take from the earth and atmosphere only as long as we keep them free of pollution.

What is Environmental Pollution?

The term *'pollution'*, derived from the Latin word *Pollutionem* from *polluere, pollutus* (*pol* = per or through + *luere* = to wash, meaning to defile or to make dirty which is to be cleansed), is the act of polluting the environment. It is defined in various ways :

1. In generalized form environmental pollution means the pollution of air, water and soil by various pollutants, which are released into the environment in large concentration by the activities of man. These pollutants are injected in such concentration in the environment that these have disrupted the natural ecosystems of the biosphere.
2. According to Odum, E. P., pollution is an undesirable alteration in the physical, chemical or biological characteristics of our air, land and water that harmfully affect the human life and the desirable species or that may waste or deteriorate our raw material resources.
3. Pollution is the unfavourable alteration of our environment, largely because of human activities.
4. According to the report entitled "Pollution, Nuisance and Nemesis" (1972) : Pollution is the deliberate or accidental contamination of the environment with man's waste.
5. Pollution may be defined as "matter in the wrong place" or "anything released into the environment which degrades its quality".
6. Pollution is defined as a deviation from the natural composition of a part of the environment, resulting in adverse effects on life.
7. According to U.S. President of the Science Advisory Committee : Pollution is the unfavourable alteration of our surroundings, wholly or largely, as a by-product of man's actions, through direct or indirect effects of changes in the energy pattern, radiation levels, physical and chemical constitution and abundance of organisms.
8. Pollution is an undesirable change in physical, chemical or biological characteristics of air, water and soil that may harmfully affect human, animal and the plant life, industrial progress, living conditions and cultural assets (National Academy of Science, U.S.A., 1966).
9. According to T. G. McLaughlin, pollution is the introduction of surplus energy or waste matter into the environment by man's activities, which directly or indirectly cause hazard to man and his environment.
10. According to Edwards (1972), pollution is the release of harmful substances or energy into the environment by man in quantities that damage either his health or resources, or both.
11. Pollution as defined by National Environmental Research Council, 1976, is the release of substance and energy as waste products of human activities, which results in harmful changes within the natural environment.
12. In view of Dixon, D. M. (1972), pollution includes all those activities, conscious or unconscious, of human beings and the result thereof, in the short or long term, and his ability to derive full benefit therefrom.
13. According to Dasmann, R. F. (1975), Pollution is the accumulation of substances in the environment — or at rates of flow which exceed the capacity of ecosystems to either neutralize or disperse them — to harmful levels.

WHAT ARE POLLUTANTS?

It is well-known that every human society — be it rural, urban, industrial or most technologically and scientifically advanced — dispose-off certain kinds of waste products as well as by-products, which, when introduced into the environment in quantities so large that they affect the normal functioning of the ecosystem, are collectively known as pollutants. If they are not efficiently assimilated, decomposed or otherwise removed by the natural, biological and physical processes of the biosphere they impose adverse effects on flora, fauna and human beings. Undoubtedly, all powers pollute. In principle, the materials which cause pollution of the environment are pollutants.

Some other definitions of pollutants are :

1. Pollutants may be anything including a chemical, geochemical substance or a biological product that deteriorates our natural environment.
2. According to Holister and Porteus, a pollutant is a physical, chemical or biotic component, which adversely affects the environment by altering growth rate of species, interferes with the food chain, health, comfort, amenities or property value of man.
3. Odum defined pollutant as the residue of substances we make use of and throw away.
4. Holdgate (1971) defined pollutant as "something that is present in the wrong place, at the wrong time, in the wrong quantity".
5. Pollutant is a harmful solid, liquid or gaseous substance present in such concentration in the environment, which tends to be deleterious for the whole living biota (Environmental Protection Act, [1986] of India).
6. Unserviceable or residue of things which we manufacture, use and throw away are regarded as pollutants. In general, pollutants are residue of substances made by us, used by us and even thrown away by us as waste products which pollute the environment in one way or the other.

Types of Pollutants

The common pollutants of developed and developing countries are :

1. *Deposited Matter :* Soot, smoke, tar, dust, grit etc.
2. *Gases :* CO, CO_2, Nitrogen oxides, SO_2, chlorine, fluorine, NH_3, O_3 etc.
3. *Chemical compounds :* Aldehydes, Hydrogen fluorides, Arsines, Phosgenes, Detergents etc.
4. *Toxic Metals :* Lead, Mercury, Cadmium, Iron, Zinc, Nickel etc.
5. *Economic Poisons :* Insecticides, Herbicides, Fungicides, Rodenticides, Nematicides and other biocides.
6. *Fertilizers :* Urea, Potash, Ammonium sulphate, Ammonium phosphate etc.
7. *Radioactive substances :* Ar 41, Cobalt 60, Iodine 131, Caesium 137, Krypton 85, Strontium 90. Plutonium 239 etc.
8. *Noise*
9. *Heat*
10. *Sewage*
11. *Photochemical oxidants :* Ozone, PAN, PBzN, Aldehyde, Ethylene etc.

Classification of Pollutants

Pollutants (the creators of pollution) can be classified in a number of ways : For example :

A. On the basis of physical properties

On this basis, pollutants are classified into 4 categories. They are solid pollutants, liquid pollutants, gaseous pollutants and pollutants without weight.

B. On the basis of forms they exist in in the environment after their release

On this basis, pollutants can be classified as primary and secondary pollutants.

1. Primary Pollutants

These substances are emitted directly from an identifiable source. These pollutants exist as such after being added or released into the environment. Example are particulates, sulphur compounds, oxides of nitrogen, CO, halogen compounds, organic compounds, radioactive compounds etc.

2. Secondary Pollutants

These are substances derived from primary pollutants by chemical reactions or pollutants produced in the environment by the interaction among two or more pollutants or by the reaction with normal atmosphere constituents with or without photo-activation.

Example : Primary pollutants such as hydrocarbons and nitrogen oxides, particularly in the environment, react in presence of sunlight to form a group of nitrous compounds like peroxyacetylnitrate (PAN) as the secondary pollutant. Ozone, formaldehyde, photochemical smog are also examples of secondary pollutants (shown in Table 2.1 next).

C. From Ecosystem Point of View

From ecosystem point of view, the pollutants may be classified as *biodegradable* and *non-biodegradable* pollutants (Odum 1971).

1. Biodegradable Pollutants

The pollutants which can be decomposed by natural processes are called biodegradable pollutants. Examples : Domestic sewage, heat, noise etc. If organic waste drained out from residential buildings is easily degraded by microbes, these become useful for the ecosystem. The domestic sewage can be readily decomposed by natural processes or by engineered systems (such as municipal sewage treatment plant). Problems arise with the biodegradable pollutants when their input into the environment exceed the decomposition or disposal capacity.

Most, if not all, of the gaseous air pollutants, after being emitted, undergo various transformations in form in the atmosphere. The major processes that convert the pollutants from their primary form at emission to other secondary forms are oxidation, dissociation and dissolution as shown below :

Table 2.1 : Majour Processes involved in conversion of pollutants

Sl. No.	Reactions	Primary Pollutants	Secondary pollutants
1.	Oxidation	SO_2 H_2S Nitric oxide Peroxyacylnitrate (PAN) CO Some organic radicals	SO_3 SO_2 Nitrogen oxides Ozone CO_2 PAN
2.	Dissociation	Nitrogen oxides Oxygen (molecular) Hydrocarbons and aldehydes	Nitric oxide + O_2 Oxygen (atomic) Organic radicals
3.	Dissolution	CO_2 Hydrogen fluoride Nitrogen dioxide Nitrous acid (HNO_2) Sulphur dioxide Sulphur trioxide	H_2CO_3 Hydrofluoric acid Nitrous/Nitric acid Nitric acid (HNO_3) Sulphurous acid Sulphuric acid (H_2SO_4)

2. Non-Biodegradable Pollutants

There are many pollutants such as aluminium, iron, mercury salts, phenolic compounds, organochlorine such as DDT etc., which are usually not present in the environment. These either do not decompose (degrade) or do so very slowly or partially and thereby pollute the environment. Such pollutants are harmful even in low concentrations. These pollutants not only accumulate, but are also often biologically magnified or amplified as they move in biogeochemical cycles along food chains. They may also react with other compounds present in the environment to produce even more toxic additional products. Their high concentrations may even prove fatal. Such pollutants decrease the rate of photosynthesis by affecting planktons (responsible for about 70% O_2 source on earth), thus disrupting the natural balance existing between oxygen and CO_2.

In nutshell, the pollutants which cannot be decomposed are called nonbiodegradable pollutants. In the environment, they represent biological magnification or amplification. Also, they frequently react with other compounds in the environment to produce additional toxins.

Characteristics of Pollutants

The chemical substances fulfilling most, if not all, of the following criteria are likely to be serious environmental pollutants :

(1) Large production (2) Usage involving leakage (3) Toxicity (4) Persistence (5) Accumulation.

Principal Consequences of Environmental Pollution

Environmental pollution is the act of introduction of extraneous substances or energy into the environment by man that induces unfavourable changes. These changes may affect man directly or indirectly by endangering his health, harming his living resources and ecosystem, or by interfering with legitimate uses of the environment.

Environmental pollution causes (1) health problems by affecting human health and lives (2) economical problems by affecting (the value of) human property and materials (3) ecological problems by disturbing a balanced ecosystem, interfering with the conservation of natural resources and threatening the existence of some species; and (4) aesthetic problems by generally affecting human senses.

Causes of Environmental Pollution

The following are the main causes of environmental pollution :

1. Human Activity

The principal cause of environmental pollution is man himself. His activities, his civilization, culture etc. bring about environmental pollution. Hence pollution is said to be a necessary evil in the present age.

2. Over-Population

Southwick (1976) stated that over-population causes pollution because more people produce more sewage, more solid wastes, more fuel being burned, more fertilizers and insecticides being used to produce more food.

Current estimate indicates that an average annual rise of world population is by 2%. This percentage in real figures means the addition of about one million people every five days or five people every couple of seconds. If this trend continues unchanged, then, in another 1,000 years, the earth will have 100 people living on every square meter of its area — land or water!

3. Urbanization

Growing population leads to a greater concentration of people in the living areas. People move to urban areas (cities) abandoning rural settings in search of employment, comfort and facilities. As a result, cities are being over loaded with population that they can barely hold or support. A thickly-populated area is the home of large number of vehicles, reservoir for solid wastes with poor sanitary conditions and management problems.

Typically, a city of 1 million inhabitants require 1,800 metric tonnes of food, 2,700 metric tonnes of coal, 2,500 metric tonnes of oil, 2,400 metric tonnes of natural gas and 900 metric tonnes of motor fuel every day for its people. In turn, the inhabitants of the city will produce, each day, 1,800 metric tonnes of garbage and 45,000 metric tonnes of sewage.

According to this hypothetical model, which probably represents best an industrial society, a city of 1 million people uses approximately 8,500 mt of fossil fuels (coal, natural gas, petroleum) everyday. They produce, on an average, approximately 400 mt of CO, 135 mt of each of SO_2 and particulate matter, and 90 mt. of each of nitrogen oxides and hydrocarbons. All these pollutants are produced just by one million people. It shall not be out of place to mention that the metropolitan cities in India hold populations in the order of several million (Table 2.2) :

Table 2.2 : Populations of Metropolitan Cities in India in 2000

Sl. No.	Cities	Population in million in 2000	Population in million in 1975
1.	Mumbai	16.6	9.3
2.	Kolkata	15.8	8.5
3.	Chennai	12.5	7.8
4.	New Delhi	11.7	7.1

India incidentally happens to have one of the largest urban populations in the world of about 132 million people. Approximately 50% of the urban population in India is confined to nine major cities — Ahmedabad, Bangalore, Mumbai, Kolkata, Chennai, Ludhiana, Hyderabad, New Delhi and Kanpur.

Owing to over-crowding, diversified activities and multitude of sources, cities have inherently been susceptible to population problems. The concentrations of several of the air pollutants are higher in the urban atmospheres compared to rural areas (Table 2.3) :

Table 2.3 : Air pollutants and climatic changes produced by cities (Peterson 1969)

Sl. No.	Component	Comparison with rural environments
1.	**Air Pollutants**	
	CO	25 times more
	CO_2	10 times more
	Particulate matter	10 times more
	SO_2	5 times more
2.	**Temperature**	
	Annual mean	1 - 1.5°F higher
	Winter minima	2 - 3°F higher
3.	**Relative humidity**	
	Annual mean	6% lower
	Winter	2% lower
	Summer	8% lower
4.	**Cloudiness**	
	Clouds	5 - 10% more
	Fogs, winter	100% more
	Fogs, summer	30% more
5.	**Radiation**	
	Total on horizontal surface	5 - 20% less
	Ultraviolet (winter)	30% less
	Ultraviolet (summer)	5% less
6.	**Wind speed**	
	Annual Mean	20 - 30% lower
	Extreme dusts	10 - 20% lower
	Calms	5 - 20% lower
7.	**Precipitation**	
	Amounts	5 - 10% more
	Days with < 0.2	10% more

4. Industrialization

Human needs are never-ending. Discovery of new products and production of the needy luxuries — to suit the changing life-styles — are accomplished by the processes of industrialization. Industrialization is also the key to the economical development of a nation. Paper mills, sugar factories, detergent factories, cement factories, chemical factories, fertilizer factories, oil plants, rubber factories, blast furnaces, distilleries, oil refineries etc. cause pollution to a greater extent.

Industries, during the processing or manufacturing of intermediate chemicals and end products, generate waste materials and useless by-products as well. The production and processing steps in industrial units often result in the wastage of 1 - 10% of the quantity of parent chemicals. Chemicals may also enter the environment through spills during their use, transportation or disposal. Each industry is associated with an emission of one or other types of pollutants or potential pollutants directly or indirectly (Table 2.4) :

Table 2.4 : Industries and their Principal Pollutants

Sl. No.	Type of Industry	Principal pollutants emitted
1.	Aluminium ore reduction plants	Fluorides, particles
2.	Caustic and chlorine plants	Chlorine
3.	Ferro-alloy plants	Particles
4.	Green Iron and steel factories	Smoke, particles
5.	Lime plants	Particles
6.	Non-ferrous smelters	Sulphur oxides, particles, various metals
7.	Phosphate fertilizer plants	Ammonia, fluorides, particles
8.	Petroleum refineries	Hydrocarbons, odours, particles, smoke, sulphur oxides, CO, NO_2, HCHO, Gasoline
9.	Cement plants	Particles, Sulphur compounds
10.	Steel mills	Carbon monoxide, fluoride, particles, smoke
11.	Dye industry	Phosgene or Carbonyl chloride
12.	Refineries	Hydrogen sulphide
13.	Sulphuric acid industry	SO_2 and SO_3
14.	Nitric acid plants	NO_x (oxides of Nitrogen)
15.	Hydrochloric industry	Chlorine vapours, chlorine monoxide, HCl and unreacted hydrocarbons
16.	Power plants	SO_2

It seems important to mention here that as every process on the earth requires water (an universal solvent), the industries consume a lot of water too, which is another negative approach for future. For example, a paper and pulp industry requires about 150 m^3 of water for every tonne of its product manufactured. Water consumed by different industries varies widely and is represented in Table 2.5 :

Table 2.5 : Estimated Industrial Water Consumption

Sl. No.	Production Facilities	Water consumed (m^3/tonne ofproduct)
1.	Soap	2
2.	Wood	19
3.	Tanneries	67
4.	Pulp and paper	150
5.	Iron and steel	250
6.	Soda	320
7.	Woollen fabrics	580

The untreated effluents by industrial processes are primarily responsible for the water pollution problems. It is estimated that more than 2.6×10^{11} kg of hazardous chemical wastes are being generated annually by the manufacturing units in U.S.A. alone, and that only 10% of these wastes are disposed of in an environmentally safe manner. Each manufacturing unit uses a (some) specific chemical(s) in its processes and hence the resulting untreated wastes have the same type of contaminants in them. For example, Table 2.6 provides a list of industries and the major trace elements detected in their effluent wastes. The Indian industrial sectors are ranked as the tenth biggest in the world in gross industrial output.

5. Conveyance

Automobiles like road vehicles, trains, aircraft etc. cause air and noise pollution.

6. Smoke

Smoke comes into the environment from various industries and houses.

7. Pesticides

Various pesticides viz. insecticides, fungicides, herbicides, molluscicides, and other biocides cause pollution directly or indirectly.

8. Fertilizers

The synthetic fertilizer urea (NH_2CONH_2) required for the boosting up of agriculture crops also cause pollution.

9. Sewage

It is a mixture of liquid pollutants emerging from industries and houses. The domestic sewage contains urine, fecal matters, waste detergents, kitchen wastes etc.

10. Radioactive Substances

The radioactive substances and their radiations cause pollution problems.

Table 2.6 : Major trace elements in industrial effluents

Type of industry	*Trace elements in effluents*								
	As	Cd	Cr	Cu	Pb	Mn	Hg	Se	Zn
Alkali/chlorine	■	■	■	■		■	■	■	
Batteries	■			■		■		■	
Electrical			■	■		■	■		
Electroplating		■	■					■	
Explosive	■			■	■		■		
Fertilizer	■	■	■	■	■	■		■	
Leather		■							
Mining and metallurgy	■	■	■	■		■	■	■	
Motor vehicles	■	■		■		■	■	■	
Pharmaceuticals	■					■			
Paints and dyes	■	■	■	■		■	■		
Pesticides		■				■		■	■
Petrochemical	■	■		■		■	■	■	
Petroleum refining	■	■	■	■				■	
Pulp and paper						■			
Steam generation			■					■	
Textiles		■							

Estimation of Pollution Levels

The pollution level of a region can be compared to another when estimates of the influencing factors (population, industrialization, and urbanization) are available. The concentration of pollution in a region is given by the following expression (Thring, 1974) :

Pollutant levels in a region $\propto$:

$$(n/A)\ G(c + w) - t - C_f$$

where

n = number of people living in the region

A = area of the region

t = half-life of pollutants in the region and of people in years

G(c + w) = amount of goods consumed and wasted per head per year

C_f = carelessness factor (which has a maximum value of 1 if no trouble is taken to reduce the production or emission of pollutants into the environment).

In India the two major contributing factors to the increasing pollution problems are the increasing human population and industrial productivity, which have been increasing by about 2 - 3% annually.

Types of Pollution

Pollutants need not be material substances. Noise may be a pollutant and, so much so, electromagnetic waves too! However, man is the principal source of pollution today, because there is no way for nature to decompose various man-made materials and return their elements to the cycle of nature. This substance will just remain and will cause whatever harmful effects they can, until they are somehow dispersed or diluted so that their action is no longer harmful.

Thus pollution may be *natural* (which originates from natural processes such as forest fires, natural organic and inorganic decays, volcanic eruptions, earthquake etc.) or *artificial,* which originates due to activities of man i.e. anthropogenic sources. However, it would be rather better to classify *pollution* either according to the *environment* (air, water and soil) in which it occurs or according to the pollutants by which pollution is caused, when it may be of a number of types. For example :

1. Metallic pollution
2. Industrial pollution
3. Pesticides or biocide pollution
4. Soap and detergent pollution
5. Acid rain pollution
6. Sewage pollution
7. Chemical pollution
8. Noise pollution
9. Thermal pollution
10. Smoke pollution
11. Gaseous pollution
12. Effluent pollution

13. Oil pollution
14. Silt pollution
15. Drug pollution
16. Radioactive pollution
17. Smog pollution
18. Dust pollution
19. Ozone pollution
20. Electromagnetic pollution.

Important types of pollution have been discussed herein at appropriate places .

POLLUTION IN RELATION TO PUBLIC HEALTH

Human beings are probably most concerned about the direct effects of pollutants upon their own health; even though these effects are not necessarily the most dangerous over the long run. It has been recognized that many human diseases are air-borne as well as water-borne. Industrial workers may be exposed to various type of occupational hazards following inhalation of particulates/dust.

Some occupational diseases are given below which affect respiratory tract and the lungs of people :

Sl. No.	*Name of disease*	*Caused by*
1.	Silicosis	Due to silica
2.	Byssinosis (asbestosis)	Due to asbestos
3.	Anthracosis (Black lung disease)	Due to coal dust
4.	Bagassosis	Due to sugarcane dust
5.	Tobaccosis	Due to tobacco
6.	Siderosis	Due to iron
7.	White lung disease	Due to textile dust

Table 2.7 : Common air pollutants, their sources and pathological effects on man

Pollutants	*Where they come from (source)?*	*Pathological effect on man*
1. Aldehydes	Thermal decomposition of fats oil, or glycerol	Irritate nasal and respiratory tracts
2. Ammonia	Chemical processes — dye-making; explosives; fertilizer	Inflames upper respiratory tracts passages
3. Arsine	Processes involving metals or acids containing arsenic soldering	Breaks down red cells in blood, damage kidneys; causes jaundice
4. CO	Gasoline motor exhausts; burning of coal	Reduces oxygen-carrying capacity of blood
5. Chlorine	Bleaching cotton and flour; many other chemicals processes	Attacks entire respiratory tract and mucus membranes of eyes; causes pulmonary edema
6. Hydrogen cyanide	Fumigation; blast furnaces; chemical manufacturing; metal plating	Interferes with nerve cells; produces dry throat, indistinct vision, and headache
7. Hydrogen fluorides	Petroleum refiniing; glass etching; aluminum and fertilizer production	Irritate and corrode all body passages
8. Hydrogen sulphides	Refineries and chemical industries	Smell like rotten eggs; cause nausea; irritate eyes and throat
9. Nitrogen oxides	Motor vehicle exhausts, soft coal	Inhibit cilia action so that soot and dust penetrate far into the lungs
10. Phosgene (carbonyl chloride)	Chemical and dye manufacturing	Induces coughing, irritation and sometimes fatal pulmonary edema
11. Sulphur dioxides	Coal and oil combustion	Cause chest constriction, headache, vomiting and death from respiratory ailments
12. Suspended particles (ash, soot, smoke)	Incinerators — almost any manufacturing	Cause emphysema; eye irritations and, possibly, cancer

Pollution in relation to public health may be grouped into four main sections :

1. Air pollution and public health
2. Water pollution and public health
3. Pesticidal pollution and public health
4. Radiation pollution and public health.

1. Air Pollution and Public Health

Severe air pollution adversely affects public health and causes various fatal diseases in them. For example, as mentioned above, there occurs lung disease in workers exposed to occupational hazards — such as *Black lung disease* among coal miners who are inhaling mine dust for many years, or Byssinosis (asbestosis) among pipe filters and insulation workers, exposed to airborne asbestos fibres. As listed in Table 2.7, a variety of air pollutants have been reported to cause many human diseases like chronic bronchitis, emphysema, lung cancer, pollen allergies, especially in city-dwellers. Mountain et al (1968) reported that air pollution due to particulate matter and CO in New York City caused breathing problems in children under 8 years of age. Becker et al (1968) reported that in many American cities along the Eastern Seaboard, increasing frequencies of bronchitis, sore throat, cough, wheezes, eye irritation, and general ill-health in public occurred as air pollution level increased.

Important Common Air Pollutants and their Effects on Human Health

1. Carbon monoxide

CO exhibits its toxic symptoms mainly by inactivating the hemoglobin (Hb) in blood. Whereas the combination of CO with Hb leads to the formation of carboxyhemoglobin (COHb), oxygen normally combines with Hb to form oxyhemoglobin (OHb) :

(a) Under normal conditions

$O_2 + 2Hb \rightarrow 2OHb$ (Oxyhemoglobin)

(b) In the presence of excess CO

$CO + Hb \rightarrow COHb$ (Carboxyhemoglobin)

The association of O_2 with Hb to produce HbO_2 and its consequent dissociation serves to supply the tissues with oxygen necessary for their life. Normally blood contains about 0.5% of COHb. But when CO is freely available as oxygen, the hemoglobin shows an affinity for CO approximately 250 times more than for oxygen, thus favouring the formation of COHb. The formation of carboxyhemoglobin reduces the amount of Hb available for performing its vital role of supplying O_2 to the tissues.

The principal adverse effects of CO on us are :

(i) The adverse effects of CO in human body are unique. It is an asphyxiant gas which causes injuries to vital organs.

(ii) On inhalation, CO reacts with Hb forming stable COHb, which restricts transport of O_2 from lungs to cells.

(iii) The classical symptoms of CO poisoning are headache, fatigue and dizziness. These symptoms appear at COHb level of 10% which corresponds to atmospheric concentration of 100 ppm of CO; when 50% Hb is transformed to COHb, life may cease.

(iv) At moderately high levels of exposure, CO shows synergistic toxic effects with H_2S and NO_2 separately.

(v) At 100 ppm, CO causes difficulty in breathing and irritation of mucus membrane. It results in unconsciousness and impairs visual perception.

(vi) CO is extremely dangerous to those persons who are exposed to its 250 ppm level. It shall not be out of place to mention that cigarette smoke contains about 400 - 450 ppm of CO. The percentage of COHb in the circulating blood of smokers increases positively with the number of cigarettes smoked. The blood COHb of non-smokers ranges from 0.5 to 1.5% whereas that of smokers averages about 5%. Smokers who consume one pack of cigarettes per day average 6.3% COHb in their blood and an additional pack of cigarettes a day contributes another 1.4% COHb. Traffic policeman at busy street-crossing may also be exposed to higher levels of CO.

(vii) At 300 ppm level, CO causes throbbing headache followed by vomiting and collapse.

(viii) Common clinical complaints caused by CO exposure at 500 ppm may be choking, chest pain, irritability, anginal pain etc.

(ix) CO is fatal over 1,000 ppm causing hypoxia, paralysis and death in few hours.

2. SO_2

Sulphur dioxide is a highly irritating gas which adversely affects public health. *London* or *Sulphurous Acid Smog* — called the killer smog — had been associated with increase in SO_2 content in air. The principal adverse effects of SO_2 on human health are :

(i) Even at 2.5 ppm level, SO_2 causes spasm of the smooth muscles of the bronchioles present in lungs. It is absorbed by nasal system leading to swelling and stimulated mucus secretion. It particularly hits the aged ill persons.

(ii) At 4.5 ppm level, SO_2 contracts the muscles that encircle the bronchioles with the result that these tubes get narrowed.

(iii) In highly polluted area it may cause watery nasal discharges, sneezing, productive cough, dyspnea and post-nasal sticky discharge.

(iv) Vehicular exhaust consisting of higher level of SO_2 induces desquamation, peeling of the surface epithelium in the mucosa and windpipe.

(v) Emphysema, cough, eye irritation, lung cancer, obstructions in breathing, asthma and allergic infections are other acute effects of SO_2.

3. Ozone

Ozone, a natural constituent of the atmosphere, is composed of three oxygen atoms. The word 'ozone' (in Greek, *ozein* means 'to smell') was introduced by Shoenbein in 1840 to characterize a chemical species, which had a pronounced odour. This bluish, stinky and explosive gas is formed when an oxygen molecule (O_2) dissociates thermally. Each atomic oxygen (O) combines with a molecule of oxygen (O_2) to form an ozone (O_3) molecule. Exposure to ozone, at ground levels, is because of its use as a disinfectant for air and water bleaches, textiles and oils. Ozone is a key component of oxidant smog and is produced as a result of electrical discharges, occurring with lightning during thunderstorms or with the sparking of electrical appliances. Ozone is also present in vehicle exhaust.

The principal adverse effects of ozone on humans are

(i) Ozone is an extremely reactive molecule. Exposure of low levels of ozone accelerates the ageing of lung tissues due to oxidation of certain proteins.

(ii) Ozone is also associated with visual impairment, loss of immune system and lung cancer.

(iii) Chronic bronchitis and emphysema, a destructive lung disease, might be due to ambient ozone exposure.

(iv) 1 ppm of O_3 causes accumulation of inflammatory cells at the site of lung injury. Severe lung injury is associated with edema and hemorrhage.

(v) An increased concentration of ozone brings about change in DNA and RNA i.e., nucleic acids.

(vi) The capacity of lung phagocytes, which normally fight bacterial infections, is also affected by O_3 — resulting in increasing incidence of respiratory infections.

(vii) Ozone has been reported to be a strong irritant. Even at 9 ppm, it causes pulmonary edema.

4. Hydrocarbons

In India, automobiles constitute the principal sources of hydrocarbons (C_nH_{2n+2}) and are emitted by vehicles as a result of evaporation or incomplete combustion. The hydrocarbons other than methane known as NMOC (non-methane organic compounds viz; aromatic hydrocarbons like benzene, toluene, benzpyrine and PAHs), especially the olefins, can react with NO_x, CO, PAN, PBN, PPN, H_2O_2, aldehydes and ketones to produce photochemical smog and ozone. The relative reactivity of various hydrocarbons with hydroxyl radical to produce photochemical smog are :

- CH_4 1
- C_2H_6 33
- Benzene 180
- Propane 270
- n-Butane 370
- Toluene 750
- n-hexane 790
- 1, 2, 4 Trimethyl benzene 4,170
- Benzo-a-pyrene 8,750

PAHs, often adsorbed on particulate matter, are risky portion of vehicular emissio·

The principal adverse effects of hydrocarbons on human health are :

1. Hydrocarbons are carcinogenic in nature i.e., these cause cancer in humans affecting cell growth, DNA and RNA.
2. Excess of hydrocarbon emission increases mucus secretion as a result of which respiratory tracts are blocked and the patient coughs regularly.
3. PAHs with other pollutants cause synergistic effects.
4. NMOC, viz. hydrocarbons like benzene and toluene, produce extreme fatigue, weakness, headache, edema, drowsiness and symptoms akin to drunkenness.
5. Cyclic hydrocarbons affect nervous system, pupil dilatation and cause loss of coordination.
6. Methane and benzpyrene present in gasoline exhaust in absence of O_2 create narcotic effects on human beings.

5. Photochemical Smog

Nitrogen oxides in combination with unsaturated hydrocarbons, CO, O_3, H_2O_2, and organic peroxides in presence of sunlight lead to the formation of photochemical smog. Photochemical reactions are initiated by ultraviolet light in air. PBN (peroxybenzoyl nitrate) is also produced in photochemical smog in presence of NO_2 and olefins. The smog may have a 50 ppb of PAN also. PBN is 100 times more toxic than PAN and 200 times as powerful as HCHO.

Metropolitan cities of India (Delhi, Mumbai, Kolkata, Chennai) are suffering from this smog because of abundant sunshine, atmospheric temperature inversion and growing vehicular emission.

The principal adverse effects of photochemical smog on humans are :

1. Photochemical smog is characterized by yellowish gray haze in the atmosphere. This smog alone, with ozone, may irritate the eyes and nose.
2. Symptoms like nasal block, discharge of sneezing, hacking cough, reduction, or changes of airway obstruction may persist in highly vehicular areas.
3. Studies also revealed prevalence of chest problems and upper airway morbidity in areas with higher levels of oxidants in Mumbai and Delhi.
4. Occurrence of cold, cough and sputum and breathlessness are closely related to peaking of vehicular pollution.
5. Photochemical smog promotes respiratory problems, reduces visibility, alters various blood parameters and aggravates diseases like headache and bronchitis.

6. Particulate Matters

Particulate matters or particulates or aerosols are dispersed solid or liquid substances with a diameter ranging between approximately 0.02 and 100 micron. These are present in vehicle emission and sometimes as smokes (e.g., diesel particulates). Particles of dust soot, fumes and mists are potentially dangerous for human health as :

1. Particulates, due to their small size, penetrate deep into the alveoli of lungs for indefinite period of time and damage lung tissues.
2. Particulates may slow down the ciliary's function, thereby inhibiting the removal of deleterious substances in the mucus flow and cause bronchitis.
3. Soot particles deposit in nose and throat.
4. Insoluble particulates, which cannot be phagocytized by WBCs, pass through the alveolar walls into lymph channels and accumulate in specific organs exerting toxic actions.
5. Silicosis, a chronic disease of lung is caused by inhalation of dust containing free silica.
6. Particulates create mutagenic and carcinogenic effects in humans by carrying deleterious chemicals like lead compounds, metals as Cd, Be, Hg, As, and hazardous organics adsorbed onto the small particulates such as PAHs (polycyclic aromatic hydrocarbons).
7. Chronic exposure to vehicular exhaust containing arsenic leads to the so-called Black Foot Disease in metropolis.

7. Nitrogen Oxides (NO_x)

Nitrogen Oxides (NO_x) are largely emitted by automobiles and the electric power industry in developed countries. In metropolitan cities, vehicular exhaust is the most important source of nitrogen oxides. NO_x may be present in the atmosphere in the following forms : NO, N_2O, NO_2,

N_2O_3 and N_2O_5. Nitrogen Oxides are the second-most abundant atmospheric pollutants and are extremely dangerous to human health.

NO is biochemically inert and not extremely toxic. NO_2 is five times as toxic as NO. Threshold limit value (TLV) of NO and NO_2 are 25 and 5 ppm, respectively. NO is not a major health hazard, but it does oxidize to NO_2 :

$$NO + \frac{1}{2} O_2 \rightarrow NO_2$$

The following are the adverse effects of NO_x on human health :

1. The health effects of NO_2 vary with the degree of exposure. An exposure to 50 to 100 ppm of NO_2 causes inflammation of lung tissues for 30 - 50 minutes for a period of 5 - 8 weeks.
2. NO_2 has irritating effects on mucus membrane.
3. Higher doses of NO_2 cause bronchitis and respiratory problems. An exposure to 150 - 200 ppm of NO_2 results in "bronchiolitis fibrosa obliterans", a dangerous disease, which occurs within 3 to 5 weeks after the exposure.
4. 500 -600 ppm of NO_2, after 2 to 10 days of exposure, results in death of the victim.
5. NO_2 also causes edema followed by death.
6. NO_2 is known as pulmonary irritant whose excess concentration causes pulmonary hemorrhage.
7. NO_2 poisoning results in Silofilter's disease when NO_2 is produced by the fermentation of ensilage-containing nitrates.
8. Death often results by irritation of gases containing NO_x from burning celluloid and nitrocellulose films.
9. Cigarette smokes readily develop lung disease as the cigarettes contain 330 - 1,500 ppm of NO_x.
10. NO_2 also lowers the resistance to influenza and irritates eyes.
11. Nitric, nitrous acid and several nitrates cause respiratory, digestive and nervous ailments.
12. NO gets attached to Hb and reduces O_2 transport efficiency of blood.
13. Higher levels of NO_x cause inflammation, internal bleeding, pneumonia, O_2 deficiency, lung cancer etc.

2. Water Pollution and Public Health

Water, the wonderful natural resource, is extremely essential for survival of all living beings. But today clean water has become a precious commodity and its quality is threatened by numerous sources of pollution. Several waterborne infectious diseases are directly related to polluted water. In addition, the aquatic food chain acts to concentrate several toxic substances as it ascends from micro or through various predators. Organochlorine and organomercury pesticides, PCBs and some radioactive pollutants are concentrated in this way.

Well water contaminated by nitrates from fertilizer run-off poses a hazard to health, particularly for infants.

Municipal waste is the principal contributor of water pollution. A recent report from Water Pollution Research Laboratory, London, indicates that domestic sewage contains trace quantities of toxic metals such as Cu, Cr, Zn, Pb, Co, etc.

The toxic and pathological effects of some heavy metal water pollutants which affect public health have been listed in Table 2.8 :

Table 2.8 : Pathological effects of heavy metal water pollutants on human

Metal	*Pathological effects on human*
1. Mercury	Abdominal pain, headache, diarrhea, hemolysis, chest pain and brain damage.
2. Lead	Anemia, nausea, vomiting, loss of appetite, convulsions, damage of brain, liver and kidney.
3. Arsenic	Disturbs peripheral circulation, mental disturbance, liver cirrhosis, hyperkeratosis, lung cancer, ulcers in gastrointestinal tract, kidney damage and carcinogenic effects.
4. Cadmium	Diarrhea, growth retardation, bone deformation, kidney damage, testicular atrophy, anemia, injury of central nervous system and liver, hypertension.
5. Copper	Hypertension, uremia, coma, sporadic fever.
6. Barium	Excessive salivation, vomiting, diarrhea, paralysis, colic pain.
7. Manganese (essential trace element)	Chronic exposure leads to neurological problems and disorders.

Contd...

Metal	Pathological effects on human
8. Zinc	Vomiting, renal damage, cramps.
9. Selenium	Damage of liver, kidney and spleen, fever, nervousness, vomiting, low blood pressure, blindness and carcinogenic effects.
10. Hexavalent Chromium	Nephritis, gastrointestinal ulceration, impairment in central nervous system functioning, cancer.
11. Cobalt	Diarrhea, low blood pressure, lung irritation, bone deformities, paralysis.

Loss of Public Recreational Areas

Beaches are closed when the bacteria count of water is too high, indicating pollution by fecal matter. Beaches are also closed when the water contains deleterious substances which give off foul odour. In heavily contaminated areas where algae have thrived and died, rotting mats of floating debris wash-up on the shore. Solid wastes sometimes wash-up also, and some oil spills have caused spectacular damage to valuable beach properties. Water containing considerable organic matter generally becomes depleted in oxygen as microbes that feed on dead and decay materials themselves die-out! Decay consumes oxygen and desirable fish needing relatively high oxygen level either die out or go elsewhere. Mercury pollution or pollution by PCBs also reduces the recreational value of a fishing lake. Resort owners, commercial fisherman and the communities that are benefited from such recreational activities suffer considerable economic damage from different types of water pollution

Aquatic oil pollution and public health

Oil pollution in marine water also affects public health critically in the following ways :

(i) Paraffins like methane and ethane are asphyxiants i.e., they cause acute suffocation. Some paraffins act as central nervous system depressants. Liquid paraffins can remove oil from exposed skin causing dermatitis and pneumonia in lung tissues.

(ii) Breathing higher concentrations of unsaturated cycloparaffins can result in irritation and anesthesia.

(iii) Aromatic thiophenes, benzothiophenes and mercaptans are lethal as they damage liver and kidneys.

(iv) Crude oil contains sulphur compounds, small amount of nitrogen, little olefins, metals like iron, nickel and vanadium. These are extremely lethal. Carbonyl sulphide is dangerous poisonously. Actually it dissociates to hydrogen sulphide, which acts on central nervous system resulting in death mainly from respiratory paralysis.

Disease-Causing Agents in Water

Water has been a potential carrier of deleterious inorganic and organic materials, non-biodegradable matters and pathogenic microbes which can endanger public health and life. The potable water contaminated with municipal sewage is the root cause of dangerous diseases in public life.

The waste-water coming from slums, hotels, restaurants, residential areas and animal waste, human fecal matters, organic wastes dissolved compounds as proteins, fats, carbohydrates, pathogens like bacteria, algae, viruses etc. which pollute water bodies directly or indirectly are nuisance to public health.

Effects of Pathogens

1. Parasites are considerably harmful for humans. Eggs of nematodes, hookworms and tapeworms occur mostly in crude sewage. When such sewages are discharged into water bodies without treatment, contamination of water takes place with resultant danger to human and aquatic life .
2. Mainly, drinking contaminated water or swallowing food transmits the enteric diseases. The pathogens most frequently transmitted through water cause infections of intestinal tract like typhoid, paratyphoid, amebic dysentery, cholera, polio and infectious hepatitis.
3. The disease-causing organisms present in the feces of infected persons get ultimately mixed with the water supply — spreading chronic diseases.
4. The guinea worm, responsible for dracontiasis, is transmitted through open village-wells and ponds infested with the copepod intermediate host.
5. Several human diseases, whose epidemics recurrently detriment human population, occur when the populace get contaminated by water.
6. Intestinal heliminths like *Ascaris* and *Trichuris* are also water-borne.

Infections transmitted from man to man

Man's infectious diseases which spread especially through water and food are :

1. Typhoid fever, bacillary dysentery, cholera, poliomyelitis and hepatitis occur from contaminated water and food.
2. Staphylococcal and streptococcal infections may occur from water and food or also by contact and inhalation.
3. Coxsackie and Echovirus diseases occur from water through ingestion.

3. Pesticide Pollution and Public Health

Generally, pesticides are the organic chemicals used to control unwanted and dangerous species of plants and animals. These are the economic poisons employed to regulate the impact of various pests upon our life and economy. Pesticides include fungicides, insecticides, herbicides, nematicides, molluscicides etc.

Effects of Pesticides on Human Health

Pesticides acutely affect man, animal, plants, soil, as well as the aquatic biota. Toxicity towards the aquatic flora and fauna is an important criterion for a pesticide to be considered as a hazardous pollutant. Pesticides affect public health in a variety of ways :

1. Higher concentration of DDT in body parts and blood of human beings causes anxiety, tension, cancer, mutations, stress reactions, congenital in uteri, malformations and impotency.
2. Central nervous system is the target of DDT poisoning in man, which ultimately leads to death. It is soluble in lipids and accumulates in the body fats.
3. The most threatening point is that DDT concentrates and accumulates in the food chain. It is continuously recycled in living organisms. The level of DDT estimated in human tissues lies in range of 5 to 10 ppm. According to an estimation, DDT concentration in man's body fat varies from 3.3 g/m^3 in U.K. to 25g/m^3 in India.
4. Clayson has suggested that low levels of DDT may cause cancer, high blood pressure and cirrhosis of the liver. DDT also interferes with signal transmission at the synaptic gap.
5. Insecticides like BHC, Aldrin, Dieldrin, Chlordane and Endo-sulfan are extremely toxic to human beings. These are considered to affect the vital organs, heart, brain, kidneys and liver — producing chronic disturbances.
6. Pesticide 2, 4, 5-T consists of a highly toxic component called dioxin or TCDD which is teratogen and liable to cause congenital malformations.
7. Organophosphates like Ethion, Fenthion, Dimethoate, Trithion, Monocrotophos, Dazinon, Dursban, Phosdrin, Thionazin and Metasyotox inhibit the production of cholinesterase at the junction between adjoining nerve cells; consequently cholinesterase breaks down acetylcholine secreted by nerve cell axons. Excessive accumulation of acetylcholine interferes with the nerve impulse transmission.
8. The organophosphorus insecticides such as parathion, malathion and TEPP may be absorbed by the lungs, eye,membrane and skin to toxic amounts. This excessive absorption leads to larger accumulation of acetylcholine in the body disurbing the normal functioning of blood.
9. Chronic accumulation of pesticide plays a major role in liver and kidney malfunctioning, secretion of excess of amino acids in human blood and urine, blood abnormalities as well as deformations of brain tissues.
10. The danger of long-term consumption of pesticide residue in food is far more serious than acute poisoning from the viewpoint of national health.
11. Pesticides are also reported to cause chronic impotency in man.
12. Pesticides poisoning in Kerala (1951) resulted in death of 112 people while in Indore (1968) 35 cases of malathion poisoning were reported. On Dec. 3, 1984 Bhopal Gas Tragedy was the worst-ever pesticidal MIC accident in history. This carbaryl (a popular carbamate pesticide) leakage caused an increased risk of sleeping, digestive problems, vision

problems, sterility, kidney and liver infections as well as brain damage. According to Pearson (1985), pesticide- reported deaths in developing countries are estimated as 10,000 per year while two million persons suffer from acute pesticide poisoning.

4. Radiation Pollution and Public Health

The deleterious effect of radiation upon human is due to its ability to ionize and ultimately destroy the organic molecules of which body cells are composed of. There are two important factors upon which the extent of damage depends. These are : the energy and the type of radiation. The energy is usually expressed in rads (a rad corresponds to the absorption of 100 ergs or 10^{-8} Joules of energy per gram of tissue). The overall biological effects of radiation, expressed in rems, is found by multiplying the number of rads by an appropriate factor for the particular type of radiation :

Number of rems = n × Number of rads

where n is 1 for β, γ and X radiation and 10 for α radiation or high-energy neutrons. Some of the effects to be observed when a man is exposed to a single dose of radiation at various levels are given in Table 2.9 :

Table 2.9 : Effect of exposure

Dose (rems)	*Probable effect on man*
0 — 25	No observable effect
25 — 50	Small decrease in white blood cell count
50 — 100	Lesions, marked decrease in white blood cells
100 — 200	Nausea, vomiting, loss of hair
200 — 500	Hemorrhaging, ulcers, possibly death
500+	Fatal, leading to immediate death

Small doses repeated over long periods of time can have extremely serious consequences.

Many of the early workers in the field of radioactivity, for example Madam Marie Curie, suffered from cancer because of chronic over-exposure to radiation. Cancer has also been developed even after 60 years of initial exposure. Recent studies have revealed an abnormally large number of **leukemia** (blood cancer) cases among the survivors of the nuclear bombs dropped at **Hiroshima and Nagasaki** in August 1945.

Radiation may also cause mutations in plants and animals by bringing about changes in chromosomes. Similar genetic effects can also arise in human beings as well. This is confirmed by increased frequency of congenital defects in children born to the survivors of Nagasaki and Hiroshima atomic bomb explosions about 60 years ago (1945)! Perhaps the most disturbing aspect of this problem is that there appears to be no lower limit of **tolerance level** below which the genetic effects of radiation become negligible. Even a small increase in the background level of radiation (now estimated to range from 0.08 to 0.2 rems/year) can be expected to produce a proportional increase in undesirable mutations. For this reason an upper limit of 0.17 rem/year above background has been fixed as maximum dosage to which major population groups can be exposed. This compares to 0.001 rems/year resulting from the operation of nuclear power plants.

The β-rays are electrons moving at high velocities and their velocities range from 0.3 to 0.99 that of light. This has been proved by their deviations in electric and magnetic fields and by their charge to mass (e/m) ratio. The α-rays are particles bearing a positive charge of 2. Their velocity is about 0.05 of that of light. The γ-rays are electromagnetic radiations and not deflected by electric and magnetic fields. They are similar to X-rays but usually have shorter wavelengths.

The energies of α, β and γ-rays are usually measured in units based on eV. An electron volt (eV) is the energy acquired by an electron when it falls through a potential of one volt. Thus

$$1\text{eV} = \text{charge of electron} \times \text{one volt}$$
$$= 1.602 \times\times 10^{-19} \text{ coulomb volt.}$$

The dangers of external exposure to γ-rays are in general much greater than those due to α and β rays.

Biological Magnification of Radioactive Elements

A number of radioactive elements get concentrated in food chain. This is known as *Biological magnification* or *Biological amplification*. An example is P-32, which was found to be concentrated as much as 500 time above levels in the water fish downstream from the Hanford Atomic Power Plant on the Columbia river. Blue gills and crappies in the same river were found to contain 20,000 - 30,000 times the concentration in water of P-32, while filamentous algae were found to have up to 100,000 times the water level. P-32, Zn-65, Fe-59 and I-131 are the other radioactive isotopes that can be concentrated by the living organisms. These are the sources of radiation exposure to the general public when certain foods are eaten. Fish may contain P-32, oysters and clams may contain Zn-65, and milk may contain I - 131.

Two radioactive substances are particularly dangerous because of their tendency to accumulate in the human body :

(i) Radio-strontium 90 — taken into body by breathing or more likely by swallowing with food and water; it tends to collect in bones (half-life 30 years), so causing a serious prolonged exposure to body.

(ii) Radioiodine 131, if assimilated, concentrates in the thyroid gland.

Revision Questions

A. Long-Answer Questions

1. Write an essay on Environment Pollution.
2. Define Pollutants. Give their types and classify them on different basis studied by you.
3. Describe in detail the various causes of environment pollution.
4. Write an essay on pollution in relation to public health.
5. How air pollution affects public health? Discuss.
6. How water pollution affects public health? Discuss.
7. How pesticide pollution affects public health? Discuss.
8. How radiation pollution affects public health? Discuss.

B. Short-Answer Questions

1. Give various definitions of pollution.
2. What are pollutants? Differentiate between primary and secondary pollutants, with examples.
3. Briefly describe the biodegradable and non-biodegradable pollutants.
4. Describe the various causes of environmental pollution in brief.
5. Make a list of common air pollutants, their sources and pathological effects on man.
6. Describe CO and SO_2 pollutions and their effects on human health.
7. Describe the nitrogen oxides (NO_x), photochemical smog and particulate matter pollutions and their effects on human health.
8. Briefly describe the pathological effects of heavy metals as water pollutants on man.

3

Pesticide Pollution

SYNOPSIS

Definition, Classification, Synthetic insecticides (Organochlorines, Organophosphates, Carbamates, Pyrethroids, Dinitro-Phenols), Herbicides, Fungicides, Other pesticides (Rodenticides, Molluscicides, Piscicides), Uses of pesticides into the environment, Effects of pesticides, Fate of pesticides, Salient features of pesticides, Control of pesticidal pollution, Pesticides in integrated pest management.

Definition

According to the Oxford English Dictionary, a pest is described as a troublesome or destructive organism and the suffix '*cide*' implies "slaughter of". Thus a pesticide may be defined as any substance intended for killing pests. In other words, a pesticide is any substance or mixture of substances intended for preventing or controlling any unwanted species of plants and animals.

Actually, pesticides are one of the biocides. A biocide is any substance used with the intention of killing living harmful organisms. Generally pesticides are also known as plant protective agents or crop protection products.

Classification

Pesticides can be classified in various ways :

1. By target organism (Table 3.1)
2. Mode of action (Table 3.2)
3. Type of application (formulation) (Table 3.3)
4. Class of chemical (Table 3.4)

Table 3.1 : Classification of Pesticides by Target Organism

Class	*Target Organism*
Acaricide	Used against mites and ticks, members of the Acaride
Algaecide	Used against algae
Avicide*	Kills or discourages birds
Fumigant	Used to kill soil pests and sometimes weeds
Fungicide	Used against fungi
Herbicide	A weed-killer in strict sense, but defoliants and silvicides are often called herbicides
Insecticide	Used against insects
Miticide	(see Acaricide)
Molluscicide	Used to kill or control molluscs and other invertebrates (e.g., predatory snails, sluge, oyster, drill, lamprey)
Nematicide	Used against nematodes, tiny roundworms that feed on decaying matter, roots or other parts of plants
Ovicide	A chemical used against the eggs of insects, mites and nematodes
Piscicide*	Used to reduce the population of rough fish in a water body
Rodenticide	Used against mice, rats and others rodents
Silvicide	Used against bushy trees and other woody plants and herbs

*Hardly used today

Table 3.2 : Classification of Pesticides by Mode of Action and Effect Type

Cells toxicants	Inhibit important steps of the cell metabolism; e.g., photosynthesis in the case of plants or phosphorylation of ADP to ATP in some animal pests
Neurotoxic substances	Affect the function of membrane permeability of neurons or their metabolism, especially in the case of insects
Growth regulators	Artificial or natural hormones used against undesired plants (rarely insects) by accelerating or inhibiting growth in an unnatural manner
Defoliants	Remove the leaves of plants
Attractants and pheromones	Attract animals, especially insects, even over large distances, into traps; very often, they are sex pheromones
Repellents	Usually gaseous substances which deter insects or other organisms from feeding but do not necessarily kill them
Chemosterilants	Sterilize males of insects or pest vertebrates; classical method of biological pest control
Disinfectants	Destroy or inactivate harmful microorganisms
Contact effects	Uptakes of the active ingredient at the site of application by direct contact; substances that cause localized injury to the plant or animal tissue on contact
Systemic Pesticides	Compounds that are absorbed by the crop and translocated mostly upwards into the stem and the leaves, where they have contact with the target organisms
Selective Pesticides	Whereas insecticides are more or less selective, herebicides are often designed to be non-selective for preplanting applications or selective only against either grasses or dicotyledons
Protective or eradicative	Especially concerning fungicides, bactericides and viricides : prophylactic protection against the respective microorganisms versus treatment after infection
Desicant	Dries out plant before harvest

Table 3.3 : Classification of Pesticides by Types of Formulation

Sprays	Insecticides, herbicides or fungicides as emulsifiable concentrates, water-miscible liquids, water-soluble powders, oil solutions etc.
Dusts	Undiluted toxic agents, aerosol dust, etc., mainly insecticides and fungicides
Granulars	Insecticides, herbicides and algicides in the form of an inert carrier impregnated with a pesticide
Coatings	Insecticides and fungicides for seeds treatment
Aerosols	Indoor application of insecticides, repellents and disinfectants in the form of a spray
Fumigants	Vapours, liquids or gases for soil treatment with mainly nematicide and insecticide effect; also widely used in storage pest control
Baits	Practically no release of the active ingredient into the environment; the toxic substance is ingested by the target animals (insects, molluscs and rodents) after they have been attracted to the bait
Slow-release formulation	Microencapsulated insecticide compounds, paint in lacquers or resin strips for indoor use

Table 3.4 : Classification of the Main Chemical Classes of Pesticides

Insecticides	*Herbicides*	*Fungicides*
Organochlorines (= chlorinated hydrocarbons)	Criteria often used for classification ◆ In relation to the sowing or planting date of the crop to be protected	◆ Inorganic fungicides (based on sulphur, copper or mercury) ◆ Organic fungicides (e.g., Dithiocarbamates)
Organophosphates (=phosphoric acid esters) Carbamates	◆ In relation to the exposure pathway (e.g., as contact herbicide)	◆ Systemic fungicides (e.g., Benzimidazoles)
Natural and synthetic pyrethroids	◆ In relation to the mode of action (e.g., as contact herbicide Phenoxyaliphatic acids, substituted amides, Nitroanilines, substituted ureas	◆ Antibiotics (produced by microorganisms) ◆ The chemical heterogeneity of fungicides is even larger than that of herbicides
Dinitrophenols		
Others	◆ Carbamates and thiocarbamates	◆ In addition to substances applied in the field or greenhouse, a large number of fungicides for seed treatments are of great importance

On the basis of chemical nature, the pesticides may be broadly divided into two groups :

I. *Inorganic pesticides* viz., arsenicals, fluorine compound, zinc phosphide, aluminium phosphide etc.

II *Organic pesticides* viz., hydrocarbons (especially the oils) toxicants of plant and animal origin and synthetic organic pesticides.

Actually, at present there is very little use of inorganic pesticides. Mostly the synthetic organic pesticides, in bulk, are being used in developed and developing countries. Keeping this in view, synthetic organic pesticides are only being discussed in this chapter.

1. SYNTHETIC ORGANIC INSECTICIDES

About 75% of the animal species in the world are insects. Some are beneficial predators and pollinators but many are pests acting as competitors for food. Others are vectors of infections and parasitic diseases. Not surprisingly, man has tried always to control such pests. In the past, preparations containing sulphur, arsenic compounds, extract of tobacco and chrysanthemum and strychnine were used, but only the synthetic pesticides (insecticides) produced by the application of modern chemistry have been really successful. Agricultural yield has increased dramatically over the last 50 years and insecticides have played a major role in this. Unfortunately, many of the compounds used may be harmful to the environment when used carelessly. Most of the insecticides come into three classes i.e., organochlorines; organophosphates, and carbamates.

I. Organochlorine Insecticides (OCIs)

Organochlorines are also known as chlorinated hydrocarbons. These were used extensively in the mid-1940s to 1960s for agricultural pest control and for malarial control programmes. Their properties of low volatility, chemical stability and environmental persistence led to their bioaccumulation or biomagnification in the food chain of fishes, birds and mammals owing to their lipophilicity and slow metabolic degradation. It was then demonstrated that these compounds possess strong estrogenic and enzyme inducing properties which interfere with the reproductive system, although the presenting symptoms are varied and non-specific (Table 3.5). In avian species of a high trophic level like pelicans, seagulls, eagles and vultures, the adverse effects of the dichlorodiphenyl derivatives are related to induction of steroid-metabolizing enzymes and the inability of the reproductive organs to mobilize enough calcium in the production of the eggshell. This eggshell-thinning leads to cracks allowing bacteria to infiltrate with resultant death of the fetus even if there is not complete breakage of the egg in the nest.

Recently the hormone-like activity of DDD (Dichlorodiphenyl Dichloroethane) has raised concern in connection with the steady decline of sperm counts in man over the past 30 years.

Table 3.5 : Symptoms of Organchlorine Poisoning

Compound	*Acute Symptoms*	*Chronic Symptoms*
Dichlorodiphenyls viz., DDT; DDD; Methoxychlor; Prolan etc.	Paraesthesia, ataxia, dizziness, headache, nausea, vomiting, fatigue, lethargy, tremor etc.	Anorexia, weight loss, anemia, tremor, weakness, hyperexcitability, anxiety etc.
BHC and Lindane	Headaches, tremor, convulsion	
Cylodiene compounds viz., Aldrin; Dialdrin; Endrin; Isodrin, etc.	Dizziness, headache, nausea, vomiting, hyperexcitability, hyper-reflexia, myoclonic jerking, convulsion	Dizziness, headache, hyperexcitability, muscle twitching, myoclonic jerking, anxiety, insomnia, irritability, epileptiform convulsions
Others viz., Chlordecone, Mirex etc.		Arthralgia, ataxia, slurred speech, blurred vision, irritability, loss of memory, weakness, tremor, decrease of sperm count

Classification of Organochlorine insecticides

The organochlorine insecticides can be divided into three groups as listed in Table 3.6 :

(i) Dichlorophenylethanes (DDT and related compounds)
(ii) Cyclodiene and related compounds
(iii) Hexachlorocyclohexanes and others.

Although there is no structure common to all organochlorines, they are all characterized by one or more chlorine atoms positioned around one or more hydrocarbon rings. Members of each group of organochlorines share similar or identical compositions although they may have very different three-dimensional structures and shapes. These isomers may differ significantly in their toxicities and other characteristics.

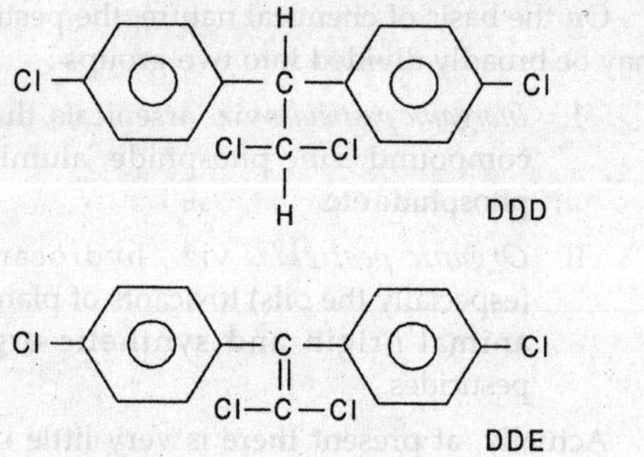

DDD

DDE

Important Metabolites of DDT

DDT was first synthesized by O. Ziedler, a German chemist, in 1874 but its insecticidal property was reported by Paul Muller in 1939.

Table 3. 6 : Chlorinated Hydrocarbons

Sl. No.	*Group*	*Example*	*Salient Property*
(i)	Dichlorophenyl-ethanes	DDT* Chlorobenzilate Dieofol Methoxychlor*	These are highly soluble in fat and thus can easily penetrate the outer integument of insects. This very property poses a threat to human beings too
(ii)	Cydodienes and related compounds	Aldrin Chlorodane* Dieldrin Endosulfan* Endrin Hepatachlor Toxaphene Isodrin	Cyclodiene type insecticides are cyclic halogenated hydrocarbons. These compounds undergo an initial epoxidation forming respective epoxides. These oxidation products generally are more deleterious to biological system than their parent compounds — a phenomenon known as *bioactivation.* The stable epoxides represent the major metabolites of the cyclodienes, which are stored as such in the adipose tissues
(iii)	Hexachloro-cyclohexanes and others	Hexachlorobenzene* Lindane* Chlorthatonil Dalapon Mirex Quintozene Pentachlorophenol	Highly persistent and carcinogenic

***Note** * Indicates that a profile for this compound is precisely included ahead in this chapter (2 compounds from each section)*

(i) Dichlorophenyl Ethanes

1. DDT (p,p : Dichloro Diphenyl Trichloro-ethane)

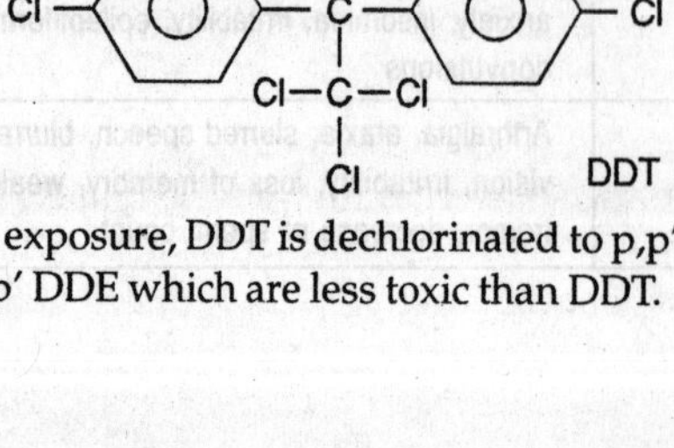

DDT

On exposure, DDT is dechlorinated to p,p' DDD and p,p' DDE which are less toxic than DDT.

Properties

DDT is a highly effective insecticide at broad spectrum. It acts as a contact and stomach poison having longer residual action. It is insoluble in water, fairly soluble in organic solvents and highly soluble in xylene. It is stable under most conditions and possesses a very low vapour pressure. DDT, once introduced into the environment, keeps circulating for many years.

Uses : It is used for cotton, soyabean and peanut pests. It is also used for mosquito control. Its persistence in the environment causes accumulation in food chain. DDT use has been eliminated in U.S. though it is still applied in many regions throughout the world.

Permissible limit in freshwater — 0.001 μg/L

Toxic Effects

- It affects the nervous system and causes violent agitation in insects followed by paralysis and death. However, it acts very slowly.
- It represents the best example of biological magnification as represented in an aquatic ecosystem (Table 3.7) where concentration of DDT continuously increased from water to birds :

Table 3.7 : Biological Magnification of DDT in a Aquatic Ecosystem

Component	*DDT Concentration (ppm)*
Aquatic body	10^{-6}
Zooplanktons	10^{-2}
Minnows	10^{-1}
Needlefish	1.00
Fish-eating birds	10.00

- The oral Ld_{50} (96 hours) of DDT for rats is of 250 mg - 300 mg/kg. A dose of 10 mg/kg produces illness in man.
- The half life of DDT in humans range from 0.5 - 0.7 years. Not only does the accumulation pose a hazard of unknown consequences, but also carries with it the potential of being transferred and secreted in human milk.
- DDT possesses carcinogenic property, as it leads to tumour formation in liver of mouse and rat. Evidence for hepatic carcinogenecity in human is inadequate.
- DDT inhibits Calcium ATPase in CNS.
- Large doses of DDT may produce a moderate leucocytosis and a decrease in hemoglobin without a decrease in the concentration of red cells.

Symptoms of DDT Poisoning in Man

A. Acute Symptoms

Paresthesia, ataxia, dizziness, headache, restlessness, nausea, vomiting, fatigue, hyper-excitability, tremor etc.

B. Chronic Symptoms

Anorexia, weight loss, anaemia, tremor, weakness, hyperexcitability, anxiety, leucocytosis, etc.

2. Methoxychlor

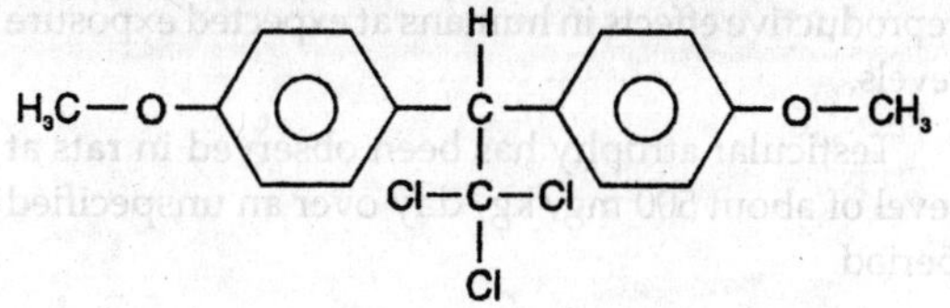

It is also known as methoxy DDT. Its chemical name is 1,1,1 : Trichloro -2, 2-bis methoxyphenyl ethane.

Properties

Pure methoxychlor is a colourless crystalline solid. Technical methoxychlor is 88 - 90% pure. Its vapour pressure is very low. Melting point is 77°C. It is not accumulated in fatty tissues or excreted in milk as DDT is. Therefore, methoxychlor is preferred to DDT for use on animals, in animal feed, and in barns. Since it is more unstable, it has less residual effect.

Major products of its breakdown in a neutral solution are anisoin, anisil and also p, p-dimethoxydichloroethane (DMDE). Its half-life in river water may be 5-10 hours but in soil it approximately 120 days.

Uses : It is a highly popular organchlorine insecticide and is a DDT substitute derivative.

Permissible limit in freshwater — 0.003μg/L

Toxic Effects

- Actually methoxychlor is practically non-toxic via the oral route, with reported (96 hours) oral LD_{50} values of 5,000 - 6,000 mg/kg in rats, 1,850 mg/kg in mice and 2,000 mg/kg in hamsters. The lowest oral dose that can cause lethal effects in humans is estimated to be 6,400 mg/kg and the lowest dose through the skin that produces toxic effect in humans is 2,400 mg/kg based on behavioural symptoms.
- Symptoms of high acute exposure include CNS depression, progressive weakness and diarrhea. Extremely high doses may cause death within 36 - 48 hours.
- High doses of methoxychlor (88 - 90% pure) or its metabolites may have estrogenic or reproductive effects. In rats, dietary doses of about 125 mg/kg/day reduced mating

and many did not produce litters. Rats fed doses of about 50 mg/kg/day had normal fertility and fecundity but their offspring had abnormal reproductive functioning.

It is unlikely that methoxychlor may cause reproductive effects in humans at expected exposure levels.

Testicular atrophy has been observed in rats at level of about 500 mg/kg/day over an unspecified period.

Ecological Effects

- Methoxychlor is slightly toxic to bird species with reported acute oral (96 hours) LD_{50} values of greater than 2,000 mg/kg in the mallard duck and sharp-tailed grouse.
- Methoxychlor is very highly toxic to fishes and aquatic invertebrates. Reported 96 hour LC_{50} values are less than 20μg/l for cutthroat trout, book trout and Atlantic salmon. Reported LC_{50} values (96 hours) are between 20 - 65 μg/l in Rainbow trout, Goldfish, Channel catfish and Yellow perch.
- Methoxychlor is nontoxic to bees.

(ii) Cyclodienes

1. Chlordane

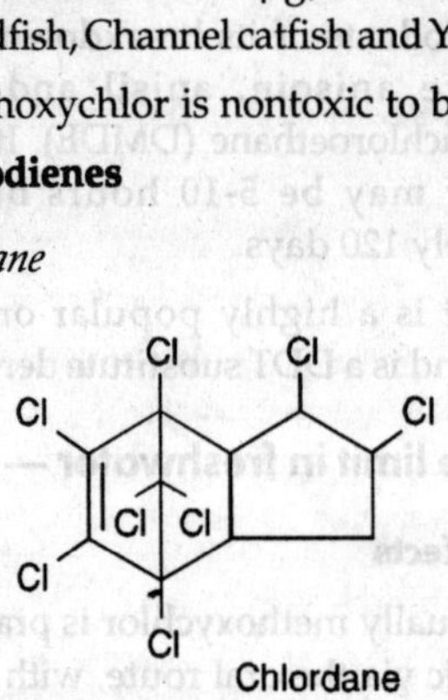

Chlordane

It is also known as Kilex Lindane or Kypchlor or Octachlor or Toxichlor. Chlordane is no longer distributed in the US since 1988 because of concern about the risk of cancer. The only commercial use still permitted is for ant and termite control.

Properties

Chlordane is a persistent cyclodiene insecticide. It kills insects when ingested and on contact. It is metabolized to oxychlordane. Consequently oxychlordane is stored in fat and is excerted in the milk of lactating animals. It does not appear to be a residue of concern, at least in plants. It is a neurotoxicant.

Uses : It is very much effective against termites and ants.

Permissible limit in freshwater — 0.01μg/L

Toxic Effects

Acute toxicity

It is moderately to highly toxic through all routes of exposure. The oral (96 hours) LD_{50} for Chlordane in rats is 200 - 700 mg/kg, in mice 145 - 430 mg/kg, and in rabbits 20 - 300 mg/kg.

Symptoms usually start within 45 minutes to several hours after exposure to its toxic dose. Convulsion may be the first sign of poisoning or they may be preceded by nausea, vomiting and gut pain. Initially poisoning victims may appear agitated or excited, but later they may become depressed, uncoordinated, tired or confused. Other symptoms include headache, dizziness, vision problems, irritability, weakness and muscle twitching. In severe cases respiratory failure and even death may occur.

Chronic toxicity

Liver lesions and changes in blood serum occurred in rats exposed to 1.0 mg/l chlordane. Increased kidney weights occurred in rats exposed to 10.0 mg/l. Animal studies have shown that consumption of chlordane caused damage to the liver and the CNS.

Reproductive effects

Chlordane affects reproduction in test animals. Fertility was reduced by about 50% in mice injected with Chlordane at 22 mg/kg once a week for 3 weeks. However, data reveal that reproductive effects in humans are unlikely at expected exposure levels.

Mutagenic effects : Chlordane is weakly mutagenic or nonmutagenic.

Carcinogenic Effects : It is a probable human carcinogen.

Organ toxicity

CNS, liver and blood disorders have been reported in human exposed to chlordane. It may also cause blood diseases, including aplastic anemia and acute leukemia in rats.

2. Endosulfan

Trade or other names for endosulfan products include Afidan, Endocel, Malix, Thiodan and Thiotox. It is a acarcide of the cyclodiene sub-group. Technical endosulfan is composed of a mixture of two molecular forms (isomers) of endosulfan, the alpha and the beta isomers.

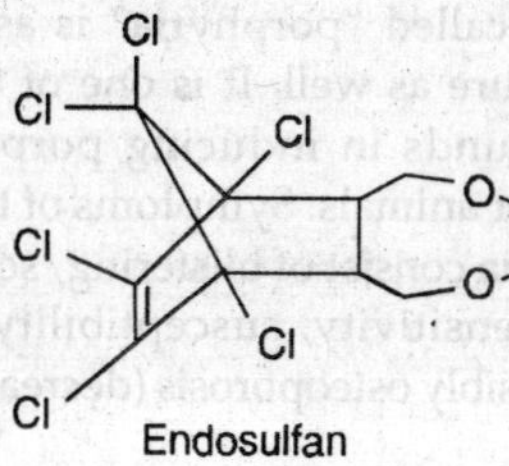

Endosulfan

Properties

It is a brownish crystalline solid, insoluble in water but soluble in xylene. It is a stomach and contact poison with slight fumigant action. It is not compatible with alkaline materials. The alpha isomer is considered to be more toxic than the beta isomer. Its average half life in soil is 50 days.

Uses : Endosulfan acts as a poison to a wide variety of insects and mites on contact. Although it may also be used as a wood preservative, it is used primarily on a wide variety of food crops, including tea, coffee, fruits and vegetables as well as on rice, cereals, maize, sorghum or other grains.

Permissible limit in freshwater — 0.01μg/L

Toxic effects

Acute Toxicity

Endosulfan is highly toxic via the oral route, with reported oral (96 hours) LD_{50} values ranging from 18 - 160 mg/kg in rats, 7 - 36 mg/kg in mice and 77 mg/kg in dogs. It is reported not to cause skin or eye irritation in animals. Stimulation of the central nervous system is the major characteristic of endosulfan poisoning.

Symptoms reported in actually exposed humans include incoordination, imbalance, difficulty in breathing, gazing, vomiting, diarrhea, agitation, convulsions and loss of consciousness. In an accidental exposure, sheep and pigs grazing on a sprayed field suffered a lack of muscle coordination and suffered from blindness.

Chronic toxicity

In rats, oral doses of 10 mg/kg/day caused high rates of mortality within 15 days, but doses of 5 mg/kg/day caused liver enlargement over the same period. Reports reveal that administration of this dose over 2 years in rats also caused reduced growth and survival, changes in kidney structure and blood chemistry.

Reproductive effects

Oral doses for 15 days at 10 mg/kg/day in male rats caused damage to the seminiferous tubules and lowered testes weight. It is unlikely that endosulfan may cause reproductive effects in humans at expected exposure levels.

Teratogenic effects

An oral dose of 5 - 10 mg/kg/day in rats in a three generation study resulted in bone deformity in the offspring. Teratogenic effects in humans are unlikely at expected exposure levels.

Mutagenic effects

Endosulfan is reported to be a mutagenic to bacterial and yeast cells. Evidence suggests that exposure to Endosulfan may cause mutagenic effects in human if exposure is great enough.

Carcinogenic effects

Reports reveal that Endosulfan is not carcinogenic.

Organ toxicity

Data from animal studies reveal that the organ tissues most likely to be affected include kidney, liver, blood and the parathyroid gland.

Effect on Birds

Endosulfan is highly to moderately toxic to bird species, reported oral (96 hours) LD_{50} values in mallards range from 31 - 243 mg/kg.

Effect on fishes

Endosulfan is reported to be highly toxic to various fish species. The reported 96 hour LC_{50} values in μg/l for some fishes are :

Rainbow trout 1.5
Fathead minnow 1.4
Channel catfish 1.5
Blue gill sunfish 1.2

Effects on Non-Target Species

Endosulfan is moderately toxic to bees, however, it is relatively nontoxic to other beneficial insects such as parasitic wasps, lady-bird beetles, etc.

(iii) Hexachlorocyclohexanes

1. Hexachlorobenzene

Trade names for hexachlorobenzene (HCB) products include Perchlorobenzene, Bentcure and Granero.

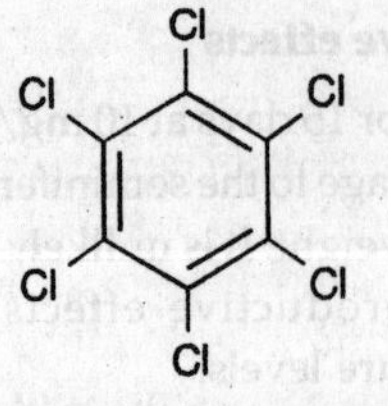

Hexachlorobenzene

Note : This compound should not be confused with hexachlorohexane (HCH), which has historically been referred to as benzene hexachloride (BHC) and is also commonly known as *lindane*, a name proposed after Van der Linden, a German chemist, who isolated this isomers of HCB.

Properties

It has strong musty odour. its melting point is 226°C. It is a colourless crystalline solid at room temperature. HCB is a highly persistent compound with reported field-half lives in the soil environment ranging from 2.7 - 7.5 years. Evaporation is rapid while it is on soil surfaces, but considerably less so when it is mixed into the soil. It may be degraded both aerobically and anaerobically. It has a low water solubility. It has been banned from use in U.S.

Permissible limit in freshwater — 0.01μg/L

Uses : HCB is a selective fungicide used as a seed protectant, especially on wheat, to control common and dwarf bunt. It may be used with or without other seed treatments, fungicides and/or insecticides.

Toxic effects

Acute toxicity

HCB is slightly to practically nontoxic via the oral route of exposure. The reported acute oral (96 hours) LD_{50} values are 3,500 mg/kg in the rat, 4,000 mg/kg in the mouse and 2,600 mg/kg in the rabbit.

Chronic toxicity

Dietary doses to rats of approximately 50 mg/kg/day killed 95% of the females and 30% of the males within 120 days. Almost all survived when the dose rate was reduced to approximately 5 mg/kg/day. Rats receiving doses of 25 - 50 mg/kg/day showed nervous system effects such as tremor, hyperexcitability and lethargy, skin eruption, as well as increases in weights of liver, kidneys, spleen and lungs.

A syndrome called "porphyria" is associated with HCB exposure as well. It is one of the most effective compounds in inducing porphyria in humans and other animals. Symptoms of the HCB-induced porphyria consist of blistering/scarring of the skin, light sensitivity, susceptibility to skin infection and possibly osteoporosis (decreased bone calcium content).

Reproductive effects

In a four generation reproduction study of rats at doses of approximately 8 mg/kg/day, decreased fertility was observed. However, it does not appear that exposure to HCB at normal levels will cause reproductive effects in human population.

Teratogenic effects

Doses of 80 mg/kg/day for one or more days of pregnancy reduced fetal birth weights and caused a slight increase in 14th ribs in rats. The potential for HCB to cause birth defects in human population is likely to be small at common levels of exposure.

Mutagenic effects

Available data reveal that it is unlikely that HCB is mutagenic.

Carcinogenic effects

HCB caused increased numbers of tumours per animal in a long-term study where the lowest dose tested was 4 mg/kg/day. Increases in tumours of the lung, thyroid and spleen were noted. However, the potential for HCB to cause carcinogenic effect in human at normal levels of exposure is not known.

Organ toxicity

Available data from animal tests indicate that the nervous system, liver, kidneys, spleen and lungs are the target organs affected by exposure to HCB.

Ecological effects

Effects on birds

HCB is slightly to moderately toxic to bird species.

Effect on aquatic organisms

HCB is slightly toxic to fish species, with reported (96 hours) LC_{50} values :

Channel catfish	11 - 16 mg/l
Bluegill	12 mg/l
Fathead minnow	22 mg/l
Coho salmon	>50 mg/l

Available data reveal that HCB possesses a significant potential for bioaccumulation.

Effect on Non-Target Species

HCB is nontoxic to bees.

2. Lindane

Trade or other names for lindane include BHC, (benzene hexachloride), Gammexane, Hexachlorocyclohexane, (HCH), gamma - HCH, Linox, Triple six ($C_6H_6Cl_6$) etc.

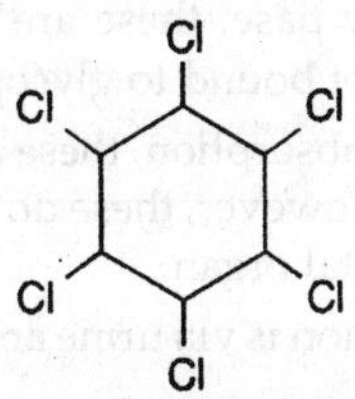

Lindane

Note : Lindane or hexachlorocyclohexane (HCH) has historical and widely been inappropriately referred to as 'benzene hexachloride' or "BHC". This compound should not be confused with hexachlorobenzene or HCB.

Properties

Lindane is a moderately toxic compound. It possesses high potential to cause cancer; hence it is no longer manufactured in US. It is a grayish or browny amorphous solid and has a strong musty odour. It is highly persistent in most soils, with a field half-life of approximately 15 months. When sprayed on the surface, the half-life was typically much shorter. Its melting point is approximately 113°°C.

Uses : Lindane has been used on a wide range of soil-dwelling and phytophagous insects. It is commonly used on a wide variety of crops, in warehouses, in public health to control insect-borne diseases and (with fungicides) for seed treatment. Lindane is also presently used in lotions, creams and shampoos for the control of lice and mites (scabies) in humans. There are 4 isomers of HCH — alpha, beta, delta and gamma. However, gamma HCH (gamma hexachlorocyclohexane) i.e., lindane, has been shown to be the insecticially effective isomer.

Permissible limit in freshwater — 0.01μg/L

Toxic effects

Acute toxicity

Lindane is a moderately toxic compound via oral exposure, with a reported (96 hours) oral LD_{50} 88 - 190 mg/kg in rats; 100 - 127 mg/kg in guinea pigs. Effects of high acute exposure to lindane may include mental/motor impairment, excitation, toxic convulsions, increased respiratory rate and/or failure, pulmonary edema and dermatitis. Other symptoms in human are more behavioural in nature, such as loss of balance, grinding of the teeth and hyperirritability. Most acute effects in humans have been due to accidental or intestinal ingestion.

Chronic toxicity

No observable effects over periods of up to 2 years have been found in mice, rats and dogs at doses of 1.25 mg/kg day. Doses of 40 - 80 mg/kg/day were rapidly fatal to dogs in a study over 2 years.

Reproductive effects

In rats, doses of 10 mg/kg/day for 138 days resulted in marked reductions in fecundity and litter size. In general, lindane has been found to be slightly estrogenic to female rats and mice, and also caused the testes of male rats to become atrophied. It is unlikely that lindane will cause reproductive effects at the expected low levels of exposure in human population.

Teratogenic effects

Though doses as low as 0.5 mg/kg/day over four months in rats decreased growth in offspring, it appears that lindane is unlikely to cause developmental effects at expected low levels of exposure in human population.

Mutagenic effects

Most tests on microbes and mice have shown no mutagenicity due to lindane exposure and it is unlikely that lindane would pose a mutagenic risk in humans at normal exposure levels.

Carcinogenic effects

Studies reveal that rodents may suffer from liver tumor from high doses of the gamma-isomer i.e., lindane. The available evidence is contradictory and does not allow assessment of the potential for carcinogenic effects in humans from lindane exposure.

Organ toxicity

Data from animal tests indicate that lindane may affect the CNS, liver, kidneys, pancreas, testes and nasal mucus membranes.

Ecological effects

Effects on birds

Lindane is moderately to practically nontoxic to bird species, with a reported (96 hours) LD_{50} of more than 2,000 mg/kg in the mallard duck.

Effect on aquatic organisms

Lindane is highly to very highly toxic to fishes and aquatic invertebrate species. Reported (96 hours) LC_{50} values range from 1.7 - 90 µg/l in trout (rainbow, brown and lake), coho, salmon, fathead minnow, blue gill and yellow perch.

Effects on non-target species

Lindane is highly toxic to bees.

Mechanism of toxic action of organochlorines

The chlorinated hydrocarbons are stimulants of the nervous system. Their mode of action is similar in insects and humans. They affect nerve fibers, by disturbing the transmission of the nerve impulse. More specifically, the members of this group of pesticides disrupt the sodium/potassium balance that surround the nerve fibers. The result of this imbalance is a nerve that sends transmissions continuously rather than in response to stimuli.

Despite the similarity of many of the compounds within each of three subgroups, the individual toxicities vary greatly. The compounds also vary greatly in their ability to be stored in tissue. For example, the structure of methoxychlor is very similar to DDT, but its toxicity is far lower, as is its tendency to accumulate in fatty tissue. Storage in fatty tissue is a strategy that the body uses to remove toxic materials from active circulation. Fatty storage prevents the toxic agent from reaching in the circulation and consequently on various organs like liver, kidney etc.

DDT type insecticides interact with the neuronal membrane by altering the membrane permeability (transport) for potassium and sodium and the calcium mediated processes. By inhibiting these functions, the repolarization of the nerves is distributed resulting in hyperexcitability.

Cyclodienes and cyclohexane compounds have a central nervous system stimulating mode of action. These compounds antagonize the neurotransmitter, gammaminobutyric acid (GABA), permitting only partial repolarization of the neuron and, thus, uncoordinated nervous excitation.

Owing to their lipophilicity, organochlorine compounds are partitioned and stored largely in adipose tissues, where they are biologically inactive. There is an equilibrium between body fat and free circulating compounds. Redistribution and mobilization of fat, for example, due to disease, ageing or fasting, may result in mobilization of stored organochlorines in quantities that may lead to manifest toxicity. Owing to the different patterns of use of organochlorine insecticides in industrialized and developing countries, the pattern of distribution of the residues in human fat (including breast milk) differs between countries.

Absorption and Fate of Organochlorines

1. These compounds are insoluble in water but soluble in lipids — so absorption from the gut is very poor.
2. In oily base, these are rapidly absorbed and get bound to glycoproteins.
3. After absorption, these are stored in body fats. However, these do not accumulate in any vital organ.
4. Excretion is via urine and milk.

II. Organophosphate (Organophosphorus) Insecticides

Organochlorine insecticides have largely been replaced by organophosphate (OP) compounds. The OP compounds are essentially polar, hydrophilic (water soluble) and are derivatives of the mineral acid i.e., phosphoric acid, having generally one — rarely two — atoms of phosphorus. Its one or more hydrogen atoms are replaced by alkyl groups. In some cases oxygen atoms are replaced with sulphur atoms (Table 3.8). The sulphur derivative of phosphoric acid may not themselves be active as pesticides, and are activated only by the enzymatic transformation taking place within the organisms. These reactions usually involve the replacement of one or more of the sulphur atoms by oxygen in the phosphoric acid portion of the molecule. They are very toxic even in small quantities, exhibiting anti-cholinesterase activities. OPIs inhibit cholinesterase at the neuromuscular junction.

Needless to mention that the OP compounds are structurally similar to acetylcholine, a potent neurotransmitter. At the enzyme site, they compete with acetylcholine and bind with cholinesterase, thus inhibiting its normal function.

OP insecticides, in general, are less persistent than the organochlorines, a property due to which they are preferred to the organochlorines. But this property also brings their repeated application during the crop period.

R_1 and R_2 = alkyl groups

X = diversified form of structure containing OH

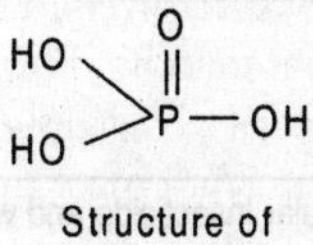

Structure of Phosphoric acid (H_3PO_4)

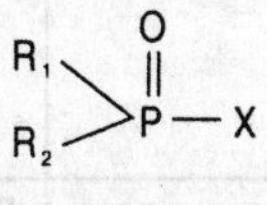

Generalised structure of OPIs

The first OPI to be introduced was TEPP, which was developed in Germany during World War II (1939-45) as a substitute for nicotine. The OPs are esters, amides, or other simple derivatives of phosphoric and thiophosphoric acids. Most OPIs are esters of alcohol with a phosphorus acid or anhydrides of a phosphorus acid with some other acids.

The most commonly used OPIs are listed in Table 3.8. The OPIs act as stomach and contact poisons as fumigants and as systemic insecticides for nearly every type of insect control :

Table 3.8 : List of commonly used OP compounds, their structure and applications

OP compounds/ chemical *name*	*Structure*	*General content*
1. TEPP (Tetra-ethylpyro-phosphate	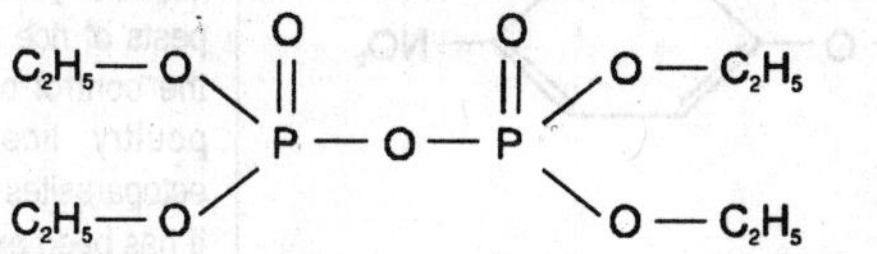	It is a higly toxic OPI. Extremely toxic to mammals. It is not a systemic insecticide. It has been used to control aphids and red spider mites in agricultural and ornamental crops.
2. Malathion*	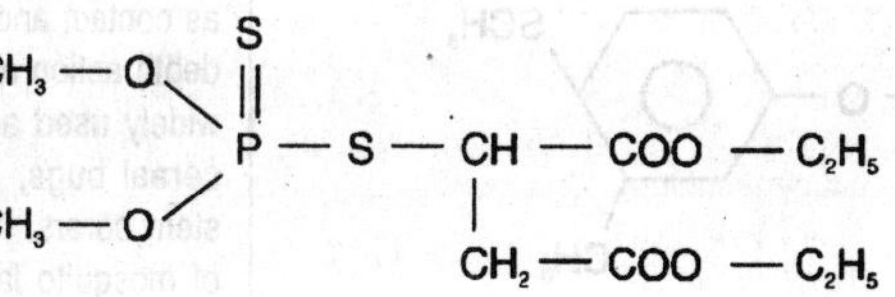	It is one of the safest insecticides. It kills insects by contact or vapour action and is also a stomach poison. It is widely used for the control of many insects. It is a non-systemic insecticide and acaricide of low mammalian toxicity. It is extensively used as a home garden insecticide. It is an indirect inhibitor of cholinesterase. It brings about reduction in develop fingerlings of freshwater fishes (Shukla, J. P. and K. Pandey, 1988).
3. Parathion or Thiophos	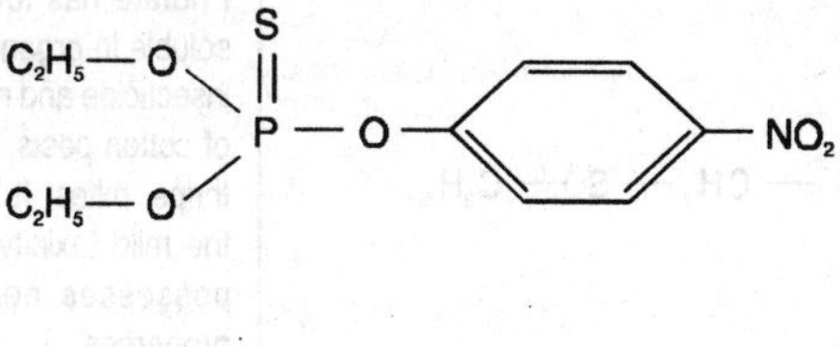	It is most widely used of all the OPIs. Upon oxidation it converts into paraxon which is the typical example of bioactivation. It is highly toxic to mammals. It is used for the control of a variety of insects such as aphids, mites, beetles, leafhoppers, leaf miners, and other insect pests found on fruits, cotton, vegetables and forage crops. It also controls several soil insects such as wireworms, rootworms and syphilids. It is 100 times more toxic than malathion.
4. Methyl parathion	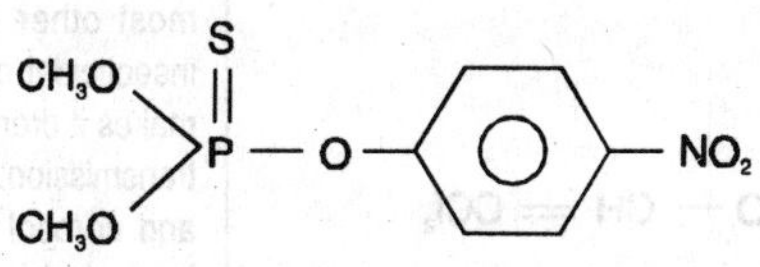	It has the same toxicity to insects as parathion and generally controls the same group of insects i.e., aphids and beetles. It ranks second in pesticide consumption in US. It is less toxic to mammals in comparison to parathion.

Contd...

OP compounds/ chemical name	*Structure*	*General content*
5. Diazinon	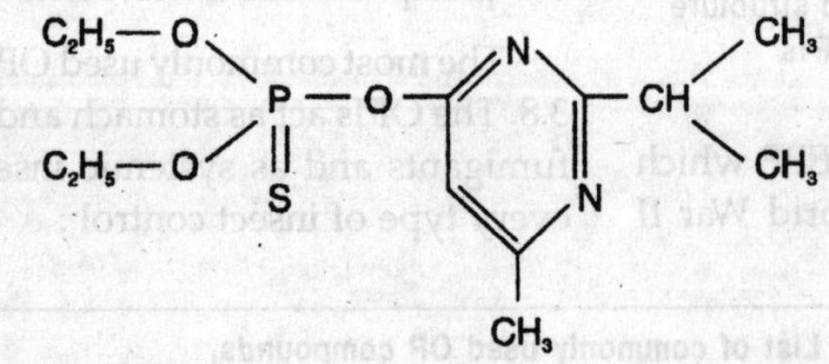	Diazinon is a popular insecticide and widely used today. It does not persist for lengthy periods in either animal or plant tissue. It is less toxic than parathion. It has a wide spectrum of insect-killing power and is used to control household pests, various vegetable and forage crop insects and DDT-resistant flies. It also possesses nematicidal property.
6. Fenitrothion	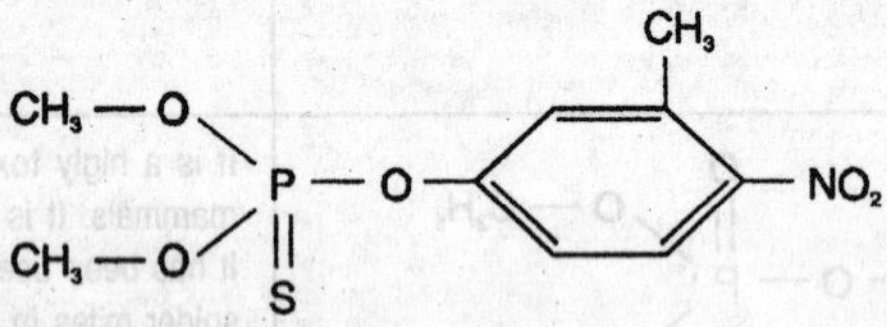	It is a broad spectrum insecticide with a much lower acute mammalian toxicity than several similar insecticides. It is insoluble in water but soluble in organic solvents. It acts as contact and stomach poison. It is especially used for the control of insect pests of rice and is also recommended for the control of mosquito larvae, bedbugs, poultry lice, fleas, ticks and other ectoparasites of livestock and pet animals. It has been extensively used as a substitute of parathion and DDT in forest pest control.
7. Fenthion	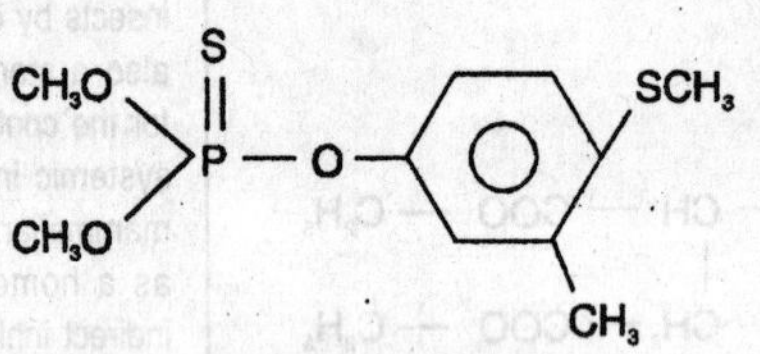	Fenthion is a colourless liquid which acts as contact and stomach poison possessing depth action and long residual effects. It is widely used against fruitflies, leafhoppers, cereal bugs, mango nut weevil and rice stem borers. It is also used for the control of mosquito larvae and bed-bugs.
8. Formothion	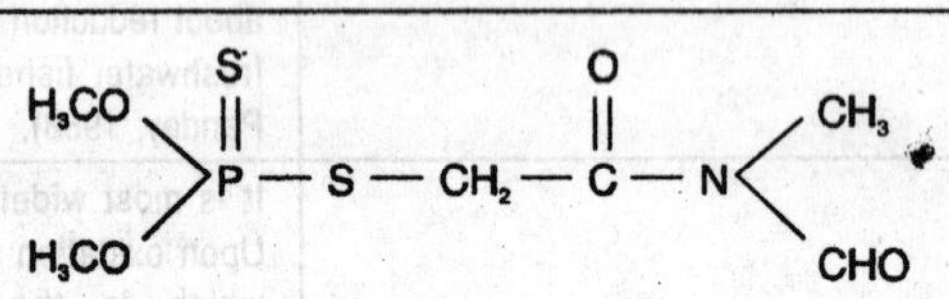	Formothion acts as contact and systemic insecticide and acaricide. It is extensively used against various sucking insects and mites of crops. It is a yellowish viscous oil or crystalline solid.
9. Phorate or Thimet	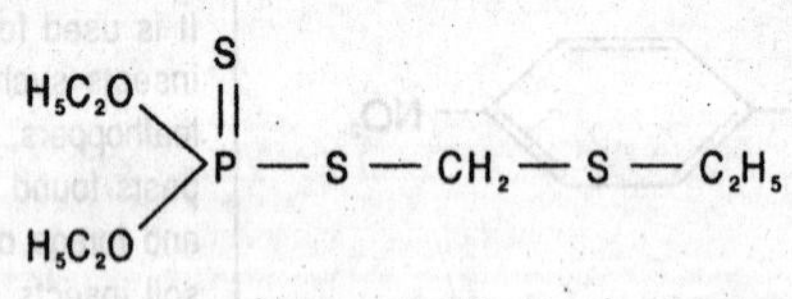	Phorate has low solubility in water but is soluble in organic solvents. It is a systemic insecticide and miticide. It is used for control of cotton pests, as well as for flea, beetles, thrips, mites. It is not phytotoxic except for the mild toxicity to cabbage plants. It also possesses nematicidal and acaricidal properties.
10. Dichlorvos (DDVP) (0,0-dimethyl 2,2-dichloro vinylphosphate)	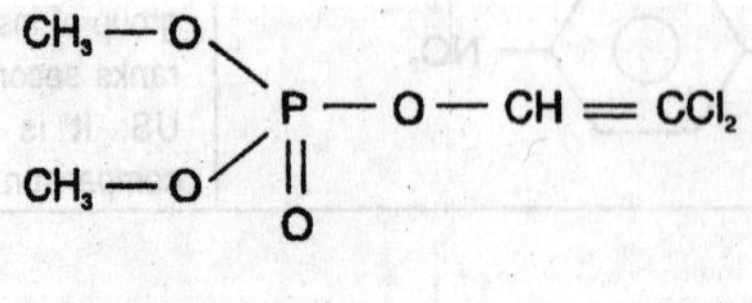	Dichlorvos (DDVP or vapona) differs from most other OPIs in that it produces an insecticidal concentration of a vapour that makes it promising for the control of malaria transmission. It is extensively used in baits and aerosol formulation for the control of household insects of public health importance such as flies and mosquitoes and also for the control of insects of stores. It is moderately toxic to fishes and highly toxic to bees.

Condt...

11. Phosphamidon or Dimecron	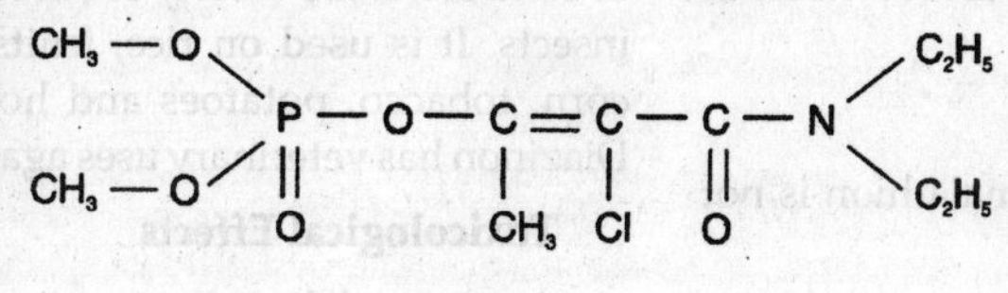	Phosphamidon occurs in two isomeric forms, α and β. Its insecticidal property is mostly attributed to β isomer. It is widely used for the control of yellow borer and other sucking pests of rice.
12. Monocrotophos	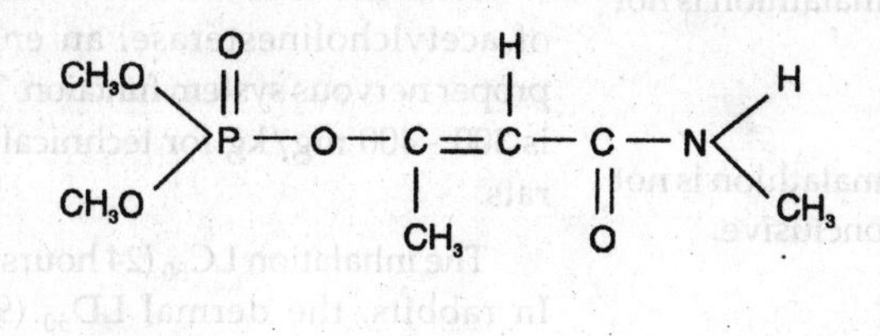	Monocrotophos acts as stomach, contact and systemic poison. It is miscible with water, acetone and alcohol. It is a degraded in plant and animal species by hydrolytic and oxidative mechanisms. It is highly toxic acting as ChE inhibitor. It is excreted rapidly and does not accumulate in the body. It is extensively used against thrips, aphids, leaf miners, leaf webbers and mites of crops.
13. Schradan (OMPA) octamethyl-pyrophosphor amidate	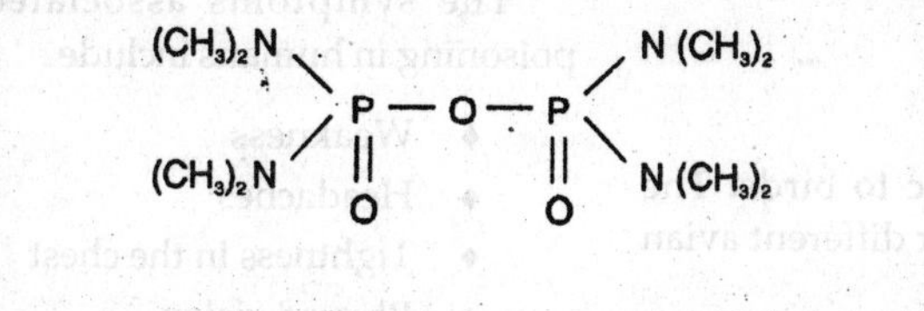	Schradan was the first OPI to be studied as a systemic insecticide. Its systemic action was first discovered by kii Kenthol and Schader in early 1940s. It kills sucking insects without killing their predators. Schradan is a colourless, odourless liquid, miscible with water and most organic solvents except petroleum hydrocarbons.

***Indicates its precise profile described below**

1. Malathion

Chemical structure : Shown in Table 3.8

Trade Names

Malathion is also known as Carbophos, Mercaptothion and Maldison.

Properties : It is a non-systemic wide-spectrum organophosphate insecticide. It is of low persistence in soil, with reported field half-lives of 1 - 25 days. Its life in air is of about 1-5 days. It is moderately bound to soils and soluble in water. Technical malathion is a clear, amber liquid at room temperature. Its melting point is 285°C and adsorption coefficient is 1,800. It is an indirect inhibitor of cholinesterase.

Uses : It is widely used as home garden insecticide. It is suited for the control of sucking and chewing insects on fruits and vegetables. It is also used to control mosquitoes, flies, household insects, animal ectoparasites viz., head and body lice.

Toxicological Effects

Acute toxicity

Malathion is slightly toxic via the oral route, with reported oral (96 hours) LD_{50} value of 100 mg/kg on the rat.

Symptoms of the acute exposure are :

- Numbness
- Tingling sensation
- Headache
- Dizziness
- Tremor
- Nausea
- Sweating
- Blurred vision
- Abdominal cramps
- Respiratory depression
- Slow heartbeat
- Incoordination

Higher doses may cause convulsions or fatality.

Chronic toxicity

Rats fed dietary doses of 5 - 25 mg/kg/day over 2 years showed no symptoms apart from depressed cholinesterase activity.

Reproductive effects

Rats fed high doses of 240 mg/kg/day during pregnancy showed an increased rate of newborn mortality. Low doses caused no reproductive effects in rats. It is not likely that malathion may cause reproductive effects in human under normal circumstances.

Teratogenic effects

Current evidences reveal that malathion is not teratogenic.

Mutagenic effects

Available evidences reveal that malathion is not mutagenic.

Carcinogenic effects

Available evidences reveal that malathion is not carcinogenic, but the data are not conclusive.

Organ toxicity

Malathion affects CNS, immune system, adrenal glands and circulatory system.

Ecological effects

Effects on birds

Malathion is moderately toxic to birds. The observed acute oral LD_{50} values for different avian species are :

In pheasants	167 mg/kg
In chickens	525 mg/kg
In mallards	1,485 mg/kg

Effects on fishes

Malathion has a wide range of toxicity (highly to slightly toxic) in fishes :

In walleye (96 hours) LC_{50}	.06 mg/1
In brown trout (96 hours) LC_{50}	.1 mg/1
In cutthroat trout (96 hours) LC_{50}	.28 mg/1
In fathead minnows (96 hours) LC_{50}	8.6 mg/1

Effects on non-target species

Malathion is highly toxic to honey-bees.

2. Diazinon

Structure : Dazzel, Spectracide, Nucidol and Basudin.

Properties

Diazinon is a non-systemic organophosphate insecticide. It is a colourless to dark brown liquid. It has a flashpoint of 180°°F (82°C). Its adsorption coefficient is 1,000. It has a low persistence in soil. The half-life in soil is 2 - 4 weeks and in animals about 12 hours. Metabolism and excretion rates for diazinon in human is also about 12 hours.

Uses : It is used to control cockroaches, ants, silver fishes and fleas in residential and non-food buildings. It also used in home gardens and farms to control a wide variety of sucking and leaf-eating insects. It is used on rice, fruits trees, sugarcane, corn, tobacco, potatoes and horticultural plants. Diazinon has veterinary uses against fleas and ticks.

Toxicological Effects

Acture toxicity

Toxic effects of diazinon are due to the inhibition of acetylcholinesterase, an enzyme needed for proper nervous system function. The LD_{50} (96 hours) is 300 - 400 mg/kg for technical grade diazinon in rats.

The inhalation LC_{50} (24 hours) in rats is 3.5 mg/l. In rabbits, the dermal LD_{50} (96 hours) is 3,600 mg/kg.

The symptoms associated with diazinon poisoning in humans include :

- Weakness
- Headaches
- Tightness in the chest
- Blurred vision
- Salivation
- Sweating
- Nausea
- Vomiting
- Diarrhea
- Abdominal cramps
- Slurred speech

Death has occurred from both dermal and oral exposures at very high levels.

Chronic toxicity

Chronic effects have been observed at doses ranging from 10 mg/kg/day to 1,000 mg/kg/day for rats. Inhibition of RBC cholinesterase occurred at lower doses in the rats.

Actually, diazinon itself is not a potent cholinesterase inhibitor. However, in animals, it is converted to diazoxon, a metabolite that is a strong enzyme inhibitor.

Reproductive effects : No data are currently available.

Teratogenic effects

One study has revealed that injection of diazinon into chicken eggs resulted in skeletal and spinal deformities in the chicks. However, the data on teratogenic effects due to chronic exposure are inconclusive.

Mutagenic effects

Some tests have suggested that diazinon is mutagenic.

Carcinogenic effects

It is not considered carcinogenic.

Ecological effects

Effects on birds

Birds are significantly more susceptible to diazinon than other wild lives. LD_{50} values for birds range from 2.75 - 40.8 mg/kg.

Effects on fishes

Diazinon is highly toxic to fishes as it may be justified by its LC_{50} value (96 hours) in rainbow trout, which is 2.6 - 3.2 mg/l. However, diazinon does not bioconcentrate significantly in fishes.

Effects on non-target species

Diazinon is highly toxic to bees.

Classification of Organophosphate Insecticides

OPIs are classified into two broad groups on the following basis :

1. On the basis of their activity
2. According to the manner in which OPIs exert their insecticidal action

1. **On the basis of their activity** : On this basis OPIs are grouped into two :

(i) **Direct action OPIs** : These compounds act by directly inhibiting the cholinesterase enzyme, e.g., Fenchlorvos, Dichlorvos, Disopropyl fluorophosphate (DFP), Tabun, Serin, Soman, Tetraethylpyrophosphate (TEPP), Mevinphos, Disulfoton, Chlorothion, Triazophos, Chlorpyriphos, Bidrin and Anilophos.

(ii) **Indirectly acting OPIs** : These insecticides undergo microbial conversion into metabolites which inhibit cholinesterase enzyme e.g. :

Compound	*Active metabolite*
Malathion	Malaoxon
Parathion	Paraoxon
Fenthion	Fenoxon
Fentrothion	Fentroxon

2. **According to the manner in which OPIs exert their insecticidal action** : On this basis OPIs are grouped into two :

(i) **Contact poisons** : These organophosphorus insecticides exert their insecticidal action like the chlorinated hydrocarbon insecticides and are called contact poisons e.g., paraxon, parathion, malathion, EPN, methyl parathion.

(ii) **Selective systemic insecticides** : Such compounds are absorbed into the sap of plants, remaining active and soluble for a reasonable period and are toxic to the plant pests but not to their predators e.g., mipafox, dimethoate, demethon schradan.

Mechanism of Toxic Action of OPIs

The OPIs, in general, are less persistent than the organochlorines, a property due to which they are preferred to the organochlorines. In principle, OP compounds react with the active site of AChE (a serine hydroxyl group). In other words, all OP compounds are anticholinesterase agents i.e., they inhibit the enzyme cholinesterase (ChE) at the neuromuscular junction.

During normal muscle contraction, the cholinergic nerve fibers of autonomic system liberate acetylcholine (ACh) at the neuromuscular junction. The ACh excites muscle fibers and, consequently, muscle contractions take place. Very soon ACh is splitted by ChE present at the neuromuscular junction into choline and acetyl CoA.

Actually, splitting of ACh by ChE prevents its accumulation at the neuromuscular junction and thus assists in muscle relaxation after each contraction.

In its usual course of action, the ChE first gets acylated by losing one H^+ and then fast recovers by using one molecule of metabolic HOH.

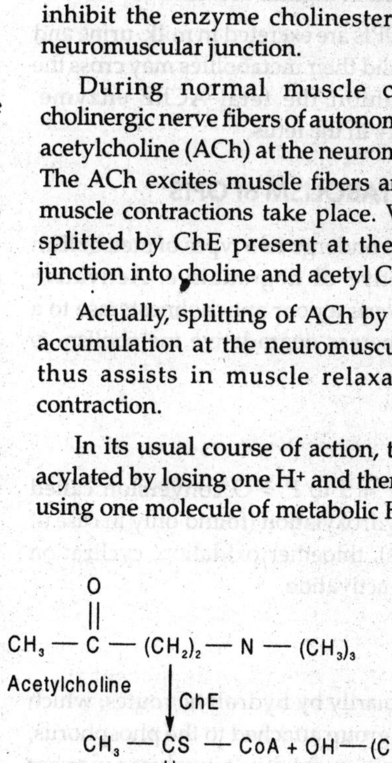

Actually, the OP compounds are structurally similar to ACh. At the enzyme site, they compete with ACh and bind with ChE, thus inhibiting its normal function. The OP compounds phosphorylate the ChE and form a stable complex. Ultimately the enzyme fails to recover quickly. These events bring about an accumulation and, consequently, increase of ACh at the neuromuscular junction, which, in turn, respectively, excites muscle fibers and produces repetitive muscle contraction that causes tremors in the body muscles of the insect — and the insect finally dies of exhaustion.

Absorption and Excretion of OPIs

Mostly, OPIs are highly lipid-soluble except few like ecothiopate, and are rapidly absorbed practically by all routes including GIT, skin, mucus membrane and lungs. Dermal absorption is highly influenced by the solvent used. The extent of absorption is more if these are applied in organic solvents.

Usually, the OPIs are excreted in milk, urine and bile. Some OPIs and their metabolites may cross the placenta and inhibit the fetal AChE enzyme, producing toxicity in the fetus.

METABOLISM of OPIS

OPIs basically undergo two types of metabolism — either activative or degradative. Activative metabolism converts a poor anticholinesterase to a stronger one, whereas degradative metabolism is the reverse.

Activation

It involves P = S to P = O conversion called desulfuration, hydroxylation (found only in case of phosphoramides), thioether oxidation, cyclization and microsomal activation.

Degradation

It occurs primarily by hydrolytic routes, which leave an anionic group attached to the phosphorus, thereby reducing its positivity. It involves a number of enzymes such as phosphatases, carboxylesterases and amidases. The most common hydrolysis is by phosphatases which convert insecticides to their degradation products e.g.

Sumithion → Ethyl sumithion
Dichlorvos → Dimethyl phosphate
Mevinphos → Dimethyl phosphate
Malathion → Demethyl malathion.

Organophosphate containing carboxamide group are cleaved by amidases

e.g. Dimethoate → dimethoate acid
Imidan → Phthalamic acid

In compounds containing carboxyester group cleavage occurs by carboxylesterase

e.g. Malathion → malathion acid

Some other metabolic processes may also play a role in degradation of OPIs.

Reductive degradation

These reactions result in severe lessening of the positivity of phosphorus. The commonest mechanisms are hydrolytic which introduce an anion.

N-Dealkylation and N-Hydroxylation

The N-derivatives of amino or amidic organophosphates differ considerably in their anticholinesterase activity. The removal or modification of N-substituents leads to activation, degradation, or neither. N-dealkylation is catalysed by hepatic microsomes and preceded by hydroxylation of the alkyl group.

Non-enzymatic reactions

Organophosphates undergo certain non-enzymatic reactions which alter the biological properties of these compounds in important ways. The most crucial among these reactions are hydrolysis and isomerization.

Hydrolysis

All the organophosphates can be hydrolysed at a rate directly related to the alkalinity and this reaction determines the half-life of an OPI.

For example

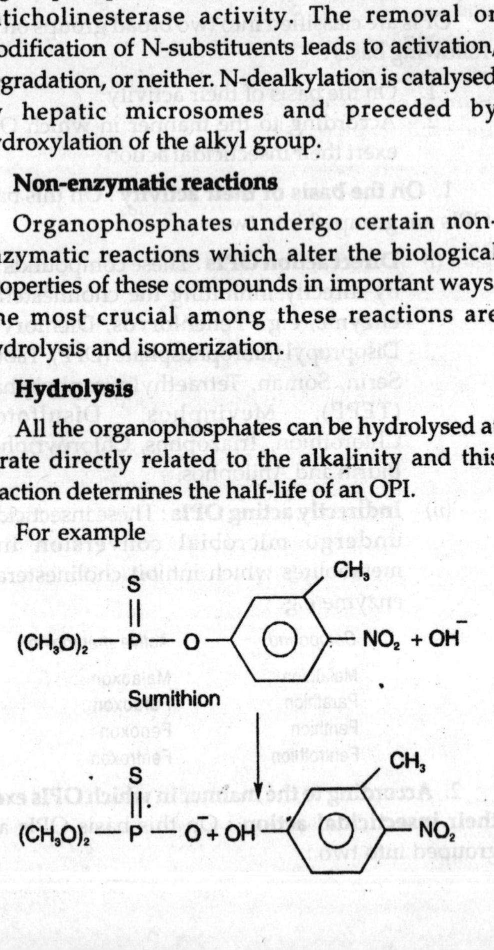

Certain amino acids, hydroxamic acids, chlorine, inorganic phosphates, copper and molybdenum ions promote the hydrolysis of organophosphates.

Isomerization

The properties of the isomers of certain organophosphates differ in salient ways from those of the parent compound.

e.g., S-ethyl parathion, an isomer of parathion and

S-alkyl malathion, an isomer of malathion.

These isomers are responsible for the direct antichlolinesterae activity, increased susceptibility to hydrolysis and increased potency against cholinesterase.

Reactions of Different Enzymes with OPIs

Acetylcholinesterase

Most organic phosphorus pesticides bind with the esteratic site on the AChE molecule. The reacting group of the esteratic site is the hydroxyl group of serine. The bond between phosphorus atom of OPI and the esteratic site of AChE is more stable than the bond between carboxyl carbon of ACh and the same enzyme site. Thus the phosphorylated enzyme is inhibited because its active site is occupied and, therefore, is incapable of hydrolysing ACh.

Carboxylesterase

Carboxylesterase present in plasma hydrolyse organophosphate compounds to products that are entirely inactive as cholinesterase inhibitors, thus reducing their toxicity. Thus OPIs serve as substrates for carboxylesterases and are detoxified by them.

Inhibition of other enzymes

Organophosphates phosphorylate several other enzymes including acid phosphatases, aliesterases, lipases, trypsin, chymotrypsin, succinoidase, ascorbic acid oxidase, dehydrogenase and sulphydryl enzymes. The reaction with these enzymes is generally slower than that with cholinesterases.

The potentiation of certain OPIs by others is partly due to inhibition of aliesterases.

Parathion decreases the activity of β-glucuronidase in liver and choline acetyl transferase in CNS which is responsible for the synthesis of ACh.

Microsomal enzymes

OPIs inhibit a wide range of microsomal enzymes in contrast to CHIs which are microsomal enzyme inducers.

Transferases

Organic phosphorus molecules may split not only by carboxyesterase and mixed function oxygenases but also by certain transferases including glutathion-s-alkytransferase and glutathion-s-aryltransferase.

III. Carbamates

Carbamates were developed into commercial pesticides in the 1950s. Although members of this family of chemicals are effective as herbicides and fungicides, they are most commonly used as insecticides. Some carbamates are also used as nematicides.

There are approximately 25 carbamate compounds currently in use as pesticides or pharmaceutical agents. Perhaps the most widely applied carbamate insecticide is carbaryl, which is especially utilized for lawns and gardens.

Pesticidal carbamates are analogs of carbamic acid (NH_2COOH), in which the carbonyl hydrogen and the amino hydrogen(s) are substituted with an alkyl, phenyl or other aromatic structures as shown below :

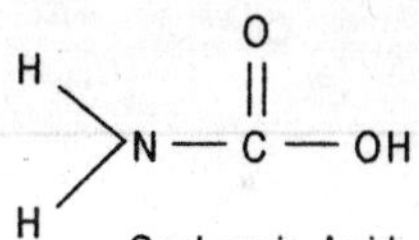

Carbamic Acid

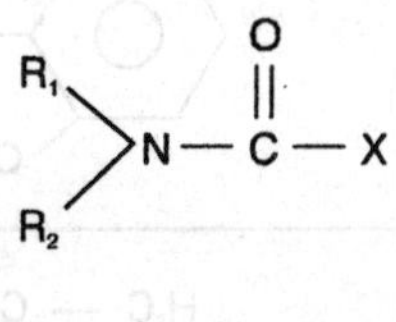

General structure of carbamates, where R_1 and R_2 are alkyl groups and X = a diversified form of structure

In simple words, carbamates are derivatives of carbamic acid. These are colourless compounds which are crystallic at normal temperature. Table 3.9 lists the important carbamates commonly used. Most pesticidal carbamates are lipophilic, a characteristic imparted to them by non-polar substitution groups. Actually, carbamates are not broad-spectrum insecticides and they show erratic patterns of selectivity to insects. They are widely used for the control of ectoparasites in both large and small animals. Because of extreme toxicity they are recommended only for limited use.

Table 3.9 : Commonly used carbamates, their structure and salient properties

Sl. No.	Name	Structure	Salient Properties
1.	Aldicarb	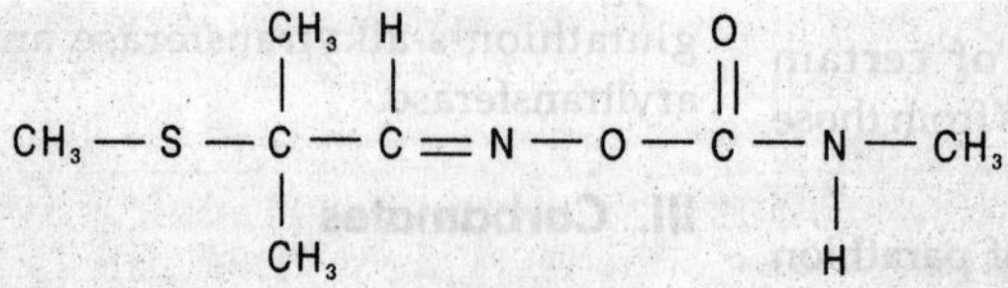	It is 2-methyl-2-(methylthio propionaldehyde)-0-1- (methyl carbomyl) oxime. It is a systemic carbamate insecticide, acaricide, and nematicide. It is widely used for the control of a number of insects, mites and nematode pests. Its trade name is Temik. It is highly toxic. Its acute oral and dermal LD_{50} (96 hours) for rat are 6.2 and 3,200 mg/kg, respectively
2.	Carbaryl	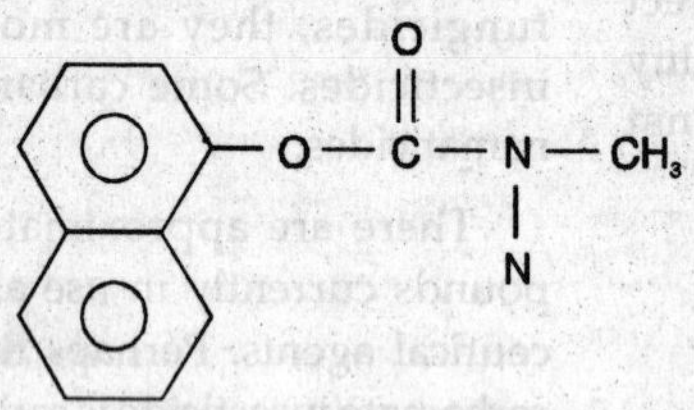	Its trade name is Sevin or Sevimol. Carbaryl is 1-naphthyl, N-methyl carbamate. It acts as contact and stomach poison with longer residual action. It is used against cotton pests and pests of apple and pear. It is moderately toxic to mammals. Its acute oral and dermal LD_{50} values (96 hours) for rat are 400 and 7,500 mg/kg, respectively. It is also used as molluscicide. It is practically non-toxic to wild bird species but lethal to bees and other beneficial insects
3.	Carbofuron	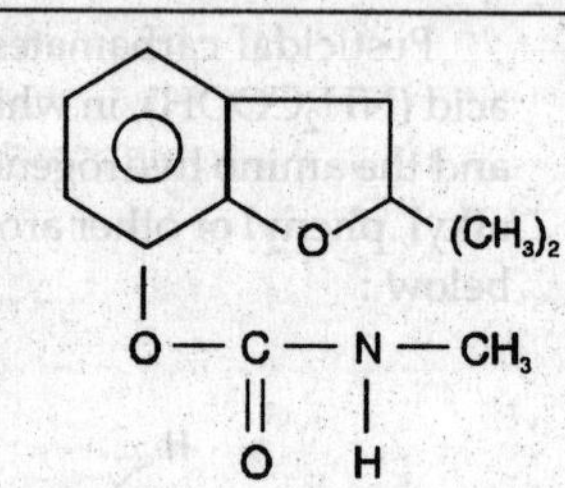	Its trade name is Furdan. It is 2, 3-dihydro 2-2-dimethyl 7-benzofuranyl methyl carbamate. It is a systemic poison used widely for the control of sucking pests and nematodes
4.	Propoxur	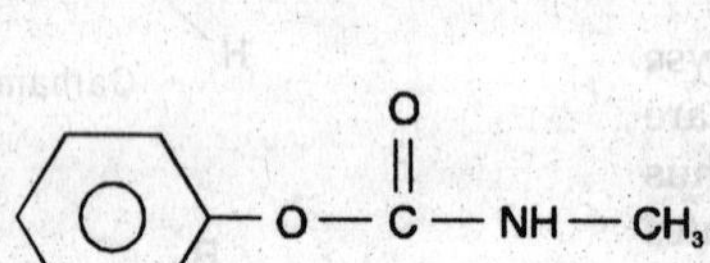	It is 2-isopropoxy phenyl N-methyl carbamate. Its trade name is Baygon. It is a broad-spectrum insecticide; extensively used for the control of household and public health pests. Its acute oral LD_{50} value (96 hours) for rat is 80-175 mg/kg
5.	Pyrolan	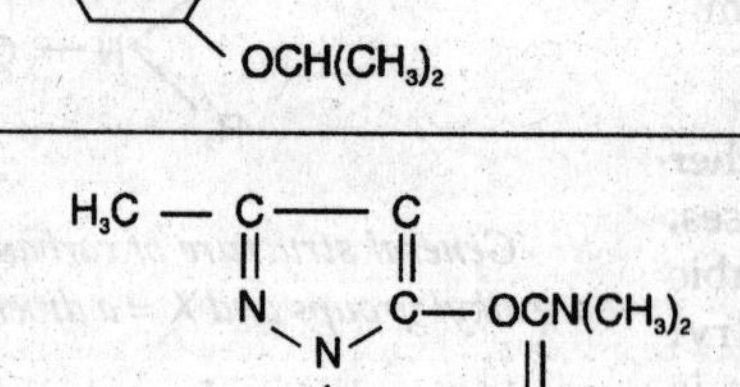	It is a heterocyclic carbamate. It is a systemic insecticide, effective against aphids and fruit flies

Mechanism of Toxic Action of Carbamates

The primary way that insecticidal carbamates work on both target and non-target species is through the inhibition of the enzyme acetyl-cholinesterase. Acetylcholine (ACh) is a substance that transmits nerve impulse from a nerve cell to a specific receptor such as another nerve cell or a muscle cell. Acetylcholine, in essence, acts as a chemical switch. When it is present (produced by the nerve cell) it turns the nerve impulse on. When it is absent, the nerve impulse is discontinued. The transmission ends when the enzyme acetylcholinesterase breaks down the ACh into choline and acetic acid. Without the action of this enzyme, ACh builds up at the junction of the nerve cell and the receptor site, and the nerve impulse continues. Carbamate insecticides block (or inhibit) the ability of this enzyme, acetylcholinesterase, to break down the ACh and end the nerve impulse.

Carbamate inhibition of acetylcholinesterase is a reversible process. Estimate of the recovery time in humans range from immediate up to 4 days, depending on the dose, the specific pesticide, and the method of exposure. The breakdown of carbamate compounds within an organism is a complex process and is dependent on the specific pesticide structure. In nutshell, the mode of action of carbamates, like OPIs, is the inhibition of **acetylcholinesterase**. In fact, carbamates are potent reversible inhibitors of cholinesterase. In addition, all the carbamates inhibit aliesterase of insects, but they kill insects and mammals entirely by cholinesterase inhibition. The mechanism of inhibition of AChE and its reactivation is shown in Fig. 3.1 taking an example of carbaryl, a widely used carbamate :

Absorption and Excretion of Carbamates

Carbamates are well-absorbed through skin, lungs and the gastrointestinal tract and widely distributed in tissues. They are rapidly metabolised and are excreted in urine, mainly as sulphate or glucuronide conjugates within 24 hours. A small amount is also eliminated via feces and milk. Unlike the OPIs, most of the carbamates produce low dermal toxicity.

Symptoms of Carbamate Toxicity

In general, carbamates produce toxicity similar to that of organophosphates, but poisoning is less severe and the effects do not last long.

Symptoms of acute poisoning are typically cholinergic with salivation, lacrimation, miosis, convulsions and death.

Chronic toxicity produces signs of neuromuscular type characterized by incoordination, ataxia, recumbancy and prostrating but there is no demyelination or damage to nerves. Carbaryl has been reported to be teratogenic in dogs (Small et al, 1968).

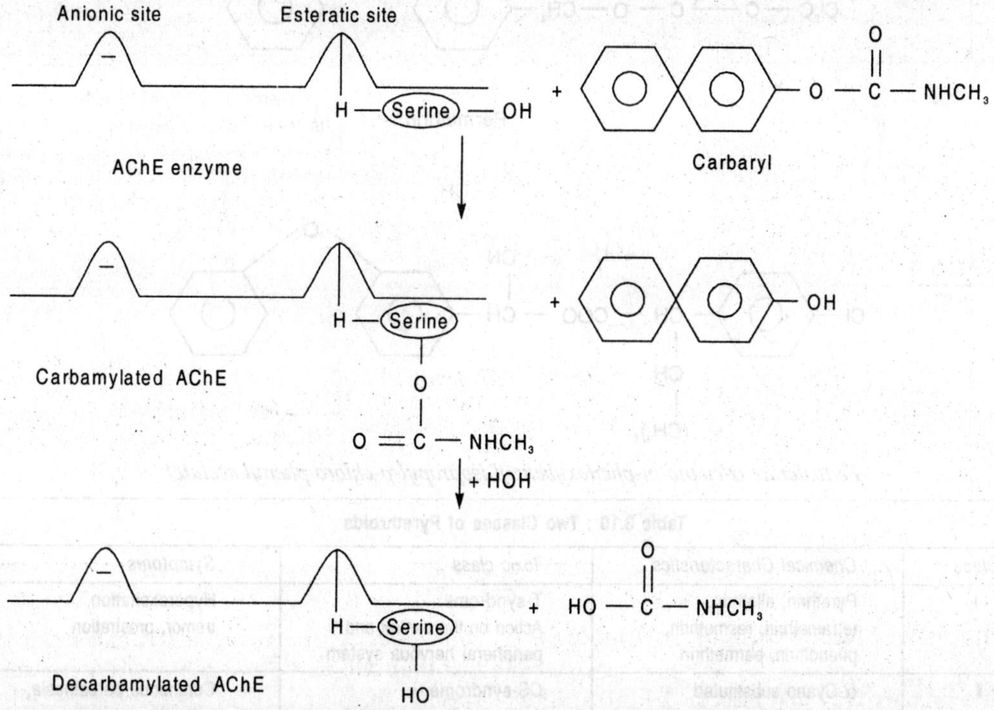

Fig. 3.1 : Inhibition of acetylcholinesterase enzyme by Carbaryl

IV. Pyrethroid Insecticides

Synthetic pyrethroids are currently the most widely used pesticides. Historically they were extracted from *Chrysanthemum cinerariaefolium* flowers which contain the insecticidal esters (pyrethrins). Being very potent insecticides with low mammalian toxicity (due to their rapid bio-transformation), they have gained a strong market share. Based on the characteristics of poisoning, pyrethroids are distinguished into two classes, as shown in Table 3.10.

Members of both the classes of Table 3.10 are rapidly decomposed on exposure to metabolizing systems such as soil microbes. Hence, they are not much important from environmental pollution and related health hazard view point.

With natural pyrethrum, allergic sensitisation occurs but rarely any other toxic effect. Inappropriate use and careless handling of pesticides has resulted in some cases of human intoxication. Most prominent symptoms are burning, itching, tingling sensation of the skin, paresthesia upon dermal contact and after ingestion, epigastric pain, nausea, vomiting, headache, vertigo, blurred vision, muscular fasciculations, convulsions and seizures.

Mechanism of Toxic Action of Pyrethroids

The Type I esters affect the sodium channels in nerve membranes with prolongation of sodium influx causing repetitive neuronal discharge and prolonged after potential but no severe membrane depolarisation. The Type II α-cyanoesters lead to greater and more prolonged sodium influx with persistent membrane depolarisation and eventually nerve blockade. Whereas the first group exerts its main effects on synaptic transmission causing hyperexcitability and tremor, the second group shows its first effects on the sensory nervous system. Other actions include inhibition of Na^+/Mg^{2+} ATPase (adenosine triphosphatase) and alteration of calcium and chloride ion homeostasis.

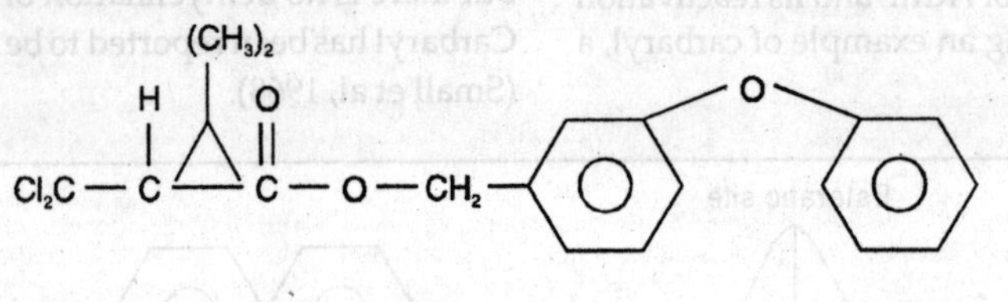

Permethrin

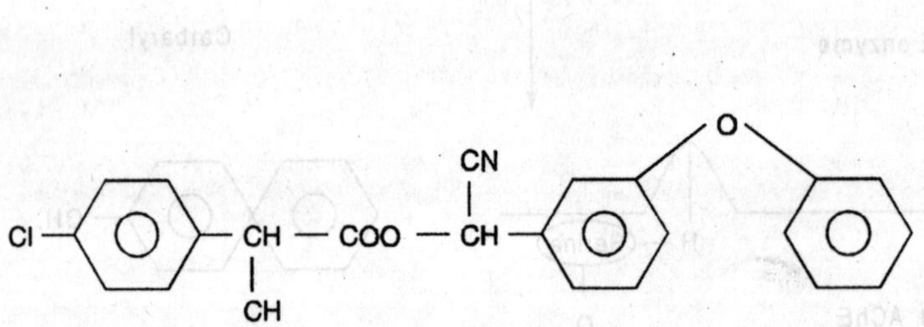

Fenvalerate α-cyano-m-phenoxybenzyl isopropyl-p-chloro-phenyl acetate)

Table 3.10 : Two Classes of Pyrethroids

Class	*Chemical Characteristics*	*Toxic class*	*Symptoms*
I	Pyrethrin, allethrin, tetramethrin, resmethrin, phenothrin, permethrin	T-syndrome Action on the central and peripheral nervous system	Hyperexcitation, tremor, prostration
II	α-Cyano substituted cypermethrin, deltamethrin, fenvalerate	CS-syndrome Action on the mammalian central nervous system	Cutaneous paresthesia, salivation, chronic seizures, dermal tingling

Toxicokinetics of Pyrethroids

Pyrethrins are lipophilic and are readily absorbed by oral, dermal or inhalation routes. Because of their rapid metabolism, the distribution of pyrethroids is relatively unimportant. Plasma and live esterase enzymes rapidly hydrolyse the ester linkage. The hydrolysed insecticide is less toxic. Metabolism also occurs readily in GIT, so oral toxicity is low. Metabolites of pyrethroids are conjugated with glysine, glucuronic acid and sulphate and are eliminated in urine.

V. Dinitrophenols

These are insecticides as well as acaricides. These kill mite eggs and are also used as weedicides. Example Dinitrocresol. It is highly toxic to mammals.

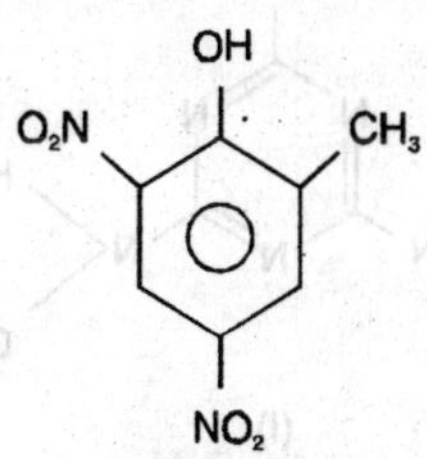

Dinitrocresol

2. HERBICIDES (WEEDICIDES)

The pesticides employed to kill weeds (unwanted plants) are called herbicides or weedicides. In other words, herbicides are the chemicals employed to control the weed plants. In early 20th century various inorganic salts viz., copper salts, sodium salts, organic compounds, etc. were used to control weeds, but they were costly and practically ineffective. Most of the herbicides at present are organic compounds characterized by high physiological activity and effectiveness at relatively low rates of use.

Categories of Herbicides

Principal categories of herbicides or weedicides are :

(i) Selective contact herbicides
(ii) Non-selective contact herbicides
(iii) Residual herbicides
(iv) Translocated herbicides.

Selective contact herbicides (weed-killers) are those which are employed for destroying weeds without injuring crops. Example 2, 4-D.

Non-selective herbicides (weed-killers) e.g. Simazine, Reglone, Glyphosate, etc. are used for cleaning up roads, canals, railway tracks, irrigation canals etc. from all types of unwanted vegetation.

Residual herbicides are applied to the soil at the time of sowing where they remain active for several weeks and check the growth of weeds in competition with the germinating crop.

Translocated herbicides are sprayed on leaves and are capable of moving to all parts of the plants through the internal translocation mechanism.

Actually, the herbicides kill the plants primarily by contact with plant tissue (contact herbicides), or move within the plant tissue from the site of application to the sites of action (systemic herbicides).

Chemically herbicides can be classified into inorganic and organic compounds (Table 3.11) :

Table 3.11 : Classification of herbicides

Class	*Example*
(a) Inorganic	
Acids	Arsenicals, sulphuric acid
Salts	Copper sulphate
(b) Organic	
Oils — diesel	Aromatic oils
Amides	Propachlor, bensulide, alachlor
Benzoics	Dicamba, chloramben
Dipyridiliums	Paraquat, diquat
Carbamates	Propham, metham, barban
Nitroanilines	Trifluralin
Nitriles	Dichlobenil, bromoxynil
Phenols	Dinoseb, Nitrofen, DNP, DNOC
Phenoxys	2,4-D; 2,4,5-T, MCPA, silvex
Thiocrbamates	Diallate
Triazines	Atrazine, simazine, propazine
Triazoles	Amitrosle
Uracils	Bromocil, terbacil
Substituted Ureas	Diuron, linuron, monuran, fenuran
Unclassified	Picloram, calcium cynamide

Some of the commonly used weed-killers are :

(i) Phenoxy Herbicides

Phenoxy herbicides are used against a wide variety of broad-leaved weeds to protect crops such as maize, sorghum, cereal grains, fruit trees and some vegetables. Phenoxy compounds have varying degrees of selectivity for certain types of broad leaved weeds and crops. These compounds have very little effect on grasses.

Among various derivatives of phenoxyacetic acids, 2,4-D salts and esters and 2,4, 5-T are widely used as selective herbicides :

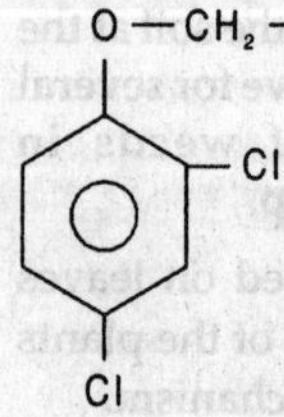

2, 4-D

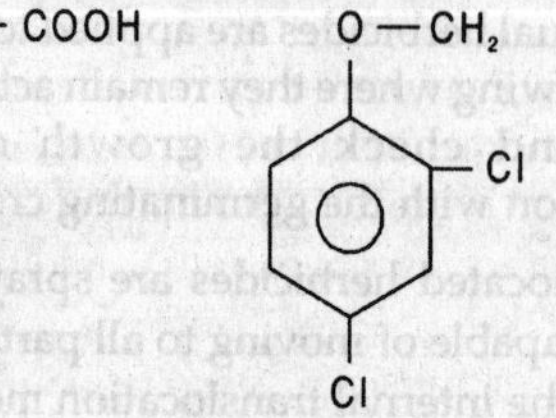

2, 4-D-Sodium salt

(2,4-Dichlorophenoxy acetic acid)

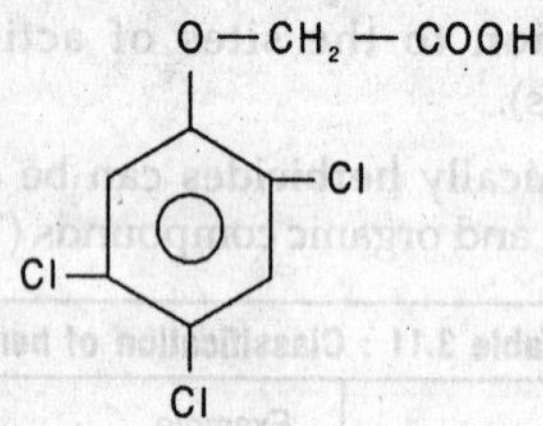

2, 4, 5-T

(2,4,5-Trichlorophenoxy acetic acid)

These herbicides act as growth regulators, hence also termed as *hormone weed-killers*. 2,4,5-T is also used as military defoliant. 2,4-D has low soil persistence. The half-life in soil is less than 7 days. Contact must be avoided since these produce neuromuscular disorders, paralysis etc. Actually the phenoxy herbicide compounds affect the peripheral nervous system in mammals.

Mechanism of Toxic Action of Phenoxy Herbicides

The phenoxy herbicides such as 2,4-D are taken up by the roots or absorbed through the leaves and then distributed throughout the plant. They accumulate in the active growth regions at the tips of the stems and roots where they disrupt normal cell growth. The phenoxy herbicides act like plant growth hormones (auxins). They stimulate the growth of old cells and the rapid expansion of new cells. This rapid growth in cell size, without normal cell division, effectively crushes the plant's water and nutrient transport system, the phlem, in the active growing regions.

(ii) Triazine derivatives

The compounds of class of triazine herbicides possess a six-numbered ring, with 3N atoms replacing alternatively. These are systemic as well as contact herbicides. Some chlorine substituted derivatives are *simazine, atrazine* and *propazine*. The derivatives happen to penetrate into plants mainly through their roots and, therefore, are used as soil herbicides. Such derivatives inhibit the growth of plants and their leaves become chlorotic which indicates the retardation of photosynthesis. Recent researches revealed that triazines inhibit the photolysis of water and the Hill reaction of photosynthesis. These derivatives have longer persistence in the soil and have moderate toxicity to humans and other mammals. Atrazine and simazine are valuable selective weed-killers, especially for broad-leaved grasses; however, atrazine is the best-known example of triazine herbicides.

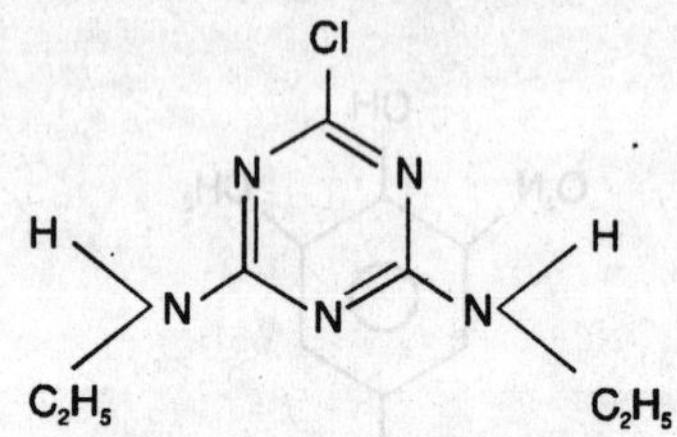

(I)

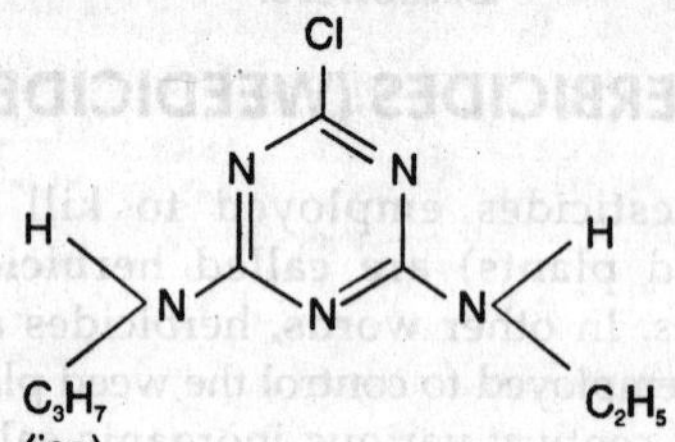

(II)

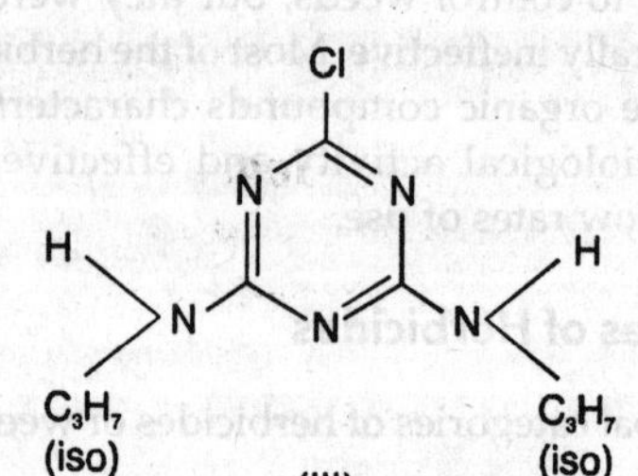

(III)

I. (Simazine) 2-chloro-4, 6-di (ethyl amino) 1,3,5-Triazine

II. (Atrazine) 2-chloro-4-ethyl amino-6-(isopropylamino) 3,5 Triazine

III. (Propazine) 2-chloro-4, 6-di(isopropylamino) 1,3,5 Triazine

(iii) Bipyridyls

This group of herbicides are widely used for ground clearance prior to sowing. It includes two important compounds, diquat and paraquat, which are no selective contact herbicides and kill the weeds even in small doses. These generally kill dicotyledon weeds and show mammalian toxicity.

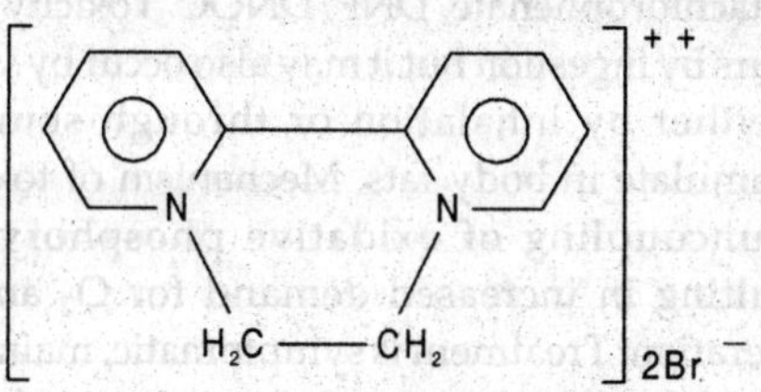

Diquat (1,1-ethylene-2,2 bipyridilium dibromide)

(iv) Ureas

Diuron is used as a pre-emergence herbicide. It has a low acute and chronic toxicity. The LD_{50} (96 hours) for male rat is around 3,400 mg/kg.

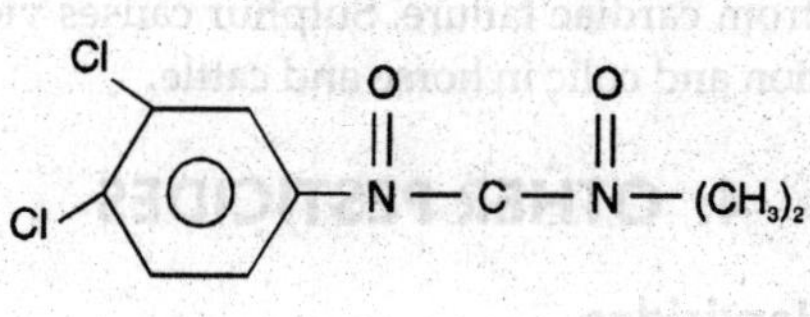

Diuron

Toxicity of Herbicides

The recently developed organic synthetic weed-killers have selective toxicity for specific plants and are safe for mammals. But other less selective ones cause fatal poisoning. These are described below :

1. **Arsenicals (e.g., Sod. Arsenite, arsenic trioxide, mono sodium mono arsenate (MSMA) and di sodium mono arsenate (DSMA)** : These are highly toxic to livestock in recommended doses. These are used as cotton defoliants and have high environmental persistence. Symptoms are profuse watery diarrhea, colic, dehydration, depression and weak pulse.

2. **Amides (e.g., Bensulide, propanil, alachlor, propachlor, diphenamid)** : These are toxic at very high doses. Prominent symptoms are anorexia, weight loss and muscular weakness, fibrosis. Oral intake causes gastritis, renal impairment, CNS excitement and death.

3. **Methyl uracil compounds (e.g., bromocif, terbacit)** : These cause mild toxicities at high doses. Symptoms include bloating, incoordination, depression and anorexia.

4. **Phenyl or substituted urea compounds (e.g., diuron, monuron, fenuron, linuron)** : Chances of toxicity are rare under normal use : These induce hepatic microsomal enzymes and alter metabolism of other xenobiotics. Symptoms include general depression, muscular weakness and anorexia.

5. **Triazines (atrazine, cyanazine, propazine)** : Poisoning from triazines is not common. Large dose ingested accidentally can cause toxic effects or death. Symptoms are muscular weakness, anorexia and posterior paralysis.

3. FUNGICIDES

Pesticides (biocides) employed to protect agricultural crops, edible and commercial products against fungal diseases/infections are called fungicides. These are also applied on seed, grains, plants, soil or wood for preventing fungal growth. Some fungicides are applied before the onset of fungal infection and some are applied after infection of fungal pathogens.

Based upon the mode of action against pathogens, fungicides are grouped into two :

I. Protective fungicides
II. Eradicative (curative) fungicides.

Protective fungicides are applied before fungal infection in the crop takes place. Most of them have a specific mode of action and disturb different processes of the cells. Some common protective fungicides are alkyl mercury, phenyl mercury, organothiocarbamates such as Thiram, Ziram, Zineb, Karathana, Dineb, Captan etc.

Eradicative or curative fungicides penetrate and move within the tissues of that plants. Example : benzo-imidizoles (benomyl carbendazin).

With respect to the distribution relative to plants, the fungicides are further divided into two types :

I. Contact fungicides
II. Systemic fungicides.

Contact fungicides act on the pathogens by direct contact with them. In other words, they do not penetrate into plants but remain on pathogen surface.

Systemic fungicides are assimilated by the plants, prevent infection and exterminate the pathogens.

Dithiocarbamates form the major group of fungicides, largely employed by the farmers. Of the other fungicides, hexachlorobenzene is of great concern and interest.

In general, fungicides may be considered broadly under two heads :

I. Agricultural fungicides
II. Industrial fungicides.

Agricultural fungicides are used in the control of plant diseases in 3 ways :

I. Protection against infection
II. Eradication of active or latent infections, and
III. Destruction of loci of infection in the vicinity of crops.

Agricultural fungicides are also used for the disinfection of seeds, tubers and fruits. Some mercury and copper compounds are used as industrial fungicides.

All the fungicides have a low toxicity to other phytons, aves and mammals. However, in long term, some fungicides or their products containing metals are known to accumulate in food chains. The main hazard to livestock occurs from their use as seed dressings.

These fungicides are classified into three main groups :

I. Organomercurials

The best example of this group is phenyl mercuric chloride acetate. Toxicity is due to mercury which causes CNS stimulation in cats and dogs producing symptoms of blindness, convulsions and hypersensitivity. Posterior weakness, tremors, incoordination and death due to destruction of central neurons. Lesions include degenerative changes in central neurons and nephrosis.

In cattle and poultry, organomercurials cause CNS depression producing symptoms of depression anorexia, incoordination, tremors, recumbancy, dyspnea, hypersalivation and mucosal hemorrhage. Lesions include congestion and hemorrhage in brain, enlargement of kidney and lymph nodes, hemorrhagic enteritis and hepatic necrosis.

Neurotoxicity by organomercurials is due to inhibition of sulfhydryl and disulphide groups of enzymes and due to reduction of glutathione. Treatment is done with mercury chelators such as dimercaptosuccinic acid, mercapto dextran, glutathione or pinicillamine orally which form a complex with mercury in the gut and prevent its absorption. Vitamin E also protects from toxicity.

II. Chlorophenols and Nitrophenols

These include such compounds as Pentachlorophenate, DNP, DNOC. Toxicity mostly occurs by ingestion but it may also occur by vapours — either by inhalation or through skin. They accumulate in body fats. Mechanism of toxicity is by uncoupling of oxidative phosphorylation, resulting in increased demand for O_2 and heat generation. Treatment is symptomatic, mainly O_2 is supplied and body temperature is lowered.

III. Thiaram and Sulphur

Symptoms of Thiaram toxicity in poultry are weakness of legs, diarrhea, weight loss, decreased production and abnormal eggs. In sheep, symptoms are anorexia, dyspnea, lethargy, weight loss and death from cardiac failure. Sulphur causes violent purgation and colic in horse and cattle.

4. OTHER PESTICIDES

I. Rodenticides

The chemical compounds applied to control rodent population are rodenticides. Most widely used rodenticides are sodium fluroacetate and warfarin.

The former is an extremely hazardous rodenticide that results in hyperstimulation of the central nervous system and interferes with heart functioning. Since it is highly persistent, it can be transferred in food chains. Warfarin acts as an anticoagulant — depressing levels of prothrombin, a blood protein essential for blood clotting. Repeated intake of warfarin, therefore, eventually results in death from internal hemorrhaging.

The common rodenticides which may cause accidental hazards are mentioned ahead. Actually, the toxicological problems from the described compounds is primarily acute accidental.

1. **Warfarin** : It is the most effective rat poison. Its mechanisms of action is due to anticoagulant effect and capillary damage (Lisella et al, 1971). Small

repeated doses are more toxic than a single large dose. Symptoms are lameness, diarrhea and sudden death. Lesions include multiple hemorrhage throughout the musculature.

2. **ANTU** : Ruminants are relatively resistant to *alpha naphthyl thiourea* (antu) toxicity. Mechanism of action is by increasing permeability of lung capillaries leading to pulmonary edema. Symptoms are vomiting, breathing trouble, cyanosis, incoordination, prostration, coma and terminal convulsions.

3. **Red Squill** : It is safest rodenticide for general use because it is unpalatable and induces vomiting. In small doses it induces convulsions and in higher doses it causes cardiac arrest. Lesions include gastritis and enteritis. Kidneys, liver, lung and myocardium show degenerative changes. Quinine sulphate is used to reduce cardiac irritability.

4. **Alphachloralose** : Poisoning occurs in dogs and cats. It acts by reducing the body temperature. Symptoms are hyperexcitability, aggressiveness, incoordination, profuse salivation, convulsions and coma.

5. **Fluroroacetates** : (Sodiumfluoroacetate and fluoroacetamide) : These are among the most potent rodenticides. Mechanism of action is by inhibiting aconitase enzyme which is responsible for metabolism of citrate, thus cellular respiration is blocked. Brain and myocardial cells are affected most severely (Murphy, 1980). Dogs exhibit symptoms of CNS, ruminants and horses show cardiac abnormalities, cats and pigs show both effects. Lesions are cyanosis, dark blood, organ congestion, pulmonary changes and enteritis.

6. **Zinc phosphide** : It is non-selective in action, so is hazardous to animals. It acts by producing phosphine gas in GIT, which causes severe gastrointestinal irritation. Symptoms are vomition of blood, dyspnea, abdominal pain, ataxia, prostration, convulsions and coma. Lesions include pulmonary congestion and edema, congestion of liver and kidneys and gastroenteritis.

7. **Strychnine** : It is a potent rodenticide. Mechanism of action is antagonizing the central inhibitory neurotransmitter glycine, leading to convulsions. Symptoms are titanic convulsions, muscle tremors, increased heart rate and respiration rate, paralysis and death due to respiratory failure.

8. **Reserpine** : In dogs it produces symptoms like diarrhea and severe depression.

9. **Crimidine** : It produces convulsions immediately in poultry.

10. **Pindone** : It has anticoagulant action like warfarin. In large doses it causes fatal vascular collapse.

11. **Noxbromide** : It is a selective rodenticide. Its toxic action is specific to rats only. The receptor sites for noxbromide in the peripheral vascular smooth muscles are assumed to exist uniquely in the rat (Rosazowski, 1965). So it is not of much concern as far as poisoning to man or domestic animals is concerned.

II. Molluscicides

Pesticides used to kill molluscs, especially snails and slugs, are called molluscicides. Examples of molluscicides are Molucid, Sodium pentachlorophenate, Bayluscide (Bayer 73 niclosamine), Copper sulphate, Triphenyltin acetate, etc.

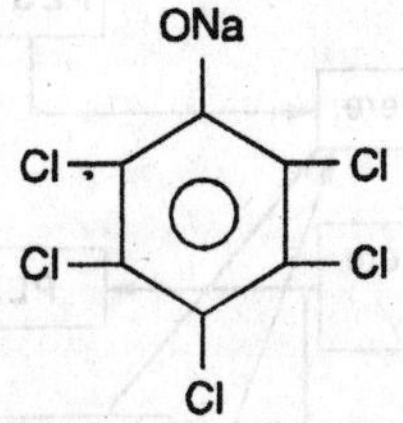

Sodium Pentachlorophenate

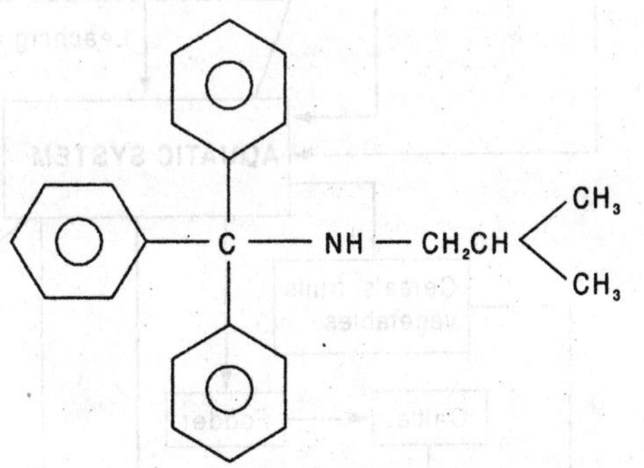

Molucid

(Isobutyl-triphenyl-methylamine)

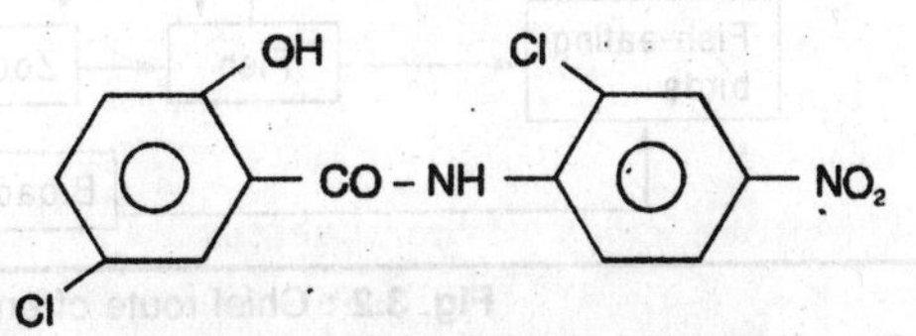

Bayluscide (Bayer 73, niclosamine)

The most commonly used molluscicide is Bayluscide but its position is being challenged by N-tritylmopholine (Frescon).

Certain aquatic herbicides viz., acrolein gramoxone (paraquat) and diquat are of particular interest in that they have bivalent properties. Actually, these are directly not deleterious to snails, but also decline the snail population by eliminating aquatic vegetation which forms the natural shelter for various snail species.

Boyce et al (1967) reported that the most effective molluscicide is Frescon, which has very little environmental effects on non-target organisms.

III. Piscicides

The chemical compounds which are used for the control of undesirable fish species or trash fishes are termed piscicides. Example — Toxaphene, Antimycin A and TFM (Trifluromethylnitrophenol).

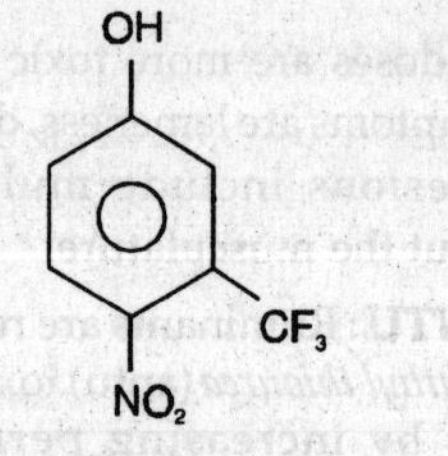

TFM

(3-Trifluromethyl-4-nitrophenol)

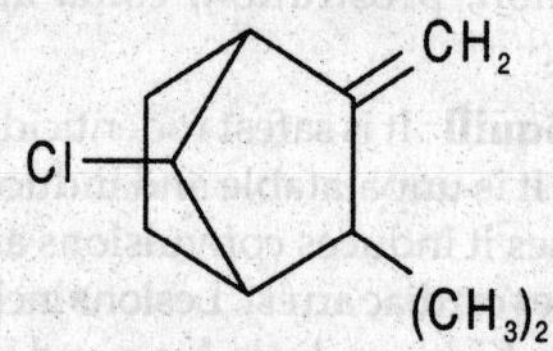

Toxaphene

Note : Piscicides are hardly used today

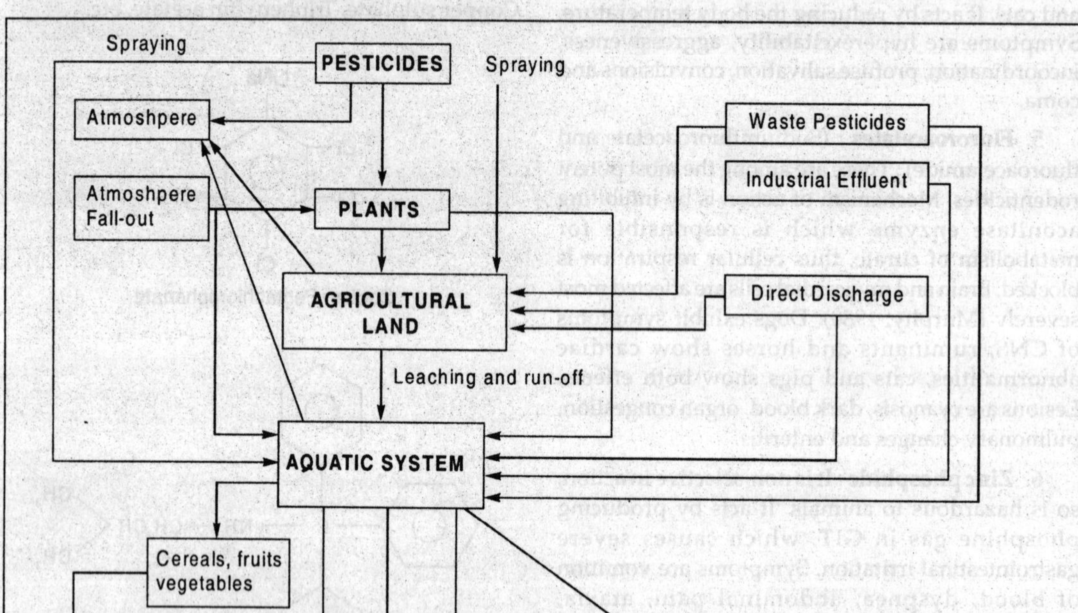

Fig. 3.2 : Chief route of entry and pathways of pesticide pollution

Uses of Pesticides

Pesticides are used for the following purposes :

(i) In agriculture for the protection of crops from various pests.

(ii) In public health programmes for the control of vectors of various diseases.

(iii) In control of household and garden pests.

(iv) In control of ectoparasites of domestic animals and even the human beings.

(v) In industry and commercial establishments.

(vi) In paints, glues etc. for the protection of furniture and other wooden materials.

(vii) In weed control.

Pathways of pesticides into the environment (Environmental contamination by pesticides)

There are various ways through which the pesticides get into the environment viz., direct spraying, run-off from atmospheric fall-out, industrial effluents and direct discharges into aquatic environment as shown in Fig. 3.2.

The main ecosystems that may be polluted by pesticides are lakes, ponds, estuaries, oceans, crops/ agricultural lands and forests.

Effects of Pesticides

Application of pesticides has now become a necessary evil of the modern agro-technology and, perhaps, man cannot stop their uses. Although the use of pesticides is directed against pests, they also exert effects on a wide range of non-target organisms, including desirable animals and plants. The adverse effects of pesticides are :

1. Environmental pollution

Varieties of pesticides such as chlorinated hydrocarbons, viz., DDT and others, are highly persistent. They can rapidly pass through a single ecosystem or from one ecosystem to another. For example, DDT may be washed away from a sprayed cotton field by water run-off and eventually be carried to a river and finally to the ocean. Pesticides like DDT, endrin, dieldrin etc., may gradually slip downward into ground water and eventually contaminate the drinking water supplies. When water evaporates from an irrigated field or from a lake or river, the pesticide molecules may co-distill, i.e., may be carried into the atmosphere along with the evaporating molecules of water. In a spraying field, DDT and other chlorinated hydrocarbons adhered to soil particles may be carried away along dust erosion and then eventually return to land, lake, river and ocean many thousands of kilometers away from the site of application. As the different components of the biosphere are functionally interrelated, pesticide pollution has spread into all of them. In the fatty tissue of seals and penguins of the Antarctic, Moran et at (1973) reported the presence of trace amounts of pesticides.

Although pesticides furnish several benefits, they entail a number of risks and problems. One is their toxicity or their potential for inducing tumors, cancers or birth defects. The World Health Organization estimates that 500,000 pesticide poisoning cases occur annually in the world and that 1% are fatal (5,000 deaths/year). In the early years of modern pesticides (1940 - 1960), pesticide manufacturers and users evaluated the safety of a pesticide almost solely in terms of its short term effects on people or on plants and animals of obvious economic interest. Some pesticides, however, have been found to cause cancer or tumors in laboratory animals. Many scientists believe that positive animal test results indicate a possible cancer hazard to people. Moreover, the tumor-causing agents must be considered to be cancer-causing agents as well; this view is supported by many scientists but questioned by others.

When it appeared that some pesticides might cause cancer, another pesticide-related problem emerged. The property of persistence was desirable, provided people were not killed or made ill, since the more persistent a pesticide, the less frequent the need for applications. Long persistence, however, also means more exposure of a population to any harmful effects and more time to destroy non-target organisms. The worldwide antimalarial programme still finds DDT its most valuable weapon, because DDT is inexpensive, easily made, nonlethal to humans, and highly persistent.

2. Effects on non-target animals

Because of the magnification, a number of non-target animals accumulate non-biodegradable pesticides from the corresponding environment. The impact of pesticide on non-target organisms may be direct or indirect, but almost always deleterious. Pesticides may produce morphological, behavioural, biochemical and physiological alterations. The most common pesticidal impact on non-target organisms are behavioural and physiological sterility, some other impairments or even death occurs. The biocides affect the non-target populations similar to that of a target population. This is particularly significant when the non-target population proves to be a beneficial insect such as pollinator or the natural enemy of any important pest are often more susceptible to insecticides than the pests themselves.

In many cases, harmful effect of a pesticide is difficult to detect and demonstrate, because the pesticides do not seriously affect the adults but at certain embryonic stages during development. Mckim and Benoit (1971), Pandey and Shukla (1984) reported that the juvenile stages of fishes are very much sensitive to a variety of chemical toxicants. Burdick et al (1964) observed that as long as the DDT concentration in trout's egg was 3.0 ppm or less, some fries could not survive, but above 5.0 ppm, mortality of the young fry was 100%.

Many pesticides may adversely affect the reproduction of fish-eating birds and other higher animals (Pimental, 1971).

The herbicides viz., 2, 4 4-D and 2, 4, 5-T at a relatively high doses lower the reproduction in chickens, while chickens exposed to Thiram produce soft shelled eggs of abnormal shape.

Biocides can adversely alter the growth rate of numerous organisms. For example, Dieldrin has been reported to decrease the growth rate of female white-tailed deer.

Besides, the pesticides may also produce some behavioural alterations in non-target organisms. Sublethal concentration of DDT causes trout to forget most of their learned evidence responses. Mosquito fish exposed to relatively lower concentration of DDT (01 - 20 ppb) tend to prefer more saline water than usual for the species. Even the fingerlings of *Colisa fasciata* were reported to produce increased mucus secretion, fast movement and gulping of more air under a sublethal concentration of an organophosphorus insecticide — malathion (Shukla and Pandey, 1988).

3. Effects on insect's predators

Pesticide does not selectively kills the pest only, but also catuses deleterious impact on predatory forms that have been keeping the pest species at reasonably low levels.

4. Development of resistance in insects

A drawback of repeated applications of a given pesticide results in the development of resistant or immune strains of the target pest species. The development of such genetic resistance in pest population is well-known. For example, when DDT was first applied in 1945, the mosquito population appeared to be under control. However, within 6-7 years, a DDT resistant population developed, so a new chemical, ethlyparathion, was applied. By 1961, ethylparathion was no longer effective, and another new pesticide, methylparathion, was introduced, which also became ineffective in 1963, as did Flenthion in 1968.

Of great significance is the fact that this resistance is passed on from generation to generation. It is, therefore, suggested that earlier pesticides should be replaced through a new pesticide each year to minimize the hardship for pest control.

5. Impact of pesticides on human health

Strobel et al (1981) reported that common man who is not occupationally exposed to DDT and other pesticides opts DDT into his body when eating DDT-contaminated food items. Thus almost all human beings are probably taking DDT into their bodies every day with food they eat and the air they breath and some DDT is stored in body fat (3 ppm-90 ppm or bit more). Small amounts of DDT may also be stored in other tissues. It may be stored in liver, tissues of brain, gonads, kidney and blood at concentrations of only 1% of that in fat. Miller (1980) reported that pesticides, when stored in organs, may lead to cancer, softening of the brain tissues, high B.P., stroke, and cirrhosis of the liver. Clayson (1962) suggested that low levels of DDT in our bodies sooner or later may cause cancer.

The observed DDT residue in the fat of Indian population ranges from 12.6 ppm for Gujarat to 31.0ppm for Delhi (Gupta, 1985).

There are several reports of mass poisoning due to careless handling of pesticides in different parts of India. In 1957, pesticide poisoning in Kerala caused death of 102 persons. In Indore, out of 35 cases of Malathion poisoning reported during 1967-1968, five died. In a foggy and windy night on December 3rd 1984, thousands of men, women and children died due to MIC from Union Carbide factory at Bhopal. Actually, Methyl Isocyanate is an intermediate or raw material for the manufacture of several pesticides of the carbamate group of insecticide being manufactured by Union Carbide. A report revealed that within a week total human mortality was about 10,000, one thousand people became blind while more than one lakh people continue to suffer from various respiratory disorders. The most vulnerable parts of the body to MIC were eyes and respiratory organs.

A medical survey held hundred days after the exposure revealed that out of 250,000 people exposed, 65,000 were subjected to severe medical disability (respiratory, eye, gastrointestinal, neuromuscular symptoms) and 45,000 to mild to moderate medical disability. People suffering from breathlessness, sleeping and digestion problems were incapable of carrying on even light physical labour. Women were badly affected, the worst victims being pregnant women. In addition to the loss of human life, a large number of cattle and birds also met the same fate.

Further, high infant mortality has been reported from places where high residues of DDT were found in human milk. A study from Punjab state of India revealed high levels of DDT and BHC in breast milk samples.

The chronic impact of pesticides is much less obvious but much more widely spread throughout the human population. Different people respond to pesticides in different ways. Many are not affected adversely, while some people are very sensitive and may be adversely affected even by very low doses of pesticides.

Common symptoms of pesticidal poisoning in man are :

1. Apprehension
2. Paresthesia of tongue, lips etc.
3. Irritability
4. Dizziness
5. Headache
6. Vomiting
7. Convulsions
8. Ataxia
9. Tremors
10. Weakness
11. Neuromuscular disorders
12. Paralysis etc.

Fate of Pesticides

From the public health view-point the biochemistry of pesticides is of considerable concern. Biochemical processes constitute the mechanism by which pesticides in the environment are degraded and detoxified.

Among the pesticides, the biological action of DDT on the environment has been most extensively studied. Like many other insecticides, the central nervous system is its main target. DDT dissolves in fat tissue and is accumulated in the fatty membrane surrounding nerve cells. Most likely, it interferes with the transmission of nerve impulses along the axons, which are projections connecting nerve cells. Finally, it leads to the disruption of the central nervous system killing the target insect (pest).

While DDT is fairly stable and persists in the environment, the other groups — organo-phosphate and carbamates — degrade quite rapidly in the environment. The latter react with O_2 and HOH and is decomposed within a few days in the environment. The end products are, however, not toxic.

Salient Features of Pesticides

Human beings, along with their domesticated and other animals, are directly or indirectly exposed to the growing use of pesticides. The persistent pesticides, however, increase the chances of biomagnifications. Studies on effect of pesticides on non-target animals can be of the following type :

1. Acute toxicity — Oral, dermal, inhalation, irritation etc.
2. Sub-acute — Oral, dermal, inhalation etc.
3. Chronic toxicity — Carcinogenicity, mutagenicity, reproduction, neurotoxicity etc.
4. Occupational exposure — The sub-acute and chronic exposures are the direct effect of contaminated food, air and water.

Certain important characteristics of pesticides as mentioned below determine their mode of action, potential etc.

1. Stability and persistence

Since the pesticides finally find their way into the aquatic bodies, their stability and persistence in aquatic system is a function of their chemical structure. Pesticides range in stability from few hours to few years. For example Malathion (an organo-phosphate) breaks down in few hours whereas Diaginon and Methoxychlor takes few days; Comphechlor few months and DDT persist for years.

Persistence in the aquatic system is highest for organochlorine insecticides, intermediate for organophosphate and carbamate insecticide and least for herbicides. The highly persistent pesticides produce a potential hazard to the productivity of aquatic systems, because the aquatic biota will be exposed to residue for considerable periods even after only a single contamination. However, if such contamination is repeated or occurs continuously, there is a potential for accumulation of more persistent pesticides in certain components of an aquatic system.

Shukla, J. P. and Pandey, K. (1982 and 1988) reported that juvenile stages of fishes in general and planktons (zoo and phyto) in particular are sensitive indicators to a variety of chemical residues in aquatic environment.

2. Toxicity

High toxicity to the aquatic fauna or flora is an important criterion for a pesticide to be considered as a toxic pollutant of aquatic systems.

Pesticides differ in toxicity to different aquatic organisms and even different development stages of the same species. Some pesticides are extremely toxic to fish at a low concentration while to aquatic invertebrates even at a lower concentration. For example, LC_{50} (96 hours) of DDT to rainbow trout is 0.018 ppm (Edward, 1976); LC_{50} (24 hours) value of malathion to *Channa gachua* is 0.525 ppm (Dudhat and Bapat, 1985); LC_{50} (96 hours) value of suqin to Rasbera daniconius and ticto is 0.35 and 0.171 ppb, respectively (Wagh and Shareef, 1985).

Toxicity may be acute and may kill the organism relatively quickly, or chronic — showing gradual effects on activity, feeding, development, reproduction and general physiology. The former is much more obvious because mortality of a large number of fishes soon attracts attention, while the latter does not become obvious.

3. Solubility

Differential water solubility is an important factor which can increase or decrease the potential of a pesticide as pollutant or toxicant of aquatic systems. For example, solubility of organochlorines in water is about 30%, organophosphates 20-60% and of herbicides like 2,4-D is 65%. Herbicides like Diquat and Dalapon are approximately 70-80% soluble in water (Edwards, 1976).

4. Adsorption

Most pesticides tend to carry electric charges that enable them to be adsorbed onto soil particles in much the same way as inorganic ions. Herbicides are much more strongly adsorbed into soil with a high exchange capacity.

5. Degradation

In addition to soil adsorption, pesticides undergo major structural changes that completely change their toxic properties. Some of these changes are biochemical, photochemical and inorganic chemical reactions with water or other inorganic substances in soil.

Undoubtedly all the organic pesticides degrade — some very slowly and others rapidly.

6. Bioconcentration and Biomagnifications

Most of the organochlorine insecticides are capable of accumulating into the tissue of animals and plants from the water-body. They often reach to a concentration many thousand times greater in the biota than in water itself. However, the process of bio-concentration is quite complex and based upon such factors as the time of exposure, rate of uptake, metabolism within the organism, rate of excretion, potential for storage and physiological condition of the organism.

The phenomenon of bioconcentration from the medium in which organisms persist is sometimes confused with that of biological magnification.

Illustration of biological magnification

The eventually accumulated non-biodegradable pesticides in an organism at a particular trophic level with a greater concentration than its preceding trophic level is referred to as biological magnification. Significantly, animals at the top of the food chains accumulate the maximum residues. This concept is widely accepted. The publication of 'silent spring' by Carson (1962) is an over-simplification of the whole process, and there are many other possible explanations for the presence of higher residue in predators than in their prey, food chains being only one factor (Rosenberg, 1975).

Biological magnification may be explained through following example.

Many pesticides applied to control harmful insects are either chlorinated hydrocarbons viz., Endrin, Toxaphene, Dieldrin, Aldrin etc. are more stable than organophosphorus insecticides like Parathion, Malathion, Chlorothion etc. For that reason they may contaminate the aquatic environment with sublethal amounts which tend to accumulate in the tissues of aquatic animals

(Nichlson, 1964). This accumulation of chlorinated hydrocarbons may reach such high levels in organisms at the summit of the aquatic food web that they are unable to reproduce or they may be poisoned outright. The widespread use of one such compound, DDT, has already aroused considerable concern in recent years and some countries, including India, have legally banned its use. Fig. 3.3 shows the biological concentration of DDT through an aquatic food chain. Animals belonging to the higher trophic levels in the food chains such as seals and birds may have up to 55 g/m^3 of DDT in their body's fatty tissues. Estimation of DDT in human body fat was made in 1964 and these varied from 3.39 g/m^3 in England to 25 g/m^3 in India.

	10.0	← DDT absorbed by fish-eating birds.
	1.0	← DDT absorbed by needle fish which feed on minnows
	10^{-1}	← DDT absorbed by minnows which feed on zooplankton
DDT in ppm	10^{-2}	← DDT absorbed by zooplankton
	10^{-3}	
	10^{-4}	
	10^{-5}	
	10^{-6}	← DDT entering the aquatic ecosystem at about this level

Fig. 3.3 : Biological magnification of DDT through the food chain

Control of Pesticidal Pollution

The following methods must be followed in using pesticides :

1. The non-selective persistent pesticides such as DDT must be phased out of use.
2. Only selective pesticides must be used.
3. Pesticides must be used in small quantities.
4. Repeated pesticide application should be stopped.
5. Pesticide education should be given to public and farmers.
6. Researches on pesticides should be in progress.

PESTICIDES IN INTEGRATED PEST MANAGEMENT

This is a lecture note delivered by G. C. Sachan of Pantnagar on 11.12.2001 at Deptt. of Zoology, DDU, Gorakhpur University, Gorakhpur [UP]

Among the Third World countries India was the first who imported DDT and BHC for the control of insects pests, soon after its independence. In India DDT was first manufactured in 1952. Among the south Asian countries our country is the largest producer of pesticides with the production of 88,751 MT in 1998 - 99. It has second place to China and 12th place in the world. Up to 2000, 155 pesticides have been registered under Insecticide Act, 1968. It includes 57 insecticides, 44 fumigants, 3 acaricides, 1 each soil sterilant, molluscicide and nematicide.

Our country has 4% of the world cropped area with a share of 1.7% of the world pesticide consumption. The world's pesticide consumption is about 3.1 million tones, of which 24% is consumed in USA, 45% in Europe and 20% in developing countries.

During the last four decades, the consumption of pesticides has increased tremendously, from 154 MT in 1953 - 1954 to 80,000 MT in 1994 - 95. After this the pesticide consumption declined steadily to the present level of 54,135 MT (1999 - 2000). This decline in the use is mainly due to banning of hard insecticide like that of chlorinated hydrocarbon, availability of high potency insecticide and adoption of IPM strategies by the farmers. Highly potent insecticide also played very important role in the reduction of pesticide consumption. The insecticide of chlorinated hydrocarbons were required in kilograms 1-1.5/ha, organophosphate in grammes (250-500/ha), synthetic pyrethroids (50 g/ha) in case of fenvalerate and cypermethrin) and still lower (10g/ha) dose is required in case of deltamethrin. This indicates the 100 - fold reduction in the dose of newer pesticides.

In our country maximum share of the pesticide is highest in cotton (39%) followed by rice (35%), cereals/millets (17%), and others (9%). During 1979, this consumption pattern was cotton (54%) followed by rice (17%), cereals/millets (6%) and others (23%). Globally the pesticide consumption pattern is slightly changed. Though cotton has disproportionately higher share, even globally. The world consumption of pesticide share is maximum (27%) in horticultural crops followed by cotton (24%), rice (17%), maize (7%) and others (25%). It is clear that rice, vegetables and fruits are also getting good share but the crops like barley, gram, jute, mustard, soyabean, sunflower and tobacco is even less than 1%.

Up to 1995-96 the major groups of pesticides in agriculture were insecticide (80%) followed by fungicides (10%), herbicides (7%) and others (3%). This consumption pattern has changed in 1999-2000. It was 60% in insecticides, 21% in fungicides, 14% herbicides and other 5%. From 1995 to 1999-2000 the type of insecticides also changed. The percentage of chlorinated hydrocarbon from 40 to 14.5%, carbamates from 15 to 4.5%, synthetic pyrethroids from 10 to 5%, whereas the use of organophosphate increased from 30 to 74%. Increase in the consumption of natural pesticides 2% has also been registered during this period. Natural insecticides include mainly neem and Bt formulations.

In Indian agriculture the average pesticide consumption was very low in 1953-54 (1.2g/ha) which increased to 377 g/ha in 1985-86 and to 431 g in 1992-93. After this a decline trend was noticed. It was 392 g/ha in 1995-96, 320 g/ha in 1997-98 and 288 g/ha in 1999-2000. These values are much more lower than the consumption of 17 kg/ha in Taiwan, 12kg/ha in U.S.A. In India the average consumption of pesticides is very low but several States/Union Territories show very high consumption of pesticides than the national average. Among them are Haryana, Punjab, Delhi and Pondicherry where consumption is more than 800 g/ha. A significant decline in the consumption pattern has also been observed in some of the states. Tamil Nadu, which used 10,500 tonnes in 1993-94 is now using 2,882 tonnes (1999-2000). Similarly, in Andhra Pradesh, the pesticide consumption decreased from 14,500 to 7,000 tonnes during this period.

Among the 53 insecticides registered monocrotophos, endosulfan, malathion, methyl parathion, phorate, quinalphos, dimethoate, chlorpyriphos and carbaryl account for 80% of the total insecticides being used. Mancozeb, sulphur, copper, oxychloride, copper sulphate, carbendazim and thiram among 44 fungicides constitute 86.8% share. Sulphur and copper fungicides still constitute 49.6% of the total fungicides used in the country. Mancozeb is the major organo compound which accounts for 25% of the total fungicides used followed by Carbendazim (7.4%) and Thiram (3.8%). In case of herbicides, out of 33 registered, only 14 are being used. The top five herbicides (isoproturon, butachlor, 2,4-D, anilophos, atrazin) account for 86.4% of the total herbicides used in the Indian.

Excessive and injudicious use of insecticide has created lot of problems like development of resistance in insects to insecticides. So far more than 504 arthropod species have developed resistance to various insecticides in the world. In our country about 7 insect pests of agricultural importance have developed resistance to insecticides. In Andhra Pradesh the injudicious use of broad spectrum insecticides, particularly synthetic pyrethroids, have created havoc to cotton-growers. The American boll worm, a major insect pest of cotton, have developed resistance to these chemicals. To kill these resistant species higher doses.cannot be used due to several consequences. Other problems due to misuse of insecticides are biological magnification, particularly of highly persistent insecticides of chlorinated hydrocarbons at each trophic level of food chain, pest resurgence (due to death of natural enemies, improvement in nutritional status of the host plant and direct effect of sub-lethal doses of insecticides on the body of the insect), outbreak of secondary pests like mite due to use of pyrethroids/carbaryl, minor pests becoming major pests (sorghum midge), environmental pollution, deleterious effects on non-target organisms like parasites/parasitoids, predators, pollinators etc. and health hazards to men and animals.

In IPM there is minimum use of pesticides which is used when absolutely necessary for effective control and restoration of the environment. Secondly, there is no need of 100% insect control to prevent the crop losses. The 100% control is neither desirable, possible nor feasible. The best time of pesticide application is based on the Economic Threshold Level values. For key pests these values have been computed. It has been reported that 10-15% of the dust and 25-30% of the spray are deposited on plant surface and only less than 1% reaches the target body. Thus there is vast scope for decreasing the amount of pesticide by improved method of application.

The misuse of insecticides can be avoided by using them as seed treatment for the control of insects attacking seeds/seedlings, seedling dip for the control of rice pests at seedling pests, root zone application for rice and other pests, whorl application for the termite in wheat/sugarcane etc., infusion and injection methods for the control of mango shoot gall maker in mango and other pests, attract and kill devices, electrostatic spraying, controlled droplet application, need base, timely application etc. In addition to these improved application technology, proper selection of pesticides is of utmost importance. Pesticides are selected on the basis of some parameters. They are selectivity ratios of different insecticides. In addition to this, different insecticides are rated for different health hazards. They are, in general, known as Pest Management Rating. In these ratings, toxicity to human being, persistence of insecticide in soil, toxicity to bees, fishes, birds etc. are pooled together and averaged. The overall rating indicates the suitability of that insecticide in IPM.

Pesticides are a viable component of IPM. Sustainability in agriculture for feeding the increasing human population cannot be achieved without the input of pesticides in agriculture, but their use must be judicious.

Table 3.12 : Pattern of Pesticide Consumption in India

Year	*Insecticide*	*Fungicide*	*Herbicide*	*Others*
1995-1996	80%	10%	7%	3%
1999-2000	60%	21%	14%	5%

Table 3.13 : Pattern of Insecticide Used in India

Insecticides	*Shifted from 1995-96 To 2000*	
Organochlorines	40%	14.5%
Organophosphates	30%	74%
Carbamates	15%	4.5%
Synthetic Pyrethroids	10%	5%
Neem and Bt	—	2%

Table 3.14 : Pesticide Consumption on Major Crops in India

Crop	*Cropped Area%*	*Pesticide Used%*
Cotton	5	54
Rice	24	17
Vegetables and fruits	3	13
Plantation crops	2	8
Cereals, pulses and oilseeds	58	2
Sugarcane	2	3
Others	6	3

Table 3.15 : Average Consumption of Pesticides in Selected Countries of World

Country	*Consumption in gm/hectare*
U.S.A.	2,500
Europe	3,000
Japan	12,000
Taiwan	17,000
Argentina	960
Mexico	750
India	570

From the above lecture, one may come to the conclusion that Integrated Pest Management (IPM) is a philosophy of pest management rather than a specific, defined strategy. It combines physical, cultural, biological and chemical control and the use of resistant varieties. It is ecologically based, relying on mortality factors, including natural enemies and weather, aimed at controlling pests below the economic injury level (EIL) and is based around monitoring of pest and natural enemy abundance. A considerable investment of time and effort is required to determine the optimum strategy, but there are long term environmental and economic benefits.

Revision Questions

A. Long-Answer Questions

1. Define pesticides, their uses, features and control.
2. Describe in detail by making a profile of one organochlorine insecticide studied by you.
3. Describe in detail by making a profile of one organophosphate insecticide studied by you.
4. Write an essay on the carbamate pesticides.
5. What are herbicides? Describe important herbicides and their mode of toxicity.
6. Make a list of important organochlorine insecticides and give the mechanism of their toxic action.
7. Make a list of the important organophosphate insecticides and give the mechanism of their toxic action.
8. Write an essay on pesticides in integrated pest management.

B. Short-Answer Questions

1. Describe briefly :

(i) Types of pesticides by target organisms
(ii) Types of pesticides by mode of action and effects
(iii) Types of pesticides by type of formulation (application)
(iv) Principal classes of pesticides.

2. Short notes on :

(i) Organochlorine insecticides
(ii) Organophosphate insecticides
(iii) Carbamate
(iv) Pyrethroids
(v) Dinitrophenols.

3. Explain in short :

(i) Herbicides
(ii) Fungicides
(iii) Rodenticides
(iv) Molluscicides.

4. Describe briefly :

(i) Impact of pesticides on human health
(ii) — Degradation of pesticides
(iii) Salient features of pesticides
(iv) Biological magnification.

5. Explain :

(i) Mechanism of organochlorine toxicity
(ii) Mechanism of organophosphate toxicity
(iii) Mechanism of carbamate toxicity
(iv) Mechanism of toxic action of pyrethroids
(v) Mechanism of toxic action of phenoxyherbicides.

6. Short notes on :

(i) DDT
(ii) Methoxychlor
(iii) Chlordane
(iv) Endosulfan
(v) Hexachlorobenzene
(vi) Lindane.

7. Explain in short :

(i) TEPP
(ii) Malathion
(iii) Parathion
(iv) Diazinon
(v) Fenitrothion
(vi) Methyl parathion
(vii) Monocrotophos
(viii) Dichlorovos.

8. Write short notes on :

(i) Aldicarb
(ii) Carbaryl
(iii) Carbofuron
(iv) Pyrolan
(v) Dinitrophenols
(vi) Phenoxy herbicides.

4

Air Pollution

> **SYNOPSIS**
>
> *Definition, Sources, Principal types of air pollutants, Common air pollutants, their sources and ill-effects, Occupational diseases, Major air pollution episodes, Ecological effects, Units for air pollutants, Prevention and control of air pollution, Biological indication for air pollution, Odourous air pollutants*

Introduction

Air pollution is considered to be one of the most dangerous and common type of environmental pollution that has been reported in most industrial towns and metropolitans of abroad and India such as New York, London, Tokyo, Mumbai, Delhi, Kolkata, Chennai, Jaipur, Hyderabad, Nagpur, Ahmedabad, Kanpur etc. One can live without clean air for only two minutes! *A clean air* is defined as the air occurring in areas sufficiently distant from places of human activities and other abnormal influences. Air Pollution has become one of the vital and serious environmental problems of the modern society.

Definition of Air Pollution

Air pollution refers to the *undesirable change occurring in air causing harmful effects on man, domestic species, flora etc.*

WHO defines *air pollution* as "limited to situations in which the output ambient atmosphere contains material in concentration which are harmful to man and his environment".

Air pollution is generally disequilibrium condition of air caused due to the introduction of foreign elements from natural and anthropogenic (man-made) sources to the air so that it becomes deleterious or injurious to biological communities.

Precisely, air pollution can be defined as "the imbalance in quality of air so as to cause adverse effects on the living beings existing on earth".

Air pollution may broadly be defined as the presence of one or more contaminants like dust, smoke, mist, and odour in the atmosphere which are injurious to human beings, plants and animals.

Normally, we breathe 22,000 times a day inhaling approximately 16 kg of air. It is, therefore, essential that we must know enough about the air around us and the means to keep it clean.

An average composition of clean dry air is shown in Table 4.1 :

Table 4.1 : Average composition of clean dry air near sea level

Group	Components	Mol. Formula	Volume Fraction
Major	Nitrogen	N_2	78.09%
	Oxygen	O_2	20.94%
Minor	Argon	Ar	0.93%
	Carbon Dioxide	CO_2	0.032%

Contd...

Group	Components	Mol. Formula	Volume Fraction
Trace	Neon	Ne	18.0 ppm
	Helium	He	5.2 ppm
	Methane	CH_4	1.3 ppm
	Krypton	Kr	1.0 ppm
	Hydrogen	H_2	0.5 ppm
	Nitrous Oxide	N_2O	0.25 ppm
	Carbon Monoxide	CO	0.1 ppm
	Ozone	O_3	0.02 ppm
	Sulphur Dioxide	SO_2	0.0002 ppm
	Nitrogen Dioxide	NO_2	0.001 ppm
	Ammonia	NH_3	0.01 ppm

Sinks of Atmospheric Gases

Sink is a medium, which is capable of retaining and interacting with a long-lived pollutant, though not necessarily indefinitely. Oceans as well as vegetations are the most important sinks for most atmospheric gases. For example, SO_2, CO_2, O_2 find oceans as the ready sink. Vegetations (plants) are important sinks of CO_2, N_2 and O_2. Limestone acts as the sink for atmospheric acid, which fixes the sulphate as part of the wall composition :

$$H_2SO_4 + CaCO_3 \rightarrow CaSO_4 + HOH + CO_2$$

Sources of Air Pollution

Air pollution is largely confined to the troposphere or lower atmosphere, with some penetration into the stratosphere. Air is never found absolutely clean in nature. Air pollution started from the very moment when the primitive man knew to make fire. Air pollutants are substances causing damage to target or receptor. The target may be man, animal, plant, building or material which may be adversely affected by pollutants.

Major sources of air pollution may be grouped into two :

1. **Natural sources** such as *Volcanic eruptions* releasing poisonous gases like SO_2, H_2S, CO etc., forest fires, natural organic and inorganic decays, marsh gases, depletion of sand and dust, cosmic dust, aero-allergens or pollen grains of flowers, soil debris, and fungal spores etc. All these produced are injurious to health. Green plants through evaporation release huge amount of CO_2. The effect of CO_2 in increasing the earth's temperature is commonly referred to as "Greenhouse Effect".

2. **Anthropogenic sources** such as increase in population, deforestation, burning of fossil fuels and fires, emissions from vehicles, rapid industrialization, agricultural activities and wars are the major causes of air pollution. Principal anthropogenic sources of air pollution are represented in Table 4.2(I)

Table 4.2(I) : Main Anthropogenic Sources of Air Pollution

Source	Examples	Pollutants
Dust generating process	Grinding, crushing, screening, milling, demolishing etc.	Mineral and organic particulate matter
Combustion	1. Fuel burning	Oxides of nitrogen, oxides of sulphur,
	2. Motor vehicles	Carbon monoxide, smoke, flyash, organic vapour,metal oxides, particulates etc.
Manufacturing process	1. Smelters, steel mills, non-ferrous industries etc.	Metal fumes, lead, arsenic and zinc fluorides, oxides of
	2. Chemical plants, petroleum refineries, pulp mills, cement factories, fertilizer plants etc.	sulphur, fluorides, organic vapour, odours, H_2S, particulates etc.
Agriculture activities	1. Crop spraying, dusting, pest and weed control.	Organic phosphates, chlorinated hydro-carbons, arsenic etc.
	2. Field burning	Flyash, soot, smokes etc.
Nuclear energy activities	1. Ore processing such as crushing, grinding, screening etc.	Uranium and dust beryllium
	2. Nuclear fission and nuclear reactors	Argon-41
	3. Nuclear device testing and atmospheric explosions	Radioactive fallout

Outdoor and Indoor Air Pollution

Air pollution may also be defined as the quantitative alteration of atmospheric constituents or addition of one or more contaminants in the outdoor atmosphere for such duration and in such quantities that are injurious (deleterious) to human health, his belongings or plants and animal life.

Pollution of the atmosphere indoors (inside enclosed) buildings, home, residence, office etc. is called indoor air pollution. Typical atmospheric pollutants encountered mainly indoors or outdoors, according to their sources, are listed in Table 4.2(II). Human beings spend about 75-80% of their time indoors; thus the degree of contamination of the indoor air is of considerable importance.

Table 4.2(II) : Typical Sources of Air Pollutants Grouped by Origin

Pollutants	*Sources*
Group I. Sources predominantly outdoor :	
Lead	Automobiles and industrial emissions
Ozone	Photochemical reactions in the atmosphere
Sulphur oxides	Fuel combustion, smelters
Pollens	Trees and other plants
Organic substances	Petrochemical solvents, natural sources, vaporization of unburned fuels
Group II. Sources both indoor and outdoor :	
Carbon dioxide	Combustion and metabolic activity
CO and nitrogen oxides	Fuel burning and tobacco smoke
Organic substances	Volatilization, combustion, metabolic action, pesticides
Polycyclic hydrocarbons	Fuel burning, tobacco smoke
Suspended particulate matters	Resuspension, condensation of vapour and combustion of products
Water vapour	Biological activity, combustion, evaporation
Group III. Sources predominantly indoor :	
♦ Aerosols	
♦ Allergens	
♦ Ammonia	
♦ Asbestos and synthetic fibres	
♦ Mercury	
♦ Organic substances	
♦ Radon	
♦ Formaldehyde	
♦ Consumer products	
♦ House dusts	
♦ Metabolic activity, cleaning products	
♦ Fire-retardant, acoustics, thermal or electrical insulation	
♦ Particle board, insulation, furnishing, tobacco smoke	
♦ Spills in dental care facilities or laboratories, thermometer breakage	
♦ Adhesives, cookings, cosmetics, solvents	
♦ Building materials (concrete stone)	

Principal Types of Air Pollutants

The principal types of air pollutants are :

1. **Gases** : The important gaseous pollutants in the air from different sources are CO_2, CO, SO_2, NO_x, H_2S, Cl_2, Fluorine etc.

2. **Chemicals** : Important chemicals which pollute the air are — Arsenic (As) from coal, oil furnaces, glass manufacturing and fertilizer plants (from HPC tower — Pandey and Shukla, 1979); benzene (C_6H_6) from refineries and motor vehicles; Cadmium (Cd) from smelters, burning waste, coal and oil furnaces; formaldehyde (HCHO) from motor vehicles and chemical plants; HCl from incinerators; HF from fertilizer plants and smelters; Hg from coal and oil furnaces and smelters; Manganese (Mn) from steel and power plants; Nickel (Ni) from smelters, coal and oil furnaces; Lead (Pb) from automobiles, smelters; SiF_4 from chemical plants.

3. **Secondary Air Pollutants** : Air pollutants directly released from any source into the environment are called primary air pollutants. For example — Soot released from unburned fuel, benzopyrene (hydrocarbon) released from cigarette smoke, CO, lead.

Secondary air pollutants are formed from primary pollutants under specific conditions. Thorndike (1976) has given following examples of secondary pollutants :

1. SO_2 :

$$2H_2S + 3O_2 \rightarrow 2SO_2 + 2H_2O$$

2. H_2SO_4 mist :

$$2SO_2 + O_2 \rightarrow 2SO_3$$
$$SO_3 + HOH \rightarrow H_2SO_4$$

3. O_3 :

$$NO_2 \xrightarrow{\text{sunlight}} NO + O; O + O_2 \rightarrow O_3$$

4. Smog :

$$\text{Hydrocarbons} + NO_2 \xrightarrow{\text{sunlight}} \text{PAN} + \text{HCHO} + \text{Others}$$

Actually, smog or so called photochemical smog is a harmful mixture of smoke + fog. It consists of a mixture of primary and secondary pollutants, of which PAN and O_3 are considered to be hazardous.

5. NO (Nitrogen monoxide) : It is formed by dissociation of NO_2 to atomic oxygen (O) and NO :

$$NO_2 \rightarrow O + NO$$

6. NO_2 (Nitrogen dioxide) : It is formed in sunlight from NO :

$$NO + O \xrightarrow{\text{sunlight}} NO_2$$

7. OH (Hydroxyl radical) : It is formed in sunlight from hydrocarbons and NO_x. It reacts with other gases to form acid droplets.

8. HNO_2 (Nitrous acid) : It is formed from NO_2 and water vapour.
9. HNO_3 (Nitric acid) : It is formed from NO_2 and is a major component of acid rain.

4. **Particulate Matters** : Particulate matters or particulates are fine solids or liquid droplets (suspended in air) with a diameter ranging between 0.02 - 1.00 μm. From health viewpoint, particles in the range of 0.02 to 10 μm are of significance. The large sized particulates are *grit*—having a diameter over 500 μm.

Particulates in the air may be viable or non-viable. The viable particulates are the minute living organisms that are dispersed in the atmosphere viz., microbes like bacteria, fungi, moulds, algae etc. Among these, viable algae are most commonly found in the atmosphere as shown in Table 4.3. Fungi in air would be allergic to humans (Table 4.4).

Non-viable particulates of importance are formed either by the breakdown of larger materials or by the condensation of minute particles and droplets.

Table 4.3 : Viable Algae in the Atmosphere

Sl. No.	Genus	Class
1.	Amphora	Bacillariophyceae
2.	Anabaena	Bacillariophyceae
3.	Calothrix	Chlorophyceae
4.	Chlora	Chlorophyceae
5.	Chrysocapsa	Chrysophyceae
6.	Heterococus	Chrysophyceae
7.	Microspora	Myxophyceae
8.	Novcula	Myxophyceae
9.	Oscillatoria	Myxophyceae

Table 4.4 : Some Allergenic Fungi in Air

Sl. No.	Name
1.	Alternaria
2.	Aspergillus
3.	Botrytis
4.	Fusarium
5.	Helminthosporium
6.	Penicillium
7.	Phoma

Non-viable particulates in the atmosphere are of four principal types : (1) Mists, (2) Smoke, (3) Fumes, and (4) Dusts. Mists and fumes are the liquid particulates whereas dust and smoke are solid particulates.

A. **Mists**

These are liquid droplets suspended in air with a diameter <10 μm. Example of mist may be the portion of insecticides and herbicides that actually miss their targets and travel through the air. Mists are formed by the condensation of vapour. For example

$$SO_2\,(\text{gas}) \xrightarrow{22°C} SO_2\,(\text{liquid})$$

B. **Smoke**

Smoke denotes very small soot particles produced by burning and combustion of organic matter. In other words, small sized particulates are smoke (gas-born solids) with particles usually less than 2.0 μm in diameter. Oil smoke, tobacco smoke and carbon smoke are typical examples of this type of particulate emission.

C. **Fumes**

These are suspended solids in air less than 1.0 μm in diameter. The condensed vapours — fumes of metals — are the well-known particulates of this type. Examples of this category also include metallurgical fumes and alkali fumes.

D. **Dust or Aerosols**

These are generally solid, rarely liquid, particles suspended in air or some other gas with a diameter of 1.0 - 100 μm. Non-viable dust particulates in the atmosphere consist of ground limestone, sand tailings from floatation, pulverized coal, cement flyash and silica dust.

Anthropogenic (man-made) dusts in the general environment originate chiefly from transportation, industrial and agricultural activities. Moreover, workers employed in specific occupation are exposed to high amounts of dusty particles. The following is a comprehensive list of the type of dusts emitted at work places and the type of occupations specifically associated with such emissions :

Sl. No.	Type of dusts	Occupations
1.	Metallic dusts	Grinders, polishers, tool and instrument makers, jewellers, brass workers, printers, compositors.
2.	Mineral dusts	Glass cutters, stone, marble and cement workers, diamond cutters, potters, plasterers, moulders and lithographers.
3.	Vegetable/fibre dusts	Cotton ginning, textile, flax, linen, hemp, paper manufacture, weavers, spinners, knitting, jute, woodworkers.
4.	Animal and mixed dusts	Hatters, silk, wool, carpet, rag and shoddy workers, hair mattresses, rug makers etc. workers.

Further particulates may be inorganic or organic type, called inorganic and organic particulate matter (OPM). OPM originates mainly from combustion of fuels, automobiles and vegetation. PAH (polycyclic aromatic hydrocarbons) like chrysene, benzo fluoroethane, benzo (a)-pyrene, benzidine etc. are some organic particulates of carcinogenic nature.

E. **Radioactive substances**

Gamma particles and radioactive nuclides like C-14, strontium 90, caesium-137 etc.

F. **Hydrocarbons**

Transport through evaporation and exhaust emission of unburnt gasoline accounts for the major part of hydrocarbon pollutants. These are chiefly benzene, toluene and benzopyrine etc. discharged by automobiles and industries.

G. **Fluorocarbons**

These come from industries, refrigeration, air conditioning, sprays etc. Examples of chlorofluorocarbons are listed in Chapter 17.

H. **Toxicants**

There are several toxicants other than heavy metals. These include PAN, acetic acid, chloroform, methylchloride etc.

I. **Metallic Pollutants**

These mainly include Al, Be, Pb, Cd, Ni, As, Sn, V, Ti, Mn, Hg, Fe etc. These are present in air as solid particles or liquid droplets or gases. Mostly metallurgical processes and automobiles produce these.

J. **Photochemical Oxidants**

These include SO_x, NO_x, PAN, PBzN, peracetic acid, acetyl peroxide etc. These are released mostly by automobiles.

K. **Hydroxypollutants**

These are formed in sunlight from hydrocarbons and NO_x. These form acid droplets with other gases.

L. **Asbestos**

Asbestos is the only naturally occurring fibrous mineral. This non-metallic substance possesses a unique property of resisting heat transfer even at high temperatures.

Common Air Pollutants, — Sources and *Ill-Effects* on Man

Various types of pollutants viz. gaseous, particulates, metals etc. and their ill-effects are precisely illustrated below in a tabular form—

Sl. No.	*Pollutant*	*Source*	*Ill-effects on man*
1.	SO_x such as SO_2, SO_3	Coal and oil combustion, electric power plants, smelting of metallic ores containing sulphur, petroleum refineries etc.	Causes chest constriction, headache, vomiting, and death from respiratory ailments
2.	H_2S	Refineries and chemical industries	Causes nausea, irritates eyes and throat, 200 ppm concentration causes depression of CNS and 2,000 ppm causes death within a few minutes
3.	NO_x such as NO_2, N_2O and NO, (NO and N_2O are considered as potential atmospheric pollutants)	Soft coal, automobile exhausts, fertilizers and explosive industries	Pulmonary fibrosis and Pulmonary emphysema i.e., distension of the lungs (if exposed to 10 - 40 ppm of NO_2)
4.	CO	Burning of coal, gasoline, automobile exhausts, industrial operations and waste disposal	Reduces O_2 carrying capacity by the formation of carboxyhaemoglobin (COHb). Inhalation of 1,000 ppm of CO quickly causes death due to inactivation of Hb. CO shows synergistic toxic effects with H_2S and NO_2 separately
5.	CO_2	Burning of fossil fuels, deforestation etc.	Causes mild narcotic effects, stimulation of the respiratory centres and leads to asphyxiation. Indirectly it is responsible for the greenhouse effect

Contd...

Sl. No.	Pollutant	Source	Ill-effects on man
6.	HCN	Blast furnace, fumigation, chemical manufacturing, metal plating etc.	Interferes with nerve cells, throat dryness, indistinct vision, headache etc.
7.	NH_3	Explosive, dye making and fertilizer plants	Inflames upper respiratory passages
8.	Phosgene or carboryl chloride	Chemical and dye making	Induces coughing, irritation and fatal pulmonary edema
9.	Aldehydes	Thermal decomposition of oils, fats or glycerol	Irritate nasal and respiratory tracts. HCHO is a potential carcinogen
10.	Arsines	Process involving metal or acid containing arsenic soldering	Damage RBC, kidneys and cause cancer
11.	Particulates viz. mist, smoke, fumes, dusts	Every manufacturing process	Cause emphysema, eye irritation, and possibly cancer
12.	Fluorides	Phosphatic fertilizer manufacturing plants, aluminium reduction process, steel and glass works	Respiratory failure, fall in B.P. and general paralysis
13.	Aluminium	Canning, food packaging industries	Adverse effects on CNS, especially brain, causing neurological diseases, especially Alzheimer's disease
14.	Aresenic as Arsenic (III) oxide (As_2O_3)	Smelting of ores containing As, coal burning. Also released in the form of weed-killers	Arsenic is a metalloid and not a metal and large doses cause gastrointestinal disorders. Smaller quantities may be carcinogenic for skin and lung. Arsenic crosses placental membranes and is a known animal teratogen
15.	Beryllium	Coal burning, nuclear fuel processing, rocket fuels	Large exposure even at low level causes a lung disease called "berylliosis" — which is often fatal
16.	Cadmium	Electroplating wastes and from burning of Cd containing materials like automobile tyres, plastic etc.	Enzyme inhibition at low level causes hypertension and at high level degenerative bone disease. It also causes the following : Aminoaciduria (excretion of amino acid in the urine); Glucosuria (excretion of blood sugar in the urine); Hypercalciuria (excretion of excess calcium in the urine); Osteoporosis (decalcification of the skeleton) and proteinuria (excretion of proteins in the urine)
17.	Lead	Leaded gasoline, paints and varnishes	Weight loss, constipation, tooth loss, nausea and brain damage
18.	Mercury	Electrochemical industry, certain fungicides	Large doses of Hg causes brain damage, often fatal

Occupational Diseases

The disease prevailing during occupation of any worker is called occupational disease. For example, one of the worker in a pharmaceutical company, who works in the boiler section of the company, had nausea and vomiting periodically. In addition, he coughed, wheezed and was short of breath and weak. His job was to clear and descale the boilers from inside. It was realized that this job involved close contact with V_2O_5 (vanadium pentaoxide). This compound is known to affect the gastro-intestinal and renal system, causing nausea, vomiting, stomach and kidney problems. It may further lead to bronchitis, pneumonia and pleurisy. Industrial workers may be exposed to various types of occupational hazards following inhalation of particulates/dust. Important occupational diseases and their causative components are :

1. Silicosis → Due to Silica (workers working in cement and rice mills)

2. Byssinosis → Due to Asbestos (workers working in Asbestos factory)

3. Anthracosis or Pneumoconoisis → Due to coal dust (workers working in the mines)

4. Bagassosis → Due to sugarcane dust (workers working in the sugar factory)

5. Tobaccosis → Due to tobacco (workers working in tobacco industries)

6. Siderosis → Due to iron (workers working in iron industries).

Emissions of Major Industrial Air Pollutants

Industries are by far the largest single emission source for polluting the environment. Some of the major pollutants' emitting sources are :

1. **Sulphuric Acid Industry** : Sulphuric acid and other industries release large quantities of SO_2 and SO_3 in the air.

2. **Nitric Acid Plants** : These normally discharge NO_x in the air which are proved serious pollutants.

3. **Hydrochloric Industry** : It releases toxic chlorine vapours, chlorine monoxides, HCl and unreacted hydrocarbons. Needless to mention that Cl_2 is now charged with being carcinogenic.

4. **Fluoride Industry** : It discharges offensive fluorides, HF and cryolite.

5. **Chloroalkali Plants** : These emit Cl_2, H_2, CO and CO_2. All these are gaseous pollutants.

6. **Iron and Steel Industry** : Emissions from these industries include metal oxides, SO_x, CO and CO_2.

7. **Phosphoric Acid Plants** : The main emissions are P_2O_5, phosphoric and phosphorous acid and gaseous fluorides.

8. **Petrochemical Refineries** : These emit CO, NO_2, HCHO, kerosene particulate, gasoline and fuel oil.

9. **Thermal Power Plant** : These are the major sources of SO_2.

10. **Catering Industry** : It emits aldehydes, hydrocarbons, CO, CO_2 and several obnoxious gases.

11. **Combustion Engines** : Emissions from internal combustion engine and fuels are —

(i) SO_2 : 60-120 lbs/ton of fuel

(ii) Particulates : 30-150 lbs/ton of fuel

(iii) Organic compounds : 20-30 lbs/ton of fuel

(iv) Natural gas

(v) Automobile exhaust like CO, NO_x, C_xHy, C, Pb particles, olefins and carbonyl compounds.

12. **Radioactive Natural Sources** : The radioactive nucleus which occurs in igneous rocks and soil are Uranium series ($_{92}U^{238}$), Thorium series ($_{90}Th^{232}$) etc.

13. **Incineration of Solid waste** : The solid refuse contains the following toxic pollutants — SO_x, NO_x, CO, CO_2, NH_3, Hydrocarbons, Benzpyrene, Organic acid and particulates.

Major Air Pollution Episodes

All air pollution episodes are of great concern because they provide evidence by which one can evaluate the deleterious effects and gravity of air pollution.

Some important air pollution episodes are :

1. **Meuse Valley of Belgium Episode**

The first case of air pollution in last century happened in the Meuse Valley of Belgium in December 1930. Reports reveal that about 60 people died in Meuse Valley due to smoky fog over industrial areas.

2. **Killer Smog Episode In Donora, Pennsylvania (U.S.A.)**

This episode took place in Donora, Pennsylvania (U.S.A.) in 1948 along the Monongahela river, in which 20 people were killed and hundreds fell ill. It was also due to thick smog formation because of strong inversion temperature on December 26, 1948.

3. **Deadlier Smog Episode in London (U.K.)**

This episode took place in London in December 1952, in which about 5,000 people died of respiratory failure. Mortality records reveal that about 4,000 excess deaths occurred in Greater London.

The mixture of smoke + fog + SO_2 affected London very badly for several centuries, after the introduction of coal as a fuel. It is chemically a reducing mixture and, hence, the London smog is also called *reducing smog*.

4. **Poza Rica Air Pollution Episode**

It resulted from an industrial accident at a refinery in November 1950. Mortality reports revealed that about 25 people died and more than 300 became ill. In this episode a huge amount of H_2S spread over the ground around the refinery and crept into adjacent homes.

5. **22 States Air Pollution Episode**

It is also known as Episode 104, which blanketed all or part of 22 states east of 'Mississippi', with air pollution haze in August 1969. Several people suffered from diseases associated with cardio-respiratory system due to polluted haze.

6. **Los Angeles (California) Air pollution Episode**

It was a serious type of air pollution during 1944. It was characterized by reduced visibility, eye irritation and plant damage. It was caused by photochemical smog. Further, it was noticed that rubber articles showed severe cracks in the presence of high concentration of such type of smog:

7. **TCDD (Tetra Chloro Dibenzo-p-Dioxin) Episode at Seveso, Italy (July 1976)**

In this air pollution episode, an explosion took place in a chemical plant manufacturing herbicide (2,4,5 Trichlorophenoxy acetic acid i.e., 2,4,5,-T). A white cloud of poisonous gas consisting of TCDD whose structure is cited below came out and engulfed the nearby town of Seveso. TCDD is a potent poison for human. The building, ground and soil were contaminated. After three weeks of accident the Italian government evacuated about 800 people from the worst affected areas. However, the dioxin pollution continued to spread and particulates were deposited over widespread areas. About 200 people including children suffered from skin disease and some of them from liver troubles. About 1% of the babies born after the accident were premature deformed and also.

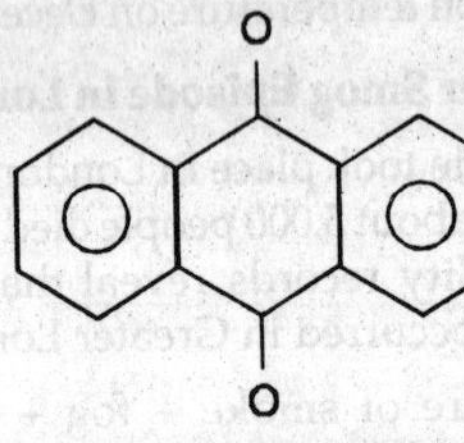

Structure of Dioxin

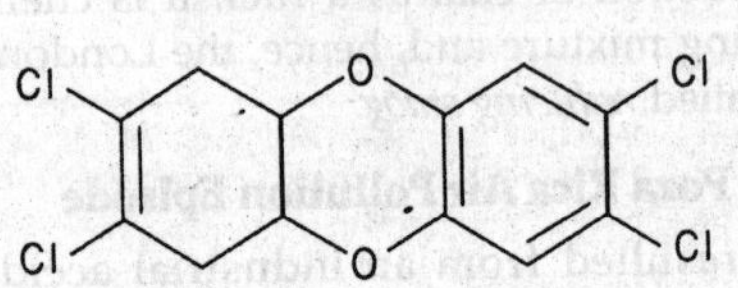

Structure of TCDD
(2,3,7,8-tetra chlorobenzo-p-dioxin)

8. **Bhopal Gas Tragedy (Toxic Gas Leak in Bhopal)**

Toxic gas leak in Bhopal is popularly known as "Bhopal Disaster".

Date : Early morning chilly windy hours of December 3, 1984.

Venue: Bhopal (M.P.), India

Source : Union Carbide Factory (Manufacturer of carbaryl, a carbamate pesticide) using methyl isocyanate (MIC), having molecular formula CH_3NCO.

Properties of MIC: Highly reactive, highly toxic, highly volatile, highly penetrable and low solubility. It can react with almost all biomembrane structures and this produce irreversibles and long-lasting damages to human body system.

Death: About 3,000 people within 24 hours and 10,000 people within a week.

Seriously Affected People : About 200,000.

Gas responsible: MIC

Mechanism of Synthesis of Carbaryl (General outline)

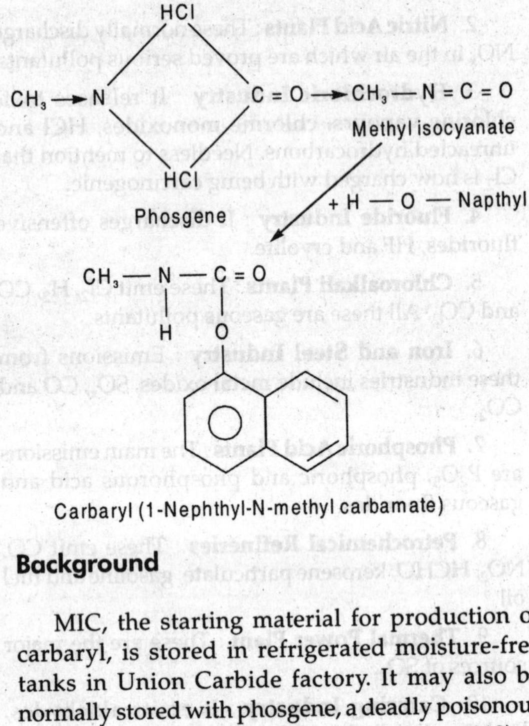

Carbaryl (1-Nephthyl-N-methyl carbamate)

Background

MIC, the starting material for production of carbaryl, is stored in refrigerated moisture-free tanks in Union Carbide factory. It may also be normally stored with phosgene, a deadly poisonous gas (used during World War I — 1914 - 1918) to prevent polymerization of the former. MIC is very sensitive to moisture and is itself a volatile liquid, b.p 43-45°C. In the preparative stage, MIC is always associated with unreacted $COCl_2$ (about 2%). In contact with moisture, it instantly produces $COCl_2$. The threshold limiting value (TLV) for MIC and for $COCl_2$ is 0.1 ppm. Workers exposed to MIC suffer from chest tightness and breathing troubles due to irritation of the respiratory tract. Since MIC is associated with $COCl_2$, the combined effect becomes fatal within 24 hours for most victims. Immediate symptoms of $COCl_2$ poisoning are bronchospasm, coughing, constriction and pain in chest. Normally 30% of the victims die in the first 24 hours and those surviving usually die from broncho-pneumonia.

Mass Panic and Death Drama

On 3rd December 1984 between 12.30 - 1.00 a.m. hopeless people awoke up coughing violently and with eyes burning as if chilli powder had been sprayed into them. They fled from their homes, some with families and many without families. They used whatever transport they could get. By 3.0 a.m. the main streets were jammed. The streets were full of foul smell due to vomiting. Those who fell were trampled by the crowd. The worst victims were the children — unable to move and breathe; they simply suffocated and expired.

The MIC emitted from the high-tech factory of the multinational Union Carbide spread over some 40 sq km and affected people seriously as distant as 5-8 km downwind.

About 200,000 people were seriously affected. A bigger tragedy could have taken place but for the two lakes which came in the way of the gas cloud.

The railway station was adjacent to the factory. Thousands of people left for distant stations away from Bhopal. The deputy chief power controller stayed on his work at the station while his wife and 15 year old son died in the neighbouring railway colony. He died later on along with the station master who alerted all nearby stations to stop trains coming for Bhopal. Next morning hundreds of sick and writhing people were found all around, on platforms, railway tracks and office rooms.

At the nearby Hamidia Hospital (1,200 beds) the first patient with eye trouble reported at 1.15 a.m. Within 5 minutes there were a thousand and by 2.30 a.m. there were 5,000 suffering from not just eye ailments but also from respiratory problems. The doctors were puzzled — nobody knew what to do and Union Carbide did not provide any useful information. Next morning thousands lay dead on roads and in homes. Corpses with distended bellies were beginning to rot, attracting vultures and dogs. More than 2,000 lay dying in hospitals. An awkward scene was that of carcasses of hundreds of dead cattle and animals all over the gas affected area, swollen up to sizes of elephants. By about 1.0 a.m. on the 4th December the hospital was crowded by 25,000 patients. The floor was splattered with blood and stinking with vomit.

The death drama continued for about 2 weeks. Within a week about 10,000 people died, 1,000 people became blind while more than 2 lakh people continued to suffer from various disorders.

A medical survey held 100 days after the exposure revealed that out of 250,000 exposed, 65,000 were subjected to severe medical disability (respiratory, eye, gastrointestinal, neuromuscular, gynecological symptoms) and 45,000 to mild to moderate medical disability. People suffering from breathlessness, sleeping and digestive problems were incapable of carrying on even light physical labour and unable to earn a living. Women were very badly affected, the worst victims being pregnant women.

9. Chernobyl Nuclear Disaster (April 1986)

World's worst nuclear disaster took place on 28th April (1986) at an atomic reactor at Chernobyl in the Ukarine area of former Soviet Union, resulting

Table 4.5 : Major Chemical Disasters in India After Bhopal Gas Tragedy

Month / Year		*Place*	*No. of people killed / injured*		*Nature of Accident*
Dec.	1985	Shri Ram Food and Fertilizers, Delhi	2	500	Oleum gas leak
July	1987	Bhorari Industrial Area, Pune, Maharashtra	—	420	Sulphur trioxide gas leak
Jun.	1987	Kaluprabhat, Bhubaneshwar, Orissa	4	NA	Ammonia gas leak
Nov.	1987	Beharampur, Ahemedabad, Gujarat	—	5,000	Oleum gas leak
May	1988	Deese town, Ahemedabad, Gujarat	19	53	Acid tanker overtuned
Oct.	1989	Chlorinators India, Ganjam, Orissa	6	100	Chlorine gas leak
Nov.	1990	Indian Petro Chemicals Ltd., Negothane, Maharashtra	35	50	Explosion in Gascracker plant
June	1991	Gwalior Rayon and Silk Mills, Gwalior, Madhya Pradesh	9	100	Explosion in filling unit
Nov.	1991	Dahanu, Maharashtra	100	19	Chemical explosion from overtuned tanker
Apr.	1992	Naya Bazar, Delhi	12	—	Chemical explosion in warehouse
Sep.	1992	National Fertilizer Plant, Panipat, Haryana	11	11	Ammonia gas leak
Mar.	1993	Century Rayon, Sahad, Maharashtra	11	123	Sulphuric acid gas leak
Oct.	1994	Jhansi, Utter Pradesh	30	25	Explosion in a factory

in clouds of radioactive smoke over a large area, causing death of thousands of people.

Further, some of the major chemical disasters in India after Bhopal gas leak have taken place which are summarized in Table 4.5.

Ecological Effects of Air Pollution

Major ecological effects of air pollution have been described as episodes of air pollution in this chapter and the rest — for example, acid rain, ozone depletion, greenhouse effect and smog pollution — is illustrated in Chapter 17 in detail.

Effects of Air Pollution on Plants

Salient aspects of air pollution and their effects on plants are being precisely illustrated below in tabular form :

Pollution Type	*Source*	*Effect*
Acid rain	Heavy industry such as electricity generation, steelworks and smelters	May cause some direct damage, especially to evergreen trees. Lowers the soil pH which mobilizes metals. Aluminium in particular may reach phytotoxic levels.
CO	Incomplete combustion of carbonaceous materials	Most plants are insensitive to CO levels but at higher concentration (100-10,000 ppm), it causes leaf drop, leaf curling, reduction in leaf size, premature ageing etc. CO is supposed to inhibit cellular respiration in plants by reacting with cytochrome oxidase enzyme system.
Dust	Cement works, road construction and quarries	Large particles shade the surface of leaves. Small particles can clog the stomata. Causes an overall reduction in growth.
Fluoride	Aluminium smelting and brickworks	Leaf edges become necrotic. Accumulates in the leaves and interferes with respiration and photosynthesis. Fluoride burns the tip of the leaves.
Heavy metals	Industrial wastes, leaded petrol	Most plants are not damaged directly by heavy metals. However, many plants can accumulate them in their leaves so making them toxic for human consumption.
HCl	Chemical manufacturing plants, combustion of coal, paper, plastic and chlorinated hydrocarbons	It is responsible for the abaxial glazing of leaves caused by collapse and plasmolysis of epidermal cells. High concentrations give rise to necrotic lesions.
H_2S	Decaying vegetation and animal material, volcanic discharges, coal pits and sewers, sulphur containing fuels	Leaf lesions, defoliation and reduced growth.
Hydrocarbons	Incomplete combustion of fuels, motor vehicles, petroleum refineries, agricultural burning, motor fuel marketing, manufacture of explosives and cracking of natural gas in petrochemical plants etc.	Yellowing and occasional necrosis of leaves, chlorosis of floral buds, inhibition of terminal growth, shortening of internodes, thickness of stem, lack of apical dominance, reduced growth, dry sepal diseases of orchids and decrease in amount of chlorophyll and carotenoids.
Nitrogen Oxides	Heavy industry and car exhausts	Visible damage to plants is rare.Causes overall reduction in growth and leaf injury.
NH_3	Refrigerators, pre-cooler systems of cold storage, manufacture of anhydrous ammonium fertilizers and nitric acids etc.	Induces bleaching of leaves, rusty spots on leaves and flowers, reduction of root and shoot growth, browning and softening of fruits and reduction of the rate of seed germination etc.
Ozone	Formed when car exhaust fumes react with sunlight	Damages leaf cell walls. May lead to magnesium deficiency. Causes stomata to close and leaves to wilt.
Sulphur dioxide	Burning coal, iron and steel works	Contributes to acid rain. Causes chlorosis of leaves. Impairs photosynthesis and increases respiration.

Units For Air Pollutants

The several units for expressing air pollutants and air quality parameters are :

1. ppm. by volume or μgm^3 (microgram per cubic meter) for gases and vapours.
2. μgm^3 — For particulate matter
3. Km — For visibility
4. M^3/min (cubic meter per minute) — for emission and sampling rates.
5. mmH — for pressure.
6. °C (Degree Centigrade) — For temperature

Prevention and Control of Air Pollution

Actually, prevention of air pollution is not so simple. It is impracticable to provide at reasonable cost. All the growing needs and amenities of modern life are causing some air pollution. In other words, air pollution has become a necessary evil of development. For example, it is difficult to drive a car or scooter without causing some air pollution. Similarly, it is difficult to run an industry without causing some air pollution. It is not possible to run a thermal plant without fouling the atmosphere.

It is, however, possible to prevent air pollution without undue cost by —

(i) Careful planning and setting of industries.
(ii) Better design equipment, and
(iii) Better operation of the equipment.

The objective of air pollution control should be to reduce the harm of a potential air pollutant. This is a multi-step process (Fig. 4.1), which indicates an effective air pollution control programme.

If a pollutant is spotted to be potentially hazardous based on the results of toxicological assessments, its control should genuinely be attempted. It is not uncommon, at least with highly carcinogenic environmental agents, that a modification of the human desire (or wants) initiates an effective pollution control programme. If man is not stringent enough to continue the use of a particular chemical despite its threatening qualities against him and his environment, pollution control is simple — 'Find an alternative'. Technological advancement has always been able to come up with almost everything that man has desired. But if there is no alternative to a chemical in use then efforts need be undertaken to control its entry into the environment.

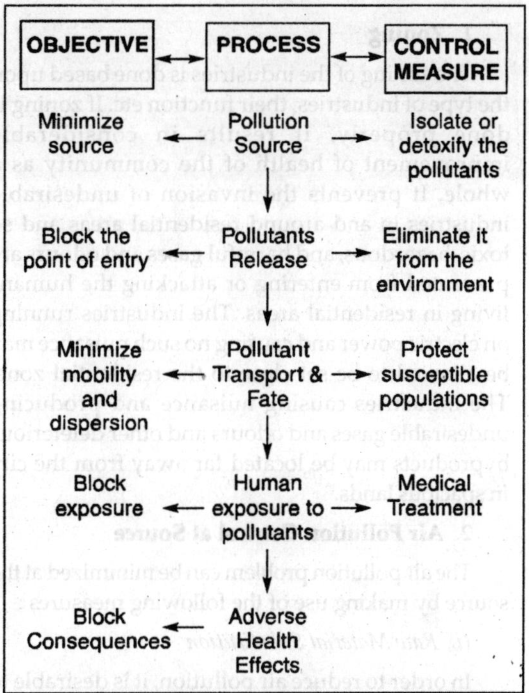

Fig. 4.1 : Planning pollution control measures

The air pollutant can either be isolated at the point of release or its production and converted for another purpose — thus keeping it within restricted areas; if that chemical (example — a byproduct or a final product) is highly toxic and not useful, then its release into the environment has to be blocked and the chemical be destroyed by specific processes in restricted areas. A systematic pollution control would begin with a search for an alternative for a process or a substance in use, if adequate control measures cannot be taken to prevent the entry of the substance into the environment.

General Methods of Air Pollution Control

Following are the general methods of controlling air pollution :

1. Zoning
2. Air pollution control at source
3. Installation of controlling devices and equipment
4. Air pollution control by high stacks or chimneys to discharge the pollutants at higher altitudes
5. By planting trees and growing vegetation.

1. **Zoning**

The zoning of the industries is done based upon the type of industries, their function etc. If zoning is done properly, it results in considerable improvement of health of the community as a whole. It prevents the invasion of undesirable industries in and around residential areas and so toxic, hazardous, and harmful gases and odours are prevented from entering or attacking the humans living in residential areas. The industries running on electric power and causing no such nuisance may be allowed to be set close to the residential zone. The industries causing nuisance and producing undesirable gases and odours and other deleterious byproducts may be located far away from the city in spacious lands.

2. **Air Pollution Control at Source**

The air pollution problem can be minimized at the source by making use of the following measures :

(i) Raw Material Substitution

In order to reduce air pollution, it is desirable to substitute the raw material (chemical) if it results in pollution by another one, which is less polluting (Table 4.6) :

Table 4.6 : Examples of Air Pollution Control by Chemical Substitution

Chemical	*Less toxic substitute(s)*	*Application*
Benzene	Toluene, Cyclohexane	Solvent, Chemical, raw material
Lead	Titanium, Zinc	Interior paint pigments
Beryllium	Calcium phosphate	Phosphorus for Fluorescent lamps
Asbestos	Glass fibers	Insulation

(ii) Redesigning Equipments or Changing the Operational Products

Release of toxic substances into the environment from a manufacturing unit or due to usage patterns can also be controlled. Toxic emissions can be reduced by redesigning equipments or changing the operational procedure (Table 4.7) :

Table 4.7 : Examples of pollution control by reducing the release of pollutants

Original Process	*Modified Process*	*Example of application*
Spraying	Dipping	Painting of parts and products
Diesel Engine	Battery powered electric engines	Delivering vehicles

In some cases, equipment alterations such as the use of floating roof tanks rather than vented tanks can cut down the evaporation losses. New type of equipment, for example the basic O_2 furnaces which are replacing the open hearth furnaces in steel industry, pose much less air pollution problems.

3. **Installation of Controlling Devices and Equipment**

Because of large number of industries, different types of gases are released into the atmosphere along with particulates and become major source of air pollution. In order to prevent these pollutants into the environment, control devices have been used depending on the collection property and capacity of the device and nature of the process used by the particular industry.

Based on the method of removal three important types of dust collectors have been used :

(i) Internal separators
(ii) Wet collection devices
(iii) Electrostatic precipitators.

(i) *Internal separators*

These separate the dust particles present in the gases. These are manufactured in various sizes and shapes. The commonly used internal separators are :

(a) Gravity settling chamber
(b) Cyclones
(c) Fabric filters.

(a) ***Gravity settling chambers***

These comprise of large chambers in which dust is separated from the gas by reducing the velocity of the gas. As a result, dust particles settle down in the chamber (Fig. 4.2). Coarse particles are thus removed. Gravity settling chambers are capable of removing only the large sized particles in the range of 25-30 μm diameter. The settling time required for smaller particles is so long that the use of this type of equipment is impractical in such cases.

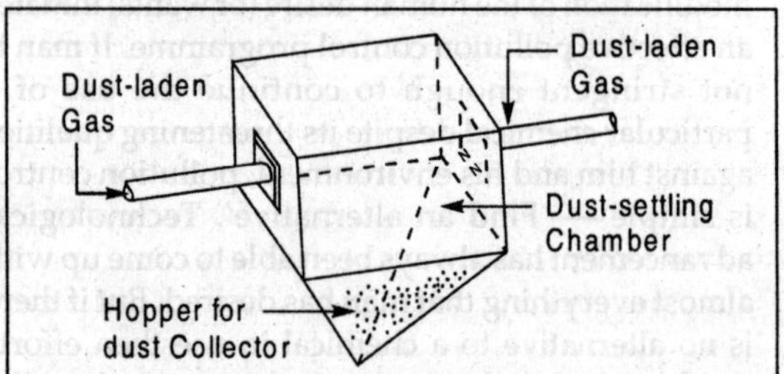

Fig. 4.2 : Gravity settling chamber

(b) ***Cyclones***

The principle of cyclone collector (Fig. 4.3) is that when a gas travels in a double vortex, the particles contained in it are separated.

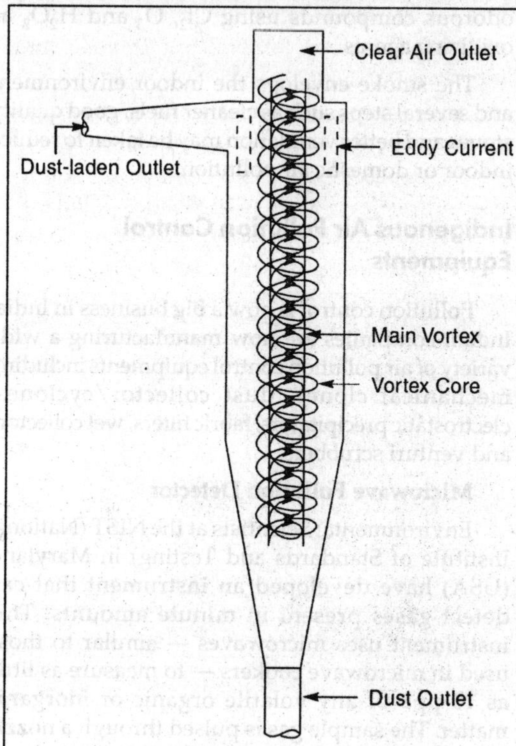

Fig. 4.3 : A cyclone

A cyclone collector consists of a tight circular spiral that produces centrifugal force, as a result of which the suspended particles in the flowing gas are forced to move outward to a wall where they are collected. This process removes all small particles (diameter 5-30 μm) efficiently.

(c) ***Fabric filters***

In fabric filters the gas with the dust is allowed to pass through a fabric to which the dust gets attached. If the gas is flowing at low velocity and contains considerable amount of large particulates, these settle down as a result of sedimentation. Due to electrostatic charges, fine particles are also attached to the fabric.

Bag filter (Fig. 4.4) is a device of removing the dust particles from the waste gases of the industrial processes. It is in the form of a tubular medium made of woven or felted fabric. These bags are fitted with hopper at bottom. The gas enters from the bottom through the hopper. The diameter of the bag is 1 m and height is about 7 to 10 m. Its collection efficiency is about 99%. The heavy particles are settled down in the hopper due to gravity. Finer particles are deposited on the fabric of the bag. The bags are cleaned by blowing compressed air in the reverse direction. It is necessary to have low gas velocities of the order of 1-3 meter per minute.

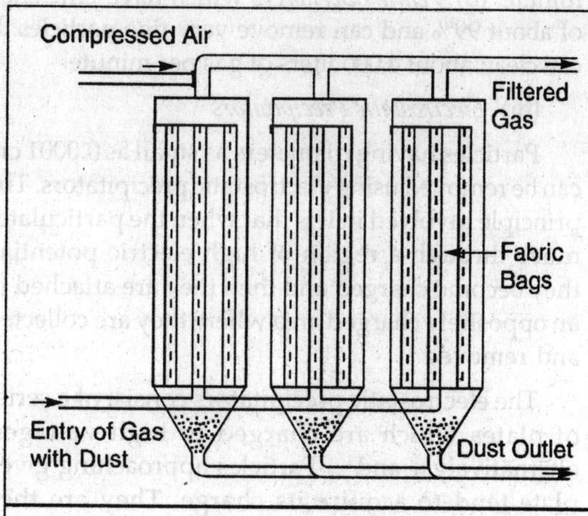

Fig. 4.4 : Bag filter

(ii) *Wet Collection Devices*

In such devices, a liquid is utilized to assist in the removal of solid, liquid or gaseous contaminants. The effectiveness of these devices depends primarily upon the frequency of contact and the degree of interaction between the liquid phase and the contaminants to be removed.

In precise form, devices using mixed phases of gases and liquid are known as wet washer and scrubbers. These are the collection devices in which the particles are washed out of the gas flow by a water spray. The aim is to transfer suspended particulate matters in the gas to the scrubbing liquid which can be readily removed by the gas cleaning device. This leaves the gas clean to pass onwards to the process for which it is being used, or alternatively to be discharged to the atmosphere. Extreme care should be taken to see that the collected waste water does not become a source of water pollution. It may need special settling tanks, chemical flocculators or filtration units.

The most common wet collectors in use are (a) *Cyclonic Scrubbers* which have an efficiency of 90% and can remove dust particles of 5.0 μm size. It is capable of cleaning about 2,000 liters of gas per minute. (b) *Venturi Scrubbers* which have efficiency of about 99% and can remove very fine particles. It can clean about 4,000 liters of gas per minute.

(iii) *Electrostatic Precipitators*

Particles having diameters as small as 0.0001 cm can be removed using electrostatic precipitators. The principle involved in it is that when the particulates move through a region of high electric potential, they become charged and then they are attached to an oppositely charged area where they are collected and removed.

The electrostatic precipitators consist of a series of plates which are charged to high voltages, alternatively + and –. Particles approaching given plate tend to acquire its charge. They are then attracted to the surface of the next plate, from which they fall into the hopper below. Thus particles pickup charge as they pass between plates and are precipitated on plates of opposite charge. The potential across the plates is around 50,000 volt. The high voltage in the wires produces billions of electrons and bombards the gas molecules, which become –vely charged. The +ve ions return to the –ve end electrode and gain electrons while the –ve ions combine with the dust particles and make them –vely charged. The negatively charged dust particles collect at the positively charged plates.

The efficiency of electrostatic precipitators is 99.9% and it is capable of cleaning 150,000 liters of gas per minute at a temperature of 600°C.

4. **Air Pollution Control by Stacks**

It is a well-known fact that when the quantity of pollutants released is much and the air possesses a limited capacity of absorption, then the pollution becomes serious. If the pollutants are carried away to long distances or taken to high altitudes, they get reduced in concentration by *diffusion* or / and dilution. Diffusion of a pollutant in air actually depends mainly upon atmospheric temperature, speed and direction of wind. Keeping the above factors in view, the pollutants are taken to high altitudes by means of stacks. The height of chimney stack should be raised to such a height that when the residual particulates are dispersed, they get spread evenly over a wide area. So the height as well as the diameter of the stack required is designed to keep the ground level concentration within permissible limits.

5. **Air Pollution Control by Planting Trees and Growing Vegetation**

In order to reduce the spreading of air pollutants coming out from industrial sources, planting of pollution-resistant tress and growing vegetation around the industry has been recommended.

In addition to above methods, the odours of gases can be absorbed by passing them through beds of activated charcoal or sand or soil. The odours can also be controlled by the oxidation of the odorous compounds using Cl_2, O_3 and H_2O_2 as oxidizing agents.

The smoke envelops the indoor environment and several steps such as cleaner fuels, good quality stoves and better ventilation may be taken to reduce indoor or domestic air pollution.

Indigenous Air Pollution Control Equipments

Pollution control is now a big business in India. Indian companies are now manufacturing a wide variety of air pollution control equipments including mechanical clones, dust collector/cyclones, electrostatic precipitators, fabric filters, wet collectors and venturi scrubbers.

Microwave Pollution Detector

Environmental Scientists at the NIST (National Institute of Standards and Testing) in Maryland (USA) have developed an instrument that can detect gases present in minute amounts. This instrument uses microwaves — similar to those used in microwave cookers — to measure as little as 10 ppb of any volatile organic or inorganic matter. The sample gas is pulsed through a nozzle into a detection cavity where it is cooled down to –272°C. At this temperature, molecules can be distinguished from one another through their light absorption patterns. The device can be used to monitor automobile emissions as well as indoor air quality.

Biological Indications Indicators for Air Pollution

Indication means "pointing out". Biological indication or bioindication is defined as the use of living matter as qualitative measures of the environmental status (in present context the air pollution).

Bioindicators can be classified into two main categories : (1) *Indicators,* and (2) *Accumulators.* Indicators are also called 'ecological indicator species' and accumulators 'chemical monitoring species' or collector organisms. At times, some organisms may act as both indicator and an accumulator for a specific pollutant. But the organisms may be different in their usefulness; a good indicator, sometimes, is not a good accumulator, and vice versa. Both animals and plants are sensitive to air pollutants.

Woodiwiss (1964) reported that the disappearance of following macro-invertebrates indicate the pollution level :

1. Stoneflies
2. Mayflies
3. Caddiflies
4. Amphipods
5. Isopods
6. Midges
7. Oligochaetes.

Plants as Air Pollution Indicators

Some plants are especially sensitive to air pollutants and these develop toxic symptoms as a result of exposure to specific pollutant (Table 4.8). These might be used to identify in general terms the pollutants in prevalence. Because of their non-specificity in response, it is often preferred to have a specifically sensitive species, that might be able to perform the function of signalling whenever the atmosphere levels of the air pollutants in question exceeds the normal limits. In Table 4.9 are identified some plant species that serve as indicators of exposure to some air pollutants.

Table 4.8 : Toxic Symptoms produced by different pollutants in plants

Element	*Toxic Symptoms*
Al	Mottling of the leaves
Co	White dead spots in leaves
Cd	Yellow-white chlorotic spots between the leaf veins which change with age to red brown
Cr	Yellow leaves with green veins and dried dead patches on lower leaves frequently beginning on the tip, purple stem
Fe	Stunted tops or sometimes stubby roots
Hg	Bleaching of the perianth and stamen injury
Mn	Chlorotic leaves, curling and dead areas on leaf margins
Mo	Stunted growth, yellow-orange coloration
Ni	White dead spots on leaves or chlorotic leaves, abnormal forms
O_3	Red or brown spots on the upper surface of leaves
PAN	Collapse of leaf cells
SO_2	Light spots on the margin or near the veins, killing of tissues
Zn	Chlorotic leaves with green veins, white dwarfed forms; dead areas on leaf tips

Table 4.9 : Some plant species that serve as indicators of exposure to specific pollutants

Indicator Plants	*Pollutants*
Dusheri Mango	SO_2 fumes
Gladiolus	Fluorine
Alfalfa	SO_2, PAN
Pine	SO_2, O_3, F, Metals
Tomato	Ozone
Tobacco	Ozone
Pinta Beans	PAN

Moreover, lichens and mosses serve as excellent accumulators of a wide range of pollutants in the environment.

Odourous Air Pollutants (Odour Pollution)

The study of smell (odour) is termed 'osmics' and an odourous substance is called an 'osmogen'. The odourous materials are usually volatile, losing molecules to the atmosphere continuously.

The production of odours by chemical substances seems to be mainly confined to thermo-volatile organics.

Sources of Odourous Pollution

There are two main sources of odourous pollution :

(1) Natural sources

(2) Anthropogenic sources (Man-made sources).

1. Natural Sources

The important natural sources and types of odour emitted are :

1. Bluegreen algae viz. *Anabaena circinalis* → Earthy odours
2. *Synura petersenii (algae)* → Fishy and cucumber-like odours
3. *Trichoderma viridae* (fungus) → coconut-like odour.

Several other microbial species also emit different odours.

2. Anthropogenic Sources

Volatile organics and unsaturated molecules of various industrial chemicals escaping into the atmosphere are also responsible for malodorous air. Various industrial operations emit different kinds of odourous substances or liberate products that emit malodours. Table 4.10 gives a list of such industries and their odourous by-product or wastes :

Table 4.10 : Industrial operations and odorous pollutants associated with them

Sl. No.	Industry	Odourous Substances
1.	Chemical Manufacturing	NH_3, H_2S, Alcohols, Amines, Aldehydes, Phenols, Esters, Mercaptans, Cl_2, Chlorinated organics etc.
2.	Coke ovens	Sulphurous, ammonical and phenolic compounds
3.	Fertilizer	Bone meal, organic nitrate compounds, NH_3 etc.
4.	Food Products	Dairy wastes, fish, chocolate flavours, cooking odours, meat products
5.	Petroleum	Sulphurous compounds from crude oil, creosols
6.	Pulp and Paper	Sulphurous compounds,
7.	Pharmaceuticals	Biological extracts and wastes, spent fermentation liquors
8.	Soap	Animal fats, perfumes
9.	Tanneries	Hair, flesh, hides
10.	Laboratories	Cl_2, H_2S, SO_2 etc.

The odour concentration at a point is dependent on several factors such as its distance from the height of the emission source, odour emission rate, air velocity and directions. Some typical pollutants, their odours and the odour thresholds are given in Table 4.11. The threshold concentration of odourous pollutant is the concentration at which its presence in air is not perceived by the human nose.

Table 4.11 : Odours and odour thresholds of important pollutants

S. No.	Chemical	Characteristic Odour	Odour Thresholds (ppm)
1.	Methanol	Sweet	100.0
2.	NH_3	Pungent	46.8
3.	SO_2	Pungent, musty	0.47
4.	Cl_2	Bleach, pungent	0.314
5.	H_2S	Rotten egg	0.00047

Health Risk

In general, children are considered to be more tolerant to malodours than adults. The problem due to odorants is more aesthetic than economical. The health effects of odours are extremely hard to quantify as such but exposure to odorants, in general, causes nausea and appetite loss. Continued exposure may result in the loss of sense of smell.

Revision Questions

A. Long-Answer Questions

1. Write an essay on air pollution and human health.
2. Define air pollution and its sources. Describe principal types of air pollutants.
3. Give the names of common air pollutants, their source and ill-effects on man.
4. Describe five important air pollution episodes studied by you.
5. Give an account of the prevention and control of air pollution.

B. Short-Answer Questions

1. Explain briefly :

(i) Outdoor and Indoor air pollution
(ii) Sinks of atmospheric gases
(iii) Primary and secondary air-pollutants.

2. Write short notes on :

(i) Particulate matters
(ii) Occupational diseases
(iii) Bhopal Gas Tragedy
(iv) TCDD
(v) Units for air pollutants.

3. Explain briefly :

(i) Cyclone collector
(ii) Fabric filters
(iii) Electrostatic precipitators
(iv) Biological indicators for air pollution
(v) Microwave pollution detector
(vi) Odourous air pollutants.

4. Write short notes on :

(i) Ozone depletion
(ii) Acid rain
(iii) Greenhouse Effect
(iv) Killer Smog.

5

Water Pollution

SYNOPSIS

Introduction, Definition, Sources, Categories, Biological magnification, BOD, Water pollution control

Introduction

Next to air, water is the most important substance for the existence of life on earth. Water accounts for about 70% of the mass total of our body. It is an essential constituent of all flora and fauna and forms about 75% of the matter of the earth's crust. Water is distributed in nature in different forms, such as rainwater, river water, spring water and mineral water. Rainwater is the purest form of naturally occurring water. The sum total of all the available water on the surface of earth is called hydrosphere.

Water is an excellent solvent because (a) It has high dielectric constant, (b) It is hydrogen bonded, and (c) It is polar a molecule and has dipole moment. Much of the heat generated by metabolism in our bodies is removed by evaporation of water through the pores of the skin. Water, the most abundant and wonderful natural resource, has become a precious commodity today and its quality is threatened by numerous sources of pollution.

The term pollution is derived from the Latin word *pollutus* — *pol* (before) + *lutus* (wash), i.e., before washing. Water contains impurities and, hence, the term pollution is applied to indicate an act of contamination or making foul the natural waterbodies such as rivers, wells, lakes, streams, seas. Salient definitions of water pollution are given next.

Definitions of Water Pollution

Water is the soul of nature; its pollution will perish the whole world. Water pollution may be defined in a number of ways :

1. Undesirable changes occurring in water, which adversely affect the life activities of human and domesticated species.
2. Alteration in physical, chemical and biological characteristics of water which may cause harmful effects on human and aquatic biota.
3. Addition of excess of undesirable substances to water that makes it harmful to man, animal and aquatic life, or otherwise causes significant departures from the normal activities of various living communities in or around water.
4. Any adverse change in conditions or composition of the water so that it becomes less suitable for the purpose for which it would be suitable in its natural state.
5. In reality, the term water pollution refers to any type of aquatic contamination between two extremes :

   (i) A highly enriched, over-productive biotic community such as a river or a lake with nutrients from sewage or fertilizer (*cultural eutrophication*), or

(ii) A body of water poisoned by toxic chemicals which eliminate living organisms or even exclude all forms of life.

Sources of Water Pollution

The well-known sources of water pollution are :

1. Domestic sewage
2. Industrial effluents
3. Agricultural discharges
4. Fertilizers
5. Detergents
6. Toxic metals
7. Silts
8. Oils
9. Thermal pollutants
10. Radioactive materials
11. Pesticides.

1. Domestic Sewage

Domestic sewage is usually a cloudy dilute aqueous solution containing organic matter, human feces, urine, soap, minerals and dirty used up water of houses. Generally, domestic sewage is released into the rivers, on the bank of which most of the cities are situated.

Municipal waste is the principal contributor of water pollution. A recent report from Water Pollution Research Laboratory, London, indicates that domestic sewage contains trace quantities of toxic metals such as Cu, Cr, Zn, Mn, Pb, and Ni. The decomposable organic matters in the domestic sewage exert an oxygen demand on the receiving waters. Organic matter in the sewage includes fatty acids, esters, amino acids, amides, amino sugars etc.

Tremendous quantities of sewages are produced due to extensive population density. Most municipal sewages receive no treatment before they are discharged in water bodies. Sewage treatment deposits the suspended material — called sludge — at the bottom, while the liquid waste consists of ions like Ca^{++}, Mg^{++}, Na^{+}, K^{+}, NH_4^{+}, Cl^{-}, NO_2^{-}, HCO_3^{-}, SO_4^{--} and PO_4^{---} in dissolved condition.

Harmful Effects of Domestic Sewage

1. Domestic sewage makes the water extremely anesthetic.
2. Affects the aquatic flora severely.
3. Interferes with the spawning of fishes.
4. Causes depletion of DO and enhances BOD.
5. Sewage serves an excellent medium for the growth of pathogenic organisms as shown in Table 5.1 :

Table 5.1 : Water-Borne Pathogens and Disease Caused by them in Humans

Sl. No.	*Pathogen*	*Disease*
1.	**Bacteria**	
	Bacillus typhosum	Typhoid fever
	B. paratyphosum	Paratyphoid fever
	B. dysenteriae	Bacillary dysentery
	Vibrio cholere	Cholera
2.	**Protozoa**	
	Entamoeba histolytica	Amebic dysentery
	Giardia lamblia	Giardiasis
3.	**Viruses**	
	Liver-infecting virus	Infectious hepatitis
	Polio virus	Poliomyelitis
	Viruses (spread by water-borne mosquitoes)	Yellow Fever
4.	**Helminths**	
	Tapeworms	Teniasis
	Schistosoma	Bilharziasis
5.	**Algae**	
	Blue green algae	Toxic effects on human health

2. Industrial Effluents

Industrial effluents discharged into water bodies contain toxic chemicals, hazardous compounds like phenols, aldehydes, ketones, amines, cyanides, metallic wastes, plasticizers, toxic acids, corrosive alkalis, oils, greases, dyes, biocides, suspended solids, nonbiodegradable pollutants, radioactive wastes and thermal pollutants from numerous industries.

The principal industries responsible for water pollution in India are chemical and pharmaceuticals, pulp and paper, sugar, distilleries, textiles, steel mills, oil refineries, fertilizer plants, soap and detergent, tanneries etc.

Some Indian rivers and freshwater streams are heavily polluted by industrial effluents (Table 5.2).

Table 5.2 : Industries and their effluents in Indian rivers

Sl. No.	River	Source
1.	Yamuna in Delhi	Indraprastha power station, insecticide factories, sewage, flyash from steel mills and automobile workshops
2.	Kali at Meerut (U.P.)	Sugar mills, soap, silk, textile, rayon, tin and glycerine industries
3.	Dajora in Bareily (U.P.)	Sewage, chemical mills, synthetic rubber factories
4.	Gomti near Lucknow	Sewage, soap, paper and pulp mills
5.	Hooghly near Kolkata (W.B.)	Paint, varnishes, chemical mills, power station, textile, vegetable ghee, oil soap, matches, jute, rayon, shella, sewage and polythene industries
6.	Damodar near Bokaro	Flyash from steel mills, fertilizers, pesticides, thermal power stations — suspended particulates
7.	Bhadra in Karnataka	Paper and pulp industries, steel factories
8.	Ganga at Kanpur	Textile mills, domestic wastes, chemical mills, surgical industries, jute, tanneries and sewage
9.	Sone (Bihar)	Cement, paper and pulp mills
10.	Godavari	Paper and chemical mills
11.	Siwan (Bihar)	Sugar mills, cement and paper industries
12.	Cauvery (Tamil Nadu)	Tanneries, distilleries, sewage, paper and rayon mills
13.	Kulu (between Mumbai and Kalyan)	Tanneries, chemical, and rayon mills
14.	Cooum (Chennai)	Automobile workshops and domestic sewage
15.	Suwao in Balrampur (U.P.)	Sugar industries
16.	Ami, Basti (U.P.) (now Sant Kabeer Nagar dist.)	Paper mill

Harmful Effects of Industrial Pollutants

1. Cause deleterious effects on living organisms and may bring about death.
2. Lower the rate of photosynthesis in aquatic flora.
3. Disinfectants added in water to control phycophytes growth and bacteria may persist in aquatic bodies and consequently may cause mortality of fishes, planktons and diatoms.
4. Free Cl_2 discharged by factories may cause fish mortality.
5. Hg in its inorganic form in water attacks mainly liver and kidneys of aquatic vertebrates whereas organomercurials traverse through the biological membrane and concentrate especially in the brain. Alkyl mercurials are much more toxic than the inorganic forms. Alkyl mercurials tend to accumulate when consumed in small quantities and eventually attack the nervous system. A concentration of 6.0 ppm in brain cells can cause irreversible brain damage. Hg is also a teratogen, capable of inducing abortion and toxic effects in human embryo. Minamata episode is a popular one in Japan revealing Hg toxicity as described ahead in this section.
6. Industrial effluents consisting of As, Pb and CN etc. cause cellular degeneration in brain, which results in frigidity, coma, stupor, and numbness in humans when they enter through food chain.
7. Effluents containing acids and alkalis make the water corrosive.
8. Heated effluents discharged into aquatic system may severely alter the aquatic ecosystem by increasing the temperature.
9. Industrial discharges impart colourful odour and turbidity to the receiving waters. They undergo putrefaction to form objectionable tastes.

It shall not be out of place to mention here that in 'holy' Ganga alone, about 315 industrial complexes are dumping their effluents. These are indirectly responsible for the ill-health of about 300 million people of northern India!

Minamata Disease

Mercury, a byproduct of the production of vinyl chloride, is used in many chemical industries. It is also a byproduct of some incinerators, power plants and laboratories (Aaronson, 1971). In Japan, illness and even death occurred in the 1950s among fishermen who consumed fishes, crabs, and shell fish contaminated with methyl mercury from the industrial city of Minamata in Japan. The main unit of this city, Shin-Nihon Chisso Hiryo Co., began to produce vinyl chloride and acetaldehyde by the catalytic conversion of acetylene, sine 1949. During this process some of the $HgCl_2$ catalyst was unknowingly converted to a methyl mercuric compound. This factory kept ejecting its effluent wastes into Minamata Bay. The people who took the fishes, crabs and shell-fish suffered from Minamata Disease. Initial symptoms of Minamata disease included numbness of the limbs, lips and tongue, impairment of motor control, deafness and blurring of vision. Cellular degeneration occurred in the cerebellum, midbrain and cerebral cortex and this led to spasticity, frigidity, stupor and coma.

In Japan in 1952, due to Minamata Disease, 17 people died and 23 became disabled permanently.

3. Agricultural Discharges

Agricultural discharges like pesticides of biodegradable and non-biodegradable nature, fertilizer, manure, slurry, animal and plant debris, soil erosion containing mostly the inorganic materials is known to cause heavy pollution to water sources. Most of the discharges adding to the soil are washed off through the rainfall, irrigation and drainage into water bodies, where they severely disturb the aquatic ecosystem. Agricultural discharges are expected to be three times more than the domestic sewage.

Harmful effects of Agricultural Dishcarges

1. Agricultural discharges containing DDT, which is non-degradable through food chain arriving in man, may cause nervous impairment. DDT is also known as cancer-promoter in humans.
2. Agricultural discharges containing endosulphan in water bodies prove to be a contact and stomach poison for man.
3. Pesticidal accumulation in fishes has also been reported.
4. Agricultural discharges in aquatic bodies disturb the ecosystem homeostasis.

4. Fertilizers

Recently NCA reported the increased application of fertilizers from 2.8 million tons in 1976 to 6.0 million tons in 1984 and 9.7 million tons in 1995. The fertilizer enhance crop yield on one hand but they disrupt the entire natural aquatic ecosystem on the other.

India[1] applies about 15-16 kg of fertilizers per hectare while the world average is 55 kg per hectare. In Netherlands the fertilizer consumption is about 71 kg/ha. However, it is not only the increasing use of fertilizers but also escalated production, which pose adverse effects on water and their living biota.

Effects of Fertilizers

Fertilizers and other agricultural chemicals create deleterious effects on man, cattle and plants as mentioned below :

A. Effects on Man (Methemoglobinemia)

The nitrate used in fertilizers enter the intestine of man through drinking water. In the intestine nitrate is reduced to toxic nitrites by intestinal bacteria. Nitrite is absorbed into the blood where it combines with hemoglobin to form methemoglobin :

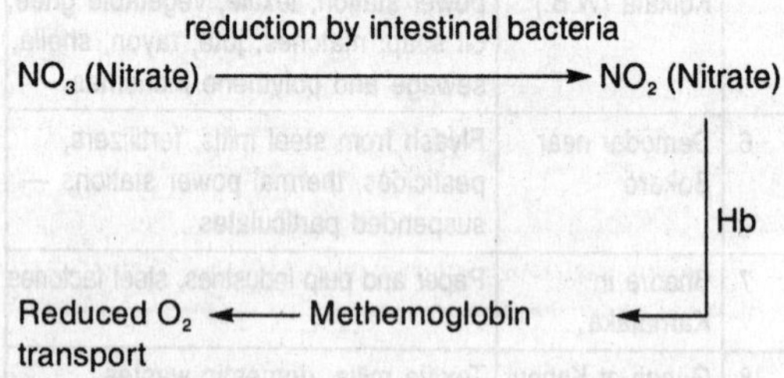

Methemoglobin cannot transport O_2. This leads to suffocation and breathing troubles, especially in infants. This disease is called "methemoglobinemia" (blue babies).

It seems important to express here that a healthy person contains 0.8% of methemoglobin, but in the diseases it reaches up to 10% in blood. Above 20% it causes headache and giddiness and above 60% it causes unconsciousness, stiffness and occupational problems. At 80% of mathemoglobin in blood, death occurs.

B. Effects on Plants

1. Protein quality of crops degrades.
2. Excessive use of fertilizers produce large sized fruits and vegetables, which are more prone to pests, insects and diseases.
3. Superphosphate may lead to Fe, Cu, and Zn deficiency in plants. Potash treatment decreases valuable nutrient like vit. C and carotene in fruits and vegetables.
4. Liming may check release and uptake of several essential elements such as Zn, Mn, Ni, Fe, Co, and Ca in plants.
5. Excessive addition of nitrates and phosphates in aquatic ecosystem makes it highly productive causing *Eutrophication.* It leads to depletion of O_2 due to excessive algal growth thereby, increasing BOD of water. It also leads to fish and other aquatic fauna mortality.

[1]The all-India average fertiliser consumption is 90 kg/ha. (INDIA 2004, p. 63)

5. Detergents

Detergents are of recent origin, used as cleaning agents and derived from surfactant (10-30%), builder (10-15%) and other ingredients. The surfactant is actually a surface active agent which dissolves partly in water and partly in organic solvent because of its dual hydrocarbon and polar character, for example, alkyl benzene sulphonates are considered as surfactants.

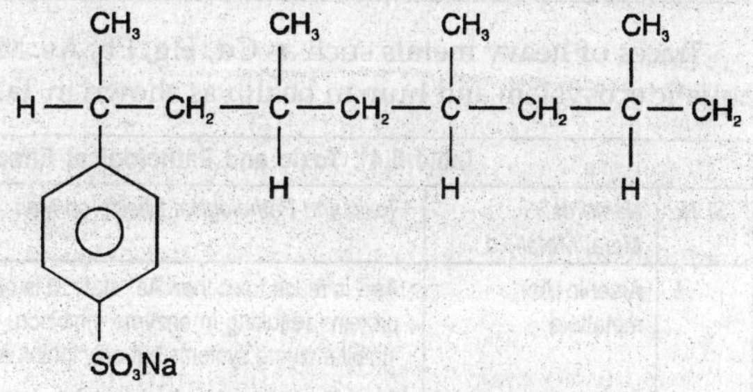

Structure of ABS (Alkyl Benzene Sulphonate)

The builder is usually a sodium phosphate (polyphosphate) of the type $Na_3P_3O_{10}$ or $Na_4P_2O_7$ acting as sequestering agent. Both the surfactant and builders of detergent create complex pollution problems in water.

Household detergents contain surface-active agents and contribute to phosphates of sodium, sodium silicates, sodium sulphate, amides and several other builders in water. Since detergents are composed of complex phosphates, they enhance the concentration of phosphates in water leading to eutrophicational problems.

Harmful Effects of Detergents

1. Act as allergens.
2. Complex formation between nitrilotriacetate (NTA) and Hg or Cd enhances the possibilities of transmission across the placental membrane (barrier) into fetus, thereby increasing the likelihood of birth defects.
3. Phosphate, the major ingredient of most detergents, favours the luxuriant growth of algae which form blooms. The extensive growth of algae consumes most of the available O_2 from water. The depletion in O_2 concentration becomes detrimental to growth of aquatic fauna which produce a foul smell upon decay. Such decomposing water are known to produce toxins as strychnine, which causes mortality in aquatic fauna.
4. Synthetic detergents destroy aquatic flora and fauna.
5. Detergents cause damage to gills of fishes and remove protective mucus from gills, skin and intestine.
6. Alkyl benzene sulphonates (ABS) may bring mortality in several species of Mayfly nymphs when they are exposed to 15 ppm for 10 days.
7. Aquatic invertebrates are severely affected by the detergents. For example, Gray fish are reduced in numbers by 15 days' exposure to 10 ppm of effluent containing detergents.

6. Toxic Metals

Toxic metals find their way in aquatic bodies from industrial processes, domestic sewage discharge, land run-off, street dust, engineering processes and fossil fuel burning (Table 5.3) :

Table 5.3 : Toxic Metals in Water and their Major Sources

S. No.	*Name of Metal*	*Major Sources*
1.	Arsenic (As)	Coal burning, impurity in phosphates, processing of sulphide ores, effluent of fertilizer plants, Arsenical pesticides, smelting of ores containing arsenic.
2.	Aluminium (Al)	Mining, manufacturing industries, canning and food packaging.
3.	Beryllium (Be)	Wastes of fluorescent lamps manufacturing industries, nuclear reactors, coal combustion.
4.	Cadmium (Cd)	Electroplating wastes, motor oils, cadmium containing industrial wastes.
5.	Chromium (Cr)	Fuel burning, textile mills, tanneries, photography wastes.
6.	Cobalt (Co)	NH_3 manufacturing waste, paint dryers' wastes.
7.	Copper (Cu)	Fertilizer plant waste, industrial effluent, municipal sewage wastes, discharge of mine tailings and flyash.
8.	Lead (Pb)	Leaded gasoline, paints, varnishes and steel industries.
9.	Mercury (Hg)	Electrochemical industry, few fungicides, vinyl chloride, manufacturing industry, Cl_2 and caustic soda industries, batteries, fossil fuel combustion etc.
10.	Manganese (Mn)	Burning of fossil fuels (coal oil), fertilizers, domestic wastes, industrial effluents etc.
11.	Selenium (se)	Rubbers, metal alloy, fossil fuel, rock phosphates, flyash from steel plants, coalfiring power plants.
12.	Zinc (Zn)	Pigment printing processes, smelting of ores, municipal refuse, motor oils, fuel additives, pesticides and fungicides.

Traces of heavy metals such as Cd, Hg, Pb, As, Mn, Co, Fe and Cr have been identified deleterious to aquatic ecosystem and human health as shown in Table 5.4 :

Table 5.4 : Toxic and Pathological Effects of Metallic Water Pollutants on Man

Sl. No.	*Name of Metal/Metalloid*	*Toxic and Pathological Effects on Man*
1.	Arsenic (As), metalloid	As^{+3} is much toxic than As^{1} state. It is because the arsenites bind strongly with the (-SH-) group of proteins resulting in enzyme inhibition, whereas the arsenates do neither bind to -SH- groups nor inhibit enzyme systems but only inhibit ATP synthesis. Arsenic causes disturbed peripheral circulation, mental disturbance, liver cirrhosis, hyperkeratosis, skin cancer, lung cancer, teratogenesis, kidney damage and ulcers in gastrointestinal tract.
2.	Beryllium (Be)	Lung disease "*Berylliosis*", heart disfunctioning, non-healing ulcers, bone tumours etc.
3.	Barium (Ba)	Colic pain, excessive salivation, vomiting, diarrhea, paralysis.
4.	Cobalt (Co)	Ingestion of excess amounts of cobalt causes polycythemia. Chronic exposure to cobalt at high concentration produces goitre. It also causes diarrhea, L.B.P., lung irritation, bone deformation and paralysis.
5.	Chromium (Cr)	Cr^{+3} and Cr^{+6} viz., Cr_2O_3, $CrCl_3$ and CrO_3 (Chromium trioxide) are biologically important. Cr^{+6} is only toxic to humans and causes dermatitis, allergic skin reactions, gastrointestinal ulcer, nephritis, teratogenesis and lung cancer.
6.	Cadmium (Cd)	It is highly toxic because there exists no homeostatic control in human body for this metal. About a third of Cd absorbed is stored in the kidney. Its excess intake causes the following disorders in man : (i) Aminoaciduria → Excretion of amino acids in the urine. (ii) Glucosuria → Excertion of blood sugar in urine. (iii) Hypercalciuria → Excetion of excess calcium in the urine. (iv) Osteoporosis → Decalcification of the skeleton. (v) Proteinuria → Excertion of proteins in the urine. It also causes hypertension, respiratory disorders, damage to liver and kidney, anemia, CNS injury, and growth retardation. **Itai - Itai Episode** — This episode occurred in Japan. In Japan, as in India, rice is the staple food. Generally, people residing in the uncontaminated parts of the world ingest about 25-60 μg of cadmium daily. But the people residing along the banks of Juntsu river in northern Japan were ingesting 100-1,000 μg of Cd every day. This is because the rice they consumed was grown with irrigation water supplies that contained 0.33-3.5 ppm of Cd. About 100 deaths took place until the end of 1965 due to Cd intake. The victims accumulated 500-600 mg of Cd in their body over the lifetime. This disorder was termed "itai - itai", literally meaning "ouch-ouch". This disease was characterized by kidney malfunction. Typically the onset of symptoms occurs only several years after the consumption of Cadmium.
7.	Copper (Cu)	The recommended daily intake of copper range is 1-2 mg. Its deficiency causes Menke's disease whereas 'Wilson disease' is its opposite. The excess intake of Cu through contaminated water etc. causes hemolysis, hepatatoxic and nephrotoxic effects, lung cancer, sporadic fever, hypertension, uremia, and coma.
8.	Mercury (Hg) [Inorganic Hg released into the environment is converted to more toxic methyl mercury compounds by anerobic bacteria]	Metallic Hg is less well absorbed by the gastrointestinal tract whereas methyl mercury absorption is about 95%. Of the salts, mercurous (Hg^{+}) salts are poorly absorbed whereas mercuric (Hg^{+2}) salts are readily absorbed from the gut. Hg, in its inorganic forms, attacks mainly liver and kidneys, whereas organo-mercurials traverse through the biological membrane and concentrate especially in the brain. Pathological symptoms include abdominal pain, headache, diarrhea, chest pain, hemolysis and abortions. Also see Minamata disease in this chapter.
9.	Manganese (Mn)	Chronic exposure to Mn^{12} and Mn^{14} may cause neurological disorders, manganese psychosis — an irreversible brain disease similar to Parkinson's disease, characterized by uncontrollable laughter, euphoria, impulsiveness, sexual excitement followed by impotence and speech disability. *Mn pneumonia results from inhalation exposure to Mn.*
10.	Selenium (se)	Damage of kidney, liver and spleen, fever, vomiting, nervousness, blindness, low B.P. and even death. Selenium is also mutganic as well as teratogenic.
11.	Zinc (Zn)	Large doses of zinc salts (220-440 mg of Zn) as $ZnSO_4$ cause gastrointestinal disorders including vomiting and diarrhea. It may also cause renal damage. There is no evidence of its teratogenic or mutagenic effects.

It is notable here that the enhanced level of heavy metals is seriously concerned because of the following reasons :

(i) Accumulate in human body.

(ii) Create sublethal and chronic effects to organisms even in minute concentration.

(iii) Produce teratogenic and mutagenic effects in human.

(iv) Cause phytotoxicity.

(v) Produce synergistic effects in living organisms.

(vi) Affect critically all aquatic bodies and their flora and fauna.

(vii) Degrade very slowly and accumulate in the food chain.

Keeping all these in view, tolerance limits for trace metals in drinking water and their recommended maximum concentration (RMC) in irrigation water is listed in Table 5.5 (source : Ayers & Westocott, 1976) :

Table 5.5 : Tolerance limits for trace metals in drinking Water

Sl. No.	Trace Metal	Symbol	Tolerance limits (ppm)	RMC (mg/l)
1.	Aluminium	Al	1.0	5.00
2.	Arsenic	As	1.0	0.10
3.	Beryllium	Be	0.50	0.10
4.	Boron	B	0.75	not known
5.	Cadmium	Cd	0.001	0.01
6.	Chromium	Cr	5.0	0.10
7.	Coblat	Co	0.2	0.05
8.	Copper	Cu	0.2	0.05
9.	Lead	Pb	5.0	5.00
10.	Lithium	Li	5.0	not known
11.	Manganese	Mn	2.0	0.20
12.	Molybdenum	Mo	0.005	0.01
13.	Selenium	Se	0.005	0.02
14.	Nickel	Ni	0.5	0.20
15.	Iron	Fe	0.1	5.00
16.	Vanadium	V	10.0	not known
17.	Zinc	Zn	5.0	2.00

7. Silts

Actually silts are the soil/sand particles suspended in the aquatic bodies. Siltation is the most widespread and damaging pollution, especially in hill streams which create high turbidities. These also obstruct the free movement of aquatic organisms, growth of fishes and their productivity. Salient deleterious effects of suspensoids and colloids on marine biota as reported by Baruton (1985) are as follows :

(i) Decreased visibility of pelagic food

(ii) Reduction in direct light penetration in water

(iii) Reduced photosynthetic rate

(iv) Less food availability

(v) Decreased availability of benthic food and plant biomass

(vi) Clogging of gill rakers and gill filaments in fishes.

8. Oils

Oil is a source of pollution in sea water. Oil pollution is due to ship accidents, loading and discharge of oil at the harbour, oil refineries and offshore oil production. It is estimated that two million tons of used lubricating oil are added every year in coastal waters. Maritime accidents due to collision, fire, explosion or grounding also result in oil release in water. International discharge of oily wastes from tank washing and accidental spillages pollute the sea water severely. From *Indira Dock* basin alone, more than 90,000 litres of waste oil was collected in 1984. A recent report shows that about 20 billion tons of wastes per year from industries, homes, farms and municipalities end up in sea.

Oil leakage from 20,000 miles (32,000 km) of pipelines which cross waterways may undergo corrosion, cracks or punctures and would lead to oil pollution in sea water.

Effects of Oil Pollution in Sea Water

Oil pollution in sea water has been an inevitable consequence of the dependence of rapidly growing population on oil based technology. The overall detrimental effects of oil pollution in marine water are :

A. Physical Effects of Oil in Water

1. Oil films on water may reduce significantly the rate of O_2 uptake by water.
2. Oil slick retards the light intensity up to 90%.
3. Smothering coats of oil kill lichens and algae along the shore lines.

B. Effect of Oil Pollution on Marine Ecosystem

1. Heavy damage to fisheries.
2. Cause lethal toxicity on aquatic flora.
3. Large-scale fish mortality as in Mumbai coast in 1981.
4. Direct oil coating unable the fishes to respire and clog their gill slits.
5. Hydrocarbons in oil get incorporated in body tissues of marine animals. These are quite stable like heavy metals and pesticides.
6. Oil pollutants may block the taste receptors of organisms.
7. Emulsified oil reaching the bottom of sea may damage flora and fauna.

C. Effects of Oil Pollution on Man

1. Oil components like aromatic thiophenes, benzothiophenes and merceptans are lethal to man as they damage kidney and liver.
2. Crude oil contains sulphur compounds, small amount of nitrogen, little olefins, metals like Fe, Ni, and V. These are extremely lethal. Carbonyl sulphide in the crude oil, after dissociation, produces H_2S which acts on CNS resulting in death — mainly due to respiratory paralysis.

D. Effects of Oil Pollution on Birds

1. Birds are especially vulnerable to damage from oil coating. The spilled oils destroy their natural insulating oils and waxes, which shield the birds from water. Consequently they lose insulation, start shivering and may freeze to death in winter. About 25,000 birds' mortality took place in *Torry Canyon Incident*.
2. Oil spilling in marine water causes abnormally low body temperature in birds resulting in *hypothermia.* Nearly 150 rare species of bald eagles also became victims when they ingested oil during Exxon Valdez accident.
3. About 1,000 sea otters died when their fur become saturated with oil by losing insulation.

9. Thermal Pollutants

Many industries use water for cooling. The resultant warmed water is discharged into rivers, stream or lakes. This brings about water pollution. Thermal pollution is also referred to as calefaction (warming). Generally the thermal pollutants include the waste chiefly from atomic, nuclear and thermal power plants. Municipal sewage also contributes to thermal pollution. The electric utilities constitute the major source of the thermal pollution of rivers and lakes. They also send heat into the atmosphere, as do in most manufacturing operations and dense, urban communities. Cooling has been the traditional and still common method of discharging waste heat from a power plant.

Major Problems due to Thermal Pollution of Water

Actually the thermal pollution of water creates two major problems :

1. The activity of biological life is more at higher temperature and, hence, as water temperature rises, there is more demand for dissolved O_2.
2. With rise in the water temperature, the amount of DO in water diminishes. Hence, at higher temperature, less amount of DO will be present in water and it may be fatal for aquatic biota.

Effects of Thermal Pollution

Thermal pollution alone poses no direct health risks to people but in aquatic system it has a profound effect on organisms as well as on water quality. If the aquatic body contains chemical pollutants, extra warmth increases their toxicity to fishes. Regarding the change in water quality, the density of water has been a maximum of 1,00,000 g/cm 2 at 4°C and then decrease slightly at higher temperature. Also, as the temperature gets increased, the viscosity of water gets decreased and the vapour pressure increases sharply. The solubility of gases in the water may also lower down with increasing temperature. These properties exert significant effects on aquatic life. Various detrimental effects of thermal pollution in aquatic bodies are :

A. Physico-Chemical Changes in Aquatic Bodies

1. Density of water decreases.
2. Viscosity of water decreases.
3. Decreases the solubility of gases in water.
4. Vapour pressure increases.
5. Decreases in dissolved oxygen.
6. Increase in BOD.

B. *Effects on Aquatic Biota due to Thermal Pollution*

1. Undesirable changes in the algal population.
2. Changes in diurnal and seasonal behaviour of aquatic invertebrates.
3. Migration of aquatic biota.
4. Excessive eutrophication.
5. Acceleration in activities of pathogenic organisms, which makes the pathogen more virulent, and fishes become less resistant. Bacteria multiply rapidly.
6. Direct fish mortality due to failure in respiratory, nervous or essential cell processes.
7. Early hatching of fish eggs.
8. Failure of trout eggs to hatch and salmon to spawn.
9. Adverse effects on spawning and reproductive mechanism during spring season.
10. Rapid settling of sediment load in water affects aquatic food supply.
11. Each type of fish has its own fatal temperature called heat death. For Sockeye Salmon fry, heat death occurs at only 75°F, (24 °C), whereas large-mouth bass can withstand temperature up to 79°F (26°C).
12. Thermal pollution also results in faster growth rates but shorter life spans in aquatic fauna.

10. Radioactive Materials

The oceanic currents carry the radioactive contaminants everywhere. Radioactive pollutants get their way into water streams from various sources such as nuclear power plants, nuclear reactors, nuclear test, nuclear installation, operations of power, processing fission and fusion product etc. In fact, hazards from radioactivity arise because radionuclides accumulated in body tissues/organs and deliver radiation dose. Potentially deleterious radioactive Cs, Cm, Bk, Np, Pu, Ru, Zr etc. are produced from neutron bombardment of atomic fuel. Once they find access into aquatic bodies, they disrupt the ecocycling process, enter the food chain and affect adversely the metabolic pathways.

Effects of Radioactive Pollutants on Aquatic Ecosystem

1. Aquatic life adversely affected.
2. Radioactive contaminants deposit on surface and ground water. This water consumed by plants during photosynthesis acts as a medium for radioactivity in them.
3. The radioactive pollutants reach to the top consumer level through food chain.
4. Ionizing radiations in aquatic bodies mainly result in cellular damage.
5. Radiation in living organisms produces hazardous chemical species like H^+, H_2O^-, H_2O^{+e-}, HO_2^{e+}, H_3O^- etc., which cause damaging effects.
6. The radioactive substances in aquatic habitat react with proteins of aquatic invertebrates and deactivate enzymes by splitting disulphide bonds. With enzyme inhibition, cell growth may continue but cell division cannot proceed.
7. Radionuclides occurring in aquatic bodies from leaching of minerals or reactors include Ra-226, Sr-90, Ba-140, I-131, Cs-137, U-236, Zr-95, Mn-44, Co-60, Kr-85, K-40 and Ru-103. Some of these fission products are insoluble in water and settle out with particulate materials and concentrate in silts and sludges. The soluble isotopes remain in solution and enter biological cycle causing chronic health effects.
8. Researches reveal that Zn-65 accumulates in oysters at high level, Fe-55 concentrates in fishes while Sr-90 accumulates in higher invertebrates. These affect their metabolism seriously.
9. Traces of radioactive substances present in water may cause cancer, leukemia, eye cataract, DNA breakage and carcinoma in human.
10. Drinking water containing Rn-222, Ra-226 and Th-232 can accumulate in humans and cause several physiological, somatic and genetic disorders. It may also cause teratogenesis and mutation defects too.

11. Pesticides

Described as a separate chapter in this book.

Categories of Water Pollution

There may be four categories of water pollution :

1. Physical Pollution of Water
2. Chemical Pollution of Water
3. Biological Pollution of Water
4. Physiological Pollution of Water

1. Physical Pollution of Water

Physical pollution of water refers to the changes in physical properties of water viz., colour, odour, taste, density, thermal properties, viscosity, turbidity etc.

2. Chemical Pollution of Water

The chemical pollution of water causes changes in acidity, alkalinity i.e. pH, dissolved oxygen (DO) and other gases in water. It may be caused either by inorganic or organic pollutants or both. The organic as well as inorganic pollutants may be biodegradable or non-biodegradable type.

(i) Biodegradable Pollutants

Biodegradability by definition is the destruction of chemical pollutants by the action of living organism. The degradation of pollutants can be accomplished by any living organisms but the microbes (bacteria, fungi and algae) play extremely important role. This is because of their catabolic versatility, species diversity and metabolic rate per unit weight. Reactions include oxidation, reduction, hydrolysis and often various types of structural rearrangement. The feasibility of occurrence of these processes is determined by the molecular structure and concentration of the pollutant, the nature of microbes in existence and the prevailing environmental conditions. In nutshell, the pollutants that can be rapidly decomposed by the natural living agents (by the action of microbes) are called biodegradable pollutants.

For a chemical pollutant to be biodegraded efficiently, the following set of criteria must be met :

1. At least one microbial species capable of metabolizing the chemical pollutant in question should be present in the media.
2. Contact between the organisms and the chemical molecules should occur.
3. The chemical pollutants must be capable of inducing the enzymes needed for its decomposition.
4. Favourable environmental conditions (optimum pH, temperature etc.) for the microbes to proliferate and for their enzymes to operate should exist.

For several chemical pollutants, these conditions are known to be met satisfactorily. In other words, there do exist various species of microbes that degrade several of the organic pollutants (Table 5.6) in the environment under normal conditions.

Easily Biodegradable Pollutants

(i) Proteins from domestic sewage, waste from creameries, canneries and slaughter houses.

(ii) Fats from sewage, soap production, food processing and wool processing.

(iii) Carbohydrates, (sugar, starch) etc. from sewage, textile mills and paper mills.

(iv) Polymers, resins, coal, oil and various other organic substances found in domestic and industrial wastes and those synthetic compounds which are non-toxic to some microbes.

Table 5.6 : Biodegradation of Organic Pollutants by Microbes

Pollutants of respective group	*Microbes capable of degrading them*
(i) Insecticides	
Dieldrin	Anaoystis nidulans Pseudomonas Actinomycetes
Endosulphan	Soil actinomycetes Bacteria Fungi
Endrin	Micrococcus Pseudomonas
Lindane	Chlamydomonas reinhardii Chlorella vulgaris Clostridium
Parathion	Bacillus subtilis Chlorella pyrenoidosa
(ii) Herbicides	
Monuron	Pseudomonas
2-4-D	Achromobactor Corynebacterium Falvobacterium
MCPA	Achromobactor Mycoplana
Daiapon	Agrobacterium Pseudomonas
DNBP	Corynebacterium Pseudomonas
(iii) Polycyclic Aromatic Hydrocarbons (PAHs)	
Benzo (a) pyrene	Cunninghamella elegans
Dioxins	Phanerochaetum chrysosporium
PCB(s)	Pseudomonas Vibrio spirillum Flavobacterium

It is important to mention here that highly branched organic pollutants are much resistant to biodegradation because increased substitution hinders oxidation, the process by which alkyl chains are broken down microbially. Other aspects of pollutants, viz., chemical structure, play key role in biodegradation; for example : position of the phenyl group, length of the alkyl chain and the presence of cyclic groups. This is evident in Table 5.7 where several pollutants have been classified according to their susceptibility to biodegradation. It shall not be out of place to mention that chlorinated hydrocarbons (DDT, endrin, dieldrin, toxaphene, aldrin, etc.) are more stable than organophosphates (parathion, malathion, chlorothion etc.). Other generalizations between biodegradation and chemical structure are summarized in Table 5.8.

Table 5.7 : Relative persistence of various synthetic organic chemicals in water

Persistent	*Slowly degradable*	*Easily biodegradable*
Alkyl benzene sulphonate	Linear alkyl benzene sulphonate	Acetic acid Cresols
Diedrin	Dibutyl phthalate	Phenols
PCBs	Biphenyl	
Dioxins		

Table 5.8 : Relationship between Chemical Structure and Biodegradability

More Biodegradable	⟶ Biodegradable ⟶	Less Biodegradable
R—CH₂—R	R—CH(R)—R	R—C(R)(R)—R
R—CH₂—OH	R—C(=O)—R	
RCHO	R—C(=O)—O—R	R—O—R
RCO_2N		
RNH_2		
– OH	Aromatic Substituents	
– COOH	– SO_3H	– Cl
	NH_2	– NO_2
	– OCH_3	

(ii) Non-Biodegradable Pollutants

The chemical pollutants and toxicants such as aluminium cans, phenolic compounds, DDT, etc. — after getting their way into the environment (water, air or land) — either do not degrade or degrade extremely slowly in the natural environment and are called non-biodegradable pollutants. Such non-biodegradable pollutants not only accumulate but are often '*biologically magnified or amplified*'. The non-biodegradable pollutants also combine with other compounds in the environment to produce additional toxins. The only possible abatement for such pollution is expensive removal or extraction from the environmental life-support system.

Pesticide persistence in soils may be defined as the time required for the pesticide level to be reduced to less than 25% of the original level applied.

Some examples are :

2,4-D	→	1 month
2,4,5-T	→	5 month
Picloram	→	18 month
Diedrin	→	3 years
DDT	→	4-5 years
Chlordane	→	5 years

It has been estimated that inorganic mercury compounds in bottom sediments of lakes and rivers may take 10 to 100 years to get converted into methylmercury, an organic form highly dangerous to animal life.

Biological Magnification or Amplification

While energy decreases and becomes more dispersed at each trophic level in the food chain, some pollutants become more concentrated with each link in a process that has now come to be known as "biological magnification" or "biological amplification".

Many heavy metals and organic insecticides concentrate biologically in the food chain. The tendency of a pollutant to accumulate in the tissue of the organisms is given by bioconcentration factor (BCF), and is expressed in L/kg :

$$\text{BCF} = \frac{\text{(Concentration of the substance in the organisms, mass/mass)}}{\text{(Concentration of the pollutant in the environment, mass/volume)}}$$

The BCF varies from species to species, and is also affected by the metabolism and elimination systems in the organisms. The BCFs of most organic pesticides are much higher than those of the inorganic pesticides. Further, it may be noted that BCFs are higher in aquatic ecosystem than in terrestrial ones. The BCFs of certain pollutants (inorganic and organic) in fishes are :

Chromium	16 L/kg
Arsenic	44 L/kg
Lead	49 L/kg
Malathion	0
Chloridane	14,000 L/kg
DDT	54,000 L/kg

DDT was a very popular pesticide in developed countries. Paul Miller won the Nobel Prize in 1948 for its synthesis. However, though its use is still widespread in developing countries, its use in the USA is banned since 1973. Biological magnification of DDT in one aquatic ecosystem is given in Fig. 5.1 :

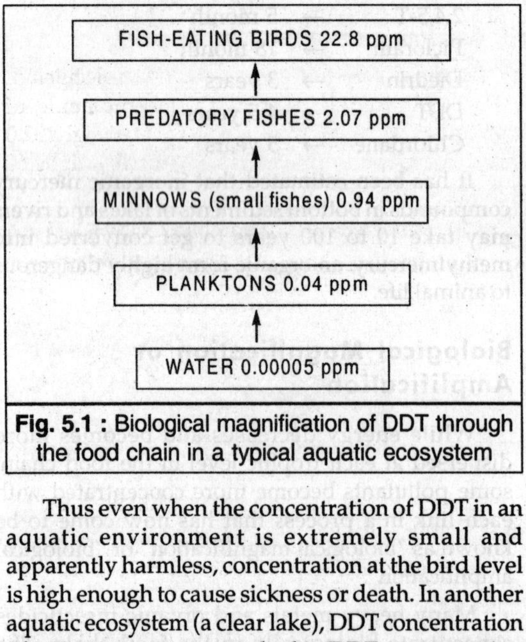

Fig. 5.1 : Biological magnification of DDT through the food chain in a typical aquatic ecosystem

Thus even when the concentration of DDT in an aquatic environment is extremely small and apparently harmless, concentration at the bird level is high enough to cause sickness or death. In another aquatic ecosystem (a clear lake), DDT concentration of 0.02 ppm in water resulted in DDT concentration of 2,000 ppm in the tissue of predatory birds. In another aquatic ecosystem, radioactive phosphorus concentrated 2,00,000 times in duck eggs. Man himself is threatened by this process of biological magnification, being at the top of the food chain. Estimations of DDT in human body fat were made in 1964 and these varied from $3.3 g/m^3$ in UK to $25 g/m^3$ in India.

The accumulation of chlorinated hydrocarbons may reach such high levels in animals at the summit of the aquatic food web that they are unable to reproduce or they may be poisoned outright.

Another classic example of biomagnification of chlorinated hydrocarbons is shown by DDD which is used as a larvicide :

Parameter : Water → Planktons → Small fish → Predatory fish → Grebe fish

Conc. of DDD (ppm) : 0.02 → 5.3 → 10.0 → 1,500 → 1,600

The incidence of mercury pollution at Minamata Bay (Japan in 1952) is a famous episode, which indicated the potential dangers to the human from the natural process of biological amplification. Mercury (inorganic form) was discharged into the bay by one chemical company "Shin-Nihon Chisso Hiryo Co". Anerobic methane synthesizing bacteria in water converted the inorganic mercury to "dimethyl mercury", which, under acidic condition, was altered to "methyl mercury" subsequently. Methyl mercury, being soluble in water, entered the food chain and became concentrated in the shell fishes. The fishermen around the bay who consumed such fishes became sick to the extent of permanent paralysis and sensory losses and genetic disorders.

From the above discussion it may be concluded that the increasing accumulation of non-biodegradeble pollutants in higher organisms may be called *biomagnification* or *biological amplification.*

3. Biological Pollution of Water

The excretory products of endothermic (warm-blooded) bacterial population in water are main cause of biological pollution. Biological pollution is also brought about by algae, diatoms like protozoans, rotifiers, crustaceans, plant toxins, and viruses. Contaminated water supplies frequently produce infection of intestinal tract (like dysentery, cholera, typhoid and gastroenteritis), polio and infectious hepatitis.

4. Physiological Pollution of Water

Several chemical agents such as chlorine, SO_2, H_2S, ketones, phenols, amines, mercaptans and hydroxy benzene are responsible for physiological pollution of water. Chlorination of water usually converts phenol to ortho or para chlorophenol which tastes like medicine and produces offensive odours.

Oxygen Demanding Wastes

Actually dissolved oxygen (DO) is a fundamental need for the maintenance of all life forms in any water body. A water body is said to be polluted when the DO level falls below a certain minimal concentration necessary for sustaining a normal biota for that water. Oxygen demanding pollutants may remove large quantity of DO from water. DO is required maximum by fishes, then by other aquatic vertebrates and followed by invertebrates, and least by bacteria.

Oxygen demand increases as the organic matter in the water decomposes due to the death and decay of the explosive growth of blooms forming algae and related species. Generally the organic compounds are oxidized by microbes like bacteria to CO_2 and water. Consequently, aquatic body becomes O_2-deficient destroying fish habitats and leading to elimination of many desirable aquatic biota.

Biological Oxygen Demand (BOD)

BOD is not a pollutant but an indicator of water quality (pollution index of aquatic body). It has been a standardized measurement of the amount of O_2 that would be required by microbes to cause the decomposition of certain organic and inorganic matter in the water. The measurement is done under standardized condition (at 20°C and 5 days to allow the decomposition to take place). The result is called the 5-day BOD and is expressed in milligrams of O_2 per liter of water. It measures no particular substance but a family of any substance that microbes can consume (using O_2 as they do).

The BOD test as a measure of pollution degree in water was first adopted in 1912 by *"The Royal Commission on Sewage Disposal"*.

One of the most serious effects of human and industrial wastes (or effluents) is to increase the BOD of natural water supplies. Aerobic bacteria use O_2 to decompose the complex organic compounds in sewage to simpler and generally unobjectionable species such as CO_2, NO_3 ions and SO_4 ions. This process reduces the amount of DO in water, sometimes to the point where animal life is difficult to survive. If the oxygen content drops too low, anaerobic bacteria take over the decomposition process, forming rather noxious pollutants such as CH_4, NH_3 and H_2S as shown below in Table 5.9.

Table 5.9 : Decomposition products under aerobic and aerobic conditions

Aerobic Conditions	*Anaerobic Condition*
$C \rightarrow CO_2$	$C \rightarrow CH_4 + C_2H_4$
$N \rightarrow NH_3 + HNO_3$	$N \rightarrow NH_3 + NH_2$
$S \rightarrow H_2SO_4$	$S \rightarrow H_2S$
$P \rightarrow H_3PO_4$	$P \rightarrow PH_3$ + lower valent phosphorus compounds

Determination of BOD

The BOD of polluted water can be determined by measuring the amount of O_2 consumed by a sample of known volume. The water sample is first diluted with air-saturated distilled water to ensure an excess of O_2. The concentration of DO in the dilute sample is immediately determined, and again after a period of 5 days. From the decrease in concentration, BOD can be calculated :

$$\text{BOD} = \frac{\text{Number of mg } O_2 \text{ required}}{\text{Number of liters of sample}}$$

$$= \text{parts } O_2/\text{million part of sample}$$

Example : Calculate the BOD of a water sample which contains one gram of urea for every 100 liters of water. The reaction between urea and O_2 is

$$NH_2CONH_2 + 4O_2 \rightarrow CO_2 + 2NO_3^- + 2H^+ + HOH$$

Solution :

Balanced equation reveals that 4 molecules of O_2 are required to react with one molecule of carbamide, i.e. urea. One molecule of O_2 weighs 32.0 gm and one mole of NH_2CONH_2 (urea) weighs 60 gm. Now we have

$$4(32.0 \text{ gm } O_2) = 1(60.0 \text{ gm. } NH_2CONH_2)$$

$$\text{or } 4 + 32 \text{ gm} = 128 \text{ gm } O_2 - 60.0 \text{ gm } NH_2CONH_2$$

To react with 1.0 gm urea we need

$$\frac{1.0 \text{ gm } NH_2CONH_2 \times 128.0 \text{ gm } O_2}{60.0 \text{ gm } NH_2CONH_2} = 2.13 \text{ gm } O_2$$

$$= 2.13 \times 10^3 \text{ mg } O_2$$

Thus BOD = 2.13 + 10^3 mg O_2/100 liters = 21.3 mg O_2/liter

Answer : BOD of given sample = 21.3 mg/l or 21.3 ppm.

Note : A BOD test normally measures the amount of O_2 used up over a 5 day period by aerobic decomposers in certain volume of effluent at 20°C, the result being expressed in ppm. Thus a 100 ppm BOD means that 100 mg of O_2 are consumed by one liter of the test sample (e.g., effluent, sewage, water) over a 5-day period at 20°C. Increased BOD lowers the DO content in water causing the suffocation and death of aquatic flora and fauna.

Water bodies according to their BOD status

In Table 5.10 is represented a classification of water bodies according to their BOD status (BOD is higher in polluted water and lesser in drinking water).

Table 5.10 : BOD as Index of Water Quality

General condition of water	BOD (ppm)
Pure water	1
Clean water	2
Moderately clean water	3
Of doubtful cleanliness	4
Poor	5
Untreated city sewage	100 - 400
Complex Industrial Wastes	1,000 - 10,000

Limitations of BOD Test

1. Before BOD test, the pre-treatment of sewage is necessary if it also contains toxic wastes.
2. The test is applicable only in the case of biodegradable organic matter.
3. A high concentration of the active bacteria is necessary to be present in the sample of sewage.
4. Before applying BOD test, the effects of nitrifying organisms are to be reduced.
5. There is no validity of the test after the organic matter present in the sample of sewage is utilized or exhausted.
6. The time required for the test is long as well as arbitrary.

Applications of BOD Test

(i) Used as a measure for determining the strength of sewage.

(ii) Aids in finding out the amount of clear water needed for the successful disposal of sewage by dilution.

(iii) Gives information of polluting power of sewage or its nuisance value.

(iv) Gives an idea of the load of organic matter on the sewage treatment plants.

What is COD?

Chemical oxygen demand (COD) may be defined as the amount of O_2 required by organic matter in a sample of water for its oxidation by a strong chemical oxidant, and is expressed as ppm of oxygen taken from a solution of acidic $K_2Cr_2O_7$ in two hours at high temperature.

As BOD value approximates the amount of oxidizable organic matter, it is, therefore, used as a measure of degree of water pollution and waste strength. But COD value has been a poor measure of strength of organic matter as O_2 also gets consumed in the oxidation of inorganic matter such as nitrate, sulphates, reduced metal ions and also that some organic materials like Benzene, Pyridine and few other cyclic organic compounds do not get oxidized by this test. COD of sewage is higher than its BOD, because more compounds can be oxidized than can be biologically degraded.

Types of microbes, pH, presence of toxins, some reduced mineral matter and nitrification process have been the important factors that influence the BOD test. But presence of toxins and other such unfavourable conditions for the growth of microbes are not able to affect COD values. The limiting value of COD has been specified to be 250 ppm. The inorganic compunds should be removed before COD test.

Water Pollution Control

Actually water pollution has become a global problem, partly because of the population explosion and partly due to the phenomenal advancement in industrialization. India has many localized areas where water pollution already creates situations not only hazardous to health but also adverse to the economy in several ways.

Water pollution problem can be minimized by adopting the following techniques :

1. Stabilization of the waste
2. Recycling of the waste
3. Waste treatment
4. Removal of pollutants.

1. Stabilization of the Waste

It is the most relevant economical and wise way to control water pollution. The principles involved in this method include reduction of the waste at source, harvesting and removal of biomass, trapping of nutrients, fish management and aeration. It is a very economical method for the treatment of industrial effluent and domestic sewage. Industrial and domestic wastes are stored in a dilute condition in shallow ponds called *oxidation* or *stabilization ponds*. After a week, microbes and algae start flourishing. The microbes decompose the organic wastes by oxidation, and water is purified. This water is rich in nitrogen, phosphorus, potassium and other nutrients. This method was devised by NEERI at Nagpur.

2. Recycling of the Waste

Water pollution can be controlled to a certain extent by reutilizing the wastes, called recycling. In developed countries the waste water consisting of industrial effluents, domestic sewage, thermal pollutants, radioactive pollutants and municipal sullage etc. receive some sort of treatment before these get mixed into the water bodies.

Examples: 1 : The dung of cattle can be reused for the production of *gobar gas*.

2. Sewage can be reused for irrigation and pisciculture after treatment in oxidation pond. Benefits of reuse of effluents in industries is shown in Table 5.11 :

Table 5.11 : Recovery of Matter from Industrial Waste Water

Sl. No.	*Industry*	*Recoverable Matter*
1.	Distillery	Potassium salts, Yeast
2.	Dyestuff	Na and K hydroxides
3.	Electroplating	Chromium, Ni salts
4.	Fertilizer	$CaSO_4$
5.	Coke oven	NH_3, $(NH_4)_2SO_4$, Aromatics, Phenol, Naphthalene
6.	Petrochemical	Acetone, Carboxylic acid
7.	Paper and pulp	Ligno-Sulphonate, Sodium salts
8.	Rayon	Sodium sulphate, Zinc
9.	Textiles	NaOH

3. Waste Treatment

Water pollution control by sewage treatment includes the following steps :

(a) Sedimentation (b) Dilution, and (c) Storage.

(a) Sedimentation

When sewage is allowed to stand, the suspended particles settle to the bottom. So by sedimentation the suspended particles are removed from sewage.

(b) Dilution

The sewage can be diluted with water. This increases the O_2 contents and reduces BOD and CO_2.

(c) Storage

The dilution sewage is stored in a pond. This facilitates the growth of microbes. This renders further oxidation of sewage.

4. Removal of Pollutants

Various techniques involved in the removal of biological, chemical and radioactive water pollutants may be adsorption, electrodialysis, ion-exchange, reverse-osmosis etc.

The processes for the removal of various pollutants are :

(i) Removal of Phosphorus by Electrolysis

In Norway, organic sewage is mixed with 10 to 15% sea water and subjected to *electrolysis* to remove phosphorus from sewage. The removal of phosphorus reduces the eutrophication of lake and rivers. Phosphorus compounds in the sewage get precipitated as calcium or magnesium phosphate. These salts along with sludge and suspended particles adhere to $Mg(OH)_2$ at the negative pole during electrolysis. Hydrogen gas liberated during electrolysis makes the phosphate and sludge to float on the surface as scum, which can be scraped off from the top layer. Chlorine produced at the positive pole can be used for disinfection at the outlet.

(ii) Electrodialysis

Placing an electrical potential difference across the waste water produces an electric current, making the cations to migrate toward the cathode and the anions to migrate toward the anode

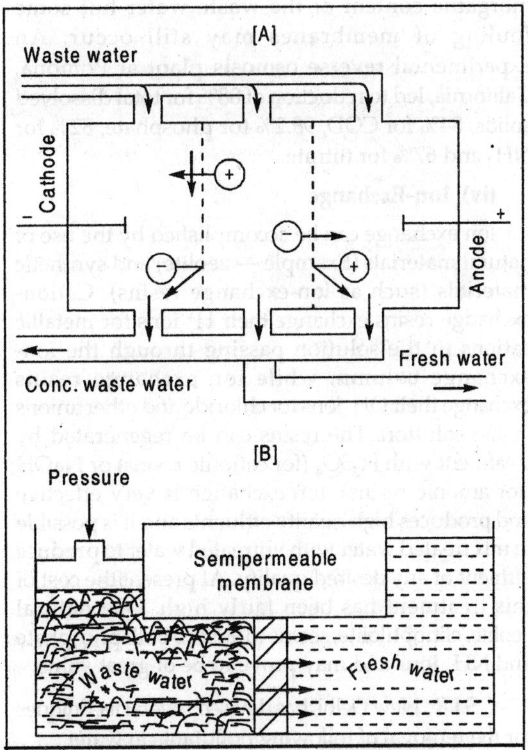

Fig. 5.2A & B : Schematic diagrams of membrane processes for water treatment. [A] Electrodialysis. [B] Reverse osmosis.

(Fig. 5.2A). Membranes (really ion-exchange resins in sheet form) permeable to only cations or only anions are used to control the migration of the ions and permit demineralized water to be taken out of the appropriate chambers. Organic molecules are not removed and they can collect on and clog the membranes.

(iii) Removal of Salts [Reverse Osmosis Technique]

Various toxic substances and salts present in the water can be removed by *reverse osmosis* by forcing the waste water through a semipermeable membrane under a pressure (Fig. 5.2B) exceeding the osmotic pressure. During the process, flow occurs in the reverse direction. The solvent is attracted while the solute is repelled. This method is commonly used to desalinate the brackish water and to purify water from sewage. Process of reverse osmosis is the opposite phenomenon of osmosis and is produced by applying pressure to the more concentrated solution and thereby forcing water to the side of the membrane.

Reverse osmosis reduces both the organic and inorganic content of the waste water but some fouling of membranes may still occur. An experimental reverse osmosis plant at Pomona, California, led to reduction of 88% for total dissolved solids, 84% for COD, 98.2% for phosphate, 82% for NH_3 and 67% for nitrate.

(iv) Ion-Exchange

Ion exchange can be accomplished by the use of natural materials (Example — zeolite) and synthetic materials (such as ion-exchange resins). Cation-exchange resins exchange their H^+ ions for metallic cations in the solution passing through the ion-exchange column, while ion-exchange resins exchange their OH^- ions for chloride and other anions in the solution. The resins can be regenerated by treatment with H_2SO_4 (for cationic resins) or NaOH (for anionic resins). Ion exchange is very effective and produces high quality effluents and it is possible to mix treated water with untreated water to produce effluent of any desired quality. At present the cost of this treatment has been fairly high. One natural zeolite, clinoptilolite, seems to adsorb both phosphate and NH_4 ions and may prove to be of great value.

C.S.I.R., New Delhi, has devised several techniques for the removal of following pollutants in water :

1. **Mercury** : Mercury can be removed from chloro-alkali effluent plants by applying mercury-selective ion exchange resin.

2. **NH_3** : It can be removed from waste water of chemical industries by ion-exchange technique as described above.

3. **Sodium Salts** : The sodium salts can be removed from industrial effluents by reverse osmosis method as described above.

4. **Phosphorus Compounds** : Phosphorus and its compounds can be removed from industrial effluents viz., reactor effluents by passage through beds of aluminium turning.

5. **Phenolic Compounds** : By the polymeric absorbents, phenolic compounds from the waste water of pulp and paper mills, tanneries, petroleum refineries and resin plants can be recovered.

6. **Decolourization of Water** : The brackish waste water from industries can be decolourized by electrolytic decomposition technique.

Specific Devices for Water Pollution Control

Important specific devices for control of water pollution are :

I. Cooling Methods

In some developed countries, thermal pollution abatement schemes are used to control water pollution. These methods include once-through cooling, cooling ponds, wet cooling towers, evaporative and dry cooling towers (Thorndike, 1976).

II. Use of Solar Power

Recently American scientists have claimed to use solar power for purifying the polluted waste water cheaply. Experiment concluded that a combination of sunlight and a catalyst such as titanium dioxide can dissociate deleterious chemical pollutants of water. Such photolytic reactions may kill pesticides like DDT, PCBs, malodorous microbes and destroy explosive solvents, dioxins and cyanides.

III. Use of Advanced Biological System

New biological methods are now being considered for polluted water treatment. The use of shallow (near about 1 meter deep) oxidation ponds allows water to be purified by the action of aerobic bacteria and algae.

These ponds use solar radiation for photosynthesis and the organic material is used for both bacterial and algal growth, greatly reducing BOD (and also coliform organisms, perhaps through the production of antibiotic substances by the algae).

The ponds may have disagreeable odours if anerobic conditions are permitted to exist but under proper conditions there is plenty of O_2 produced and the effluent may even be supersaturated with DO. The ponds eventually have to be cleaned out (perhaps every few years) and weeds must be kept under control (this is part of the reason for keeping the ponds shallow).

IV. Use of Water Hyacinth in Controlling Water Pollution

Recently in Mississippi, the US National Aeronautics and Space Administration (NASA) at the National Space Technology Laboratories used water hyacinth (*Eichhornia Crassipes*) and duckweeds (*Spirodela species, Lemna species and Wolfia species*) to upgrade waste water treatment lagoons and to treat chemical waste water.

In fact, water hyacinth (Hindi — *Jalkumbhi*) can be used as natural filters to absorb contaminated effluents from domestic and industrial sewage. Other pollutants like S, Ca, K, Fe and other minerals may be extracted from domestic sewage by harvesting the plant biomass. This harvested weed material is also a potential source of high quality protein, energy (biogas), fertilizer etc. Water hyacinth may also absorb various organic compounds such as phenols, benzene, toxaphene etc. So much so, it can remove radioactive toxicants from effluents which are actually very difficult otherwise. Further, it effectively maintains the pH of water between 6.8 - 7.8. Moreover, it inhibits the algal bloom to develop and the BOD and COD are lowered to the level at which the effluent can be satisfactorily mixed without causing any nemesis in aquatic bodies.

Process Involved in the Treatment of Sewage

The treatment of sewage consists of the following three stages :

1. **Primary or Mechanical Treatment**

In nutshell it consists of mechanical screening and sedimentation of solids which are burned or buried. In all developed countries about 30% of domestic sewage receives primary treatment. The principal function of this primary treatment is to remove suspended solids through the following steps :

(a) *Screening* : A series of screens remove large objects from the inflowing raw sewage.

(b) *Grit Chamber* : Any fairly large object that may have passed through the screens are ground up to facilitate their processing.

(c) *Settling or Sedimentation Tank* : From grit chamber the sewage is piped into settling tank where it is detained for several hours so that sand and suspended organic materials can settle to the bottom. The fluid that is left is removed from the tank, chlorinated to destroy pathogens and discharged into a stream or lake, etc.

(d) *Sludge digester* : The sediment solids are then piped to a sludge digester, where the organic matter is decomposed by microbes in the absence of oxygen. Primary treatment removes about 60% of settlable solids as sludge and about 35% of the oxygen-demanding waste. In other words, it results in 35% reduction of BOD value.

2. **Secondary or Biological Treatment**

In all developed countries about 60 to 70% of domestic sewage receives the secondary treatment. It is a more advanced treatment — primarily biological in nature — that involves intensive use of bacterial activity. Secondary treatment is often inadequate during cold weather because microbes are not very active at low temperatures. The main function of the secondary treatment is to remove much of the suspended organic material through the following steps that remain in the effluent after primary treatment :

(a) *Aeration* : The fluid from the first settling tank is piped to an aeration tank in which air is bubbled to provide maximum supply of oxygen (Fig. 5.3). This enables the aerobic bacteria to decompose the organic compounds. The liquid that accumulated at the top of second sedimentation tank is then chlorinated and discharged.

(b) *Activated Sludge Treatment*: The sludge that settles at the bottom of the second settling tank is called *activated sludge* because it bears a high density of bacteria that have become conditioned to the unique environment of the system. A part of the activated sludge is piped to the sludge digester. The remaining part is recycled back to the aeration tank and the second settling tank, where it, in turn, activates the inflowing not activated effluent. The

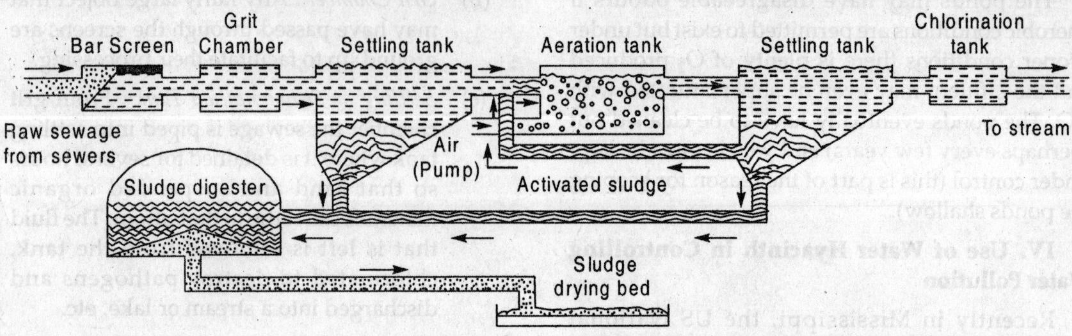

Fig. 5.3 : Secondary sewage treatment

activated sludge process was first developed by Ardorn and Lockett (1914) in UK. At the end of secondary treatment the strength of the effluent is reduced to 30 : 20 (i.e., suspended solids 30 mg/l and BOD 20 mg/l) which can be discharged into water sources. However, major part of the nitrogen and phosphorus compounds still remain in the effluent.

Trickling Filter : A second form of secondary treatment is the *trickling filter process* in which the sewage (or the effluent), after primary treatment, is sprayed by the arms of a rotating sprinkler into a filter bed of stones or clinker. The filter bed may be about 1.8 m thick and up to 60 m in diameter. The stones are coated with a slimy film containing bacteria, fungi, protozoa, insect larvae and small worms that have accumulated during the operation of filter. The sewage (effluent) containing a load of dissolved organic compounds gradually trickles downward or percolates through the stones. Some of the bacterial slime enters the trickling waste water and the bacteria decompose the dissolved organic matter removing nutrients, and energy for their growth and reproduction. Present solids are piped to a settling tank and later transferred to a sludge digester. Primary treatment combined with secondary treatment can reduce BOD to about 90%, COD to about 80%, total N_2 to about 50%, total phosphorus to about 30% and suspended matter to about 90%.

3. **Tertiary or Advanced Biological, Chemical and Physical Treatment**

Only 1 or 2% domestic sewage receives tertiary treatment, which is the most advanced phase of sewage treatment. The major function of this treatment is to reduce the load of phosphorus and nitrogen through the following steps :

(a) *Precipitation*: The waste (effluent obtained after secondary treatment) is mixed with lime (calcium oxide). The lime reacts with phosphorus-containing compounds in the effluent to form insoluble calcium phosphate ($Ca_3(PO_4)_2$), which then settles to the bottom of the settling tank and is removed.

(b) *Nitrogen Stripping* : Nitrogen is normally present in the waste water as ammonia gas (NH_3), nitrites(-NO_2), and nitrates (-NO_3). Nitrogen in the ammonia may eventually contribute to eutrophication and ammonia itself is highly undesirable in lakes and streams because it is harmful to fishes. The ammonia-containing waste water is directed into a metal tower. As this water slowly flows downward over a series of small plastic baffle plates, air is forced upward through the effluent and removes the ammonia gas from the water.

(c) *Carbon Absorption* : The main purpose of this process is a to remove dissolved organic compounds from the effluent. This is done by passing the water through a tower packed with small particles of carbon. The dissolved organic matter sticks to the carbon particles, a process sometimes referred to as *carbon polishing.*

(d) *Chlorination* : After the phosphorus, nitrogen and dissolved organic compounds are removed, the effluent is chlorinated again to destroy any pathogenic organism that might be present.

The resulting effluent at the end of tertiary treatment (Fig. 5.4) is of 10 : 10 strength (suspended solids 10 mg/l and BOD 10 mg/l). The water quality of our lakes and streams would be very good if all

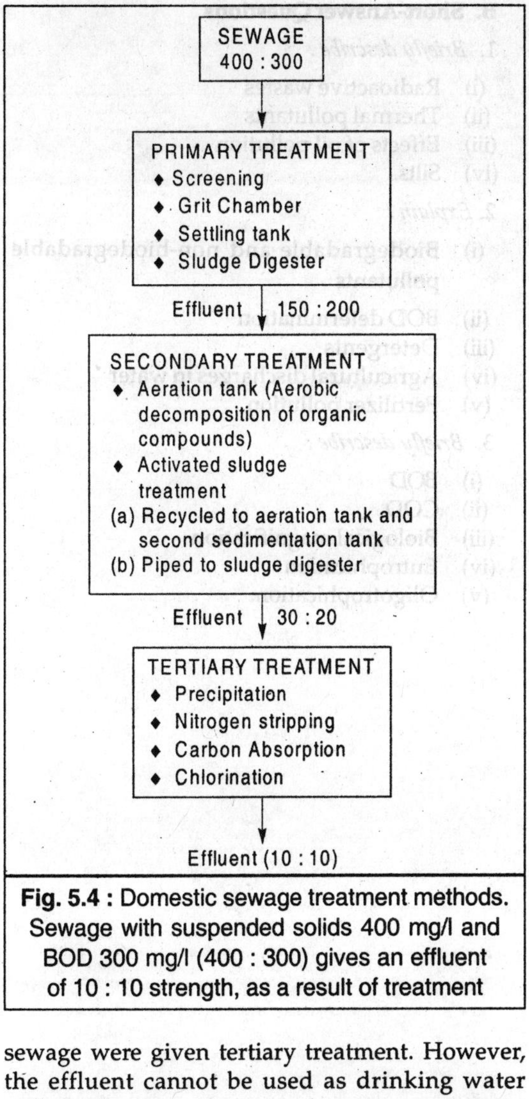

Fig. 5.4 : Domestic sewage treatment methods. Sewage with suspended solids 400 mg/l and BOD 300 mg/l (400 : 300) gives an effluent of 10 : 10 strength, as a result of treatment

sewage were given tertiary treatment. However, the effluent cannot be used as drinking water without the normal chemical treatment given to natural waters.

Approaches to Prevent and Control Water Pollution

The objective of all control legislation and management are the same, i.e., to achieve and maintain water quality at acceptable levels.

Some suggested approaches are :

1. **Licencing of Discharging Effluents for Prevention of Water Pollution**

In developed industrialized countries, the system of administrative agreement for the release of effluents into aquatic bodies, considering technological and economic conditions, have been established. The system includes consent conditions, abatement requirements, enforcement methods and severe tax penalties for violating the rules of agreed limits. Discharge license has to be issued to any polluter by any competent authority or organization like water pollution control authority, a state government or a federal agency to avoid water pollution.

2. **Economic Incentives for Control of Water Pollution**

Incentives can be applied to waste water discharges to maintain the desired quality in the receiving water. The incentive is principally based upon *"pollute pays principle"*.

3. **Environmental Impact Assessment**

This scheme may enable to prevent environmental damage prior to the construction of industrial plant. The format will have to be prepared by environmentalists and scientists.

4. **Set up of Water Pollution Control Boards**

Water Pollution Prevention and Control of Act, 1974, was adopted by 16 stages in India up to 1977 and they also constituted *Water Pollution Control Boards.* Central Water Pollution Control Board is set up in New Delhi to coordinate the work of State Boards and Union Territories. The principal functions of a State Pollution Control Board are :

(i) To educate the water consuming industries for effective water pollution control.

(ii) To conduct seminars and workshops for water pollution control.

(iii) To evolve effective and efficient methods of disposal of sewage and industrial effluents.

(iv) To perform functions from time to time as suggested by the Central Pollution Control Board.

(v) To carry out sewages of aquatic bodies for classification for the beneficial uses of water.

(vi) To encourage, conduct and participate in investigations and research programmes related to water pollution and its prevention.

Prevention and Control of Water Pollution Act was implemented in 1974 and amended in 1988. Now the Central and State Pollution Control Boards have set up "National Implementation Programmes" which could be planned in each Five Year Plan for different industries.

Revision Questions

A. Long-answer Questions

1. Define water pollution. Name main sources of water pollution. Add a note on the harmful effects of domestic sewage.
2. Give a detailed account of industrial effluents and the harmful effects produced by important industrial pollutants.
3. Make a list of important heavy metals and their effect on human health.
4. Describe the various categories of water pollution. Add a note of biodegradable and non-biodegradable pollutants.
5. Give an account of water pollution control and approaches to prevent water pollution.
6. Describe in detail the process involved in sewage treatment.

B. Short-Answer Questions

1. *Briefly describe :*
   (i) Radioactive wastes
   (ii) Thermal pollutants
   (iii) Effects of oil pollution
   (iv) Silts.
2. *Explain :*
   (i) Biodegradable and non-biodegradable pollutants
   (ii) BOD determination
   (iii) Detergents
   (iv) Agricultural discharges in water
   (v) Fertilizer pollution.
3. *Briefly describe :*
   (i) BOD
   (ii) COD
   (iii) Biological magnification
   (iv) Eutrophication
   (v) Oligotrophication.

6

Soil (Land) Pollution

SYNOPSIS

Introduction, Definition of soil (land) pollution, Sources of soil pollution, Detrimental effects of soil pollutants, Diseases caused by soil pollution, Remedial measures for soil pollution, Procedures for reducing/control of soil pollution, Soil indicator plants.

Introduction

Soil is one of the most significant ecological factors which is derived from the transformation of surface rocks. The word *soil* is derived from a Latin word *solum,* which means earthy material in which plants grow. The study of soil is commonly referred to as *soil science* or *pedology* (*pedos* = earth) or Edaphalogy (*edaphos* = soil + *logos* = study). Soil can be defined as weathered layer of the earth's crust with living organisms and their products of decay intermingled.

Soil is nothing but a substratum on which plants depend for their nutrients, water and mineral supply and anchorage. It constitutes an important medium wherein numerous animals live. In fact, the soil of a nation is its most valuable material heritage.

Definition of Soil Pollution

The undesirable changes in the land that harmfully affect the life activities is called soil or land pollution.

Sources of Soil Pollution

Soil pollution mainly results from the following sources :

1. Industrial wastes and sludges
2. Urban wastes
3. Radioactive pollutants
4. Agricultural practices
5. Chemical and metallic pollutants
6. Biological agents.

1. Soil Pollution by Industrial Wastes

Industrial waste disposal is the major problem responsible for *soil pollution.* These industrial pollutants are chiefly released from the following sources :

- Chemical industries
- Oil refineries
- Pulp and paper mills
- Tanneries
- Chemical industries
- Coal and mineral mining industries
- Drug factories
- Steel industries
- Sugar factories
- Textile industries
- Metal processing industries
- Glass factories
- Cement factories
- Engineering industries
- Petroleum industries
- Distilleries
- Pesticide industries

Various newer types of industrial wastes with the advent of technology are produced and get finally deposited in the land. These waste products are also deposited on soil, enhancing the extent of soil pollution.

Electrical, thermal and atomic power plants are also the villains to add pollutants to the soil. Actually the furnaces of these industries generate "fly ash" which severely pollutes the soil as well as air and water. Fly ash is actually unburnt brownish black substance. Various industrial effluents are either discharged into adjoining streams or dumped into the surrounding land.

Industrial wastes chiefly comprise of organic compounds along with inorganic complexes and non-biodegradable materials. These pollutants adversely affect and alter the chemical and biological properties of soil. Consequently, hazardous chemicals may enter the human food chain from the soil or water, interfere with the biological processes, and finally lead to serious deleterious effects.

2. Soil Pollution by Urban Wastes

Urban wastes comprise both commercial and domestic wastes consisting of dried sludge of sewage. All the urban solid wastes and refuse, particularly in urban areas, contribute to soil pollution. This refuse contains garbage and rubbish materials like glasses, metallic cans, fibres, plastics, paper, street sweepings, abandoned vehicles, broken iron pieces, fuel residues, containers and other rejected manufactured products.

Urban domestic wastes, though disposed off separately from the industrial wastes, can still be dangerous. This is so because these cannot be easily degraded.

Over-population, increasing consumption and, consequently, increase in pollution concentration in urban areas, have posed severe soil pollution. Also unplanned industrial progress in and around the urban areas have contributed to soil pollution problem in India to a greater extent.

3. Soil Pollution by Radioactive Pollutants

Radioactive wastes consist of several radionuclides such as Sr-90, I-129, Cs-137 and isotopes of iron which are more deleterious. Strontium-90 gets deposited in bones and tissues instead of calcium. Nuclear reactors produce waste containing Ru-106, 1-131, Ba-140, La-140, Cs-144 which Pm (promethium)-144 along with the primary nuclides Sr-90 and Cs-137. These are also produced from nuclear fission. Rainwater carries Sr-90 and Cs-137 to be deposited on the soil where these are held firmly with the soil particles by electrostatic forces. Heavy rains and also soil erosion carry away the deposited Cs-137 and Sr-90 (having half life 30 and 28 years, respectively) with the soil and clay. All these radionuclides deposited on the soil emit harmful radiations.

Recently it has been observed that some plants viz., lichen and mushroom, can accumulate Cs-137 and some other radionuclides, which ultimately may get concentrated in grazing animals.

4. Soil Pollution by Agricultural Practices

Actually, modern agricultural practices have polluted the soil to a large extent. With the advancing agro-technology today, huge quantities of pesticides, fertilizers, herbicides, weedicides and soil conditioning agents arise mainly from their excessive application rates. Although the fertilizers are used to fortify the soil, yet they do contaminate the soil with their impurities. When the fertilizers are contaminated with other synthetic organic pollutants, the water of the soil may also get polluted. Generally, fertilizers are retained by the soil and crop efficiently but there are possibilities to certain extent for the nitrates to be washed out. These nitrates may cause several undesirable effects on the water quality of the low land aquatic bodies creating numerous health hazards. Reports reveal that if PO_4 and NO_3 – concentration exceeds one and thirty parts per hundred million parts of water, respectively, it causes eutrophication.

It shall not be out of place to mention that India alone utilizes 16 kg/hectare of fertilizers (90 kg/hectare in 2003) whereas the world average is 55 kg/hectare. NCA has estimated the increased application of fertilizers — 9.7 million ton in 1976 to 6.0 million ton in 1984 and 9.7 million ton in 1995. Phosphate fertilizer consumption rose from 2.5 mt. in 1985 to 3.4 mt. in 1990. However, it is not only the increasing utilization of fertilizers but also escalated production, which create soil pollution hazards.

Other than fertilizers, use of pesticides, soil conditioners, fungicides and other chemical agents pollute the soil.

Pesticides

The need of increase food production due to increasing population density was emphasized since long, which consequently led to manipulation of land resources. Different kinds of pesticides used to control pests are causing a high stress on the natural environment.

Among pesticides the most important are the chlorinated hydrocarbons e.g., D.D.T., B.H.C. aldrin, endrin, dieldrin, lindane, chloracdane, heptachlor and endosulphan; organo-phosphates include — malathion, parathion, ethion, fenthion, trithion, dussban, dimethoate, phosdin and metasystox etc. The remnants of these pesticides may get absorbed by soil particles, which may contaminate roots of crop grown in soils.

Soil Conditioners, Fungicides and other Chemical Agents

In addition to the fertilizers, pesticides and biocides, soil conditioners and fungicides are also employed to the land system to increase and protect the soil fertility as well as to kill the hazardous insects. These chemical agents are reported to cause alterations in both agricultural and horticultural soil area. They contain several toxic metals like lead, arsenic, cadmium, mercury, cobalt etc., which, when applied to a land, will accumulate on the soil permanently — thereby introducing these chemical components into growing crops. However, researches are being carried out to synthesize pesticides of short-lived degradable residues sources so that the persistence of the pesticide residue and their degraded products on soil, food and forage crops may be reduced considerably.

5. Chemical and Metallic Pollutants

Various industries including metal, textile, paints, dyes, pesticides, soap and synthetic detergents, tanneries, drugs, cements, asbestos, rubber, petroleum, paper and pulp, sugar, steel, electroplating and glass industries pour their hazardous effluents in soil and water creating deleterious effects on living organisms.

Synthetic chemicals and fertilizers are important source of trace metals, which are added to the soil either deliberately or as an impurity, for example, As, Pb, Fe, Cd. Today various trace elements such as Ni, Co, Cu, Zn, Ba, V, Mn, As, Hg, Mo and Si are being added to the soil in one form or the other. Mn and Fe oxides have a tendency to concentrate trace metals by isomorphism replacement of ions. In many soils, 50 to 100% of soil carbon is found complexed with clay containing organic and inorganic components which affect the soil texture, its fertility and stabilization of organic matter. Presence of high levels of Na, Mg and K causes calcium deficiency in soil. Magnesium deficiency in soil has been attributed to high concentrations of Ca, Na, and K, which are added as artificial fertilizers. Plant leaves may absorb excess of sulphur in soil. Although Mg, Fe, Zn and V may be involved in photosynthesis but, if present at high levels, pose lethal effects on crop production. Metallic contaminants in soil are considered to be highly dangerous since they affect the production of atmospheric oxygen as well as living beings.

6. Biological Agents

Soil receives large amount of excreta from human, cattle, birds and other domestic and wild animals (biological agents) which constitute the major source of land pollution. In addition to these excreta, faulty sanitation, municipal garbage, waste water and wrong procedures of agricultural practices also induce heavy soil pollution. Sludge does have faults as they contain enough viruses and viable intestinal worms. In developing Western countries, intestinal parasites constitute the most serious soil pollution problems.

The pathogens that pollute soil may be grouped into three categories :

1. **Pathogens occurring naturally in contaminated soil**

Examples of this group are bacteria, fungi, algae, protozoan, rotifers, annelids, molluscs, arthropods etc.

2. **Pathogens excreted by human beings**

Examples of this group are enteric bacteria, amebae and parasitic worms.

3. **Pathogens excreted by animals**

Examples of this group are pathogenic bacteria and parasitic worms.

Detrimental Effects of Soil Pollutants

I. Effects of Industrial Pollutants

1. Industrial wastes consist of a variety of chemicals which are extremely toxic to living beings. Industries manufacturing paper, fertilizers, textiles, pesticides, steel etc. release metallic wastes, grease, oils,

plastics, fertilizers, solvents, heavy metals, suspended solids inorganic and organic pollutants and non-biodegradable materials in the soil. Consequently, these toxicants are transferred to different organisms through food chain causing a number of undesirable effects.

2. Industrial effluents, when discharged through sewage system, may poison the biological purification mechanism of sewage treatment causing several soil and water borne diseases. Most of these pathogens are not subjected to degradation and are injurious to health.

3. According to an estimate about 5 lac people die annually because of these toxic effluents. Most of the foodgrains in USA and other countries have become uneatable due to synthetic chemicals.

4. Metallic contaminants (Zn, Hg, Pb, Cd, Na, K, As, Cr, Cu, etc.) destroy bacteria and beneficial microorganisms present in the soil. Heavy metals tend to precipitate phosphate compounds and catalyze their decomposition. These metals are considered to be destructive poisons and their accumulation in soil for a long period may be highly fatal to living biota.

5. Soluble salts released by the industries are responsible for crop damage. They cause crop loss, soil texture loss etc. Salt accumulation in the soil has been a perpetual problem. The Food and Agriculture Organization (FAO) states that soluble salts are damaging 50% of the irrigated farms in the world.

6. Products of industries, such as synthetic fibers, plastics and waste paper are consigned to incineration — their emissions may contaminate air with their toxic vapours and particulates, causing air pollution. When discarded plastic packaging, materials, textiles and toys of polyvinyl chloride are burnt in soil, they emit highly toxic gases like hydrochloric acid fumes, SO_2, NO_x etc. which directly affect human health adversely.

7. Some of the trade wastes contain pathogenic bacteria. For example, pathogen Anthrax bacilli is present in tannery wastes. Chemicals either emitted into air or applied by aerosol spray ultimately reach the soil where they have significant effect on plants causing disruption.

II. Effects of Urban Waste Products

1. Dumped wastes along highways and other places may spread several chronic diseases posing a serious threat to human health.
2. Sewage appears to be an excellent medium for optimum growth of pathogenic microbes, which affect human health. For example *Vibrio cholrae* found in sewage cause cholera, *Shigella dysenteriae* cause bacillary dysentery and *Salmonella typhus* spread typhoid in man.
3. Suspended matter in sewage can blanket the soil, thereby interfering with the soil moisture.

III. Effects of Radioactive Pollutants

(Refer Chapter 7 — Radiative Pollution)

IV. Effects of Fertilizers

Modern agricultural practices i.e., the use of fertilizers, pesticides etc. are employed to increase the soil fertility and crop production. In fact, the fertilizer-enriched soil cannot support microbial flora, hence there remains poor humus in the soil.

Salient adverse effects of the use of fertilizers are :

1. Fertilized soil produces larger sized vegetables and fruits, which are more prone to pests.
2. Excessive use of nitrogenous fertilizers in land leads to accumulation of nitrate in the soil, which are transferred to human through plants.

   Nitrates, being highly soluble, proceed into drinking ground water and become deleterious when this concentration exceeds 90 ppm, causing diarrhea and cyanosis (blue jaundice) in children. In human body these nitrates or nitrites may be converted to nitrosoamines and other nitroso compounds, which are suspected to cause stomach cancer.

   Recent reports reveal that nitrogenous fertilizer has raised NO_3 levels in soils to dangerous amounts. The danger appears to be acutely severe, particularly to infants (less than 4 months old) who are more susceptible to fatal disease called *methemoglobinemia* or *blue baby syndrome* where nitrates interfere with O_2-carrying capacity of blood. It also damages respiratory and vascular system causing ultimately death in infants.

3. Potassium fertilizer in soil decreases the ascorbic acid (Vitamin C) and carotene in vegetables and fruits.

   Researches reveal that there is a 25-30% decline in protein content when corn, maize, gram and wheat crops were grown on soils fertilized with NPK fertilizers. Moreover, the subtle balance of amino acids with the protein molecule is also disrupted, degrading the carbohydrate and protein quality leading to malnutrition.

4. Excess use of fertilizer intensively reduces the ability of plants to fix nitrogen.
5. Fertilizers get contaminated with other synthetic organic chemicals, thereby polluting the soil water. These fertilizers enter the aquatic system contributing to eutrophication.

V. Effects of Pesticides

Some adverse effects of pesticides present in soil are :

1. Pesticides may adhere to soil particles causing harm to vegetation.
2. Pesticides like DDT, BHC and other organochlorines retained in soil concentrate in vegetables, crops, cereals and fruits. Consequently, these recycle in the food chain. DDT, if accumulated in human, may cause impotency besides other disorders.
3. Persons who used vegetables contaminated with 0.5 gm or more PCBs developed darkened skin, eye damage and severe acne.
4. Pesticides like DDT, endrin, dieldrin, heptachlor etc. are known to slip gradually through soil into ground water and eventually contaminate public drinking water supplies.
5. From the pesticides-contaminated soil, when irrigated water evaporates, it may also spread pesticide molecules into the atmosphere, indirectly causing ill-effects on human health.
6. Long-lasting effects of pesticides are visible in animals and man where they affect the tissues and interfere with normal metabolic activities by disturbing the enzymatic functioning in the body.
7. Herbicides and chlorinated insecticides are very potent pollutants of the soil and affect soil texture and function of the ecosystem. Many of these have long-lasting effects and if used indiscriminately may be suicidal for people. Even certain herbicides — such as dioxan — has been found to cause congenital birth defects in offspring of experimental female rodents who have ingested or inhaled certain concentration of dioxan at critical periods.
8. Organophosphate insecticides cause extreme muscular weakness, tremor and dizziness in poisoned animals.
9. Many hunting birds feeding on grains, particularly contaminated with high levels of DDT, are threatened with extinction. It has also been implicated as the cause of thin and fragile egg shells because of inhibition of enzymes and interference with hormones which regulate calcium metabolism. Vultures' declining population may be due to this reason.
10. Excess intake and consequently absorption of pesticides may lead to greater accumulation of acetylcholine in the body. Chronic absorption damages liver and kidney causing malfunctioning, excess of amino acid in blood and urine, blood abnormalities and cancer.

Diseases Caused by Soil Pollution

1. Pathogenic soil bacteria are *Mycobacterium, Salmonella typhosa, Lepospira, Pasteurella* and *B. tuberculosus* pose a serious threat to man's health. They cause infections of intestinal tract like amebic dysentery, cholera, typhoid, T.B., polio and hepatitis. Many of these are endemic to particular population.

2. Pathogenic soil bacteria are chronic disease-carrier, which are transmitted from man to soil, or vice versa, causing cholera, typhoid, bacillary dysentery, paratyphoid fever etc. Flies which breed or get in contact with the contaminated soil become carriers of disease-causing organisms.
3. Viruses are infectious agents of plants and animal cells. Common viruses present in sewage added to soil are *adenoviruses, enteroviruses, polioviruses* and *hepatitis viruses*. Derby et al have reported that about 80 viruses are most dangerous for man. Most dreadful viral disease is poliomyelitis in which body parts get paralysed due to destruction of nerve cells controlling the muscles.
4. Q. Fever is caused by *Rickettsia coxiella burneti* which is present in dust and soil. Anthrax is caused by *Bacillus anthracis*, whose spores are extremely resistant to attack. They survive in soil and animal products for years together.
5. Algae poison is considered to be the most virulent poison, which produces cirrhosis of liver.

Remedial Measures for Soil Pollution

The salient measures to meet the challenge of soil pollution are :

1. Production of Natural Fertilizers

Organic wastes contained in cattle's dung may be used for preparing compost manure and biogas rather than throwing them wastefully contaminating the soil.

2. Proper Dumping of Unwanted Materials

Excess of waste products by man and animals cause chronic disposal problems. Crude Tripping or Open Dumping is the most widely used method in Indian cities. Recently controlled tipping is used in metropolitan cities like Delhi and Mumbai etc. for solid waste disposal. The surface sources obtained then can be used for housing or sports fields.

3. Use of Biological Weapons

In opinion of Committee of U.S. National Academy of Science the natural soil micro-organisms should be used to increase crop fertility rather than the use of pesticides or insecticides. The committee have also recommended the formation of new generation of biological agents that can destroy the target insects.

4. Recycling and Reuse of Wastes

To minimize soil pollution, the wastes such as paper, plastics, glass, metal, organics, petroleum products, industrial effluents etc. should be recycled and reused.

5. Ban on Toxic Chemicals

Ban should be imposed on chemicals which are otherwise fatal to plants and animals (such as a few pesticides e.g., DDT, BHC etc.). Nuclear explosions and disposal of radioactive wastes should be banned.

6. Proper Hygienic Conditions

People should be trained regarding the sanitary habits. Lavatories should be equipped with quick and effective disposal machinery.

7. Plantation of Trees

Extensive plantation is extremely essential to prevent soil erosion.

Recently Industrial Toxicological Research Centre (ITRC), Lucknow, has established central facility for safety evaluation of pesticides to prevent their adverse effects on soil and crops.

The aforementioned remedial measures in terms of effective and integrated ecological management of soil systems are necessary to restore normal conditions.

Procedures For Reducing/Control Soil Pollution

The problem of soil pollution can be reduced to some extent by adopting the following procedures :

1. **Dumping wastes into Proper bins** : For recycling the wastes, civilians will have to dump paper. plastics and food trash into separate colour coded bins.
2. **Separate paper and plastic wastes** : Waste papers should not be mixed with plastic or glass which are difficult to recycle.

3. **Encouragement of people** : People must be encouraged for subsiding the wastes.

4. **Recycle paper use** : Use of recycled paper should be highlighted instead of use of fresh ones.

5. **Replantation of Forest** : Replantation of forests to reduce sulphur contamination of soil.

6. **Utilization of night soil** : In India, water carriage system is available only for 8-10% of the population. Majority of the urban and rural community is having dry conservancy system where the night soil is collected from pail type of lavatories, which is disgusting to people. Dumping or trenching causing soil pollution and spreading chronic disease disposes off night soil thus collected.

   National Environmental Engineering Research Institute (NEERI), Nagpur, took a realistic approach to the problem and conducted research on total utilization of night soil by the procedure given below :

Night Soil Digester

It has a masonary structure partly above and partly below the ground with a floating dome at the top to collect the gas generated in the digester. There is provision in the digester to feed night soil slurry and withdraw the digested sludge and supernatant. About 0.025 m^3/capita/day of gas with 70% methane content is produced in the digester. The gas contains a calorific value of 5,000 k.cals/m^3.

Uses

The gas can be utilized for cooking, lighting or a motive power. The dried sludge cake (moisture content 20%) is rich in nitrogen, phosphorus and potassium. The sludge acts as a good organic manure and efficient soil conditioner.

7. **Bonfire treatment** : It is a very simple technique. The red hairy caterpillar emerges as moth from the soil just after the first spell of rains. The white moth appears on a mass scale during nights and is attracted towards light and bonfires. To monitor their emergence, rain gauges and light traps are arranged in the fields. As the pests emerge, the villagers are alerted and they burn rubber scrap, thereby eradicating the pests. As rubber scrap is a pollutant, modified light traps have reflected bonfire treatment.

   This non-pesticidal group action technique also protects the soil from hazardous effects of pesticides.

8. **Reducing the need of insecticides** : Robert Plaisted and his colleagues, in 1989 at Cornell University in USA, have developed a potato species whose leaves trap any insects which happens to alight on them. If more other such species are developed they may repel the insects and thus dramatically reduce the need of synthetic chemical insecticides.

9. **Protecting crops from pests** : The scientists of China, to avoid the application of insecticides and pesticides, are now applying a pinheaded wasp to wipe out insect pests, which harm crops, as reported in May 1989.

10. **Fungi to increase soil fertility** : Fungi which enable tropical trees and plants to grow in the poorest soils have been successfully isolated by a team of Indonesian and British scientists (June 1991).

11. **Converting basic slag into nitro-phosphatic fertilizer** : Basic slag, which is the major by-product of steel industries, poses a serious problem of disposal all over the world. G. N. Pant and Pathak of the Regional Research Institute of Technology at Jamsedpur have indicated that basic slag can be used for preparing nitrophosphatic fertilizers.

12. **Bacterial fertilizer replacing chemical fertilizer** : A great deal of research on bacterial fertilizers is at present going on at national and international levels to minimize farmers' dependence on chemical fertilizers. The steep rise in the cost of energy necessary for the production of phosphatic fertilizers led scientists to intensify their research on the fixation of atmospheric nitrogen with the help of symbiotic bacteria.

    The best-known nitrogen fixing bacteria belóngs to genus *Rhizobium.* These produce nodules on the roots of legumes such as peas, alfa-alfa, soyabean and clover. *Rhizobium* strains, which are specific for each group of legumes, invade root hairs and induce root nodules that become packed with modified forms of Rhizobium cells known as *bacteriods*.

In tropical countries, blue green algae belonging to class Myxophyceae such as *Nostoec*, *Anabaena* and *Colothrix* species contribute significantly for plant growth by converting atmospheric N_2 into fertilizers. When they die, the nitrogenous matter present in their cells becomes available for the growth of other microbes and plants. In rice field, addition of blue green algae enriches the soil by 50 kg N_2/hectare. Blue green algae, in addition with nitrogenous fertilizer, always yield better growth than higher doses of nitrogenous fertilizer alone.

Soil Indicator Plants

Some plants serve as soil indicator as shown in Table 6.1 :

Table 6.1 : Soil Indicator Plants

Sl. No.	Plant Species	Soil Characteristics
1.	*Andropogon scoparium*	Sandy-loam soil
2.	*Argemone mexicana*	Flooded or disturbed soil
3.	*Butea monosperma*	Heavy alkaline soil
4.	*Capparis decidua*	Alkaline soil
5.	*Lippia nodiflora*	Nitrate-abundant soil
6.	*Prospis cineraria*	Good soil for crop plant
7.	*Salvadora oliloides*	More calcium, boron and good soil crop plants
8.	*Stellaria setucea*	Mercury minerals
9.	*Viscaric alpina*	Copper-rich soils
10.	*Viola lutea*	Zinc-rich soil
11.	*Damara ovata*	Iron-rich soil

Revision Questions

A. Long-answer questions

1. Write an essay on soil pollution.
2. Describe important sources of soil pollution and disease caused by them.
3. Describe detrimental effects of soil pollution and give procedures for reducing soil pollution.
4. Define soil pollution and describe remedial measures for it.

B. Short-answer questions

1. Briefly describe

(i) Sources of soil pollution

(ii) Detrimental effects of two soil pollutants

(iii) Soil indicator plants.

2. Write short notes on

(i) Control of soil pollution

(ii) Bacterial fertilizer replacing chemical fertilizer

(iii) Soil pollution by radioactive pollutants.

3. Discuss in brief

(i) Soil pollution by industrial wastes

(ii) Soil pollution by urban wastes

(iii) Soil pollution by agricultural practices

(iv) Soil pollution by metallic pollutants

(v) Bonfire treatment.

7

Radiative Pollution

SYNOPSIS

Introduction, Types of radiation, Sources of radioactive pollution, Radioisotopes, Half life of radio nuclides, Measurement of radiation, Effects of ionizing radiations, Mechanism of ionizing radiation reaction, Types of high energy radiations, Effects of non-ionizing radiation, Biological magnification of radioactive elements, Maximum permissible dose, Preventive measures from radiation, Important radiation episodes, Prime nuclear power stations in India.

Introduction

Radiation is also a significant source of environmental pollution. The main radiation hazard in the environment comes from ultraviolet, visible, cosmic rays and microwave radiation. The biggest hazard comes from X-rays, which account for 95% of air radiation exposure other than cosmic rays. It shall not be out of place to mention that man has been exposed to low levels of ionizing radiations from natural resources throughout his evolution. But it is only after the advent of nuclear weapons that the exposure levels have increased enormously. The radioisotopes persist much longer in the environment.

To know the consequences of pollution caused by radiation on living systems and his environment one must know what the radiation is.

Radiation, in broad sense, is a process in which energy is transferred from one place to another without being influenced by the intervening medium. Precisely speaking, radiation is energy, which travels in waves or particles. Or, radiation is the release of energy from an atom into the environment. Or, radiation is energy being propagated from one place to another through space.

Types of Radiations

There are mainly two types of radiation :

I. Ionizing Radiations

II. Non-Ionizing Radiations.

I. Ionizing Radiations

It is also called high energy radiation in waves (example — X-rays, gamma rays) or as particles (example — alpha and beta). It carries energy level so high that it causes injury to the living tissue, hence called high energy radiation. These are capable of removing electrons from atoms and attach them to other atoms, hence called ionizing radiation. Thus, ionizing radiation produces ions as it passes through the protoplasm. It is the principal cause of injury to the living matter and the damage is proportional to the number of ion pairs produced in the absorbing materials. Isotopes of element that emit ionizing radiation are known as *radioisotopes* or *radionuclides.*

The ionizing radiations are further categorized into two types :

(i) Electromagnetic radiation

(ii) Corpuscular or Particulate radiation

(i) *Electromagnetic radiations* : These are of shorter wave-length and with broad spectrum of energy. These travel long distances and readily penetrate the matter. Example — X-rays and gamma rays.

(ii) *Corpuscular* or *Particulate radiations* : These consist of the particles ejected from atoms at high speed and often with tremendous energy. These have electron, proton or neutron. Examples — alpha particles, beta particles, neutrons.

II. Non-Ionizing Radiations

The radiations viz., heat, light, radio waves that carry enough energy to excite atoms but not enough to produce ions are called non-ionizing radiations. The UV rays are the important example of non-ionizing radiations.

Sources of Radioactive Pollution

Living biota may be exposed to a variety of radiation sources which are grouped into two :

1. Natural Sources of radiation
2. Anthropogenic or Man-made radiations.

1. Natural Sources of Radiation

These sources include :

(i) Solar rays (ii) Environmental radiation (iii) Radio nuclides in earth's crust (iv) Internal radiation.

(i) **Solar rays** : Solar rays coming from the sun keep up bombardment of a steady drizzle of cosmic rays, gamma rays and heavy particles. Solar radiation consisting of cosmic rays are highly energetic particles (10^9 Mev) which reach the earth's surface from different galaxies.

(ii) **Environmental radiation** : Radioisotopes of naturally occurring radio elements emit enormous amount of radiations in the form of α, β particles and γ rays. Radioactive elements occurring mainly in the lithosphere comprise thorium, uranium, radium and isotopes of K and C.

(iii) **Radionuclides in earth's crust** : Radioactive minerals such as U-238, Th-232 and K-40 are widely distributed in earth's crust. These give rise to the phenomenon referred to as 'terrestrial radioactivity'.

(iv) **Internal radiation** : Radiations may originate within our bodies, particularly during the decay of potassium in our muscles. Besides potassium, radioactive elements such as U, Th, Sr, and C do exist but in very minute quantities in human body. Radioactive substances that emit α or β particles are referred to as "internal emitters" because these rays severely affect the living tissues upon ingestion and absorption. The radioactive substances that emit gamma rays are referred to as "external emitters" because these are highly penetrating and can produce their effects without being absorbed inside the body. In man, internal radiation values varies from 25-75 m rads per year. Some metabolically important radionuclides which emit radiation in our body include Co, Ca, I, P, C, Mn and Fe. These radioelements are constituents of human body in traces, but are not ecologically harmful.

2. Anthropogenic or Man-made Radiation

Anthropogenic sources include :

(i) X-rays
(ii) Radioisotopes
(iii) Nuclear tests
(iv) Radioactive fallout
(v) Nuclear reactors
(vi) Nuclear power plants
(vii) Radioactive ore processing
(viii) Use of radioactive materials
(ix) Electric field.

(i) X-rays

X-rays constitute approximately 15-18% of artificial radiation used in radiotherapy for diagnostic purposes. These are highly penetrating, like the gamma rays. Repeated X-ray exposure may be cumulative in the body and may also create chronic defects in the internal organs.

(ii) Radioisotopes

Isotopes of elements that emit ionizing radiation are called radioisotopes or radionuclides. Radioisotopes administered to patients during radiation therapy are now proving to be hazardous source of nuclear pollution.

(iii) **Nuclear Tests**

Researches reveal that radionuclides formed in explosion test include fission fragments such as Sr-90, I-131, Cs-137 and Ba-141 along with unused explosives and activation products.

(iv) **Radioactive fall-out**

The radioactive dust that falls to the earth after atomic explosions is called radioactive fall-out. It becomes suspended at a height of about 10-15 km above the earth's surface and may come down to soil and water in the form of radioactive rain. Fig. 7.1 shows transmission of radioactive fall-out in food chain :

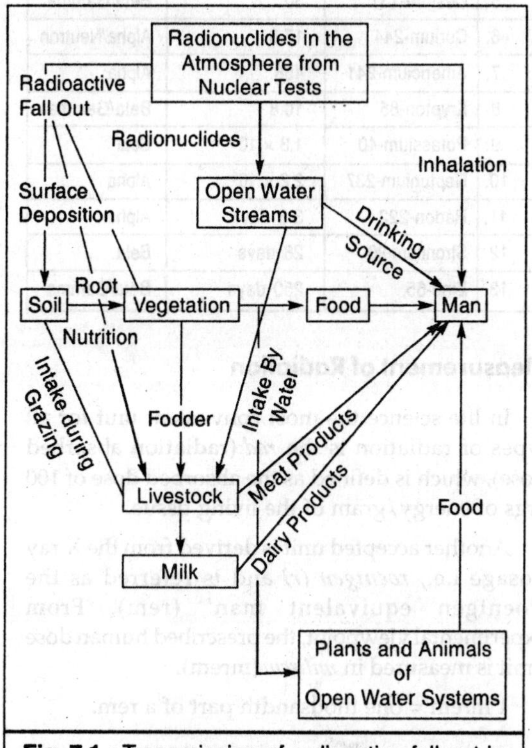

Fig. 7.1 : Transmission of radioactive fall-out in food chain of man

Radioactive fallout causing radiation pollution is mainly of two types :

(a) Early fall-out, (b) Delayed fall-out.

(a) ***Early fall-out***

If the nuclear explosion is at a very low altitude, it sucks up large quantities of soil and water affecting severely all the living biota.

(b) ***Delayed fall-out***

If the nuclear explosion occurs at a high altitude, it sucks little dirt and water. It is also known as world-wide fall-out.

(v) **Nuclear Reactors**

Release of nuclear radiations from nuclear reactors are the :

1. Fission products
2. Gaseous wastes of several nuclides comprising H-3, C-14, Ar-41, Fe-54, Kr-85, I-129, Xe-133 etc.
3. Liquid wastes containing H-3, Co-58 and other corrosion products are also produced.

However, pollution from such sources is bound to increase in future.

(vi) **Nuclear Power Plants**

The following are the radioactive wastes generated by power plants :

1. **Low-level radioactive liquid wastes** : Radioactive wastes in solution coming from power plants contaminate with aquatic life. These radioactive elements eventually find their way into human through food chain.
2. **Gaseous and particulates radio wastes** : Stack effluents from atomic power plants contain gaseous and particulates radio-isotopes such as H-3, C-14, Kr-85, I-129. Some of these radionuclides have critically long half-lives and may be distributed in the environment for several years. When man inhales these radioisotopes, they get concentrated in specific target organs posing deleterious health effects.
3. **Fission fragments** : The largest volume of radioactive wastes comes from reprocessing of irradiated fuel, viz., Co-58, Sr-90, I-131, and Cs-137, Sr-90 released from atomic power have been reported to get concentrated in the aquatic food web.
4. **Heat release** : Researches reveal that one ton of uranium may produce as much energy as 12 million barrels of oil. This nuclear fuel produces enormous heat. One gram of fissionable radioactive material releases 23,000 kw of heat causing thermal pollution in air and water bodies.

(vii) **Radioactive Ore-Processing**

Radioactive ores of uranium like pitchblende and uranites as well as thorium are applied in nuclear processes, $_{92}$Uranium^{235} undergoes natural fission and emits radiation such as α, β and γ. The important processes in the nuclear technology include mining, washing, refining, separation and milling. These cause nuclear pollution in the atmosphere. All these treatments during ore processing result in the release of radioactive gases which subsequently adsorb on the particles present in the atmosphere. Uranium and thorium ores form dusts in air posing toxic effects on living organisms.

(viii) **Use of Radioactive Materials**

Many nuclear industries wastes contain hazardous radionuclides. Some industries leave their waste products in open and allow their effluents to flow into soil or water bodies. These effluents also contain small amount of radionuclides. The population living nearby is thus exposed continuously to radiation.

(ix) **Electric Fields**

Electrical gadgets and power transmission lines generate electric fields causing environmental radiation hazard. Human population is continuously exposed to such low frequency electric fields, which are harmful enough. Studies reveal that bees turn violet near high-tension wires. Artificial magnetic and radio waves can misguide migratory birds when sunlight is blocked out completely.

What are Radioisotopes or Radionuclides?

When an atomic nucleus becomes unstable, it may undergo spontaneous disintegration or decay. The isotopes of elements that emit ionizing radiations are called *radioisotopes* or *radionuclides*. The atom that undergoes disintegration is said to be radioactive. Approximately 18 types of naturally occurring elements have been known to exhibit radioactivity. However, all elements may become radioactive when processed in a nuclear reactor.

Half-life of Radionuclides

The rate of disintegration of radionuclides is represented by a term 'half life' which may be defined as "the period of time within which half of the nuclei in a sample of radioactive substance undergoes disintegration or decay." Once a radioactive substance is released into the environment, it continues to emit radiation for a period of time commensurate with its half-life, until all the atomic nucleic have disintegrated. In Table 7.1 half-life period and type of radiation emitted by certain important radionuclides have been given :

Table 7.1 : Name of radionuclides, their half-life and type of radiation

Sl.No.	*Name of Nuclide*	*Half life Period (years)*	*Type of radiation*
1.	Uranium-238	4.5×10^9	Alpha
2.	Thorium-234	1.4×10^{10}	Alpha
3.	Radium-226	1622	Alpha
4.	Carbon-14	5,574	Beta
5.	Cesium-137	30	Beta/Gamma
6.	Curium-244	17.6	Alpha/Neutron
7.	Americium-241	458	Alpha
8.	Krypton-85	10.8	Beta/Gamma
9.	Potassium-40	1.3×10^9	Beta
10.	Neptunium-237	2.2×10^6	Alpha
11.	Radon-222	3.8 days	Alpha
12.	Strontium-90	28 days	Beta
13.	Zinc-65	250 days	Beta/Gamma

Measurement of Radiation

In life science the most convenient unit for all types of radiation is the *rad* (radiation absorbed dose), which is defined as the absorbed dose of 100 ergs of energy/gram of the living tissue.

Another accepted unit is derived from the X-ray dosage i.e., *roentgen (r)* and is referred as the 'roentgen equivalent man' (rem). From experimental viewpoint, the prescribed human dose limit is measured in *milirem* (mrem).

1 mrem = one thousandth part of a rem.

Effects of Ionizing Radiations

Radioactive contamination in the environment may be natural or artificial (man-made) and occurs when radioactive elements disintegrate, emitting α, β, and γ rays. Actually, the effects of nuclear radiation at all levels are not always deleterious to living beings. It has been observed that low doses of radiation are helpful in increasing the longevity and may reduce the incidence of cancer. Radiations at very low levels (close to that of the background) have some beneficial effects. This phenomenon, first observed in 1898 in the faster growth of irradiated phycobionts, was called as *radiation hormesis* (RH). Hormesis refers "*in small amounts*".

RH represents the stimulation by low doses of radiation that are deleterious at high levels. Evidence of RH in humans, plants and animals have actually been observed as shown below :

1. Increased longevity
2. Increased growth
3. Increased fertility
4. Reduced frequency in cancer growth
5. Increased weight and height in plants
6. Increased flowering in plants.

Despite these observations and results, the RH is still not fully accepted by majority of the scientists, perhaps due to the fact that it is tedious and rather difficult to reproduce the results or because the mechanism of RH is not very much clear. One of the possible mechanisms proposed is that low level radiation may alert an organism to adopt itself better in the evolutionary struggle. This has become a general view now that the radiation is not beneficial.

The harmful effect of ionizing radiation upon human beings is due to its capability to ionize and ultimately destroy the essential organic molecules of the body cells. The extent of damage depends upon two important factors : (i) the energy, and (ii) the type of radiation. As mentioned above, the radiation energy is expressed in *rads* (a rad corresponds to the absorption of 100 ergs or 10^{-8} joules of energy/gm of tissue). The overall biological effect of radiation, expressed in *rems,* is found by multiplying number of rads by an appropriate factor for the particular type of radiation.

Number of rems = n × number of rads

where n is 1 for β, γ and X radiation and 10 for α radiation or high energy neutrons. Some of the effects to be absorbed when a man is exposed to a single dose of radiation at various levels is represented in Table 7.2 :

Table 7.2 : Effect of Exposure on Human Beings to a Single Dose of Radiation

Dose (Rems)	*Probable Effect*
0-25	No observable effect
25-50	Small decrease in white blood cell count
50-100	Lesions, marked decrease in white blood cells
100-200	Nausea, vomiting, loss of hair
200-500	Hemorrhaging, ulcers, possible death
500+	Fatal, leading to immediate death

Small doses of radiation repeated over long periods of time can have extremely serious consequences. Cancers have been developed even after 40 years of initial exposure. Many of the early workers (e.g. Madame Curie) in the field of radioactivity developed cancer because of chronic overexposure to radiation.

Detrimental effects of radiation came into knowledge in 1895 when W. Roentgen put his hand between X-ray tube and a fluorescent screen. These rays penetrated deeply and the bones cast a deeper shadow than the flesh. H. Becquerel, a French physicist, discovered in 1903 naturally ray-emitting elements and shared the Nobel Prize with Madame Marie Curie.

M. Curie discovered radium which is 25,000 times more lethal than arsenic, the potency of radiation toxicity was realized by the death of M. Curie when she died of *Leukemia.* Radiation poses a wide range of syndromes and symptoms causing several adverse somatic and genetic effects.

Somatic or body effects occur within an individual's lifetime and appear as skin damage, eye cataracts, liver damage, spleen damage, thyroid damage and decreased fertility. Laboratory studies on certain mammals — rabbit, rat etc. revealed that radiation can even induce cancer in certain organs such as lymph glands, mammary glands, lungs, ovaries, skin, besides leukemia, i.e., blood cancer.

Sensitivity against radiation varies with the age (fetuses and newborn are much sensitive) and with the type of tissues or organs radiated. However, somatic or genetic damage may not be apparent in one generation, but can appear in offspring and the subsequent generations. Since radiation affects the reproductive organs, the gametes produced contain deleterious gene mutation which are inherited. Within the gametes the genetic mechanism of the chromosomes is damaged by ionization and breakage.

Extensive studies have been conducted on the Japanese who survived the atomic bomb explosion at Hiroshima and Nagasaki in 1945, and their offsprings have confirmed the effects of radiation damage. In a survey conducted in 1976 in India of 13,000 people who lived in an area with a high thorium-234 soil content, it was found that incidence of severe mental retardation in children was 3 in 1,000 or 0.3% of the survey sample as compared to 1 in 1,000 i.e., 0.1% in the neighbouring area of normal background radiation. Chromosome

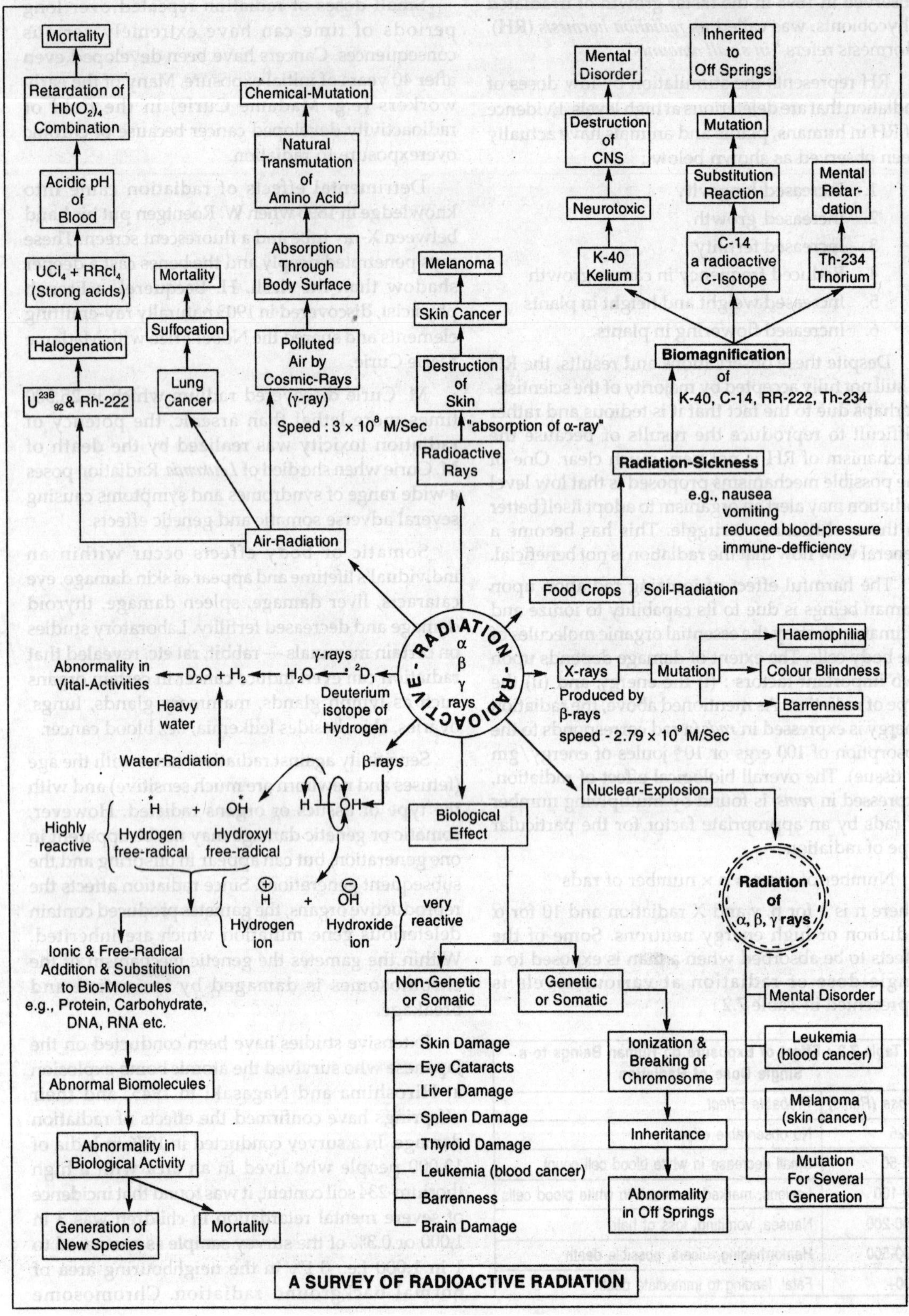

A SURVEY OF RADIOACTIVE RADIATION

damage was observed and more cases of mongolism were reported. It has been estimated that X-ray probably cause between 2,500-100,000 genetically defective birth and 200-300 leukemia (blood cancer) deaths annually in India.

Embryos of various animals as well as of man are more sensitive to the deleterious impact of radiation because they contain rapidly growing/ dividing cells in all parts of their body. The degree of sensitivity varies with the embryonic stage, the younger embryos being the most sensitive because their cells grow and divide most rapidly as shown in the following histogram :

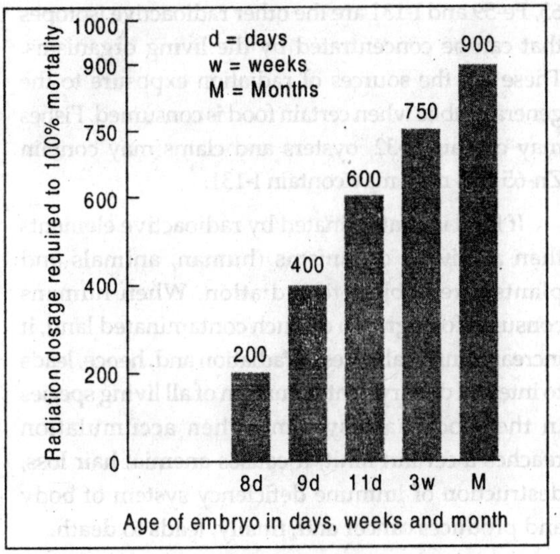

When we consider the effects of radiation at the ecosystem level, it has been found that radiation stress leads to reduction in species diversity.

Radiation stress can also alter key population interaction such as pre-predator equilibrium (Awerbench, 1958) or bring on a pest erruption. Radioactive substances of the soils, ocean and other environments where man does not actually live, nevertheless, affect man's vital life-support system. Any radioactive substance with a long half-life introduce into the environment anywhere in the biosphere is likely to find its way into man's body. Therefore, looking after the man radio-biologically, one must take maximum care of the ecosystem (Odum, 1971). Overall radiation hazard is shown in the chart on page 120.

Types of High Energy Radiation

Different types of high-energy radiation are as follows :

A. Alpha (α) particles

Actually these are positively charged energy particles and consist of two protons and two neutrons. α-particles move at comparatively low velocity, therefore, these rays are less penetrative and destructive as compared to ß-particles. With a positive charge, any negative charged ion that may lie near their pathway deflects these. The outer layer of human skin will deflect and absorb practically all the α-radiation to which it is exposed. It moves with 1/10th of speed of light and has high ionizing power.

B. Beta (β) particles

These are fast-moving electrons emitted from the radioactive isotopes at a high energy. Being a high-energy electron with a negative charge, these are deflected by positive charges in the matter through which they pass. These are very much powerful in penetrating capacity. The velocity of β-particles is close to that of light (3/4 of the speed of light) hence β-rays are more penetrating than α-rays but have lower ionizing power than α-particles and, like α-particles, β-rays can cause internal damage to a person's body if inhaled or swallowed.

C. Gamma (γ) Rays

Gamma rays are short wave high-energy rays and thus represent the electromagnetic radiation. They are given off as a form of energy released from some radioactive isotopes or they may be generated from modern high voltage X-ray tubes. These rays are highly penetrating, but gradually lose energy as these pass through the object i.e., matter, in proportion to the density of the matter. Since gamma rays travel with the speed of light, these are most penetrating and cause severe damage in the living cells. Gamma rays travel with a speed that equals speed of light. Gamma and X-rays are practically of same nature.

D. Neutrons

As its name indicates, neutrons are the neutral or uncharged particles released from the atomic nuclei. It is highly penetrating because it has got no electrical charge. It is neither deflected nor slowed by passage near charged particles. It travels in a straight path until it collides with the nucleus of an atom. The cells inside an animal body receive just as much neutron radiation as is received by the skin on the outside of the body.

The neutrons prove to be very much dangerous for another reason — they can cause atoms to become radioactive. When the stable C^{12} is bombarded with neutron, some of the carbon atoms receive extra neutrons and C^{14} is created.

Mechanism of Ionizing Radiation Reactions— (Refer Chapter 18)

Mechanism of Non-Ionizing Radiation Reactions — (Refer Chapter 18)

Beneficial Aspects of Radiation (Food Irradiation)

According to a recent report of 2001, India loses approximately 10% of its food production through ill-storage, which accounts for Rs. 10,000 crore loss every year. Drying, canning, salting, heat treatment, dehydration and freezing were used previously to prevent food putrefaction. At present nuclear technology has added another process called "food irradiation" i.e., bombarding food substances with relatively low energy *gamma rays* emanating from a radioisotope such as Co-60 or Cs-137.

Following are the benefits of food irradiation :

1. It absolutely eliminates microbes from the food.
2. Irradiation diminishes microbial growth sufficiently to retard spoilage.
3. Irradiation destroys insects in cereals, grains, flour, peanuts, dried fruits, walnuts etc.
4. Irradiation inactivates pathogens and parasites that endanger health.
5. Irradiation sterilization can be used successfully to store pork, sausage, beef, chicken, canned beef, cakes, cod fish and shrimp for long durations under non-refrigerated conditions.
7. Recently DRDO scientists reported that food like *chapatti, pulao* etc. can be safely irradiated.
8. Irradiation can drastically reduce the use of food additives and fumigants, which pose genotoxic and occupational health hazards.
9. Co-60 is water insoluble so there is no risk of environmental contamination. Co-60, after its use, gets converted into non-radioactive nickel. So there is no problem of radioactive waste disposal.

Biological Magnification of Radioactive Elements

Accumulation of various radioactive elements in the food chain is known as *biological magnification.* An example is P-32, which was found to be concentrated as much as 5,000 times above levels in the water in white fish downstream from the Hanford Atomic Power Plant on the Columbia River. Blue gills and crappies in the same river were found to contain 20,000-30,000 times the concentration in water of P-32, while filamentous phycobionts were found to have up to 100,000 times the water level of P-32. Zn-65, Fe-59 and I-131 are the other radioactive isotopes that can be concentrated by the living organisms. These are the sources of radiation exposure to the general public when certain food is consumed. Fishes may contain P-32, oysters and clams may contain Zn-65 and milk may contain I-131.

If land is contaminated by radioactive elements then all living organisms (human, animals and plants) are subject to radiation. When humans consume food grown on such contaminated land, it increases internal doses of radiation and, hence, leads to internal dietary contamination of all living species in the food chain system. When accumulation reaches a certain limit, it causes anemia, hair loss, destruction of immune deficiency system of body and produces cancer and, finally, leads to death.

Maximum Permissible Doses for the Protection from Radiation

"Tolerance dose" — a dose that worker could tolerate for a prolonged period of time without any appreciable body injury was — first proposed by A. Mutscheller (1925). In 1928, ICRP (International Commission on Radiological Protection) set a standard for the tolerance dose at 100R/year. Mutscheller in 1934 reduced this dose to 60R/year. The concept of MPD (Maximum Permissible Dose) was introduced in 1960 with the fact that for somatic effects such as leukemia, bone necrosis and premature ageing etc., there was no tolerance dose below which they could not occur. According to ICRP — "The permissible dose for an individual is that dose which carries a negligible probability of severe somatic or genetic injury". MPD limit set by ICRP is shown in Table 7.3 :

Table 7.3 : MPD (Maximum Permissible Dose) limit set by ICRP

Sl. No.	Tissue/organ	Adult radiation worker (rems/year)	General people (rems/year)
1.	Gonads, red bone-marrow	5	0.5
2.	Skin	30	3.0
3.	Thyroid	30	3.0
4.	Extremities	75	7.5
5.	Other single organ	15	1.5

Preventive Measures From Radiation

Following preventive measures from radiation should be adopted :

1. Nuclear devices should never be exploded in air. If these activities are most essential then they should be exploded underground.
2. Production of radioisotopes should be minimized.
3. Minimum number of nuclear installation should be commissioned.
4. Fission reactions should be minimized.
5. In nuclear mines, wet drilling may be employed along with underground drainage.
6. Extreme care should be exercised in the disposal of industrial wastes contaminated with radionuclides.
7. Use of high chimneys and ventilations at the working place where radioactive contamination is high. It seems to be an effective way for dispersing radio-pollutants.
8. Nuclear medicines and radiation therapy should be applied only when absolutely necessary and with minimum doses.

Important Radiation Episodes

1. Hiroshima and Nagasaki 1945

The first atom bomb was exploded about 580 metres in the atmosphere over ill-fated Hiroshima on August 6, 1945. The second was detonated on 9th August 1945 507 metres high in air over Nagasaki.

At least 100,000 people were reported killed, severely injured and missing in Hiroshima alone, where the bomb virtually demolished all set-up and constructions in about 15 square km area. In Nagasaki 49,000 people were killed, injured and disappeared while an area of 6 to 7 sq. km was devastated.

The atom bomb exploded in Hiroshima used Uranium-235 with a half life period of 8.5×10^8 years, while the Nagasaki bomb had plutonium-239 as an explosive man-made radionuclide with half life of 24,000 years.

Atom bomb survivors showed higher than average leukemia rates, which peaked in 1951 (six years after exposure) but which were still higher than usual in 1966.

Despite frequent criticism of these studies in Japan, studies are being continued and are playing an important role in defining the hazards of radiation. They are especially important now when increased cancer rates or other effects might be appearing after a long period of latency.

2. Three Mile Island Nuclear Disaster of U.S.A.

This episode of nuclear disaster took place in 1979, in which the 'Three Mile Island' melt down of nuclear reactor crystallized nuclear fears. Its adverse effects have been observed till now.

3. Chernobyl — World's Worst Nuclear Disaster

This episode is also known as 'Melt Down' of Chernobyl Nuclear Power Plant in former USSR. April 28th 1986 was a sad day for nuclear power generation when a major accident took place at 1.23 a.m. in the nuclear reactor at Chernobyl, 80 miles (130 km) north of Kiev in former USSR. It resulted in clouds of radioactive smoke over a large area in Scandinavian countries, which are 2,000 km away in the Russian region itself. It caused unspecified deaths, suffering and environmental and health damage. Around 2,000 hapless people near the plant had been killed. A top Swedish nuclear official reported that water supplies to the homes of 3.5 million persons of Kiev had probably been contaminated with acute radionuclides.

Nuclear Power Stations in India

1. **Tarapur Nuclear Power Station** : This power station was operated on 1.4.1969 and was constructed in collaboration of Central Electric Company of USA. It consists of two enriched uranium reactors of the boiling water type. It is a 420 MW energy station. It is the country's first power reactor (based on enriched uranium).
2. **Kota Nuclear Power Station** : It is the second nuclear power station. It is 430 megawatt power station, situated at Kota (Rana Pratap Sagar, Rajasthan). It is natural uranium fuel reactor moderated with heavy water and started feeding power in 1973 into the power system of Rajasthan. In this power station the design and the bulk of the equipment have come from Canada.
3. **Kalpakkam Nuclear Power Station** : It is 500 MW power station located at Kalpakkam, about 100 km away from Chennai. It is popular as Indira Gandhi Centre for Atomic Research (IGCAR). It is the third power station in India. It is a fast-breeder reactor and is based on Plutonium-Uranium Carbide.
4. **Narora Nuclear Power Station in U.P.** : It consists of two atomic reactors of 235 MW each and delivers 440 MW of power. It uses natural uranium as fuel and heavy water as moderator. It is the fourth Indian Atomic Power Plant.

All the above four are based on the Canadian CANDU reactors. The fifth nuclear power station is at Kakrapur in Gujrat.

Atomic Explosions in India

India conducted its first atomic explosion on May 11, 1974 in the Pokharan deserts of Rajasthan at 3.45 p.m.. It was code named "Smiling Buddha". It was a test reaction explosion mainly meant for peaceful purposes as claimed by the Govt. of India and the Atomic Energy Commission. Further, the second explosion in Pokharan was conducted into two series on May 11 and 13, 1998 at 3.45 p.m. and code named "Operation Shakti". At present India has become the sixth nuclear power in the world.

Revision Questions

A. Long-Answer Questions

1. Write an essay on *Radiation pollution.*
2. Describe in detail about the sources of radioactive pollution.
3. Define ionizing radiation. Give the effect and mechanism of ionizing radiation.
4. Define non-ionizing radiation and give its effects.

B. Short-Answer Questions

1. Briefly describe :

(i) Ionizing radiation
(ii) Non-ionizing radiation
(iii) Radioisotopes
(iv) Measurement of radiation
(v) Preventive measures from radiation.

2. Write short notes on :

(i) Types of high-energy radiation
(ii) Beneficial aspects of radiation
(iii) Half-life period
(iv) Biological magnification of radioactive elements
(v) Maximum permissible dose for the protection from radiation
(vi) Important radiation episodes
(vii) Nuclear power stations in India
(viii) Natural sources of radiation
(ix) Anthropogenic sources of radiation.

8

Food-Additives

SYNOPSIS

General account, Commonly used food-additives and contaminants, Incidental or indirect additives. Intentional or direct food additives, Terms related to adverse reactions to food, Food-borne molds and mycotoxins, Testing of food additives.

General Account

Food additives have acquired a social acceptance in the processing and preservation in order to maintain the taste and quality of food products. These additives have now got legal and scientific sanctions because, without changing the nutritional value of the food, they keep the quality, stability and appearance of food in good order. The perishable foods which otherwise go waste are generally saved due to these additives within check the microbial development and quick oxidation, thus reducing the food wastage.

An additive is a substance not normally present in the food in question and which is added either deliberately or incidentally in order to improve the quality or to preserve the food. In other words, food additives are a group of substances which are directly ingested by humans along with food-items or food product. Actually, naturally-occurring substances are not necessarily nontoxic, for example : the causation of lathyrism, a serious neurological disease due to ingestion of *khesari dal (Lathyrus sativus)* is well known and its incidence in endemic form occurs in many parts of India and abroad as well. In India, Lathyrism — which is a crippling disease accompanied by paralysis of the leg muscle — is found to occur in Bihar, U.P. and M.P. Outside India, it has been reported to occur in Spain, Algeria, France and Italy. The Nutrition Research Laboratory at Hyderabad has done a painstaking and splendid piece of work to remove toxic factors from *khesari dal* and has evolved ways and means to remove toxic factors from this seed. Lathyrogens bring about alterations in collagen, elastin and mucopolysaccharides of connective tissue.

The existence of harmful fish toxins has been known for centuries. Teraodontoxin, the active principle of puffer fish, and saxitoxin from muscles are two of the best-known fish toxins. Vegetable cooking fat may contain 5 to 20 ppb of carcinogenic polycyclic aromatic hydrocarbons including benzopyrene.

The whisky drinker suffers in additions to benzopyrene consumption from the barley used in the manufacture of Scotch whisky. Most pulses, legumes and oil seeds including soyabeans that we eat contain various types of antitryptic factors, hemagglutinin or hemolysins and other growth-inhibitory substances. Unless they are adequately processed and detoxified they will be deleterious to human health.

A number of vegetables, especially the Brassica family and soyabean, contain goitrogens. In recent years many goitrogenous compounds in foodstuffs, for example crambe seed meal, have been detoxified by treatment with gaseous ammonia. It is well known that tapioca contains cyanogenic glycosides and steam treatment is necessary to remove these compounds. Peanut meal contains aflatoxin, which is one of most the carcinogenic compounds known today. Recently the *Central Food Technical Research Institute*, Mysore, has evolved methods of detoxification of aflatoxin containing meal by treatment with hydrogen peroxide under alkaline conditions. Unfortunately, this treatment may also affect the nutritive value of the meal to some extent.

Those of us with vegetarian tendencies will be depressed to learn that the greens we collect from the market place may add yet another burden of carcinogens. Cabbage and spinach are especially rich in phenanthrene and flouranthene. Seed foodgrains also contain variable quantities of polycyclic aromatic hydrocarbons with phenanthrene (20 to 100 ppb) as a major constituent. Coffee has been known to have a mutagenic effect on male larvae of the fruit fly as also in mammalian tissue cultures and has been reported to be hypercholestremic. In the gestating rat, caffeine has been reported to cause fetal resorption and abnormalities of various kinds. Certain oxidized and heated fats are known to produce growth retardation, serious pathological changes at the cellular and sub-cellular level, interference with mitochondrial electron transport and alterations in membrane permeability. Gossypol, present in cottonseed meal, has a growth-retarding effect, especially to nonruminants subsisting on low protein diets as has been shown by the recent work of Dr. Gopalan at the Indian Nutrition Research Laboratories. Cycad plants grow in many parts of the world and their leaves and seed are eaten by people and animals. These plants contain methylisodimethanol glucosides which, when hydrolysed, behave like their chemical relatives — the alkylnitro-samines.

Let us now consider the toxicity of food additives. Table 8.1 shows the list of commonly used food additives. They are additives added to foods with specific technological on nutritional objectives. In addition, food also contains other adventitious chemicals, as those from pesticides, from packaging materials and bacterial and chemical contaminants of various types. Some of these are listed in Table 8.2.

Table 8.1 : Commonly Used Food Additives

- Common Salt (NaCl)
- Sugar
- Spices
- Fruit extracts
- Monosodium glutamate
- Maltol
- Saccharin
- Citric acid
- Cyclamates
- Phosphoric acid
- Fumaric acid
- Malic acid
- Tartaric acid
- Adipic acid
- Lactic acid
- Succinic acid
- Vitamins (A, D, E, B_1, B_2, B_6, Niacin)
- Minerals (iron, copper, calcium)
- Amino acids (lysine, threonine, methionine, tryptophan)
- Emulsifiers
- Stabilizers and thickeners (gum arabic, guar gum, CMC, gelatin, carrageenan, agar-agar, carobbean gum, etc.
- Firming agents (Ca salts)
- Anit-caking agents (Ca silicate, $Ca_3(PO_4)_2$, Sod. Al. Silicate
- Moisture-retaining agents (Sorbitol, glycerol, propylene, glycol)
- Preservatives (Propionates, sorbic acid, Sod. Diacetate, benzoic acid, ethyl formate, ethylene and propionates oxides)
- Anti-oxidants (BHA, BHT, NDGA, tocopherol, ascorbic acid, erythorbic acid)
- Sequesntrants (Citric acid, sod. hexametaphosphate, EDTA)
- Colours (saffron, cochineal, caramel, turmeric, Red No. 2, Yellow No. 5, Yellow No. 6).
- Flavouring agents (delta-decalactone)
- Leaching or contact additives
- Solvents (ethyl alcohol, glycerol, ethyl acetate, glycerol mono di and tri acetates, isopropyl alcohol, propylene glycol, diethylethar)

Table 8.2 : Food Contaminants

- Aflatoxin (mainly from groundnut meal)
- Argemone oil (in mustard oil)
- Gossypol (in cottonseed meal)
- Solvent residues
- Pesticide residues, Agricultural chemicals
- Defoaming agents and detergents
- Polymers, Plasticizers and Packaging contaminants
- Natural chemicals

By and large the additives do not increase the nutritional value of the food, rather they only improve the appearance and sale value and also the texture and flavour. However, commonly used food additives are usually excreted from the body.

The food additives may be broadly classified in two groups :

A. The incidental or indirect additives.

B. The intentional or direct additives.

However, a third group, i.e., naturally occurring additives, may result from processing conditions, metabolic reactions and unanticipated chemical combinations. These include safrole and related compounds and contaminants as aflatoxins.

(A) Incidental or Indirect Additives

A food may contain minute traces of a chemical as a result of contact with a substance used in its production, processing or packing. Since its presence serves no useful purpose in the final food product, such a chemical may be considered to be an incidental additive, e.g., food may pick up material from a wrapper or a container, either by dissolving it out or by abrasion from the container into the food. Detergents used for cleaning dishes or food equipments could be such an incidental source.

Of greatest concern are the pesticides which are used in crop production in order to avoid destruction by insects, nematodes, viruses, fungi or other plants peril. Residues of some of these pesticides sometimes may be present on fruits and vegetables even after careful washing. It is self-evident that these chemicals are toxic to some forms of life or they would not have their protective properties for the crop. Just what hazard they cause to man is a fundamental question. Since the minimum lethal dose of these chemicals for various species of animals is well-known, the products must be used according to specified concentrations. More difficult to determine is the effect of built-up in the body when foods that contain such residues are eaten day after day for long periods of time—even a lifetime. Toxicological studies on animals throughout their life cycles, including the effects on reproduction and the next generation, are continuously being conducted to minimize such hazards.

(B) Intentional or Direct Additives

Chemicals are intentionally added to foods by the processor in order to enhance the quality of a food such as texture, colour, flavour, nutritive value, or keeping properties to gain better consumer-acceptance.

The intentional additives frequently used can be classified depending on their nature, composition and quality of preservation required :

1. Antioxidants
2. Emulsifiers
3. Enzymes
4. Flavoring agents
5. Colour and preservatives
6. Artificial sweetening agents
   (a) Saccharin
   (b) Urea derivatives.

1. Antioxidants

The warm climate, temperature and humidity are such variable factors that they boost microbial infections to contaminate the food which, in due course, gets oxidized. The oxidation of food material makes the food unfit for human consumption. The dairy products such as cheese, butter, oils and fats etc. start stinking if exposed to air for a longer period. To check or reduce the oxidation of these food products, chemical antioxidants are used, of course, with following restrictions :

(a) Such chemicals must be incorporated by the foodstuff and also must not tend any flavour of their own or the colour even during long storage.

(b) It must be fairly cheap and also active even at low concentrations. Only a little quantity of the antioxidants should be mixed with a diluting agent facilitating its proper and uniform mixing without affecting the food texture.

Not more than often, the antioxidants are also mixed with those agents which enhance their action and such substances are known as "synergists", though they have no or very little activity and only catalyse the action of antioxidant e.g., citric acid.

The natural antioxidants which are used in checking the oxidation of animal fats are oils of soyabean, rice, palm, peanut, olive, cotton seed, coffee etc. Apart from this, certain spices and other natural products—e.g., oxalic, succinic, citric, malic, tartaric and amino acids like cystein, glycine and leucine or amides and proteins — are also used as antioxidants in food preservation.

The commonly used antioxidants by food industries — butylated hydroxy anisole (BHA), butylated hydroxy toluene (BHT) and propyl gallate — been found to be the cheapest and best. The aforesaid antioxidants are frequently used to preserve chicken fat, meats, fishes, baked and fried food etc.

2. Emulsifiers

To accelerate the process of emulsification certain emulsifying agents are used to boost the desired emulsification. The most commonly used agents are glycerides of fatty acids and their esters, stearic acid, lecithin, and brominated vegetable oils. These agents, however, dilute the consistency of the products and thus, *Agar, Gelatin, Pectin, Calcium, Sodium acetate* and various starches are used to thicken food products as they possess the affinity with water.

3. Enzymes

Most harmless and without any toxic affect — the enzymes have been found to play an important role in food preservation and are supposed to be the best commercially available agents. These are *amylases, proteases, lipases, pectinases* etc.

It is known to all of us that *amylases* hydrolyse starch and this reaction has been adopted by those engaged in the syrup production from various types of starches. Likewise, various fruit juices are also treated with peptide enzymes for clearing and thereby reducing the viscosity, it also helps storage (wine and soft drinks).

The protein hydrolyzing enzymes have also been successfully picked up in dairy products e.g., cheese manufacturing and also in the preparation of bread and wines etc. Those who consume meat use raw papaya for meat tenderisation while cooking on large scale to make the meat soft and easily digestible. The animals are given intravenous injection of proteolytic enzyme before slaughter so that the enzymes may reach uniformly in each and every part of the body making the flesh soft and tender.

Since these additives are cheap and almost harmless, they are thought to be good ones.

4. Flavouring of food products

The specifically desired flavour and taste is the first and foremost prerequisite of the food products to attract large number of consumers. In order to improve the taste and increase or decrease the natural flavour of food the agents which are used are known as *seasoners, enhancers, potentiators* and *flavour integers*. Since long, scents and aromatic oils are being extracted from the leaves, flowers, fruits, barks and seeds of various aromatic plants. The most commonly used odorous plants are orange, lemon, grapes, rose, kewara, khas, vanilla etc. Even in the common houses pickles are made — a good number of raw spices are added to it to increase the taste and aroma.

Apart from vegetative plant materials a good number of comparatively stronger synthetic flavour boosters have also been synthesized. However, such chemicals are used in very small quantity (not more than 300 ppm) yet they have been found quite effective and also easier in mixing. They are usually some acids along with sugar or certain salts and chemicals viz., Di-sodium 5'-inosinate and Di-sodium 5' quanylate (intensifiers).

However, quite safe and most commonly used natural vegetable aromatic agents are coriander, ginger, garlic, onion, clove, cardamom, saffron, *tejpatta* etc. These are freely used in our kitchens and houses without any toxic risk. Some of the fruits like mangoes, melon, banana etc. make other substances aromatic.

5. Colour and Preservatives

Colours have been found to be one of the important attractants in the finished food products. Therefore, colours and preservatives are freely added in a variety of edibles e.g., icecream, sweets, candy, biscuits, soft and alcoholic drinks etc. Besides dairy, confectioneries and beverages, even the pulse like *arhar* (*Cajanus cajan*), *masoor* etc. are also coloured for an attractive food and consequently increased sale (*Phool Chap Dal*).

Though only a limited number of colours are approved for colouring the food yet people utilize most randomly even the prohibited ones. The colours which are permitted are either of plant origin or synthetic. Some of the permitted colours are *Erythrosine, Ponceau* 4 R (red), *Sunset yellow, Indigo, Carmine, Fast green, Amarnth* etc. The permitted colours are either added to some of the edibles e.g., aerated drinks, sauce, jams, syrups etc., or in good number of medicines. However, since the colours (synthetic) are not quite safe, their doses should be carefully watched and monitored because they cause toxicity in animals. The excess use of some of these colours may cause diarrhea, lowering of RBC count (Ponceau), cholesterol disbalance, growth retardation etc. A higher concentration of amaranth has been reported to be carcinogenic, reduces fertility, induces abortion and fetal deformities. The ill-effect of most of the colours is because they do not decompose during digestion. Therefore, colours must be used with caution and care should be taken to restrict to the permitted ratio.

But it is most alarming that most of the sweet-meat shops use certain nonpermitted dyes e.g., mentil yellow, auromine, orange II, rhodamine B, congo red, malachite green and methylene violet etc. which are highly toxic and cause irreparable harm to some of the most vital body organs e.g., kidney, spleen, liver, heart, eyes.

6. Artificial sweetener

The most commonly used natu. .. sweetening agent is sugar ($C_{12}H_{22}O_{11}$) and is manufactured from sugarcane, palm, sugar beet, pineapple juice etc. In addition to sucrose, other natural sweeteners are glucose, fructose, maltose, lactose etc. But since these sugars are costlier, the industrialists use certain synthetic chemical sweeteners which are, on one hand, cheaper, and, on the other hand, 100 times sweeter than sugar. These chemical sweeteners are the compounds of amides, imides, nitrites, oximes, ureas and nitroaniline groups. Though they are excreted through urine they exert extra pressure on kidneys.

The most commonly used sweetening agents are :

(1) **Saccharin** : This is a byproduct of coal-tar distillation and was developed during World War II (1939-45) by Germans and is chemically called o-sulphobenzimide. It is about 700 times more sweet then natural sucrose. Since this chemical has obtained universal acceptance it is widely used in beverages and cold drinks, toothpastes and several such products. Like other food-additives, saccharin has low acute toxicity in experimental animals. However, it may produce bladder tumors in rats.

(2) **Urea derivative** : Its commercial name is *Dulcine* and the chemical name is p-phenetyl. It is 300 times sweeter than sugar.

Besides, there are many more such chemicals in commercial use but they are not as safe as saccharin. These compounds have been reported to be toxic and cause damage to kidney, produce liver tumour and several more side-effects. Typical uses of various well-known intentional (direct) food additives by functionality is shown in Table 8.3 :

Table 8.3 : Well-known Important Direct (intentional) Food Additives by Functionality

Sl. No.	Agent's name	Description	Examples
1.	Anticaking agents and free-flow agents	Substances added to finely powdered or crystalline food products to prevent caking, lumping, or agglomeration	Glucitol, sodium ferrocyanide, silicon dioxide
2.	Antimicrobial agents	Substances used to preserve food by preventing growth of microorganisms and subsequent spoilage, including fungistats, mold and rope inhibitors	Nisin; methyl-ethyl-propyl, or butyl ester of p-hydroxybenozoic acid; sodium benzoate; sorbic acid and its salts
3.	Antioxidants	Substances used to preserve food by retarding deterioration, rancidity or discolouration due to oxidation	Butylated hydroxyanisole (BHA), butylated hydroxytoluene (BHT), propyl gallate
4.	Colours and colouring adjuncts	Substances used to impart, preserve, or enhance the colour or shading of a food including colour stabilizers, colour fixatives, colour-retention agents	Tartrazine, β-carotene, annatto, turmeric
5.	Curing and pickling agents	Substances imparting a unique flavour and/or colour to a food	Calcium chloride, glucitol
6.	Dough strengtheners	Substances used to modify starch and gluten, producing a more stable dough, including the applicable effects	Calcium bromate, baker's yeast extract, calcium carbonate

Contd...

Sl. No.	Agent's name	Description	Examples
7.	Drying agents	Substances with moisture-absorbing ability used to maintain an environment of low moisture	Calcium stearate, cobalt tallate
8.	Emulsifers and emulsifier salts	Substances that modify surface tension in the component phase of an emulsion to establish a uniform dispersion or emulsion	Phosphate esters of mono-and diglycerides, acetylated monoglycerides, calcium stearate
9.	Enzymes	Enzyme used to improve food processing and the quality of the finished food	Papain, rennet, pepsin
10.	Firming agents	Substances added to precipitate residual pectin, strengthening its collapse during processing	Calcium acetate, calcium carbonate
11.	Flavour enhancers	Substances added to supplement, enhance, or modify the original taste and/or aroma of a food without imparting a characteristic taste or aroma of their own	Monosodium glutamate, inositol
12.	Flavour agents and adjuvants	Substances added to impart a taste of aroma in food	Cinnamon p-cresol, thymol zingerone
13.	Flour-testing agents	Substances added to milled flour at the mill to improve its colour and/or baking qualities	Calcium bromate
14.	Formulation aids	Substances used to promote or produce a desired physical state or texture in food	Palm kernel oil, tallow
15.	Fumigants	Volatile substances used for controlling insects or pests	Aluminum phosphide, potassium bromide
16.	Humectants	Hygroscopic substances incorporated in food to promote retention of moisutre	Arabic gum, calcium chloride
17.	Leavening agents	Substances used to produce or stimulate production of carbon dioxide in baked goods to impart a light texture, including yeast, yeast foods, and calcium salts	Carbon dioxide, adipic acid
18.	Lubricants and release agents	Substances added to food contact surfaces to prevent ingredients and finished products from sticking to them	Mineral oil, acetylated monoglycerides
19.	Nonnutritive sweeteners	Substances having less then 2 percent of the caloric value of sucrose per equivalent unit of sweetening capacity	Aspartame, saccharin
20.	Nutrient supplements	Substances that are necessary for the body's nutritional and metabolic processes	Calcium carbonate
21.	Oxidizing and reducing agents	Substances that chemically oxidize or reduce another food ingredient, producing a more stable product	Calcium peroxide, hydrogen peroxide
22.	pH control agents	Substances added to change or maintain active acidity or basicity including buffers, acids, alkalis and neutralizing agents	Acetic acid, propionic acid, calcium acetate, calcium carbonate, carbon dioxide
23.	Processing aids	Substances used as manufacturing aids to enhance the appeal or utility of a food or food component	Carbon dioxide, ammonium carbonate, ammonium sulphate, potassium bromide
24.	Propellants, aerating agents, and gases	Gases used to supply force to expel a product or reduce the amount of oxygen in contact with the food in packaging	Carbon dioxide, nitrous oxide

Contd...

25.	Sequestrants	Substances that combine with polyvalent metal ions to form a soluble metal complex to improve the quality and stability of products	Acetate salts, citrate salts, gluconate salts metaphosphate, calcium acetate
26.	Solvent and vehicles	Substances used to extract or dissolve another substance	Acetic acid, acetylated monoglycerides, ethyl alcohol, ethyl acetate etc.
27.	Stabilizers and thickeners	Substances used to produce viscous solutions or dispersions, to stabilize emulsions	Calcium acetate, calcium carbonate
28.	Surface-finishing agents	Substances used to increase palatability, and inhibit discolouration of foods	Ammonium hydroxide
29.	Synergists	Substances used to act or react with another food ingredient to produce a total effect different from or greater than the sum of the effects produced by the individual ingredients	Acetic acid, propionic acid
30.	Texturizers	Substances that affect the appearance or feel to food	Calcium acetate

Terms related to Adverse Reactions to Food

Various terms have been used for the adverse reactions of food. Their definitions are mentioned in Table 8.4 and description of very important terms generally used viz. food allergy, idiosyncrasy, pharmacological food reactions and metabolic food reactions are described in proceeding section.

Common Terms used for the Adverse Reactions of Food

(i) **Food hypersensitivity or food allergy** : Food hypersensitivity (allergy) actually refers to a reaction involving an immune-mediated response. Such a response is generally (immunoglobulin) IgE-mediated. What generally distinguishes food allergy from other reaction is the involvement of

Table 8.4 : Adverse Reactions of Food : Definition of Terms

Sl. No.	*Term*	*Definition*
1.	Adverse reaction (sensitivity) to a food	Common term that may be applied to a clinically abnormal response attributed to an ingested food or food additive.
2.	Food hypersensitivity (allergy)	An immunologic reaction resulting from the ingestion of a food or food additives. This reaction occurs only in some patients, may occur after only a small amount of the substance is ingested and is unrelated to any physiological effect of the food or food additive.
3.	Food anaphylaxis	A classic allergic hypersensitivity reaction to food or food additives.
4.	Food intolerance	A general term applied to describe an abnormal physiological response to an ingested food or food additives; this reaction may be an immunologic, idiosyncratic, metabolic, pharmacological, or toxic response.
5.	Food poisoning	A term used to imply an adverse effect caused by the direct action of a food or food additives on the host recipient without the involvement of immune mechanisms.
6.	Food idiosyncrasy	A quantitatively abnormal response to a food substance or additives; this reaction differs from its physiological or pharmacological effect and resemble hypersensitivity but does not involve immune mechanisms. Food idiosyncratic reactions include those which occur in specific groups of individuals who may be genetically predisposed.
7.	Pharmacological food reaction	An adverse reaction to a food or food additives as a result of a naturally derived or added chemical that produces a drug-like or pharmacological effect in the host.
8.	Metabolic food reactions	Toxic effects of a food when consumed in excess or improperly prepared.

immunoglobulin, basophils, or mast cells. An allergic reaction may be manifested by one or more of the symptoms listed in Table 8.5. It is now a established fact that most allergens (antigens) in food are protein in nature.

Table 8.5 : Symptoms of 1gE-Mediated Food Allergies	
Cutaneous	Urticaria (hives), eczema, dermatitis, pruritus, rash
Gastrointestinal	Nausea, vomiting, diarrhea, abdominal cramps
Respiratory	Asthma, wheezing, rhinitis, bronchospasm
Others	Anaphylactic shock, hypotension, palatal itching, swelling including tongue and larynx

Some of the allergenic components of common food allergens are listed in Table 8.6 :

Table 8.6 : Known Allergenic Food Proteins

Food	*Allergenic Proteins*
Cow's milk	α-Lactalbumin β-Lactoglobulin Casein
Egg whites	Ovomucoid Ovalbumin
Egg yolks	Livetin
Peanuts	Arah II Peanut I
Soyabeans	β-Conglycinin Glycinin Gly mIA Gly mIB Kuniz trypsin inhibitor
Codfish	Gad cl
Shrimp	Antigen II
Green peas	Albumin fraction
Rice	Glutelin fraction Globulin fraction
Cotton seed	Glycoprotein fraction
Tomato	Several glycoproteins
Wheat	Albumin Gluten Gliadin Globulin
Okra	Fraction 1

(ii) **Food Idiosyncrasy** : Food idiosyncrasy may be defined as quantitatively abnormal responses to a food substance or additive; this reaction differs from the physiological effect, and although it may resemble hypersensitivity, it does not involve immune mechanisms. Food idiosyncratic reactions include those which occur in specific groups of individuals who may be genetically predisposed. Example of such reactions and the foods that probably are responsible are given in Table 8.7.

Probably the most common idiosyncratic reaction is lactose intolerance, a deficiency of the lactase needed for the metabolism of the lactose in cow's milk. A lack of this enzyme results in fermentation of lactose to lactic acid and an osmotic effect in the bowel, with resultant symptoms of malabsorption and diarrhea. Lactose intolerance is lowest in northen Europe at 3 to 8 percent of the population; it reaches 70 percent in southern Italy and Turkey and nearly 100 percent in southeast Asia (Gudmand-Hoyer, 1994).

Table 8.7 : Idiosyncratic Reactions to Foods

Food	*Reaction*
Fava beans	Haemolysis, sometimes accompanied by jaundice and hemoglobinuria; fatigue, nausea, dyspnea, fever and chills, abdominal pain
Chocolate	Migraine headache
Beets	Beetanuria : passage of red urine (often mistaken for hematuria)
Asparagus	Odorous, sulphurous-smelling urine,
Red wine	Sneezing, flush, headache, diarrhea, skin itch, shortness of breath
Choline-and carnitine containing foods	Fish odour syndrome; foul odour of body secretions
Lactose intolerance	Abdominal pain, bloating, diarrhea
Fructose-containing foods	Abdominal pain, vomiting, diarrhea, hypoglycemia

(iii) **Pharmacological Food Reactions** : It is also referred to as "false food allergies" (Moneret-Vautrin, 1987). Such adverse reactions are characterized by exaggerated responses to pharmacological agents in food (Table 8.8). These reactions are distinguished from other classifications because they are not associated with a specific anomaly. These, then, are common pharmacological agents acting in a very predictable manner, but at exceptionally low levels :

Table 8.8 : Pharmacological Reactions to Food

Food	*Reaction*	*Mechanism*
Cheese, Red wine	Severe headache, hypertension	Tyramine from endogenous or ingested tyrosine
Coffee, tea	Headache, hypertension	Methyloxantine (caffeine) acting as a noradrenergic stimulant
Chocolate	Headache, hypertension	Methylxanthine acting as a noradrenergic stimulant

(iv) **Metabolic Food Reactions** : Metabolic food reactions are distinct from other categories of adverse reactions in that the foods are more or less commonly eaten and demonstrate toxic effects only when eaten to excess or improperly processed.

Food-Borne Molds and Mycotoxins (Food Contaminants)

Molds are well-known to humans for centuries in the production of foods and have provided a number of fungal metabolites with salient medicinal applications. On the other hand, they may produce various other metabolites which may produce severe adverse health effect. Important among these are aflatoxins and trichothecenes. Other mycotoxins and subgroups are listed in Table 8.9.

Actually, aflatoxins are some of the most toxic and carcinogenic substances known. The comparative carcinogenic properties of aflatoxins with other well-known carcinogens are illustrated in Table 8.10 :

Table 8.10 : Carcinoganic effects of Aflatoxins in Comparison to other well-known Carcinogens

1.	Butter yellow	9ng/rat/day
2.	Ethionine	1ng/rat/day
3.	Dimethylnitrosamine	0.7ng/rat/day
4.	Aflatoxin B_1	0.006ng/rat/day

Other Contaminants of Food (Metals, Pesticides and Antibiotics)

Various metals, pesticides, antibiotics etc. accidently or intentionally may contaminate the food stuffs as summarized in Table 8.11 :

Table 8.9 : Selected Mycotoxins Produced by various Molds

Mycotoxin	*Source*	*Effect*	*Commodities Contaminated*
◆ Aflatoxins B_1, B_2, G_1, G_2	Aspergillus flavus parasiticus	Acute aflatoxicosis, carcinogenesis	Corn, peanuts and others
◆ Aflatoxin M_1	Metabolite of AFB_1	Hepatotoxicity	Milk
◆ Fumonisins $B_1, B_2, B_3, B_4, A_1, A_2$	Fusarium moniliforme	Carcinogenesis	Corn
◆ Zearalenones	Fusarium	Estrogenic effect	Corn, grain
◆ Cyclopiazonic acid	Aspergillus, penicillium	Muscle, liver, and spleenic toxicity	Cheese, grain
◆ 3-Nitropropionic acid	Artrinium sacchari, Sacharicala, Phaesospermum	Central nervous system impairment	Sugarcane
◆ Citreoviridin	Penicillium citreaviride Toxicarium	Cardiac beriberi	Rice
◆ Sterigmatocystin	Aspergillus versiolar	Carcinogenesis	Corn
◆ Penicillinic acid	Penicillium cyclopium	Nephrotoxicity	Corn, dried beans, grains
◆ Rubratoxins A, B	Penicillium rubrum	Hepatotoxicity teratogenic	Corn
◆ Patulin	Penicillium patulatum	Carcinogenesis, liver damage	Apple and apple products
◆ Ochratoxin	A.ochraceus, P.viridicatum	Balkan nephropathy, carcinogenesis	Grains, peanuts, green coffee
◆ Ergot alkaloids	Cladosporium purpurea	Ergotism	Grains

Table 8.11 : Deleterious Effects of some known Contaminants of Food (Metals, Pesticides and Antibiotics)

Sl. No.	Contaminants	Foods	Deleterious effect
1.	Arsenic	Edible fruits sprayed by lead arsenate	Chills, Cramps, Paralysis
2.	Barium	Foods contaminated by rat poison	Muscular twitching and convulsions
3.	Cadmium	Fruit juices, soft drinks	Liver and kidney damage, excessive salivation, prostate cancer
4.	Cobalt	Water, beer	Cardiac failure
5.	Copper	Acid foods in contact with tarnished copper ware	Vomiting, diarrhea, abdominal pain
6.	Lead	Some processed foods	Brain damage, Paralysis
7.	Mercury	Mercury fungicide treated seedgrains	Brain damage, Paralysis
8.	Tin	Canned foods	Colic, Vomiting
9.	Zinc	Foods stored in galvanised ironware	Dizziness, Vomiting
10.	Pesticides viz., DDT & BHC	All types of food	Damage to liver, kidney, brain, and nerves — leading to death
11.	Antibiotics	Meat from animals fed antibiotics	Drug resistance, hardening of arteries, heart disease

Testing of Food Additives

Unlike drugs, food additives are not used for their specific biological additives but are used for their specific technological effects and to ensure the maintenance of the natural quality of food. With food additives the hazard, if any, is more likely long-term rather than acute.

The problem of testing food toxicity is much more complicated than that of testing drug toxicity. The composition of the food additives, the conditions and extent of usage and their biological properties — all factors have to be taken into account in evaluating the possible hazard or ensuring the safety in use of food additives.

Apart from chemical and physical specifications the main headings under which information should be sought are :

1. Acute toxicity data by oral and by injection in rats, mice and one additional species.
2. Short term studies (feeding for 90 days). In rat or mice and one additional species — usually for a period of 10% of the life span. In this case a 1,000-fold margin of safety on non-effective levels in animals has been suggested in humans.
3. Long term (usually 2 years) tests which include studies on toxicity, with histopathological data, fertility and carcinogenicity. In this case a 100-fold margin of safety has been suggested in humans over non-effective levels in animals.
4. Metabolic effects on enzyme levels etc. Estimation of serum enzyme concentrations undoubtedly gives an excellent indication of tissue abnormalities and the acute effects of many hepatotoxic compounds. Elevation in urine enzyme levels can give some indications of kidney damage. These tests require sophisticated technologies and, apart from animal-house facilities, a good deal of inter-disciplinary collaboration between biochemists, physiologists, histochemists, electron microscopists, pharmacologists, and oncologists, in addition to analytical chemists, is necessary.

Revision Questions

A. Long-Answer Questions

1. Write an essay on use and abuse of food additives.
2. What are direct food additives? Make a list of direct food additives and their uses with example.
3. Define various terms selected to adverse reactions to food.
4. Make a list of food-borne molds and mycotoxins with their source and effects.

B. Short-Answer Questions

1. Write short notes on :

(i) Incidental food additives
(ii) Intentional food additives
(iii) Food contaminants
(iv) Commonly used food additives.

2. *Briefly describe :*

(i) Antioxidants in food

(ii) Flavoring of food products

(iii) Colour and food preservatives

(iv) Artificial sweeteners.

3. *Write short notes on :*

(i) Aflatoxins

(ii) Pharmacological food reactions

(iii) Testing of food additives

(iv) Food allergy

(v) Metals as contaminants of food.

9

Noise Pollution

SYNOPSIS

Introduction, Definition of noise and noise pollution, Sources, Principal types of noise, Measurement of noise — unit of sound, Sound level of different sources, Adverse effects of noise on human being, Ambient noise standards, Prevention and control of noise pollution.

Introduction

Just like chemical waste or smoke, noise has also been considered an important environmental pollutant and has penetrated almost every aspect of modern life. It is potentially a serious signal and a grave threat to the environmental health. The word "noise" is derived from the Latin "*nausea*" meaning disgust or a feeling of sickness of the stomach with an urge to vomit. The obiquity of noise has made many people apprehensive about its possible adverse effects on health. In general, noise is any unwanted sound occurring in the wrong place and time. Noise, though undesired, could be meaningful or meaningless. Meaningful noise such as a baby crying or a person screaming acts as a stimulus on the body, it helps to convey a message and receives one's attention or a response in turn. But meaningless noise is disturbing, destructive and annoying. The noise arising from industrial and transportation process can sometimes be so high as to impair hearing power. The problems caused by such unwanted sounds is *noise pollution.* There has been a considerable increase in noise from man-made sources during the last 100 years which is now doubling every decade.

It is interesting to mention here that plants too wish to grasp melodious sound. In an experiment shown on TV by Walter Cronkite, on October 26, 1970 at Temple Buell College in Denver healthy flowers were put in three separate glass cases. Acid rock music was piped into one case, soft Indian Sitar music was piped into the second case, and the third had no music. A CBS camera crew recorded the experiment using time-lapse photography. At the end of two weeks, flowers exposed to the rock music were dead, the group with no music was growing normally and the third ones that heard the sitar music had turned into beautiful blooms with stems and flowers reaching towards the source of the sitar sound. It may be inferred from W. Cronkite's experiment that plants too have an intelligence and prefer soft melodious sound.

Definition of Noise

Noise may be defined in a number of ways. For example

1. Noise is a sound* without value.
2. Noise is unwanted, unpleasant or disagreeable sound that causes discomfort.
3. Noise is that sound which is undesired by the recipient.
4. Noise is the wrong sound in the wrong place at the wrong time.

*Sound consists of wave motion caused by the vibration of molecules. Sound travels in pressure waves.

5. Noise is the loud voice.
6. Actually noise is the 'inferno' (hell) created by 'bel'.
7. When the sound waves are non-periodic, irregular and of short duration they produce an irritation effect and such a sound is called noise.

Definition of Noise Pollution

Noise pollution may also be defined in a number of ways. For example :

1. Hell created by undesirable sound may be referred to as *noise pollution.*
2. Noise pollution is the unwanted sound dumped into the environment.
3. Any unwanted sound signal that produces displeasing effect and interferes with human comfort, communication and health.

Sources of Noise Pollution

Noise may be either *natural* such as thunder, or *man-made*. The main sources of man-made noise in developed urban areas are scooters, motor bikes, cars, tempos, buses, trucks, tractors, loudspeakers, loud pop music, machines of factories, mills, industries, generators, trains, aircraft, supersonic aircraft, social gatherings etc.

Principal Types of Noise

Broadly, noise may be grouped into three :

(I) Industrial noise (II) Transportation noise (III) Neighbourhood noise

(I) Industrial Noise

High intensity sound or noise pollution caused mainly by the industrial processes has received much attention over the past century. The workers usually experience industrial noise created by the manufacturing units. Hearing impairment in certain occupations is well known (e.g., heavy industrial blowers). The hearing loss is related not only to the intensity of sound but also to the duration of exposure. For example, in the steel industry, the workers hearing the heavy industrial blowers are exposed to 112 dB for eight hours and suffer from the occupational pollution.

(II) Transportation Noise

Transportation noise is more important since it probably has an impact on the general public. Transportation noise could either be due to aircraft or motor vehicles and trains. The number of road vehicles like motors, scooters, buses, trucks, motor bikes, and, particularly, diesel engine vehicles, have increased enormously in the past years. That is why this form of pollution is gaining importance, particularly in large and over-crowded towns and cities.

Transportation noise can be grouped into three :

(a) Aircraft noise (b) Vehicular noise, and (c) Neighbourhood noise.

(a) Aircraft Noise

Aircraft noise is causing much more discomfort than that from road transport noise. Loud noise produced by high speed jet aircraft is not only disturbing human communication, comfort and health, but can also damage hearing permanently.

Whenever a solid body travels at a speed in excess of the speed of sound (330m/1,070 ft per second = 1,220 kph/760 mph), the sonic boom produced can be heard up to 80 km.

Sonic boom is an important aspect of aeroplane noise. The sonic boom occurs when an aircraft like *Concorde* flies supersonically overhead. The sonic boom, besides annoyance, causes damage to property as well as fright to animals. Sometimes it also creates mental anguish.

Fortunately, aircraft noise in India is not a serious problem, probably due to the limited number of flights per day at most of her airports. India has about 85 airports including four international airports — Palam in New Delhi, Meenambakam in Chennai, Dum-Dum in Kolkata and Santa-Cruz in Mumbai. Proper zonal planning keeps people in residential areas safe and undisturbed.

(b) Vehicular Noise

Vehicular noise is probably the most significant of all types of noise. Year by year the general noise levels in towns and cities increase mainly due to the increasing traffic. For example, the average noise level in North America doubles every ten years. The situation in India is not very different. For example, the number of motor vehicles in New Delhi alone has increased from 20,000 in 1950 to over

300,000 in 2001. The number of scooters, motor cycles and tempos increased from 2,500 to over 400,000 in the same period. In U.P., Kanpur tops in noise pollution, but conditions in other industrial towns are fast becoming intolerable. In general, noise pollution varies from place to place.

A survey conducted in metropolitan cities has revealed that noise level in Delhi, Mumbai and Kolkata is as high as 90 dB. Inhabitants of cities are subjected to this most annoying form of vehicular noise, which gradually deafen them.

(III) Neighbourhood Noise

This type of noise includes both i.e., indoor noise as well as outdoor noise. Indoor noise is created in the adjacent room or in the same room where noise is noticed. Examples of indoor noise are crying of babies, TV, transistor, banging of doors, conversation of the occupants etc. The outdoor noise is created from nearby streets. The largest source of outdoor noise is automobile traffic. Other examples are aeroplanes, social gatherings, loudspeaker etc.

In general, neighbourhood includes disturbances from household gadgets and community. Common noisemakers are TV, VCR, radio, transistor, musical instruments, washing machines, vacuum cleaners, mixers, coolers, loudspeakers etc.

It shall not be out of place to mention that Indians are religious minded and their every occasion, function and sentiment is manifested in a noisy manner. Thus the variety of sources of noise (e.g., loudspeaker at place of worship or marriage and birthday parties etc.) may cause disturbance and annoyance to the public.

Measurement of Noise — Unit of Sound

Actually the unit of noise is Bel, but it is comparatively large. In practice the sound intensity is usually measured in 'decibel' (dB) which is equal to 1/10th of Bel. Deci means 10, a reference to the basis of common logarithms and Bel (named after Alexander Graham Bell) and is defined as the ratio of the loudness of one sound to another. One dB is equal to the faintest sound a human ear can hear. The decibel is not an absolute physical measurement like gram, meter or volt but is rather a ratio expressed as the logarithmic scale relative to a reference sound pressure level.

The dB is thus expressed as :

$$\text{Sound Intensity Level} = 10 \log_{10} \frac{\text{Intensity Measured (I)}}{\text{Reference Intensity } (I_0)}$$

$$\text{or dB} = 10 \log_{10} \frac{I}{I_0}$$

Loudness is also expressed in sones.

One sone is equal to the loudness of 40 dB sound pressures at 1,000 hz. It is notable here that at a distance of 7 m away, this sound pressure is 78 dB; and at 14 m it is 75 dB, but at 28 m it is 72 dB.

Human ear is known to be sensitive to extremely wide range intensity from zero to 140 dB. Here zero dB is the threshold of hearing while 140 dB is the threshold of pain. By threshold it implies the lowest intensity at which stimulus gets perceptible. Some people feel discomfort even with sound of 85 dB whereas some do not feel discomfort even with sound of 115 dB. Pain is usually felt at 140 dB. A normal conversation is done at 60 dB levels while noise produced by a jet plane at take-off may exceed 160 dB. The effect on man depends upon the frequency or pitch duration of the sound. The sound pressure level has been regarded to be of greater loudness for higher pitched than for the lower pitched sound.

Sound beyond 80 dB harms hearing system and so it can be safely regarded as pollutant. The loudest noise a man can hear without much discomfort is thus about 80 dB. According to WHO, 45 dB is the safe noise level for a city. A noise of over 90 dB, heard for more than 10 millisecond, causes tympanic membrane to contract, because of aural reflex action. A noise of over 140 dB causes the middle ear muscles to change the direction of movement of ear ossicles, which makes a sudden decrease in the noise intensity reaching the inner ear. Such protective reflex actions, however, exist only for a short duration against loud noise. Thus a noise level up to 65 dB has been considered tolerable for hospital zones in international standards.

Noise levels in the four metropolitan cities of India (New Delhi, Mumbai, Kolkata and Chennai) are generally more than 90 dB.

Experience has revealed that the following environmental noise levels will be satisfactory for most people :

(I)	Close to hospitals, old people's homes etc. 35 dB at night	45 dB during day	Peak noise 55 dB

(II) Residential zones
45 dB at night 55 dB during day Peak noise 70 dB

(III) Commercial areas
60 dB average peak noise 75 dB

(IV) Industrial areas
65 dB average peak noise 80 dB

Various useful situations and their corresponding decibel levels are listed in Table 9.1 :

Table 9.1 : Sound Levels of Different Sources

Sl. No	Sound Level in dB	Sources
1.	Zero	Threshold of hearing
2.	2	Dropping of a pin on the floor
3.	10	Normal breathing
4.	30	A room in a quiet house at night
5.	40	Public library, hospitals, and hotels
6.	40-60	Quiet speech
7.	50-55	Business office
8.	40-60	Normal conversation at 1 meter
9.	60	Busy shop
10.	70	Vacuum cleaner
11.	72-82	Market noise
12.	75	Diesel concrete mixer at 15 m
13.	80	Printing press, alarm clock and marriage procession
14.	85-90	Public meeting
15.	87	Rock drill at 15 m
16.	90	Truck sound
17.	70-100	Loudspeaker and generators
18.	90-100	Food blender at 0.6 m distance
19.	90-95	Motor cycles
20.	100	Large factory and sirens
21.	110	Train whistle at 15 m
22.	120	Boiler factory and thunderclap
23.	130	Aeroplane noise
24.	150	Jet plane's take-off
25.	170	Space rocket launching
26.	180-200	Rocket engine

Hearing threshold is zero dB. In general, the sound may be grouped into :

1. Audible 0 - 10 dB
2. Very quiet 11 - 30 dB
3. Quiet 31 - 50 dB
4. Moderately loud 51 - 75 dB
5. Very loud 76 - 100 dB
6. Uncomfortably loud 101 - 125 dB
7. Painful 126 - 150 dB
8. Instantaneous damage Above 150 dB

The maximum level of noise, which neither irritates the occupants nor damages the acoustics of the building, is known as acceptable noise level inside the building.

Adverse Effects of Noise on Human beings

WHO has recommended 75 dB as the explosive limit for industrial noise. The Bureau of Indian Standards (BIS) has recommended acceptable noise level in an industrial area between 45 dB and 60 dB. The threshold limit universally accepted under the Occupation Safety and Health Act is 90 dB for 9.0 hours, 95 dB for 4.0 hours, 100 dB for 2.0 hours and 115 dB for 15 minutes per day. Above 115 dB, it crosses the threshold of pain.

In general, adverse effects of noise (unwanted sound) on health are physiological, psychological, sociological and psycho-acoustical. Continued exposure of high levels of noise may cause interference, fatigue, and temporary shift of hearing, which may lead to permanent loss of hearing.

To understand hearing loss due to noise, it is important to understand the general structure of ear and how it functions.

The ear consists of the outer, middle and the inner ear. The outer and middle ear convert sound pressures to vibrations. The eardrum is housed in the middle ear (Fig. 9.1). It behaves like the diaphragm on a microphone. Sound waves make the eardrum vibrate. The vibrations of the eardrum set vibrations in a chain of three small bones. The bones are arranged in a lever-like manner so that the vibrations in the third bone (the stapes), are greater than the vibrations set up in the eardrum.

The auditory receptors are housed in the cochlea of the inner ear. This is a coiled tube filled with fluid. The membrane makes the fluid vibrate. Inside it are cells which are sensitive to vibrations. They respond to the vibrations by setting up tiny electrical signals, which travel along the auditory nerve to the brain. The brain interprets these signals as sound.

The ear of young, audiometrically healthy adult male respond to sound waves in the frequency range of 20-16,000 Hz. Young children and women often

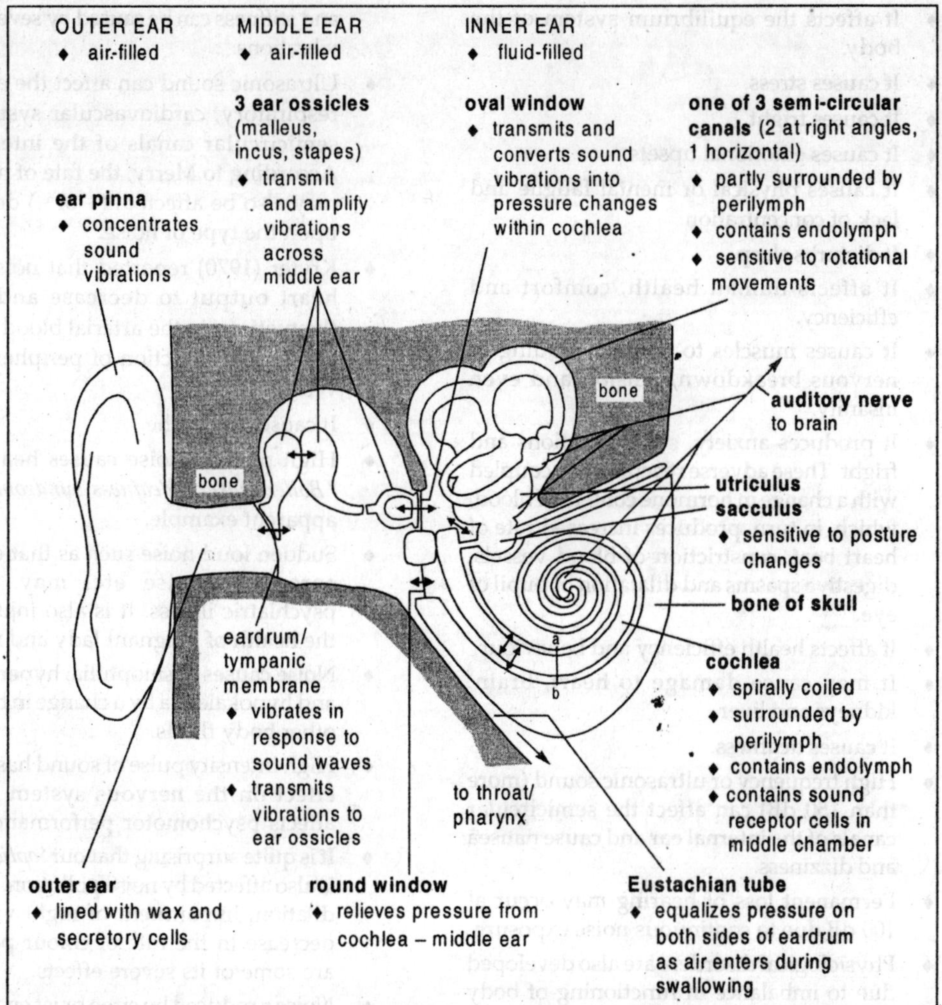

Fig. 9.1 : Vertical Section ear of human

have the capacity to respond to frequencies up to 20,000 Hz. The speech zone lies in the frequency range of 500-2,000 Hz. The ear is most sensitive in the frequency range from 2,000-5,000 Hz.

Sound beyond 80 dB actually harms the hearing and so can be safely regarded as a pollutant. A man can hear at about 75-80 dB without much discomfort. According to WHO, 45 dB is the safe noise level for a city. A noise of over 90 dB heard for more than 10 milliseconds causes the middle ear muscles to change the direction of movements of ear ossicles, which leads to a sudden decrease in the noise intensity reaching the inner ear. Such protective reflex action, however, exists only for a short duration against loud noise. Thus a noise level up to 65 dB has been considered tolerable for hospital zones internationally. The outer and middle ears ars rarely damaged by intense noise. However, explosive sounds can rupture the tympanic membrane or dislocate the ossicular chain. The permanent hearing loss that results from very brief exposure to a very loud noise is termed *acoustic trauma.* Damage to the outer and middle ear may or may not accompany *acoustic trauma.*

The most immediate and acute effect of noise pollution is impairment of hearing. A sudden loud noise can cause severe damage to eardrum. Long exposure to loud noise can cause some hearing loss that may become permanent.

Overall effects of intense noise are :

- It interferes with the communication system.

- It affects the equilibrium system of the body.
- It causes stress.
- It causes fright.
- It causes emotional upsets.
- It causes physical or mental fatigue and lack of concentration.
- It disturbs sleep.
- It affects human health, comfort and efficiency.
- It causes muscles to contract leading to nervous breakdown, tension and even insanity.
- It produces anxiety, stress reactions and fright. These adverse reactions are coupled with a change in hormone content of blood, which, in turn, produces increased rate of heart beat, constriction of blood vessels, digestive spasms and dilatation of pupil of eye.
- It affects health efficiency and behaviour.
- It may cause damage to heart, brain, kidneys and liver.
- It causes deafness.
- High frequency or ultrasonic sound (more than 150 dB) can affect the semicircular canals of the internal ear and cause nausea and dizziness.
- Permanent loss of hearing may occur at 100 dB due to continuous noise exposure.
- Physiological disorders are also developed due to imbalance in functioning of body systems and due to continuous exposure to noise. These are neurosis, insomnia, hypertension, increase in sweating, hepatic diseases, giddiness, peptic ulcers, undesirable change in gastro-intestinal activities, behavioural and emotional stresses.
- Low frequency noise of 50-60 dB affects the higher centres of brain and causes an alteration in the normal sleep pattern and prevents sound sleep.
- Constant noise and physical vibrations emitting from handheld tools may cause severe effects often described as *white fingers*, dead hands or Reynaud's phenomenon, or pneumatic drill disease. Pain, cyanosis (blue colouration) and numbness of finger result from moderate vibrations while damage to bones and joints in the hands while swelling and stiffness can be caused by severely high vibrations.
- Ultrasonic sound can affect the digestive, respiratory, cardiovascular systems and semicircular canals of the internal ear. According to Merry, the rate of heart beat may also be affected, (+ or –) depending upon the type of noise.
- Kryter (1970) reported that noise causes heart output to decrease and creates fluctuations in the arterial blood pressure and vasoconstriction of peripheral blood vessels.
- It causes headache.
- High pitched noise causes hearing loss "*Boiler-maker's deafness syndrome*" is the apparent example.
- Sudden loud noise such as that of jet and aeroplane noise etc. may produce psychiatric illness. It is also injurious for the health of pregnant lady and fetus.
- Noise causes eosinophilia, hyperglycemia and hypokalemia by a change in blood and other body fluids.
- High intensity pulse of sound has its worst effect on the nervous system and also affects psychomotor performance.
- It is quite surprising that our '*optical system*' is also affected by noise pollution. Pupillary dilation, impairment of night vision and decrease in the rate of colour perception are some of its severe effects.
- Noise produced by siren or jet engines may cause thymus gland atrophy.
- Noise also causes irritation, disinterest and affects work performance. It may accelerate or diminish work efficiency, depending upon its intensity, duration and frequency distribution etc.
- Noxious noise has been known to cause nervous disorder, headache, high B.P. and short memory.
- Effects of noise on the fetus are not fully known. However, medical scientists have found that an unborn child moves and kicks about when there is loud noise.
- Noise not only affects the biological system but also affects adversely non-living things. For example, buildings undergo physical damage by cracks, broken windows, doors and glasses etc. by sudden explosive sounds.

Overall salient effects of noise on human beings is summarized in the flow chart and Table 9.2 :

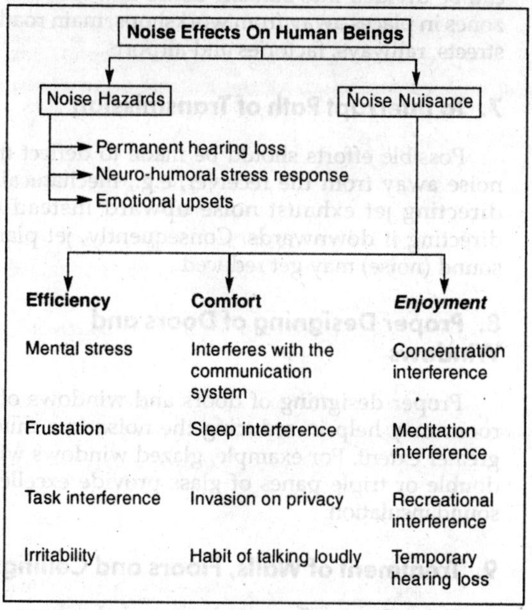

Table 9.2 : Effects of High Intensity Noise on Human Beings

Sl. No.	Noise (dB)	Observed Effects
1.	above 80	Harms hearing system i.e., hearing damage begins
2.	105	Significant change in pulse rate
3.	110	Skin receptor stimulation
4.	120	Pain threshold
5.	130-135	Nausea, vomiting, dizziness
6.	140	Increased ear trouble with pain
7.	140-145	Extreme limit of human noise tolerance
8.	150	Prolonged exposure causing burning of the skin
9.	160	Deafness and minor permanent damage if prolonged
10.	190	Major permanent damage in a short time. Suddenly evoked noise may bring death through cardiac failure.

Steady noises do not interfere with human performances unless the noise level exceeds 90 dB. Sound levels of irregular peaks — even when below 90 dB — interfere with the performance.

Effects of noise on human health depends upon its intensely as well as exposure duration. Permissible noise exposures at their specific noise intensities have been tabulated in Table 9.3 :

Table 9.3 : Permissible Noise Exposures (if unavoidable)

Sl. No.	Duration per day-hrs.	Sound level in dB
1.	8	90
2.	6	92
3.	4	95
4.	3	97
5.	2	100
6.	1½	102
7.	1	105
8.	½	110
9.	¼	115

The sounds that are generated and transmitted in air directly to human ears are called *air-borne sounds*. They possess less power, continues for long duration and is confined to places near the origin. The sounds that originate and progress on the buildings structure are known as *structure-borne sounds* or *impact sounds*. They are powerful, propagate over long distances and persist for a very short duration. Such noise is developed in solid structures and is then transmitted as *air-borne noise*.

Ambient Noise Standard

Ambient air quality standard in respect of noise in India, as given in Schedule III of the Environment (Protection) Rule, 1986, is given in Table 9.4. For industrial workers, permissible noise exposure is 90 dB for 8hr/day, 00 dB for 1hr/day and may go up to 114 dB for 2 min/day :

Table 9.4 : Ambient Sound Level Standard in India

Area	Sound level in dB	
	Day (6 am to 9 pm)	Night (9 pm to 6 am)
Industrial area	75	70
Commercial area	65	55
Residential area	55	45
Silence zone	50	40

Prevention and Control of Noise Pollution

According to a WHO report to the UN Conference on Environment : *"of all environmental problems noise is the easiest to control"*. But the question of control arises only after an awareness has been developed among the people that the noise is a public nuisance and a danger to physical and mental health and, therefore, its control is necessary.

Following measures may be taken for an effective control of noise pollution :

1. Using Ear Protective Aids

The workers working in factories having noisy machinery should be provided with the ear protection aids such as soft plastic and rubber ear-plugs, headphones etc.

2. Improving working methods

The best method of controlling noise pollution is to reduce or suppress noise at the source itself. Procedures should be planned in such a manner that it creates minimum possible noise (40-60 dB). For example, welding should be given preference to riveting. The noisy machines should be provided with such construction that they create minimum possible noise. This can be done by enclosing the noisy machines in a box-like structure having sound-absorbing materials on its surface.

3. Providing Enclosures

Providing enclosures, shields and barriers proves to be an effective and efficient method of reducing noise pollution because these can cut off the sound waves from propagating. The enclosures can be constructed from most common materials. Lead materials are good for controlling noise pollution.

4. To form Silent Zones

Prescribing noise limits for vehicular traffic can reduce noise. Formation of silent zones — especially near hospitals and schools — would be an effective measure.

5. To Plant Noise-Absorbing Trees

It has been found that plantation of trees like Neem, Coconut, Ashok, Tamarind etc. near schools, hospitals, public offices, libraries etc. can reduce the noise to the extent of about 6-10 dB.

6. Town Planning

A rational town planning can help in reducing outdoor noise pollution from factories, road traffic, heavy machinery, aeroplanes, railways etc. because vibrations from these external sources create what is known as *structure-borne* sound. The town or city can be divided into suitable zones and residential zones in places away from workshops, main roads, streets, railways, factories and airports.

7. To Interrupt Path of Transmission

Possible efforts should be made to deflect the noise away from the receiver, e.g., mechanically directing jet exhaust noise upward instead of directing it downwards. Consequently, jet plane sound (noise) may get reduced.

8. Proper Designing of Doors and Windows

Proper designing of doors and windows of a room may help in reducing the noise to a much greater extent. For example, glazed windows with double or triple panes of glass provide excellent sound insulation.

9. Treatment of Walls, Floors and Ceilings

Using floating floors and suspended ceilings can also control noise pollution. Suitable sound-absorbing materials such as acoustical tiles, perforated plywood and various porous materials are available to be fixed on walls, floors and ceilings in order to minimize the noise level.

10. To Create Awareness in Public

Educating the masses through radio, TV, newspaper and newsreel in cinema halls about the adverse effects of noise pollution will go a long way in creating awareness among the people of the need to control noise pollution.

11. To Impose Legal Restrictions

Legislative measure, especially taken during the marriage and festival functions, can prevent noise pollution.

Central Pollution Control Board (CPCB) have conducted noise pollution surveys in Delhi, Chennai, Kolkata, Mumbai, Bangalore, Hyderabad, Jaipur and Kanpur in 1989 and found that the noise levels in all these eight big cities in residential, commercial and silence zones are much above the prescribed standards — both during day as well as night.

The Govt. of India prescribed (1991-1992) noise limits for automobiles, domestic appliances and construction equipment to be adopted at the manufacturing stage.

Under the Environmental Protection Act, 1986, ambient air quality standard in respect of noise of different categories of area have been notified. Codes of practice for controlling noise from sources other than automobiles and industries have also been evolved by CPCB.

Summary For Controlling Noise Pollution

General Misconceptions

Lound sound is not dangerous, as long as you don't feel any pain in your ears.

Not true : Our threshold for pain is at about 120-140 dB, but sound begins to damage our hearing when it is above 85 dB (for an 8 hour period).

Hearing loss after sound exposure is temporary.

Not true : Some of the hearing loss will be permanent. Indication of damage is ringing and noise in the ears (called tinnitus) after sound exposure. This is a clear indication that sound exposure took place. Another indication is the difficulty to communicate on the phone and in noisy restaurants or cafetarias.

If you have a hearing loss already, you don't have to protect your hearing any more.

Not true : Hearing loss accumulates. More exposure to loud sounds leads to more hearing loss.

Hearing loss is mostly caused by ageing.

Not true : Research shows that accumulative exposure to loud sounds, not age, is the major cause of hearing loss.

Hearing loss can be repaired by medicine, surgery or hearing aids.

Not true : Certain improvements can be obtained by the use of hearing aids, however, in the case of hearing losses inflicted due to noise exposure, the resulting quality of hearing will be far from normal. So far no drug or therapy can correct noise-induced hearing loss.

Loud sound damages only your hearing.

Not true : Loud sound can change your heart rate, vision and reaction time. It may make you more aggressive and, in general, negatively affect you.

Sound systems have built-in safety features.

Not true : Most sound systems are tremendously powerful and are capable of producing sounds much louder than adjusted levels. However, these systems have no built-in protection against surges in sound due to feedback or accidents.

Revision Questions

A. Long-Answer Questions

1. Write an essay on noise pollution.
2. Define 'noise'. How is it controlled?
3. Define noise pollution. What are its adverse effects?

B. Short-Answer Questions

1. *Define noise and noise pollution.*
2. *What are the following? Describe in 50-60 words :*
   (a) Industrial noise
   (b) Transport noise
   (c) Neighbourhood noise.
3. *What is the difference between noise and air pollution?*
4. *Write short notes on :*
   (i) Decibel
   (ii) Sound level of 10 important sources
   (iii) Adverse effects of noise
   (iv) Ambient noise standards
   (v) Noise-absorbing plants.

10

Automobile Emissions

SYNOPSIS

Introduction, Kinds and sources of pollutants emitted by automobiles, Effect of important automobile pollutants on human health, Photochemical pollution, Control of vehicular pollution.

Introduction

Alarming level of pollution has shrouded the cities due to motor vehicles. In big cities of India like Delhi, Mumbai, Kolkata, Kanpur, Chennai we have now an automobile culture and we find roads and nearby places full of exhaust toxic gases. Out of the automobiles the major pollution is done by two-wheelers and three-wheelers which burn a mixture of petrol and oil and emit about 9-10 times more harmful gases than cars. Besides the emission of harmful gases, automobiles also increase the problem of noise pollution. As vehicular air pollution in this country is not monitored efficiently, it is difficult to estimate precisely the extent to which they contribute to the pollutants in the air. However, it is believed that in three metros in our country, vehicles-contribution towards air pollution is as follows :

Delhi = 64%

Mumbai = 52%

Kolkata = 30%

A study indicates that 97% of HC and 76% of CO emissions in Delhi are from vehicles running on petrol. About 70% of HC and 48% of CO in India come from vehicles having 2-stroke engines. In general, most two-wheelers and three-wheelers run on 2-stroke petrol engines. In spite of the advantages of 2-stroke engines (light, easy to manufacture, less cost, high specific output, easy cold start, and easy to maintain), the 2-stroke engines are considered to be emission-inefficient. However, NO_x emission in 2-stroke engines is only 1/3 as compared to 1/4 in 4-stroke engines. The present chapter, however, deals with the types of pollutant emitted by automobiles and their deleterious effects, particularly on human health.

Kinds and Sources of Pollutants Emitted by Automobiles

The composition and nature of air-borne pollutants produced by automobile exhaust has been reviewed by Sherwood and Bowers (1970). Although the exhaust gas, the chief culprit, pollutes the air but evaporative losses from the fuel tank and carburetor and also losses from the crank case account for a significant proportion of the hydrocarbons emitted (Fig. 10.1). If the petrol is completely oxidized in the environment, only the H_2O and CO_2 will be produced in an internal combustion engine, which can hardly be considered as pollutant, although there is considerable anxiety about the build-up of CO_2 in the environment and its possible effect on the climate. In practice, it is rather difficult to achieve 100% oxidation and, thus, CO is formed in considerable quantity.

Besides, some of the fuel remains unchanged and some is converted into other organic components. Apart from these products of incomplete or partial combustion most petrols contain lead and lead compounds which are also present in the exhaust. Further, the conditions in the combustion chamber favour the oxidation of the nitrogen in the air so that oxides of nitrogen are also formed in the engine. All these chemical products are expelled as exhaust gas. It is observed that the diesel exhausts also contain some carcinogens like benzopyrine.

The important pollutants present in the exhaust gas produced by automobiles are :

- Water vapour
- CO_2
- Unburnt petrol

→ Not considered to be pollutants

- CO
- Nitrogen oxides
- Lead compounds
- Sulphur oxides
- Carbon particles (Smoke)
- Organic compounds

→ Considered to be serious pollutants

The world average of the transport sector's share in polluted air is :

- ♦ CO → 85%
- ♦ NO_x → 47%
- ♦ HC → 45%
- ♦ SO_x → 3%
- ♦ CO_2 → 20%

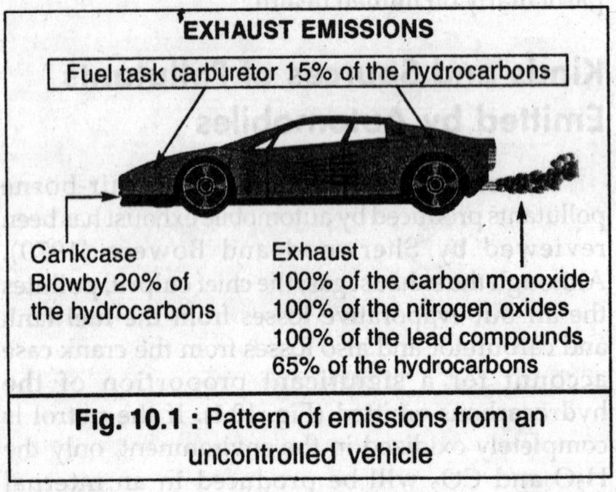

Fig. 10.1 : Pattern of emissions from an uncontrolled vehicle

It shall not be out of place to mention that under certain circumstances the products emitted by automobiles can react with each other to produce unpleasant secondary products. This frequently occurs.

Although petrol engines and diesel engines both produce similar products in their exhausts, the relative proportions present are very different — as shown in Table 10.1.

The exhaust of diesel engines contains significantly lower concentration of pollutants than the exhaust gas from petrol engines. However, to offset this, diesel engines are very liable to emit smoke and a nauseating smell if they are incorrectly operated and/or not properly maintained. Under such conditions the amount of CO and hydrocarbons produced is also increased; the CO content may then rise to 2,000 ppm. Diesel, high in sulphur content, is also responsible for high SO_2 concentration in traffic junctions of big cities.

From the above discussion, it can be concluded that another major source of air-borne pollutants, through the combustion processes, is the automobile. Pollution occurs each time a motorist presses his brake pedal. Thus vehicles on road serve as mobile sources of air pollution, contributing such pollutants as CO, nitrogen dioxide, smoke, organic vapours and incompletely burnt hydrocarbons (Table 10.2). Their contribution to air pollution is minimum, and the combustion process is nearly complete, when vehicles are driven at constant speeds, with minimum stops and start. Of course, other factors such as the engine design and the type of gasoline used also determine the quantity and profile of pollutants emitted by a car.

Table 10.1 : Comparative Composition of Exhaust Gases in Maintained Vehicles (concentrations in ppm by volume)

Type of engine	*Pollutant*	*Idling*	*Acceleration*	*Cruising*	*Deceleration*
Petrol Engines	CO	69,000	29,000	27,000	39,000
	Hydrocarbons	5,300	1,600	1,000	10,000
	Nitrogen oxides	30	1,020	650	20
	Aldehydes	30	20	10	290
Diesel Engines	CO	Trace	1,000	Trace	Trace
	Hydrocarbons	400	200	100	300
	Nitrogen oxides	60	350	240	30
	Aldehydes	10	20	10	30

Table 10.2 : Average Emission Factors of Pollution for Automobile Exhausts

Types of emission	*Kgs/1,000 liters of gasoline used*
Carbon monoxide	300.00
Hydrocarbons	25.00
Oxides of nitrogen	14.00
Particulates	1.50
Oxides of sulphur	1.00
Aldehydes	0.50
Organic acids (acetic)	0.50
Benzo(a)pyrene*	0.06

Nickel may be emitted from the Ni-added to gasoline and Ni-containing parts of automobiles. Zinc and cadmium come from the lubricating motor oils, tires and galvanized parts of the vehicles. Cadmium concentrations range from 0.07 to 0.10 ppm in diesel oils and 0.20 - 0.26 ppm in lubricating oils; cadmium in automobile tires ranges from 20 to 90 ppm. It is amazing to know of the various chemicals and materials involved in the manufacture of automobiles and keeping them on road. When discarded, several of them contaminate the environment (Table 10.3) :

Table 10.3 : Automobile Parts and Pollution Problems associated with them

Parts	*Problems*
Battery	Contains lead and HCl
Bumper	Wastes include cyanide, chromium and other heavy metals
Brake shoes	Contains asbestos
Engine	Waste per tonne of castings, 0.3 tonnes — mainly slags with some toxic contaminants
Exhaust	Contains several air pollutants : 63% lead, 20% NO_2, 23% hydrocarbons and 45% CO.
Seat textiles	Wastes include dyes, acids, solvents, greases and waxes
Gasoline tank	Serves as a source of benzene and hydrocarbon emission during fuelling
Plastic components	Toxic chemical used in the production include vinyl chloride, formaldehyde, phenols
Tyres	Toxic chemicals used in production include amines, thiurams, nitrosamines and solvents

Bond and Straub (1972) reported that diesel vehicles are important sources of nitrogen oxides, hydrocarbons and particulates (Table 10.4). Regulations proposed in the early eighties called for a reduction in particulate emission from 0.35 g/km to 0.12 g/km, and restricted the nitrogen oxide emissions from 1.2 g/km to 0.6 g/km. It had been a problem to achieve the simultaneous control of particulate and NO_x emissions efficiently.

*gram/1,000 liters

Table 10.4 : Emission Factors for Diesel Engines

Types of emission	*Emission levels* (Kg/1,000 liters of *diesel)*
Oxides of nitrogen	27
Hydrocarbons	17
Particulate matter	14
Carbon monoxide	7
Oxides of sulphur	5
Organic acids	4
Aldehydes	1
Benzo(a)pyrene*	0.1

Excessive and visible smoke is always inexcusable and is the result of bad engine maintenance. Odour, however, is a more difficult problem. No single component is responsible for exhaust odour, and complex synergisms undoubtedly occur.

Hydrocarbons are undoubtedly responsible for a considerable part of the odour, and oxygen-containing materials may also contribute. Compounds of sulphur and nitrogen may also contribute to odour although the sulphur contents of exhaust are generally so low that they are present but not detected. The concentration at which a material in the atmosphere is detected varies greatly from compound to compound — a list of these 'odour thresholds' for some materials is given in Table 10.5 :

Table 10.5 : Odour Thresholds of Possible Exhaust Components

Material	*Odour* threshold *level (ppm)*
Hydrocarbons	
Benzene	4.700
Cyclohexane	300.000
Styrene	0.050
Toluene	2.100
0-Xylene	0.200
p-Xylene	0.500
Oxygen-containing compounds	
Acetaldehyde	0.200
Acetone	100.000
Acrolein	0.200
p-Cresol	0.001
Formaldehyde	1.000
Phenol	0.050
Sulphur-containing compounds	
Dimethylsulphide	0.001

Effects of Important Pollutants Produced By Automobiles on Human Health

1. Unburnt Fuel and Secondary Products Produced from Fuel

Unburnt fuel is emitted into the environment by evaporation losses from the fuel tank and carburetor, a high proportion of the hydrocarbons in the crank-case blown and exhaust gases also contain unburnt fuel. In fact, the constituents of petrol are not regarded to be deleterious but few of them have very slight anesthetic effects in fairly high concentration.

The gaseous products released from the fuel and emitted through exhaust gases contains over 100 compounds. Most of these are hydrocarbons, acetylenes, olefins, paraffins, aromatic hydrocarbons and aldehydes. These components produce irritant action on the eyes, mucus membrane and respiratory system. Apart from the gaseous products, various polynuclear aromatic compounds are also emitted with the exhaust gas as very fine particles and these may persist in the air for long periods. Such particles are injurious to health and are known to be carcinogenic, e.g., benzopyrine.

Benzene is a constituent of crude oil; when unleaded fuel is produced, benzene is added or maintained at high levels to enhance the fuel's anti-knock property. Benzene is highly soluble in fat, and settles conveniently in fat-rich adipose tissues and bone marrow. Even a small amount of benzene can cause cancer and leukemia. On the other hand, PAHs are quickly absorbed into the gut and lungs — and thus increase the risk of being affected with bladder cancer — among truck drivers and delivery personnel by 50%. PAHs are formed during incomplete combustion of fuel; exhaust of diesel engines particularly contains high concentration of PAHs.

2. Carbon Monoxide (CO)

It is deemed as a serious toxic pollutant emitted from the engines. It has a tendency to combine with hemoglobin in the blood to form carboxy-haemoglobin (COHb) :

$$CO + Hb \rightarrow COHb$$

As a matter of fact, CO has a greater affinity for hemoglobin than O_2 and is easily absorbed even in small concentration.

Table 10.6 : Signs and Symptoms at Various Concentrations of Carboxyhemoglobin

Percent COHb	*Signs and symptoms for normal man*
0 - 10	No signs or symptoms (toxicity zero)
10 - 20	Slight headache; dilation of the cutaneous blood vessels; tightness across the forehead
20 - 30	Headache and throbbing
30 - 40	Weakness; severe headache; dizziness; dimness of vision; nausea; vomiting.
40 - 50	Syncope and increased pulse and respiratory rate; greater possibility of collapse and others as above i.e., (30 - 40%)
50 -60	Syncope; increased pulse rate; coma; intermittent convulsions, and Cheyne-Stokes respiration.
60 - 70	Coma; intermittent convulsions; depressed heart function and respiratory rate, and possibly death.
70 - 80	Weak pulse; very slow respirations; respiratory failure and death within few (4-6) hours.
80 -90	Death in less than an hour
90 +	Death within few minutes

The deleterious effects of CO is measured by the percentage of COHb formed in the blood. Table 10.6 shows the signs and symptoms at different concentrations of carboxyhemoglobin.

3. Nitrogen Oxides

Both nitric oxides (NO) and nitrogen dioxide (NO_2) are produced in the automobile exhausts. NO is produced in much larger quantity. However, NO_2 is produced in less quantity (about half of NO), but is relatively more toxic. The maximum allowable concentration for industrial exposures of 8 hours for NO is 25.0 ppm and NO_2 5.0 ppm. Normal level in city street is under 1 percent of these values. Nitrogen dioxide finally ends up as HNO_3, nitrates, or organic nitrogen compounds (under conditions of photochemical smog) :

$$3NO + HOH \rightarrow 2HNO_3 + NO$$

In the stratosphere (12-50) km above earth, NO_2 also reacts with hydroxyl radical to from HNO_3 :

$$HO^- + NO_2 \rightarrow HNO_3$$

Little is known about the deleterious effects of oxides of nitrogen on human health. However, their effects are adverse and permanent than the CO.

Dogs and rabbits exposed to a concentration of 0.5 ppm for 90 days gained weight less than normal, and monkeys, when similarly exposed, lost weight.

Mice exposed at this concentration for 3, 6 and 9 months and then exposed to *Klebsiella pneumoniae* had higher mortality rates than the control groups.

4. Lead Compounds

It has been reported that some three thousand tons of lead are emitted with the exhaust gases of automobiles, especially cars, in UK each year. This has been found to accumulate in the vegetations and soil along roadside verges. Lead could possibly enter food chains and have damaging effects far from the source of pollution.

In humans, lead enters the body either through the mouth, with things eaten or drunk, or breathed in through the nose. Far more lead is likely to be ingested than inhaled, but ingested lead is less absorbed than that which is inhaled. Poisoning by lead at high concentration has been known for a very long time.

The most common form of lead poisoning seen today is the disturbance of the gastro-intestinal system known as *lead colic*. Its symptoms are excessive tiredness; continued headaches; loss of appetite, muscular pains and nausea. These symptoms are produced only when the lead level in the blood increases above 80 mg per 100 ml of blood. As most lead in ambient air is in the form of fine particles, children are the worst sufferers of lead poisoning.

5. Smoke

The smoke emitted by automobiles consists mainly of very fine particles of carbon formed due to incomplete combustion of fuel. Smoke formation is actually related with diesel engines. Carbon particles may act as nuclei both for haze formation and in the absorption of gases e.g., CO_2 and nitrogen oxides. It has been observed that where this occurs, it is likely to cause damage to lungs.

6. Other Particulate Matters

In addition to the finely divided carbon and aromatic compounds in the exhaust, road traffic is also responsible for finely divided rubber from the tyres and for asbestos dust from the brake-linings and clutch-plates. Prolonged exposure to asbestos dust gives rise to *asbestosis* and some varieties are carcinogenic. However, the proportion produced from road vehicles is small and there is no evidence that it is a hazard to human health.

Photochemical Pollution

'Smog', as experienced in Los Angeles, is a classical example of photochemical pollution.

As initiated by sunlight, the series of atmospheric reactions between hydrocarbons and oxides of nitrogen, which lead to the formation of new products, is a most complex system. Among these products are substances termed 'oxidants' which are chemical entities detrimental to biological systems and destructive to certain materials.

In the original sense, the term 'oxidation' represents chemical reactions in which certain atoms combine with oxygen to form carbon dioxide — a typical example of an oxidation reaction.

Presently, the term 'oxidation' is more usually applied to describe the loss of one or more electrons by an atom, ion, or molecule. Conversely, a gain of electrons is referred to as *'reduction'*. Oxidation-reduction reactions cannot be separated since, in any reaction, one of the reactants is oxidized while the other is reduced. For example, in the combustion of carbon in air, carbon is oxidized while oxygen is reduced.

The combustion of coal and petroleum products such as natural gas, gasoline, and fuel oil — termed 'fossil fuels' — is an oxidation-reduction reaction which is responsible for most of the air pollution in urban atmospheres. Fossil fuels are composed principally of hydrocarbons (RH or HC) which, upon complete combustion, produce two oxides; carbon dioxide and water. These two oxides are not considered to be air pollutants, since they are relatively non-toxic and are normal constituents of the atmosphere.

The other byproducts of combustion in exhaust gases and the products of incomplete combustion lead to localized air pollution problems. Since combustion processes are normally less than 100 percent efficient, the exhaust gases contain minor amounts of the original fuel as well as minor amounts of partially oxidized fuel. Carbon monoxide, aldehydes, and unsaturated hydrocarbons (those which contain less than the maximum number of hydrogen atoms as a result of the presence of double or triple bonds) are examples of partially oxidized fuel. Small amounts of nitrogen (a major constituent of air) are oxidized — at the high temperatures characteristic of combustion processes — to nitric oxide; sulphur compounds — also found in varying quantities in fossil fuels — are oxidized to sulphur dioxide. Therefore, the combustion processes produce emission which pollute the air with oxides of carbon, nitrogen, and sulphur; also a large variety of hydrocarbons and partially oxidized hydrocarbon fragments.

When these contaminants have been released into the atmosphere, they may react chemically to produce other contaminants quite different from those originally released. Sunlight-induced oxidation processes, termed photo-oxidation, are especially important in some community air pollution problems. During day, for example, nitric oxide in polluted atmospheres is rapidly oxidized to nitrogen dioxide. Similarly, the oxidation of sulphur dioxide to sulphur trioxide is accelerated and olefins and alkylbenzenes are oxidized to form aldehydes and ketones. Formation of ozone is also observed, as well as formation of a family of compounds identified as peroxyacetyl nitrates ($RCOONO_2$). Researches suggest the presence, in such atmospheric mixture, of peroxybenzoyl nitrate, hydrogen peroxide and alkyl hydroperoxides.

Various pollutants formed during the photochemical reaction process are termed oxidants — atmospheric substances which oxidize certain reagents not readily oxidized by oxygen. Products of the photochemical air pollution process, these substances are often termed photochemical oxidants. For detail about photochemical smog, refer Chapter 12.

Control of Vehicular Pollution

Catalytic converters are recommended for the reduction of vehicular emissions. Such a converter is placed near the exhaust pipe in cars to chemically convert engine emission into environmentally benign gases. Three-way converters promote reactions which oxidize HC and CO and simultaneously reduce NO_x emission.

Such converters work best when a chemically correct mixture or fuel and air is consistently maintained by the introduction of a closed loop air-fuel ratio control system. However, no Indian manufacture, at present, uses this system, and converters become non-effective during the start up, i.e., when the emissions are maximum. Catalytic converters also require unleaded fuel.

It should be appreciated that the existing vehicular pollution is due to combined effect of :

- Bad vehicular technology
- Poor fuel quality (accounts for 30% of the problem of vehicular emission)
- Poor vehicular maintenance, and
- Non-existence of traffic planning.

To control automobile pollution vehicles need engine redesign. Emission efficiency of the vehicles should be increased. But quality of the fuel that are being supplied by the state-owned refineries has also to be improved. Crude oils, high in sulphur content, are usually cheaper, and are imported by the refineries. This results in an increase in the sulphur and carbon content in the fuel. Improvement in fuel quality would result in immediate improvement in air quality.

Experts suggest that maintenance alone can reduce emission by 40%. But only efficient traffic management, as practiced in many large cities elsewhere, can reduce the vehicular air pollution in India metro cities. The only way out of jammed road, slowly moving cars and increased emission seems to be a tough traffic management. Introduction of a Mass Rapid Transport System like Metro Rail or Suburban Rail System, control of entry of vehicles on busy roads or crowded areas, banning of old and polluting vehicles from city roads, and withdrawal of subsidy from diesel could be some of the strategies for a cleaner environment.

Revision Questions

A. Long-Answer Questions

1. Write an essay on automobile emissions.
2. Name the important pollutants emitted from automobiles and give their *effects on human health.*
3. Write an essay on photochemical pollution.

B. Short-Answer Questions

1. *Write short notes on :*
   (i) Kinds and sources of pollutants emitted by automobiles
   (ii) Photochemical pollution
   (iii) Automobile and human health
   (iv) Control of vehicular pollution.

C. Briefly describe :

(i) Emission factors for diesel engine
(ii) Contaminants of discarded automobile
(iii) CO as an exhaust toxicant
(iv) NO as an exhaust toxicant
(v) Pb as an exhaust toxicant.

11

Environmental Carcinogens

SYNOPSIS

Introduction, Characteristics of cancer, Salient Properties of cancer cells, Tumour progression, Theories for origin of cancer, Sites of infection, Effects of cancer, Types of cancer, Causes of cancer, Environmental carcinogens, Types of carcinogens on the basis of chemical nature and on the basis of mode of action, Mechanism of genotoxic and non-genotoxic carcinogens, Hormonal carcinogenesis, General mechanism of carcinogenicity, Carcinogenicity tests, Diagnosis, Treatment of cancer, Oncogenes.

Introduction

In simple words, any component of the environment which causes cancer may be termed environmental carcinogen.

A cancer is an abnormal overgrowth or tumour or enlargement of tissue. It is commonly called disease of growth. It has no coordination with the normal tissue. It is an independent, autonomous, uncontrolled growth of a tissue containing a mass of aberrant or abnormal cells. In general, cancer is a collection of multistep, and multifactorial diseases. A tumour serves no useful purpose but is harmful as it grows at the expense of the host like a parasite. Cancers now cause the second largest number of deaths in most countries.

Characteristics of Cancer and Cancer Cells

1. A tumour originates from an existing tissue or cells of the body.
2. The growth of tumour is autonomous i.e., it follows its own laws of growth; it is not regulated by those governing the tissues from which or in which it grows.
3. The tumour cells are undifferentiated and anaplastic in nature. 'Anaplasia' refers to the reversion of differentiated cells into undifferentiated or embryonic cells.
4. They possess a greater potentiality for growth and multiplication.
5. They carry out none of the functions of normal adult cells.
6. They have large and irregular nuclei.
7. They lose their contact inhibition.
8. They are irregular in shape.
9. They invade local tissues.

Salient Properties of Cancer Cells

1. **Contact inhibition** : Normal cells stop movement and division when they make contact with each other. This phenomenon is referred to as contact inhibitions. Cancer cells reveal no contact inhibitions. So when a cell loses contact inhibitions, it becomes cancerous.

2. **Nucleus and Nucleolus** : The nucleus of cancer cells is larger in size and irregular in shape. The nucleolus is also large in size and reveals hypertrophy.

3. **Chromosomal abnormality** : Most of the cancer cells are characterized by chromosomal abnormality. This is actually due to abnormal mitosis.

4. **Centrioles** : The misbehaviour of centrioles is an important feature of cancer cells. In any non-dividing somatic cells the two centrioles are close to each other and are regularly disposed so that their long axis forms a right angle. In cancer cells the two centrioles are separated by long random distances; they are randomly oriented towards each other; they also do not have any connecting fine structures.

5. **Plasma membrane** : The most visible property of cancerous cells is their invasiveness of other tissues; it is primarily a property of plasma membrane. It becomes leaky.

6. **Growth promoting factors** : The cancer cells release an excessive amount of **growth promoting factors**. They escape by leakage through the plasma membrane. After they have leaked out, they act on the surfaces of the neighbouring cells. The result is a dramatic spurt in growth and cell multiplication in the normal cells, which also become abnormal in shape.

7. **Fast glycolysis** : Cancer cells possess fast glycolytic property, thereby resulting in much acidity in the internal environment. Indirectly, acidity induces tumour formation.

Tumour Progression

The development of tumour is a progressive process involving several phases, of which three are important. They are *initiation, promotion* and *progression* as shown in Fig. 11.1 :

1. **Initiation** : It is the beginning of a cancerous growth where a normal cell begins to transform itself into cancer cell. The initiation is brought about by carcinogens, viruses, or gene mutation. It depends primarily on a specific pattern of damage to those genes which are involved in cellular homeostasis. Initiation evokes no distinct change and incites no cell-growth and proliferation.

2. **Promotion** : In this stage the change from dormancy to progressive growth occurs. The cell is stimulated to divide. It requires any simple stimulus for cell division. Continuous tissue-repair resulting from chronic initiation and the presence of initiation agents (carcinogens) promote malignancy. Tissues dependent on hormones, such as mammary glands, are especially vulnerable. Unlike initiation, promotion may be reversible.

3. **Progression** : It is the final stage in the development of cancer. It leads to faster and more independent growth. It is a wholly new event, resulting in the appearance of a new type of cancer cell distinct from the one which gave rise to it. Stable, irreversible, hereditable changes occur, during which the original histological feature of the tissue becomes progressively lost. This is most evident when the primary tumour grows by metastatic proliferation.

Theories for Origin of Cancer

Various theories have been proposed regarding the origin of tumours. These are :

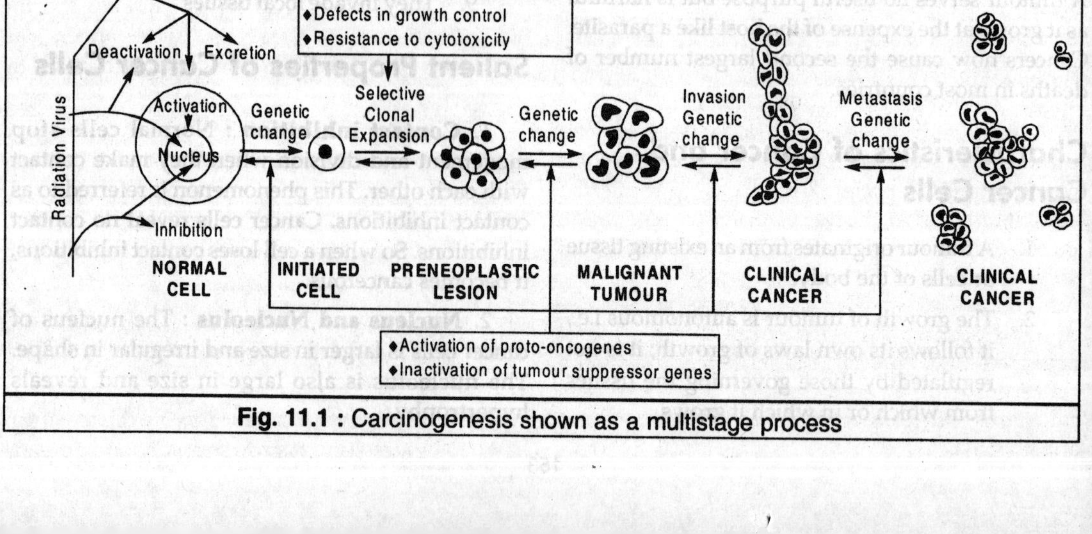

Fig. 11.1 : Carcinogenesis shown as a multistage process

1. **Conheim's theory of embryonic cell rest** : According to this theory, during the early development of the embryo, islets of cells may be misplaced and may become intermingled with other tissues. These cells acquire the power of proliferation and growth in later life.

2. **Hypothesis of Waldeyer and Thiersch** : These scientists are of view that under normal conditions there is a static equilibrium between epithelial tissues and connective tissues with respect to their proliferative capacity, one restraining or checking the other. With the advancement of age, this checking power of the connective tissues is lost and as a result the epithelium gets the upper hand, proliferates rapidly and gives rise to carcinoma.

3. **Hypothesis of anaplasia** : This theory was proposed by HANSMANN. According to this theory, differentiated cells occasionally revert to the embryonic type or anaplasia resulting in the production of more primitive type of cells. The greater the degree of anaplasia, the greater the malignancy.

4. **Theory of cell variation** : This theory was proposed by HADDOW. He states that carcinogenic agents inhibit normal cell function and that tumour-growth is the result of the process of adaptation to an unfavourable environment. The injured cells give rise to a new race of cells with a permanently increased rate of fission. The carcinogen acts as an enzyme poison resulting in the appearance of a new enzyme, which gives the cell a capacity for uncontrolled growth.

5. **Immunological theory** : This theory was proposed by GREEN. According to this theory, the carcinogens unite with special cellular components to form antigenic complexes leading to immunological reactions destroying those cells which contain the antigen. This also favours the emergence and proliferation of cells which lack it.

6. **Defective TCA cycle** : WARBURG (1956) suggested that cancer is caused owing to an injury to the cellular respiratory apparatus. The chemicals producing cancer are supposed to affect the aerobic respiration (TCA cycle) of cells. Owing to injured aerobic respiration, the rate of anaerobic respiration increases. The anaerobic respiration releases only a small amount of energy compared to aerobic respiration. This small amount of energy is incapable of maintaining the structural integrity and organization of the cells. This leads to uncontrolled cancerous growth. In a nutshell, a cell becomes cancerous when its aerobic respiration becomes anaerobic.

Sites of Infection

In man, cancers are most likely to develop in the liver, bone marrow, skin, lungs, prostate glands, stomach, intestine, colon, blood, mouth etc. In woman the commonest sites are the breast, uterus, intestine, stomach, skin, and vagina in young woman.

Metastasis

Cancerous cells tend to invade the surrounding tissues. Groups of cancerous cells grow into these tissues much as root grows into the soil. Sometimes the cancerous cells may break off from the parent growth and may migrate through the blood stream or the lymph circulation to other parts of the body, perhaps far away from the original site. Here they cause new cancerous growths. This new cancerous growth is called metastasization and the phenomenon is called metastasis.

Effects of Cancer

As the cancers grow in an unlimited and uncontrolled fashion, they become a menace to life. They interfere with bodily functions, the pressures that they create produce severe pain when the nerves are affected. Malnutrition goes hand in hand with cancerous growth. The cancerous tissues seem to draw the nutrients taken into the body; normal tissues are deprived of their fair share of these nutrients.

A cancer of the pituitary gland, i.e., hypophysis, may cause abnormal growth, and the victim may become a dwarf or a giant. When the pancreas is affected, there are often digestive disturbances and diabetic symptoms. A cancer in the brain and the spinal cord causes paralysis. Some cancers can corrode blood vessels and cause sudden death by hemorrhage.

Types of Cancer

The cancers are mainly classified in two ways :

1. On the basis of pain
2. On the basis of tissues.

On the basis of pain cancers are classified into two types. They are malignant tumour and non-malignant tumour :

(a) **Malignant tumour** : This is a cancer which gives pain; it grows rapidly and spreads from the place in which it started to the other parts of the body, nearby or away.

(b) **Non-malignant tumour (Benign tumour)** : This cancer may do or do not harm the possessor; it is confined to the area in which it originated; its growth is always slow.

Secondly, physicians classified cancers on the basis of the tissues in which they originate. A few are mentioned below :

1. **Carcinoma** : It is a tumour arising from epithelial cells. It occurs on the skin, in the lining membranes of the internal organs.
2. **Sarcoma** : Sarcomas are malignant tumours arising from any connective tissue and the allied structures of mesenchymal origin.
3. **Osteoma** : It is tumour of bones.
4. **Fibroma** : It is a tumour arising from fibrous tissues.
5. **Glioma** : It develops in the network of supporting connective tissues in the brain and the central nervous system.
6. **Melanoma** : It is a rapidly growing pigmented tumour; it originates in certain types of pigmented moles found on the skin. If these moles are irritated, they rapidly become malignant and the resulting cancers metastasize rapidly.
7. **Lymphoma** : The cancerous growths developing on the lymph nodes and the other tissues of the lymphatic system are called lymphomas.
8. **Teratoma** : It arises from cell-rest derived in the morula stage itself; later it acquires the ability for growth. It is suggested that teratoma is a suppressed fetus resulting from an undifferentiated germ cell or a fertilized polar body, or it develops by parthenogenesis. In other words, teratoma is considered an inclusion of a parasitic fetus in an individual.

Causes of Cancer

Actually it is still not clear just what causes cancer. However, there are certain contributing factors :

1. **Chronic irritation** : Chronic irritation causes cancers. There are various examples. Irritation to the tongue by broken or malformed teeth has often led to the cancer of the tongue; Cancer of the lip were quite common when clay pipes were in use.
2. **Environmental pollution** : There are evidences that cancers are more apt to strike city-dwellers than country-dwellers. It is true that the environment over cities is often polluted with the products of the combustion of coal and oil, with the motor exhaust gases. The increase in industrialization has caused a corresponding increase in cancer cases.
3. **Smoking** : Statistical studies show that heavy smoking of cigarettes cause an increase in the number of lung cancer cases. There is as yet no positive evidence that cigarette smoking alone causes lung cancer.
4. **Radiation** : It is now established that excessive radiation (X-rays, cosmic rays) causes cancers. Early investigators who used X-rays without appropriate shielding and other protective devices often fell victims to the rays, and some of them developed cancers. If the amount of radiation is allowed to increase without rhyme or reason, there may be a catastrophic increase in the number of cancer cases. Ultraviolet rays are also responsible for skin cancers. The effective wavelength has been reported to be in the range of 2,900-3,200 A^0.
5. **Chemicals** : Chemical substances added to food and medicines may have the effect of causing the development of cancers. Cancer-causing chemicals are termed chemical carcinogens or cancer-producing chemical agents. Carcinogenesis is the process involved in the development of malignant tumours (swellings). A carcinogen, therefore, is the risk factor causally related to an increase in cancer incidence or prevalence. In other form, a carcinogen is an agent that increases the incidence of any neoplasm (a new growth) — a term applied to all types of tumour, irrespective of whether it is lethal or potentially lethal. In extreme circumstances, a carcinogen may also be defined as an agent that increases the incidence of certain types of preneoplastic change. According to Pitot (1986), a carcinogen is an agent whose adminis-

tration to previously untreated animals leads to a statistically significant increased incidence to neoplasms of one or more histogenic types as compared with the incidence in appropriate untreated animals.

Carcinogens may be physical, chemical or biological agents. Important physical phenomena that are carcinogenic are UV and ionizing radiation. Circumstances involving biological agents that are judged to be carcinogenic are chronic infection with the hepatitis viruses B and C, and infection with the blood parasite *Schistosoma haematobium*, and the liver fluke *Opisthorchis viverrini*. However, there are many more chemicals that have been recognized as human carcinogens and some circumstances that entail exposures to carcinogens (without, necessarily, identifying the actual carcinogens). Those identified by the International Agency for Research on Cancer (IARC) as of May 1996 are listed in Table 11.2.

Carcinogens cause a cell or a group of normal cells to change their biological behaviour resulting in an autonomous uninterrupted growth of malignant nature.

From the aforementioned discussion, the cancer may be defined as an actively growing tissue in which normal growth controlling mechanism is impaired permanently, resulting in progressive uncontrolled growth of tissue of an autonomic nature. It is a general term used to refer to a variety of malignant neoplasms. It ultimately causes death of the living being owing to its invasive and metastasing character. Presently, cancer is one of the most dreaded diseases of the world. For the malignant neoplasm derived from tissues of mesenchymal origin the suffix "*Sarcoma*" is added to the tissue descriptor : *Fibrosarcoma, Osteosarcoma, Liposarcoma* and so on. Malignant neoplasms derived from tissues of ectodermal and endodermal descriptors : *Epidermoid carcinoma; Hepatocellular carcinoma, Gastric adenocarcinoma,* and so on.

The induction of cancer is an irreversible and often life-threatening change. The induction of cancer following exposure to certain chemical is termed as *chemical carcinogenesis* or *carcinogenicity*. It is now generally accepted that the vast majority of human cancers are caused by various environmental chemicals called environmental carcinogens.

Environmental Carcinogens

A large number of environmental toxicants like polyaromatic hydrocarbons (PAH), drugs, heavy metals, pesticides, radiations, food dyes and additives, tobacco smoke, aflatoxins, smoked beef etc. have been known to behave as "Environmental Carcinogen" by IARC as shown in Table 11.1 and 11.2 and exposure to chemical carcinogens (established/suspected) is shown in Table 11.3.

Table 11.1 : Environmental Carcinogens

Agency/Media	*Source*	*Carcinogen(s)*
Air	Industrial/Factory effluents, emission from automobiles; smoking, coal and petroleum combustion; Asbestos, Fibres; Soots; Tar etc.	Polyaromatic hydrocarbons (PAHs) viz., Benzo(a) pyrene; Chrysene; Naphthylamine etc.
Water	Industrial effluents; Oil refineries; Pesticides leaching from soil, industrial wastes dumped along river; dry cleaning solvents; laboratory effluents; Lead pipelines, Vinyl pipes; cement pipelines etc.	PAH (s) viz., Benzo(a) pyrene; Carcinogenic pesticides leaching from soil viz., DDT; Kepone; 2,4,5-T; DBCP; Asbestos; Lead tetrachloro-ethylene; Vinyl chloride etc.
Soil	Pesticides; PAHs from environmental condensate, fertilizers translocation to trees, fruits via soil.	Pesticides (DDT; BHC; Kepone, 2,4,5-T) PHAs and their alkyl derivatives, Benzo (a) Pyrene etc.
Food	(a) Beef, Fish, Meat (b) Flavanoids from fruits (c) Mycotoxins from *Aspergillus flavos* in rotten nuts, beans and food grains (d) Beverages from gas fired malt and from safrole root	Pyrolysed protein derived carcinogens viz., Tryp-P1, Tryp P-2 Kaemferol quercetin Aflatoxins B_1, B_2 and B_3 Nitrosamines and safrole
Food additives	(a) Nitrite as preservatives in meat, fish and beef (b) Food dyes (c) Sweetening agents	Nitrosamines Violet-1, Red Dye No-2; Azo dyes, Metanil yellow etc. Cyclamates and saccharive.
Life style	Tobacco chewing, cigarette smoking and drinking	CO, NH_3, Hydrazine, Vinylchloride, Nitrosamines, PAHs etc.
Pesticides	Used to kill pests, herbicides etc.	DDT, BHC, 2,4, 5-T, DBCB, Kepone, Aldrin etc.

Table 11.2 : Chemical and physical agents and exposure circumstances identified as human carcinogens by the International Agency for Research on Cancer (1996)

Agents/Groups	*Mixtures*
Aflatoxins	Alcoholic beverages
4-Aminobiphenyl	Analgesic mixtures contaning phenacetin
Arsenic and arsenic compounds[1]	Betel quid with tobacco
Asbestos	Coal-tar Pitcher
Azathioprine	Coal-tars
Benzene	Mineral-oils — untreated and mildly treated
Benzidine	Salted fish (Chinese style)
Beryllium and beryllium compounds[2]	Shale-oils
N, N-bis(2-chloroethyl)-2naphthylamine (Chlornaphazine)	Soots
Bis (Chloromethyl) ether and chloromethyl methyl ether (technical grade)	Tobacco products, smokeless (chewing tobacco, oral snuff)
1, 4-Butanediol dimethanesulfonate (Myleran)	Tobacco smoke
Cadmium and cadmium compounds[2]	Wood dust
chlorambucil	*Exposure circumstances*
1-(2-Chloroethyl)-3-(4-methylcyclohexyl)-1-nitrosourea (methyl-CCNU)	Aluminium production
Chromium (VI) compounds[2]	Auramine, manufacture of
Cyclophosphamide	Boot manufacture and repair
Cyclosporin	Coal gasification
Diethylstilboestrol	Coke production
Erionite	Furniture and cabinet making
Ethylene oxide	Hematite mining (underground) with exposure to radon
Melphalan	Iron and steel founding
8-Methoxypsoralen (methoxsalen) plus ultraviolet A radiation	Isopropanol manufacture (strong-acid process)
MOPP and other combined chemotherapy, Including alkylating agents	Magenta manufacture
Mustard gas (sulfur mustard)	Painter (occupational exposure)
2-Naphthylamine	Rubber industry
Nickel compounds[2]	Strong inorganic acid mists containing sulphuric acid
Oestrogen replacement therapy	
Oestrogens, non-steroidal[1]	
Oestrogens, steroidal[1]	
Oral contraceptives, combined[3]	
Oral contraceptives, sequential	
Radon and its decay products	
Solar radiation	
Talc containing asbestiform fibres	
Tamoxifen[4]	
Thiotepa	
Theosulphan	
Vinyl chloride	

1 This evaluation applies to the group of chemicals as a whole and not necessarily to all individual chemicals within the group.

2 Evaluated as a group.

3 There is also a conclusive evidence that these agents have a protective effect against cancers of the ovary and endometrium.

4 There is also a conclusive evidence that tamoxifen reduces the risk of contralateral breast cancer in women with a previous diagnosis of breast cancer.

Table 11.3 : Exposure to Chemical Carcinogens in the Workplace

A [Established]

Sl. No.	*Agent*	*Industries and Trades*	*Primary Affected Sites*
1.	Para-aminodiphenyl	Chemical manufacturing	Urinary bladder
2.	Arsenic	Copper mining and smelting	Skin, bronchus, liver
3.	Asbestos	Construction, asbestos mining and milling, productions of friction products and cement	Pleura, peritoneum, bronchus
4.	Alkylating agent (mechloroethamine hydrochloride and bis [chloromethyl] ether)	Chemical manufacturing	Bronchus
5.	Benzene	Chemical and rubber manufacturing, refining	Bone marrow
6.	Benzidine, beta-naphthylamine, and derived dyes	Dye and textile production	Urinary bladder
7.	Chromium and chromates	Tanning, pigment making	Nasal sinus, bronchus
8.	Isopropyl alcohol manufacture	Chemical manufacturing	Cancer of paranasal sinuses
9.	Nickel	Nickel refining	Nasal sinus, bronchus
10.	Polynuclear aromatic hydrocarbons from coke, coal, tar, share, mineral oils, and creosote	Steel making, roofing, chimney cleaning	Skin, scrotum, bronchus
11.	Vinyl chloride monomer	Chemical manufacturing	Liver
12.	Wood dust	Cabinet-making carpentry	Nasal sinus

B [Suspected]

Sl. No.	*Agent*	*Industries and Trades*	*Suspected Human Sites*
1.	Acrylonitrile	Chemical and plastics	Lung, colon, prostate
2.	Beryllium	Beryllium processing, aircraft manufacturing, electronics, secondary smelting	Bronchus
3.	Cadmium	Smelting, battery making, welding	Bronchus
4.	Ethylene oxide	Hospitals, production of hospital supplies	Bone marrow
5.	Formaldehyde	Plastic, textile and chemical production; health care	Nasal sinus, Bronchus
6.	Synthetic mineral fibers (e.g., fibrous glass)	Manufacturing, insulation	Bronchus
7.	Phenoxyacetic acid	Farming, herbicide application	Soft tissue sarcoma
8.	Polychlorinated biphenyls	Electrical-equipment production and maintenance	Liver
9.	Organochlorine pesticides (e.g., chlordane, dieldrin)	Pesticide manufacture and application, agriculture	Bone marrow
10.	Silica	Casting, mining, refracting	Bronchus

TYPES OF CARCINOGENS

Generally, the chemical carcinogens are grouped in two categories on the following bases :

I. On basis of chemical nature, and
II. On basis of mode of action.

I. On Basis of Chemical Nature

On the basis of chemical nature, carcinogens may be grouped into two categories :

(i) Organic chemical carcinogens
(ii) Inorganic chemical carcinogens.

(i) Organic chemical carcinogens

Several polycyclic hydrocarbons were isolated from active crude tar fraction in the early 1930s. The first synthetic carcinogenic polycyclic aromatic hydrocarbon was dibenz (a, h) anthracene. Later the benzo (a) pyrene (3, 4-benzpyrene) was reported in 1932. Further, their numbers increased and reached up to 50 till the end of the 20th century. Structure of some carcinogenic polycyclic hydrocarbons is shown in Fig. 11.2 and other representative organic chemical carcinogens is shown in Fig. 11.3.

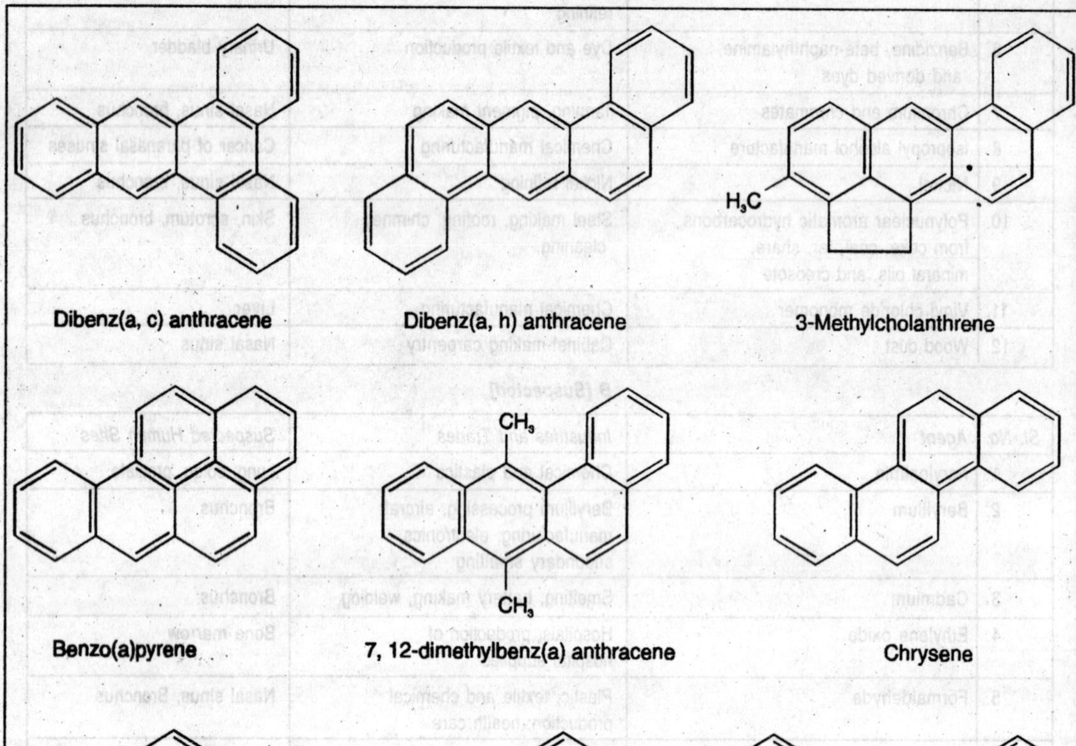

Fig. 11.2 : Chemical structures of some Carcinogenic polycyclic hydrocarbons

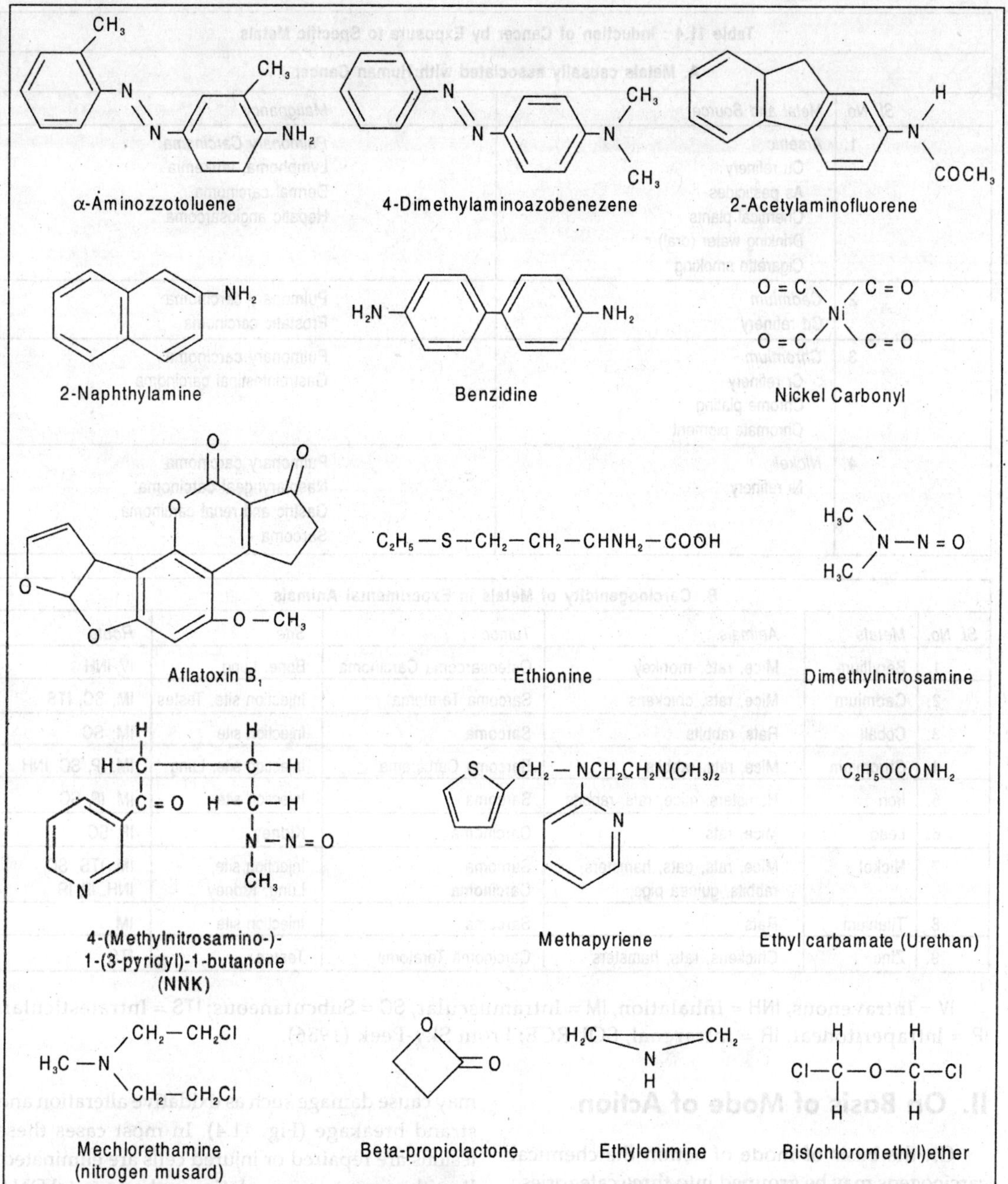

Fig. 11.3 : Chemical structures of other representative organic chemical carcinogens

(ii) Inorganic chemical carcinogens

In addition to organic compounds such as those illustrated in Figs. 11.2 and 11.3, a number of inorganic elements and their compounds have been known to be carcinogenic in both animals and humans. Table 11.4 lists metals that are proved carcinogenic in some form to humans (Part A) and experimental animals (Part B) (Sky-Peck, 1986).

Table 11.4 : Induction of Cancer by Exposure to Specific Metals

A. Metals causally associated with Human Cancer

Sl. No.	Metal and Source		Malignancy
1.	*Arsenic* Cu refinery As pesticides Chemical plants Drinking water (oral) Cigarette smoking		*Pulmonary Carcinoma* Lymphoma, leukemia Dermal carcinoma Hepatic angiosarcoma
2.	*Cadmium* Cd refinery		Pulmonary carcinoma Prostatic carcinoma
3.	*Chromium* Cr refinery Chrome plating Chromate pigment		Pulmonary carcinoma Gastrointestinal carcinoma
4.	*Nickel* Ni refinery		Pulmonary carcinoma Nasolaryngeal carcinoma Gastric and renal carcinoma Sarcoma

B. Carcinogenicity of Metals in Experimental Animals

Sl. No.	Metals	Animals	Tumor	Site	Route
1.	Beryllium	Mice, rats, monkey	Osteosarcoma Carcinoma	Bone, Lung	IV, INH
2.	Cadmium	Mice, rats, chickens	Sarcoma Teratoma	Injection site, Testes	IM, SC, ITS
3.	Cobalt	Rats, rabbits	Sarcoma	Injection site	IM, SC
4.	Chromium	Mice, rats, rabbits	Sarcoma Carcinoma	Injection site, Lung	IM, IP, SC, INH
5.	Iron	Hamsters, mice, rats, rabbits	Sarcoma	Injection site	IM, IP, SC
6.	Lead	Mice, rats	Carcinoma	Kidney	IP, SC
7.	Nickel	Mice, rats, cats, hamsters, rabbits, guinea pigs,	Sarcoma Carcinoma	Injection site, Lung, Kidney	IM, ITS, SC INH, IP, IR
8.	Titanium	Rats	Sarcoma	Injection site	IM
9.	Zinc	Chickens, rats, hamsters	Carcinoma Teratoma	Testes	ITS

IV = Intravenous, INH = Inhalation, IM = Intramuscular; SC = Subcutaneous; ITS = Intratesticular; IP = Intraperitoneal; IR = Intrarenal, SOURCE; From Sky-Peek (1986).

II. On Basis of Mode of Action

On the basis of mode of action the chemical carcinogens may be grouped into three categories :

1. Genotoxic carcinogens
2. Epigenetic or Non-genotoxic carcinogens
3. Unclassified carcinogens.

1. Genotoxic Carcinogens

The carcinogens that initiate tumour formation by chemically interacting with DNA molecules are termed genotoxic carcinogens. These chemical carcinogens, after covalently binding with DNA, may cause damage such as oxidative alteration and strand breakage (Fig. 11.4). In most cases these lesions are repaired or injured cells are eliminated. If neither event occurs, a lesion in the parental DNA strand may induce a heritable alteration, or mutation, in the daughter strand during replication. The mutation may remain silent if it does not alter the protein encoded by the mutant gene or if the mutation causes an amino acid substitution that does not affect the function of the protein. Alternatively, the genetic alteration may be incompatible with cell survival. The most unfortunate scenario for the organism takes place when the altered genes express mutant proteins that reprogram cells for the growth and multiplication. When such cells survive and undergo mitosis, their descendants also have a similar propensity for proliferation. Moreover, because enhanced cell division enhances the likelihood of mutations, these cells eventually acquire additional mutations that may further increase their growth advantage over their normal counterparts. The final outcome of this process is a nodule, followed by a tumour consisting of transformed rapidly proliferating cells (Fig. 11.4).

A small set of cellular genes are the targets for genetic alterations that initiate neoplastic transformation. Included are proto-oncogenes and tumour suppressor genes (Barrett, 1992). However, to some extent DNA repair genes are also included.

Mutation of Proto-Oncogenes. Proto-oncogenes are actually highly conserved genes encoding proteins that stimulate the progression of cells through the cell cycle (Smith et al, 1993). The products of proto-oncogenes consist of (1) growth factors, (2) growth factor receptors, (3) intracellular signal transducers such as G proteins and protein kinases, and (4) nuclear transcription factors. Transient increase in the production or activity of proto-oncogene proteins are required for regulated growth such as what occurs during embryogenesis, tissue regeneration, and stimulations of cells by growth factors or hormones. In contrast, permanent activation and/or overexpression of these proteins favours neoplastic transformation. Genotoxic carcinogens typically induce mutation of proto-oncogenes, with the mutant gene then termed as an oncogene, encoding a mutant protein. If this protein gains increased activity, it can initiate neoplastic transformation of the cell. An example of mutational activation of an oncogene protein is that of the *Ras proteins.*

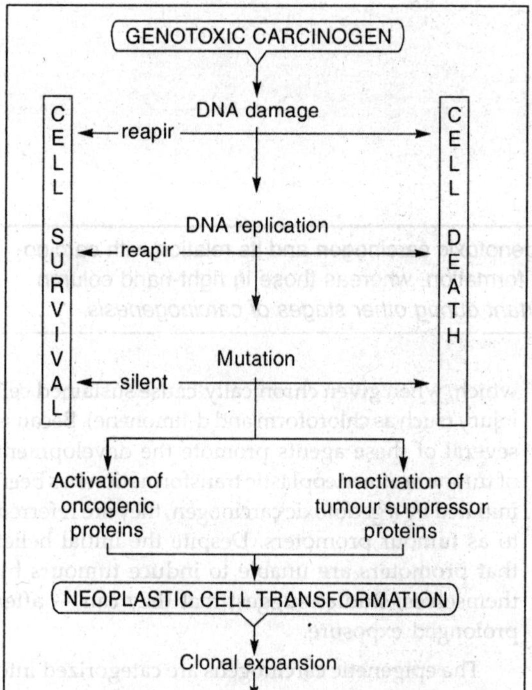

Fig. 11.4 : The process of carcinogenesis initiated by genotoxic carcinogens

Mutations of Tumour Suppressor Genes. Tumour suppressor genes encode proteins which inhibit the progression of cells in the division cycle. Uncontrolled proliferation may take place when the mutant tumour suppressor gene encodes a protein which cannot suppress cell division.

Types of Genotoxic Carcinogens

The genotoxic carcinogens are basically of two types :

(i) Procarcinogens

(ii) Direct acting or ultimate carcinogens

(i) **Procarcinogens** : These chemicals require conversion through bioactivation either directly or via an intermediate stage in order to become ultimate or proximate carcinogens. This category includes PAHs, halogenated hydrocarbons, aromatic amines, alkaloids, nitrosamines, aflatoxin B, safrole, and thioamides.

(ii) **Direct acting carcinogens** : These are actually electrophilic chemical carcinogens and can bind with DNA and other macromolecules. The examples of this group include alkyl and aryl epoxides, lactones, sulphate esters, nitrosamides, nitrosoureas, and platinum amine chelates.

Overall process of carcinogenesis, the fate of genotoxic carcinogens and the process of carcinogenesis has been depicted in Fig. 11.5 along with the mechanism of action :

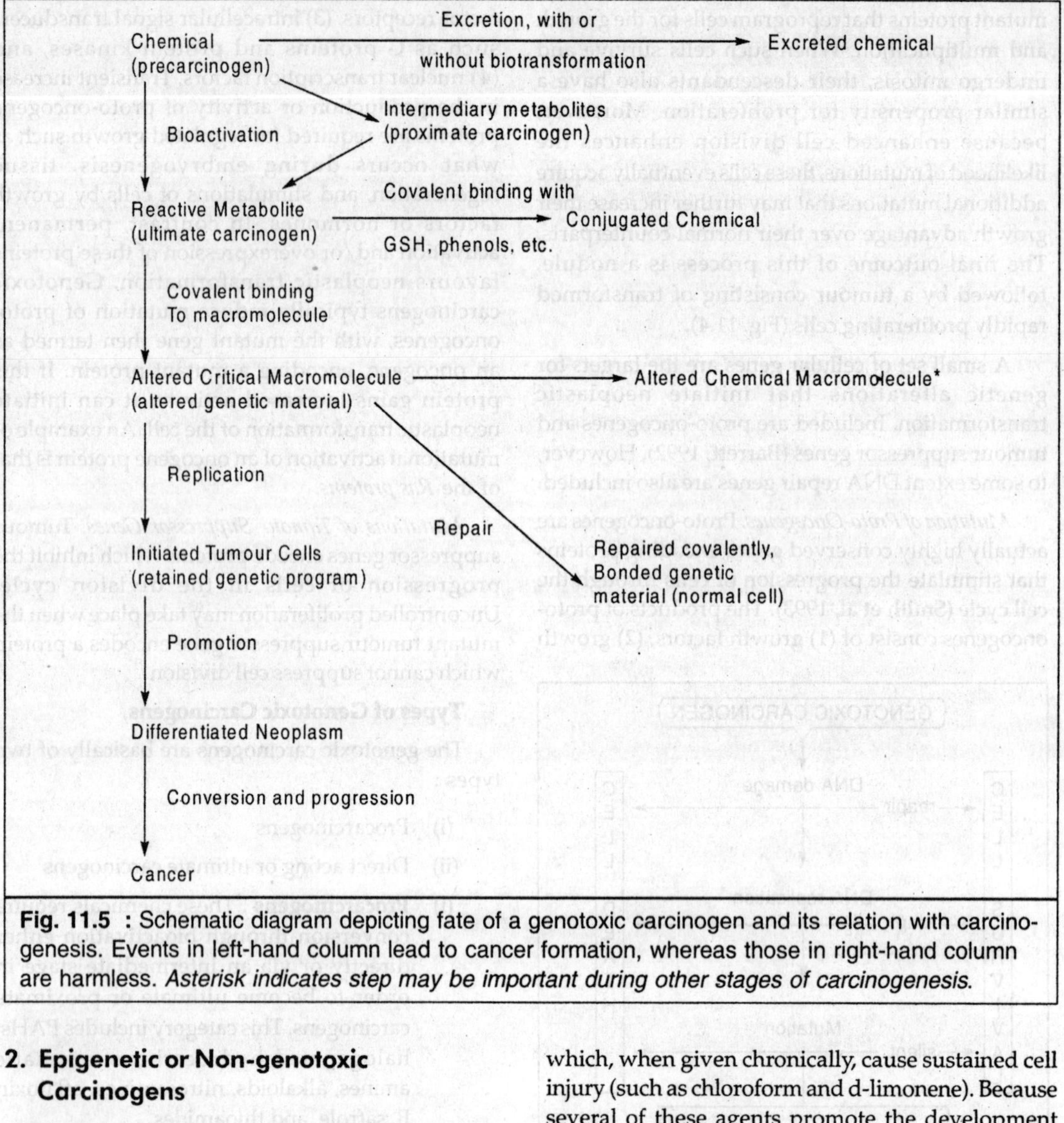

Fig. 11.5 : Schematic diagram depicting fate of a genotoxic carcinogen and its relation with carcinogenesis. Events in left-hand column lead to cancer formation, whereas those in right-hand column are harmless. *Asterisk indicates step may be important during other stages of carcinogenesis.*

2. Epigenetic or Non-genotoxic Carcinogens

Actually these chemical substances do not damage DNA but enhance growth of tumours induced by genotoxic carcinogens. These produce carcinogenicity at high and sustained exposure causing prolonged physiological abnormalities, hormonal imbalances or tissue injuries (Barret, 1992). These include (1) xenobiotic mitogens (e.g., phenobarbital, phorbol esters, DDT, peroxisomal proliferators), (2) endogenous mitogens such as growth factors (e.g., TGF-alpha) and hormones with mitogenic action on specific cells [e.g., estrogens on mammary gland or liver cells, TSH on the follicular cells of the thyroid gland, and luteinizing hormone (LH) on Leydig's cells in testes, and (3) chemicals which, when given chronically, cause sustained cell injury (such as chloroform and d-limonene). Because several of these agents promote the development of tumours after neoplastic transformation has been initiated by a genotoxic çarcinogen, they are referred to as tumour promoters. Despite the initial belief that promoters are unable to induce tumours by themselves, studies suggest that they can — after prolonged exposure.

The epigenetic carcinogens are categorized into two :

(a) Cocarcinogens

(b) Promoters.

(a) **Cocarcinogens** : They may act as synergists of genotoxic carcinogens and enhance their effects when given simultaneously. The effect of genotoxic carcinogens may be increased either by increased absorption or by increased bioactivation. The increased effects of genotoxic carcinogens can also be achieved by decreased elimination — perhaps on account of inhibition of detoxication enzymes.

(b) **Promoters** : They do not actually produce cancer themselves but facilitate the growth of dormant neoplastic cells into tumours. The promoters induce carcinogenic effects on account of promotion of spontaneously transformed cells.

The possible mechanisms of action of promoters include :

(i) Stimulation of cells proliferation through cytotoxicity produced by cytotoxins or hormonal effects.

(ii) Inhibition of intercellular communication thereby releasing the initiated cells from the restraint exercise by the surrounding normal cells; and

(iii) Immunosuppression.

Some known common promoters are tetradecanoyl phorbol acetate (TPA), phenobarbital, chlorinated hydrocarbons, butylated hydroxy anisole (BHA) and butylated hydroxytoluene (BHT).

(i) **Cytotoxins** : Some carcinogens, such as nitrilotriacetic acid (NTA), are known to produce cancer through cell proliferation resulting from cell injury and death.

Induction of cancer by the cytotoxicants can be ascribed to :

(a) Enhanced susceptibility to environmental carcinogens.

(b) Mutations during DNA replication,

(c) Aberrant methylation, and

(c) Their effects on chromosomes.

(ii) **Hormone modifiers** : Some hormones such as estradiol and diethyl stilbesterol have been known to cause cancer in animals (e.g., breast cancer in mice and endometrial cancer in menopausal human females maintained on estrogens). Herbicides, aminotriazole and certain fungicides have been known to induce thyroid tumour, possibly through interferences with the synthesis of thyroxine and thus alteration in pituitary-thyroid feedback system.

(iii) **Immunosuppressive drugs** : Various immunosuppressive drugs (such as azothioprine or 6-mercaptopurine) are given to the patients who underwent organ transplantation. These drugs have been known to promote the development to tumours (leukemia and sarcomas) in these patients — possibly through epigenetic mechanisms. Besides, some solid state materials also may induce sarcoma formation.

Solid State materials : Some plastic and metal sheets or discs implanted subcutaneously in rodents have been observed to lead to sarcoma formation after a long latent period. The shape and size of these implanted materials are important rather than their composition. Smooth materials have been reported to be more effective tumour inducers than rough and perforated sheets or discs. Asbestos and other fibrous materials, such as fine fibre glass, on reaching the pleural cavity in the animal models as well as in human-beings lead to mesotheliomas usually after long latent period.

Mechanism of Non-genotoxic Carcinogensis

Non-genotoxic carcinogens cause cancer by promoting carcinogenesis initiated by genotoxic agents or spontaneous DNA damage (Fig. 11.6). Spontaneous DNA damage, some of which give rise to mutation, generally occurs in normal cells (Barrett, 1992). The spontaneous mutation frequency per base pair in human cells is estimated to be in the range of 10^{-8} to 10^{-10}. Genotoxic carcinogens increase the frequency tenfold to 1,000-fold. Nongenotoxic carcinogens also enhance the frequency of spontaneous mutation through a mitogenic effect and by the mechanisms discussed earlier. In addition, non-genotoxic carcinogens, by inhibiting apoptosis, increase the number of cells with DNA damage and mutations. Both enhanced mitotic activity and diminished apoptotic activity brought about by non-genotoxic carcinogens expand the population of transformed cells, promoting cancer development. In nutshell, non-genotoxic carcinogens appear to act by enhancing cell division and/or blocking apoptosis.

It is easy to recognize that even epigenetic carcinogens of the cytotoxic type act in this manner. It is important to mention here that cell injury evokes the release of mitogenic growth factors such

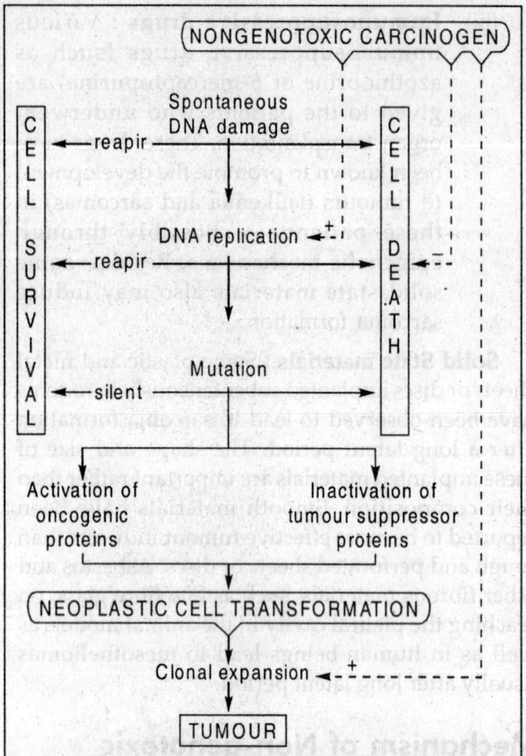

Fig. 11.6 : The process of carcinogenesis promoted by non-genotoxic carcinogens

as HGF and TGF-alpha from tissue macrophages and endothelial cells. Although these growth factors are actually instrumental in tissue repair after acute cell injury, their continuous presence is potentially deleterious because they may ultimately transform the affected cells into neoplastic cells.

It is further important to mention that even epigenetic carcinogens may exert a genotoxic effect, although indirectly. For example, chemicals causing cell injury evoke a prolonged inflammatory response, with the free radicals produced by the inflammatory cells causing DNA injury in neighbouring cells. Similarly, phorbol esters are only potent mitogens but also activators of leukocytes, which release DNA-reactive free radicals during their respiratory burst (Weiss and Lo Buglio, 1982).

3. Unclassified Carcinogens

Various compounds of uranium, radium, polonium etc. are carcinogenic, chiefly because of their radioactive property.

Numerous metals and metalloids, such as arsenic, beryllium, chromium, cobalt, lead, manganese, nickel etc., have been observed to be carcinogenic in animals and human beings under specific experimental conditions. Arsenic is exceptional among these elements because it induces carcinogenicity in human beings but not in animals.

Unclassified chemical carcinogens may be broadly grouped into two :

(a) **Peroxisome proliferators** : A number of chemicals, such as hypolepidemic drugs and certain phthalate plasticizers, induce liver cancer in rodents by increasing peroxisomes in liver cells.

(b) **Secondary carcinogens** : The term secondary carcinogens has been generally applied for substances that do not directly cause carcinogenic effect but can induce cancer by causing non-carcinogenic effects. For example, polyoxyethylene monostearate induces bladder stone at very high doses which, in turn, develops bladder tumours. This term is also used for certain genotoxic carcinogens (procarcinogens) which require bioactivation prior to their effects.

What is Hormonal Carcinogenesis?

It has been established that many hormones are also responsible for the onset of cancers.

Causal association of hormones in the development of specific neoplasms may be termed *hormonal carcinogenesis.* Furth (1975) reported that disruption of the cybernetic relationship between peripheral endocrine glands and the adenohypophysis may result in neoplasia of one of the glands involved (Fig. 11.7). One of the classic examples is the experimental transplantation of normal ovaries in the spleen of castrated rodents (Biskind and Biskind, 1944). This results in a break in the hypophysis-gonadal hormone feedback loop.

The break occurs because estrogens produced by the ovary are carried by the splenic venous system to the liver. In liver, metabolization takes place and thus estrogens are prevented from entering the general circulation to suppress the hypophysis production of gonadotropins. The excessive production of gonadotropins and their constant stimulus of the ovarian fragment in the spleen may result, ultimately, in neoplasia of the ovarian implant.

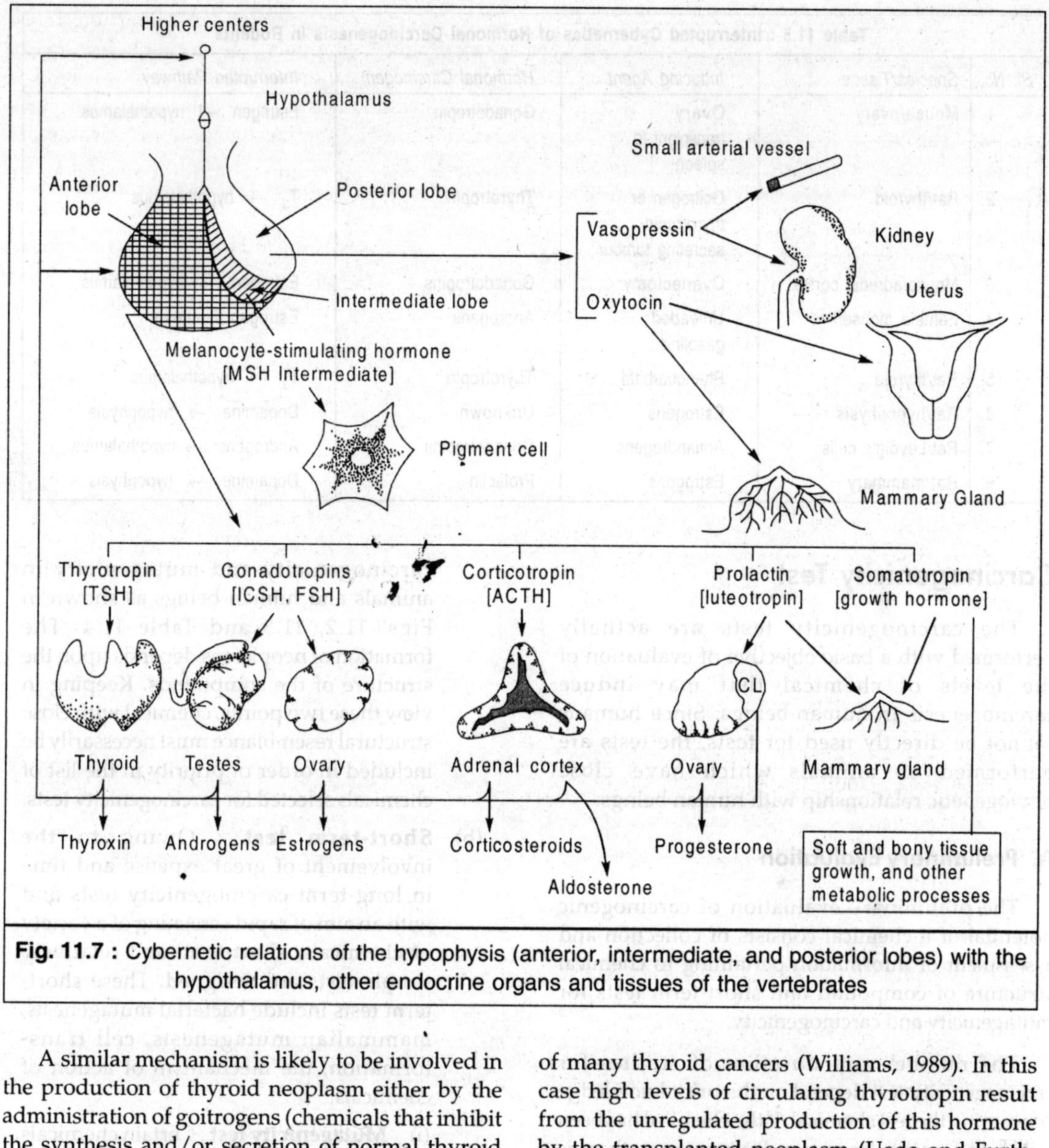

Fig. 11.7 : Cybernetic relations of the hypophysis (anterior, intermediate, and posterior lobes) with the hypothalamus, other endocrine organs and tissues of the vertebrates

A similar mechanism is likely to be involved in the production of thyroid neoplasm either by the administration of goitrogens (chemicals that inhibit the synthesis and/or secretion of normal thyroid hormone) or by a marked increase in the circulating levels of thyrotropin secreted by thyrotropin secreting pituitary neoplasms transplanted into the host. In the former instance, there is a break in the feedback loop and the hypophysis produces high levels of thyrotropin in the absence of normal feedback regulation by the thyroid hormone (Furth, 1975). In fact, in humans there is substantial evidence that this may be the mechanism of the development of many thyroid cancers (Williams, 1989). In this case high levels of circulating thyrotropin result from the unregulated production of this hormone by the transplanted neoplasm (Ueda and Furth, 1967). Thyroidectomy and neonatal gonadectomy result in the development of neoplasms of the hypophysis. Chronic administration of hypophysis growth hormone also induces a variety of neoplasms in the rat (Moon et al, 1950).

Some examples of carcinogenesis resulting from the interruption of the cybernetics of hormonal relationships shown in Fig. 11.7 are listed in Table 11.5 :

Table 11.5 : Interrupted Cybernetics of Hormonal Carcinogenesis in Rodents

Sl. No.	Species/Tissue	Inducing Agent	Hormonal Carcinogen	Interrupted Pathway
1.	Mouse/ovary	Ovary transplant to spleen	Gonadotropin	Estrogen → hypothalamus
2.	Rat/thyroid	Goitrogen or thyrotropin-secreting tumour	Thyrotropin	T_3 → hypothalamus
3.	Mouse/adrenal cortex	Ovariectomy	Gonadotropins	Estrogen → hypothalamus
4.	Female mouse/liver	Unleaded gasoline	Androgens	Estrogen synthesis
5.	Rat/thyroid	Phenobarbital	Thyrotropin	T_3 → hypothalamus
6.	Rat/hypophysis	Estrogens	Unknown	Dopamine → hypophysis
7.	Rat/Leydig's cells	Antiandrogens	Gonadotropins	Androgens → hypothalamus
8.	Rat/mammary	Estrogens	Prolactin	Dopamine → hypophysis

Carcinogenicity Test

The carcinogenicity tests are actually performed with a basic objective of evaluation of the levels of chemical that may induce carcinogenesis in human-beings. Since humans cannot be directly used for tests, the tests are performed on animals which have close phylogenetic relationship with human beings.

A. Preliminary Evaluation

The preliminary evaluation of carcinogenic potential of a chemical consists of collection and assessment of information pertaining to chemical structure of compound and short-term tests for mutagenicity and carcinogenicity.

The number and duration of studies for carcinogenicity test depend on the prior knowledge concerning the test chemical. If the chemical has been tested on a particular species for two years, there is no need of prechronic screening. Otherwise, the chemical should be tested at least for 90 days. If the chemical is suspected to cause delayed effects, more longer tests (for six months' duration) should be conducted to accurately predict the maximum tolerance limits over a two year period. In case of absence of any data on the toxicity of chemical, first short-term studies are made to evaluate lethal levels so that sublethal doses of the chemical may be fixed for long-term or chronic studies.

(a) **Chemical structure** : A number of chemicals are known to induce carcinogenicity and mutagenicity in animals and human beings as shown in Figs. 11.2, 11.3 and Table 11.4. The formation of neoplasms depends upon the structure of the compounds. Keeping in view these two points, chemical with close structural resemblance must necessarily be included in order of priority in the list of chemicals selected for carcinogenicity tests.

(b) **Short-term test** : Owing to the involvement of great expense and time in long-term carcinogenicity tests and with an aim of rapid screening of a variety of chemicals, short-term and relatively simple tests are designed. These short-term tests include bacterial mutagenesis, mammalian mutagenesis, cell transformation, the mechanism of action of chemicals.

(i) **Mutagenicity test** : Certain chemicals produce genetic changes because of their interaction with DNA. Such mutagenic chemicals resemble in their action with those of genotoxic carcinogens. The positive result of mutagenicity tests do not provide direct evidence on carcinogenic potential of chemical, but indicate that extensive and long-term testing should be initiated on the chemical. On the other hand, the negative findings do not establish that the chemical is safe.

(ii) **Short-term screening for carcinogenicity** : Bull the Pereira (1982) have suggested that short-term mutagenicity and limited carcinogenicity tests may consist of :

i. initiation and promotion of skin cancer in mouse,
ii. lung adenoma in mice,
iii. rat liver foci,
iv. cell transformation, and
v. sister chromatid exchange.

The ultimate criterion effect in carcinogenicity test is tumour formation. Certain chemicals (e.g., cocarcinogens) which yield negative result in mutagenicity test may give positive results in limited carcinogenicity tests. The positive results in more than one tests suggest the carcinogenicity potential of test chemical. Long-term studies on such chemicals are required for any inference.

B. Long-term Carcinogenicity Test

The long-term carcinogenicity test provides definite information on the carcinogenic potential of chemical tested against a particular animal. If this animal has close phylogenetic relationship with humans then the data from these studies may be extrapolated for human beings.

The procedure for carcinogenicity test is outlined :

(a) **Test animal** : Generally preferred characteristics of test animal for carcinogenicity tests are :

(i) Low incidence of spontaneous tumours,

(ii) Resemblance in rate and pattern of biotransformation with those of human beings, and

(iii) Known susceptibility with chemicals of similar structure.

Doses : The selected animals are exposed to two or three doses of test chemical. These doses are sublethal and fixed on the basis of reported acute toxicity data of the chemical (10% of LD_{50} 96 hours). The maximum tolerated dose (MTD) is often used as higher dose whereas two lower doses are usually some fraction of the higher dose. A separate untreated group of similar individuals is also maintained as control.

(c) **Route of exposure** : In case of drugs, food-additives and contaminants, the routes of exposure are the same through which the human beings are exposed or expected to be exposed. The environmental toxicants generally get entry through inhalation route. But, for the sake of convenience in exposure, these substances are usually mixed in the diet and fed to the test animals. Sometimes, the chemical is painted over the body of test animal with a purpose to examine the effect of dermal exposure. Since these tests are conducted for long duration, the chemical is exposed repeatedly.

(d) **Duration** : In order to achieve maximum duration of exposure the tests are generally started shortly after the weaning of individuals. The usual duration of test is 18 months for mice and 24 months for rats. This duration may be extended up to 24 months and 30 months, respectively, if the animals are healthy.

(e) **Observation** : The normal health of animals exposed and mortality, if any, should be examined daily. Death and moribund individuals should be segregated for gross and microscopic observations. Sick animals should also be caged separately to prevent cannibalism and spread of infection of the disease. Signs of toxicity and onset, location, size and growth of unusual tissue masses should be carefully examined and recorded regularly. All animal — including those that survive to the scheduled end of the experiment — are subjected to autopsy and almost 40 different tissues are taken from each animal for histological examination. All observations are recorded, summarized and analysed.

(i) **Body weight and food-consumption** : In order to calculate the dose of chemical exposed, the body weight and food consumption should be recorded after certain intervals. During initial phase of the test, for usually first three months and, in case of growing animals, the data are recorded weekly and thereafter biweekly.

(ii) **Functional test** : As the test relates to carcinogenic activity of chemical, the functional tests are performed regarding the health of test animals usually at the termination of the study. For mid-term evaluation of health of test animals, separate groups of animals are exposed to the respective doses of test chemicals.

(iii) **Postmortem examination** : Dead and moribund animals are subjected to gross autopsy. The surviving animals are also sacrificed at the end of test for gross morphological and histological examination of various organs and tissues to record gross abnormality and tumour growth, if any.

The following points should be considered while reporting tumours :

i. The number of tumour-bearing animals out of stock/group exposed.

ii. The number of tumors in each animals.

iii. The number of types of tumours (benign and malignant), and

iv. Onset of tumours whenever determinable.

The occurrence of unusual tumours in significant number should be considered as an important phenomenon. The stages of tumour may include a typical hyperplasia, benign tumours, carcinomas in situ, invasion of nearby tissue and metastasis in other parts of the body. The tumours of same type but at different stages should be tabulated and statistically analysed separately. The test is more reliable if it is reproducible in another species with similar results.

What are Oncogenes?

Oncogenes are genes capable of causing cancer. These were first recognised as unique genes of tumour-causing viruses that are responsible for the process of transformation (viral oncogenes).

Mechanisms of action of Oncogenes

(i) They may act on key intracellular pathways involved in growth control uncoupling them from the need for an exogenous stimulus.

(ii) The products of oncogenes may also imitate the action of a polypeptide growth factor.

(iii) The products may also imitate an occupied receptor for a growth factor.

Revision Questions

A. Long-Answer Questions

1. Write an essay on different types of environmental carcinogens.
2. Describe various types of carcinogens on the basis of their chemical nature.
3. Describe the types of carcinogens on the basis of their mode of action.
4. What are genotoxic carcinogens? Illustrate their mechanism of action.
5. What are the epigenetic carcinogens? Illustrate their mechanism of action.
6. Describe the hormonal carcinogenesis in detail.
7. Define carcinogenesis and describe carcinogenecity test.

B. Short-Answer Questions

1. Describe in brief :

(i) Environmental carcinogens

(ii) Exposure to chemical carcinogens in work-place

(iii) Important environmental carcinogens identified by IARC.

2. Write short notes on :

(i) Carcinogenic polycyclic hydrocarbons

(ii) Important organic carcinogens

(iii) Metals associated with human cancer

(iv) Metals as carcinogens.

3. Briefly describe :

(i) Genotoxic carcinogens

(ii) Epigenetic carcinogens

(iii) Hormonal carcinogenesis

(iv) Oncogenes.

12

Major Anthropogenic Global Environmental Problems

SYNOPSIS

General account, Acid rain, Eutrophication, Greenhouse effect, Smog, Ozone depletion.

General Account

The planet (earth) was formed over 4,600 million years ago from hot gases and cosmic dust. Primitive life probably evolved about 3,500 million years ago in the form of a single cell organism and bacteria. 400 million years ago life forms began to colonise the dry land. The human species evolved about 100,000 years ago. It was not until about 10,000 years ago that modern man emerged with the development of village, agriculture and the domestication of animal and plant species. From this perspective the human species may seem to be an insignificant episode in the planet's history. However, the global population is growing by 75-80 million a year and is expected to reach approximately 8 billion by the year 2025. This enormous population requires food, water, clothes, houses, medical care, transport, education, employment and recreation. The world we live in has to provide the raw materials for all these activities. The problem we face today is managing the earth's resources by utilizing man's ingenious technology, to support a huge population while not destroying the planet we live on.

The environmental awareness of the 1990s has enabled us to look at our activities and examine the effect that we are having on the planet earth. We are faced with the following major man-made global environmental problems :

1. Acid rain
2. Greenhouse Effect
3. Smog
4. Ozone depletion
5. Eutrophication.

1. ACID RAIN

Introduction

The term "acid rain" was first coined by Robert Angus in 1872. "Literally it means the presence of excessive acids in rain waters". Acid rain is, in fact, a mixture of mainly H_2SO_4 and HNO_3 where the ratio of these two may vary depending upon the relative quantities of oxides of sulphur and nitrogen emitted. H_2SO_4 is the major contributor (60-70%) to acid precipitation, HNO_3 ranks second (30-40%) and HCl third.

Acid formation takes place only at higher relative humidities (>70%).

Definition

One of the major environmental issues facing human societies at the national and international level is the problem of acid. The rainwater is always slightly acidic as CO_2 in the atmosphere gets dissolved in it. However, during recent years, it has been noted a further lowering of pH of rain water — often as low as 5.0 to 4.0. This lowering of pH is due to the dissolution of acids in the rain water. *Precipitation of oxides of sulphur (SO_x) and nitrogen (NO_x) with rain is termed acid rain.*

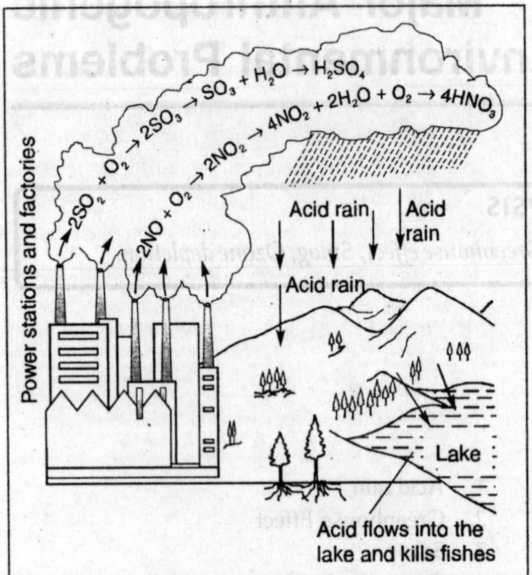

Fig. 12.1 : Schematic representation of Acid Rain

Causes of Acid Rain

Acid rain (Fig. 12.1) is caused by air pollution. When oxides of sulphur (SO_x) and oxides of nitrogen (NO_x) dissolve in rain water they form sulphuric and nitric acids, respectively. The rain water falls as acid rain. The chemical reactions for H_2SO_4 and HNO_3 formation are :

A. Reactions for sulphuric acid (H_2SO_4) *formation*

$$2SO_2 + O_2 \rightarrow 2SO_3$$
$$SO_3 + H_2O \rightarrow H_2SO_4$$

The H_2SO_4 (sulphuric acid) in combination with NH_3 as an environmental pollutant may form aerosols :

$$H_2SO_4 \xrightarrow{2NH_3} (NH_4)_2SO_4$$

B. Reactions for Nitric Acid (HNO_3) *formation*

Actually, two processes may be involved in the formation of nitric acid :

I. $2NO + O_2 \rightarrow 2NO_2$

$4NO_2 + 2HOH + O_2 \rightarrow 4HNO_3$

II. $NO + O_3 \rightarrow NO_2 + O_2$

$NO_2 + O_3 \rightarrow NO_3 + O_2$

$NO_3 + NO_2 \rightarrow N_2O_5$

$N_2O_5 + H_2O \rightarrow 2HNO_3$

From the above reactions it becomes clear that acid rain formation may take place in two phases :

(i) *Gas phase* — Where sulphuric acid and nitrogen are oxidized, and

(ii) *Liquid phase* — Where hydroxylation of oxides of S and N — (SO_x, NO_x) — takes place.

Sources of SO_x and NO_x

The main source of oxides of sulphur and nitrogen is the burning of fossil fuels in power plants based on coal and oil. This contributes more than 60% of all SO_x and 25-30% of NO_x in the atmosphere. Other source of oxides of sulphur and nitrogen are smelters.

Automobiles make a substantial contribution in large cities. Ozone is now regarded as a major factor in the formation of acid rain.

Acid Rain Problem Areas

Acidification of environment is a man-made phenomenon. The problems of acid rain are realized in North American and the Scandinavian countries where the pH of rain water has been recorded in the range of 4.4-4.0. In some regions, pH values as low as 2.8 have been recorded. In Holland, the normal pH of rainfall was anticipated to be only 3.7 in 1985. In India, low pH rain has been recorded in Mumbai (pH 4.8), Hyderabad (pH 5.7), Kolkata (pH 5.8) and Delhi (pH 6.2). Some areas like Mediterranean region and Balkans are noted as secured areas, as the soil is better and able to resist the acidification.

Role of Wind in Acid Rain

The oxides of sulphur and nitrogen (SO_x and NO_x) are swept up into the atmosphere and can travel thousands of kilometers. The longer they stay in the air, the more likely they are to be oxidized into acids. For example, a given molecule of SO_2 may remain in the atmosphere up to 40 hours, while a sulphate particle remains for three weeks. They have enough long life period and so these molecules may be transported by wind several kilometers away from their point of release.

It has been observed that a molecule of SO_2 originated in Ohio might be transported to New York. These are also wind blown from USA to Canada, or vice versa. Recently it is estimated that 87% of the sulphate in New York and New Jersey and 92% of it in New England has been carried in by long distant transport from the middle west (Wolff, 1982).

What is Wet and Dry Acid Rain?

Actually, acid rain is the one phase of acid deposition which can either be wet or dry. Acid rain, snow, dew, fog, frost and mist represent the wet form of deposition, while dust particles containing sulphates and nitrates settled on earth is called dry deposition. The wet rain is much more common.

Effects of Acid Rain

Acid rain problem is one environmental evil in India and is spreading fast in the developing countries. The most worrying aspect is that the area affected by acid deposition has been increasing day by day.

Some of the acid precipitation effects are :

Effects of Acid rain on Aquatic Ecosystem

1. Acid rain creates complex problems and their impacts are far reaching in aquatic ecosystem. There are 15,000 fishless lakes in Sweden and about 100 such lakes in the Adirondack region of USA because of increased acidity of the lake. The fishless lakes are now "fish graveyards"
2. Acid rain causes impairment in biological activity of aquatic biota.
3. Acid rain (pH below 6) causes decline in the population of phytoplanktons and zooplanktons.
4. It adversely affects nutrient cycle.
5. It causes decrease in fish food.
6. Fish population declines.
7. pH below 5 makes fish unable to reproduce.
8. It causes osmoregulatory failures in fingerlings.
9. Fecundity in fishes is highly reduced in acidified lakes.
10. Different fish species react differently to acidified lakes. Adult fish can survive in more acidic water having high concentration of aluminium than fry fish. Brook trout is most acid tolerant while rainbow trout the least. So a few critically affected lakes still have a population of mature fish.
11. Ion exchange across gill membrane in fishes is impaired causing death.
12. Low pH in lake enhances the toxicity of dissolved heavy metal ions in fishes.
13. Low pH of the lake causes damage of the leaves of aquatic plants.
14. The acidic water can also leach aluminium from the soil. So the run-off can carry dissolved aluminium to lakes, rivers and streams. It is highly toxic to aquatic animals and cause death of fishes by clogging its gills and depriving it of oxygen. Besides this, aluminium can also bind with organic particles, which are also toxic for fishes.
15. Many bacteria and blue green algae are killed due to acidification, disrupting the whole ecological balance.
16. Low pH of the aquatic environment, where fishes cannot survive, provides favourable conditions to flourish mosquitoes, black flies and other aquatic worms. Dragon fly larvae and water boatman also flourish in acidified lakes.

To sum up, acid rain causes a number of complications in ponds, rivers and lakes where it accumulates as "acid snow". In summer, rapid snowmelt gives a jolt of acid waters to lakes.

This "acid jolt" is most damaging to young fishes, algae, insects and to the food chain. In Sweden more than 2,500 lakes have their water acidic due to acid rains, and this has resulted in great damage to fish and fisheries in these lakes.

Effects of Acid Rain on Forest Ecosystem

1. In 1958, the pH of the rain at Europe was 5.0 and in 1962 the pH of rain at Netherlands was 4.5. This acid rain had damaged leaves of plants and trees and had retarded the growth of *Swedish forest*.
2. Due to acid rain, forests in *Germany, Switzerland, Netherlands and Czechoslovakia* have been damaged.
3. Acid rain causes leaching away of nutrients like calcium, potassium, iron and magnesium from the soil. Hence soil fertility is reduced.

4. Acid rain causes reduced plant growth by decreasing soil respiration and fertility.
5. Acid rain also leaches away the nutrients from the plants.
6. Acid rain reduces the photosynthetic rate of plants.
7. Acid rain decreases activity of symbiotic nitrogen fixing microbes present in the soil.
8. Acid rain also affects the mineralization of nitrogen.
9. Acid rain in Japan has damaged 5,000 sq. kms of cedar trees in Kanto plain, north of Tokyo.
10. Acid rain produces chlorotic spots on the leaves.
11. Acid rain accelerates cuticular erosion of leaves.
12. Acid deposition weakens the trees like pine, spruce, ashes, and birch which can be easily attacked by pathogens and drought.
13. Root systems are damaged by the uptake of aluminium released from the soil.
14. Lichens are adversely affected due to rise in soil sulphate concentration.

Effects of Acid Rain on Terrestrial Ecosystem

1. Acid rain causes reduced rate of photosynthesis, growth and increased sensitivity to drought and disease.
2. Acid rain has severely retarded the growth of crops such as pea, beans, radish, potato, spinach, broccoli, carrots etc.
3. The activity of symbiotic nitrogen-fixing bacteria present in the nodules of leguminoseae family inhibited, thereby decrease in the fertility of soil takes place.
4. Acidification of soil changes its biology and chemistry. Plants can easily absorb cadmium from acidified soil. High level of cadmium in plants are deleterious for animals and human beings.
5. Due to acid rain, the lichens growing on plant barks lose their hold and, therefore, fail to grow.
6. Death of lichens affects nitrogen cycle.
7. Acid rain causes leaching of aluminium and nitrates from the soil and hence root system is damaged.
8. Trees pull heavy metal ions, like Cd, Cr and Al, into their systems. These ions are immobile in many soils, but acidic water mobilizes them and causes tree damage.
9. Acid rain enhances rise in harmful microbes.
10. Acid rain affects adversely the nitrogen balance of the ecosystem.
11. Germination index of seeds slows down in acidified soils.
12. Barks of trees absorb acid and thus the protective cover of trees becomes weak, which exposes them to other diseases.
13. Acid rain kills small invertebrates of the terrestrial ecosystem.

Effects of Acid Rain on Human Beings

Acidification can play havoc with human digestive system, respiratory system and nervous system by making the person an easy prey to neurological diseases. This happens because these acids produce highly toxic compounds, which contaminate the portable water and enter man's body. Acid rains containing air pollutants contribute to a variety of safety hazards, associated with reduced visibility due to smog etc. These contaminants can be nuisance in several aspects and cause adverse health effects.

Effect of Acid Rain on Buildings

1. Acid rain causes extensive damage to building and structural components of marble, limestone, slate, mortar, etc. Limestone is attacked rapidly :

$$CaCO_3 + \underset{\text{(Acid rain)}}{H_2SO_4} \rightarrow CaSO_4 + HOH + CO_2$$

   The attack on marble is termed "stone leprosy".
2. In Greece and Italy invaluable stone statues have been partially dissolved by acid rain.
3. British Parliament building also suffered damage due to H_2SO_4 rain.
4. Due to acidity, levels of heavy metals as Al, Mn, Pb, Cd, Cr and Cu in water increases beyond the safe limit which indirectly affect the buildings.
5. Acid rains are great threat to British environment. Much of the falling snow in Britain is now highly acidic. If it does not melt it may turn into a *"pollution time bomb"*.

Episode : Taj Mahal and Acid Rain

The world's attention was drawn seriously to SO_2 emissions in India, when an oil refinery — a stationary SO_2 emission source — was planned for construction near Taj Mahal. But this refinery was badly needed by the farmers of the area, for the supply of diesel fuel for tractors and irrigation pumps. This oil refinery, lying only 40 km away from Taj Mahal, emitted about 25-30 tons of SO_2 daily. The atmospheric concentration of SO_2 in Agra have been around 1.75mg/m³. However, if the SO_2 levels in air increase, the resulting acidic precipitation may react with $CaCO_3$ present in the marbles :

$$CaCO_3 + H_2SO_4 \rightarrow CaSO_4 + HOH + CO_2$$

(present in marble) (Acid rain)

It may cause pitting in the wonderful monument that has attracted people from around the world. Thus any increase in SO_2 pollution levels in the long run may prove dangerous to this structure.

Control of Acid Rain

No effective control for acid rain has yet been devised. There is an urgent need for proper regular monitoring to provide timely warnings about acidification of our environment. Short-term control of acid deposition problem can be achieved by using lime. New York has been liming the lakes and ponds since 1959. Mr. M. C. Chawala (1989) was the first Indian citizen who filed writ petition in the Supreme Court of India regarding the protection of Taj, a wonderful monument. The Supreme Court of India immediately issued an order for Taj to be secured, especially from acid rain.

2. GREENHOUSE EFFECT

Introduction

Actually, 'greenhouse' effect is a theory which proposes that pollution by common anthropogenic pollutants such as CO_2, CH_4, CFCs, ground O_3 etc. may lead to an increased global temperature. Over the last century, especially CO_2 concentration have risen and global air temperatures have increased by 0.4-0.7°C, which supports this theory. The atmosphere formed around the earth insulates the planet from the full effects of heat loss by trapping heat in the atmosphere using greenhouse gases like CO_2, CH_4, CFCs, ground O_3, N_2O, SO_2, water vapour etc. Doubling of the atmospheric CO_2 concentration from its present level is predicted to lead to a further warming of around 3.5°C.

The concentration of CO_2 in the atmosphere has increased from about 280 ppm in 1750 to about 350 ppm in 1990 and is still rising, with the main reason for the increase being the combustion of fossil fuels. Estimates vary, but it appears likely that the concentration will rise to a mean of about 550 ppm by the year 2050.

Term Greenhouse effect coined by—J. Fourier (1827). It is also called "Atmospheric Effect" or "Global Warming".

Definition of Greenhouse Effect : According to Botkin and Keller (1982), in a greenhouse, visible light passes through the glass and heats up the soil warming the plants. The warm soil emits radiation in longer wavelengths, particularly the infrared or heat. Because glass is opaque to longer wavelength of infrared radiation, it partly reflects and partly absorbs infrared radiation. This mechanism keeps the greenhouse warmer than the outer atmosphere (Fig. 12.2).

Thus a *greenhouse* is that body which permits the short wavelength incoming solar radiation (light energy) to come in, but does not permit the long wave (heat energy) outgoing terrestrial infrared radiation to escape.

In a similar way, the earth's atmosphere bottles up the energy of the sun and is said to act like a *greenhouse,* where CO_2 — whose concentration is increasing at the rate of 0.4% per annum — acts like a glass window (Fig. 12.3 A and B). Increasing CO_2 content along with water vapours in the atmosphere transmits short wave length solar radiation but reflect the longer wave length heat radiation from warmed surface of the earth. CO_2 molecules are transparent to sunlight but not heat radiation. So they trap and reinforce the solar heat stimulating an effect which is popularly referred to as "Greenhouse Effect" (Fig. 12.4).

The *Greenhouse Effect* may, therefore, be defined as "the gradual warming or heating up of the earth's surface due to blanketing effect of man-created excess CO_2 in the atmosphere".

or

Greenhouse Effect is "the phenomenon due to which the earth retains heat."

or

Greenhouse Effect means the excessive presence of gases like CO_2, CH_4, CFCs etc. blocked in the infrared radiation from the earth's surface to the atmosphere leading to an increase in temperature, which, in turn, would make life difficult on earth in the coming future. This phenomenon is popularly known as "Global Warming".

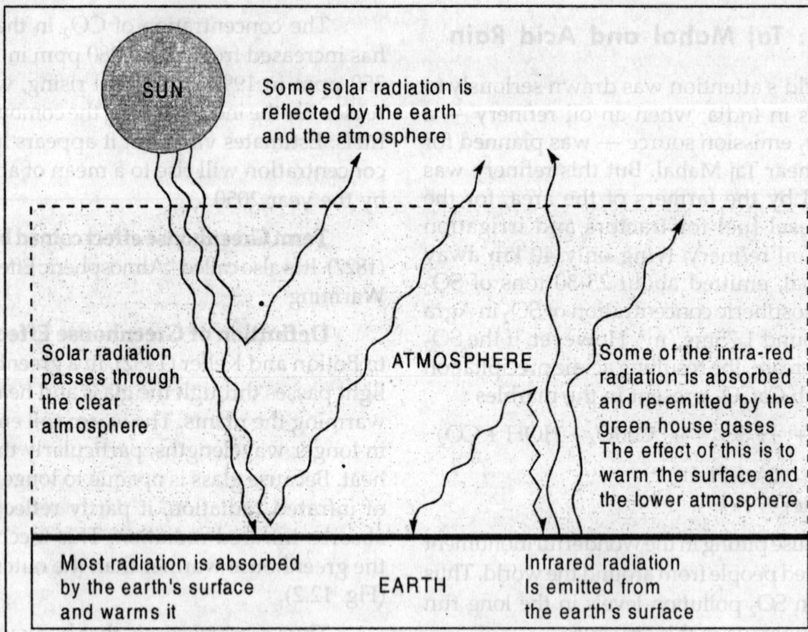

Fig. 12.2 : Simplified representation of global warming

Greenhouse Gases

The principal gases responsible for Greenhouse Effect are

1. Carbon dioxide (CO_2)
2. Methane (CH_4)
3. Nitrous oxide (N_2O)
4. Synthetic chlorofluorocarbons (CFCs)
5. Ozone (O_3) at ground
6. Sulphur dioxide (SO_2)
7. Sulphur hexafluoride (SF_6)
8. Water vapour.

Among these, CO_2 is the most common and important greenhouse gas which gets introduced into the atmosphere from fossil fuel burning, furnaces, uncontrolled industrial emissions, destruction of forests, automobile emissions, breathing of organisms etc. It shall not be out of

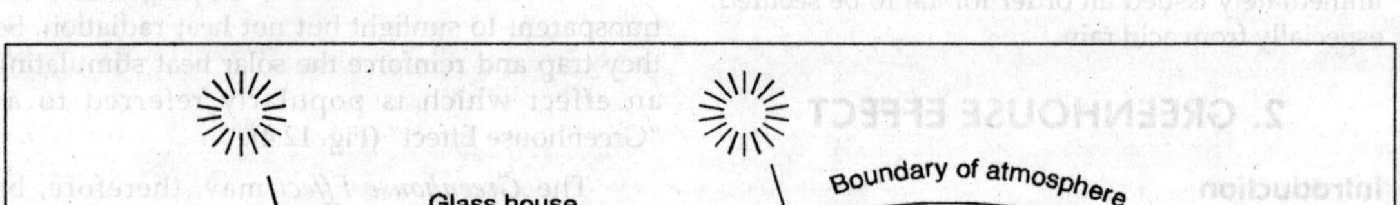

Fig. 12.3 (A) : Light rays penetrate the glass house greenhouse. They are reradiated as heat rays. The glass does not permit the heat rays to escape. So the heat remains within the greenhouse and the house remains warmer. **(B) :** Carbon dioxide of atmosphere, like glass, allows light rays to pass down. But it absorbs the infrared heat rays. An increase in CO_2 could therefore lead to an increase in the temperature of the biosphere as a whole.

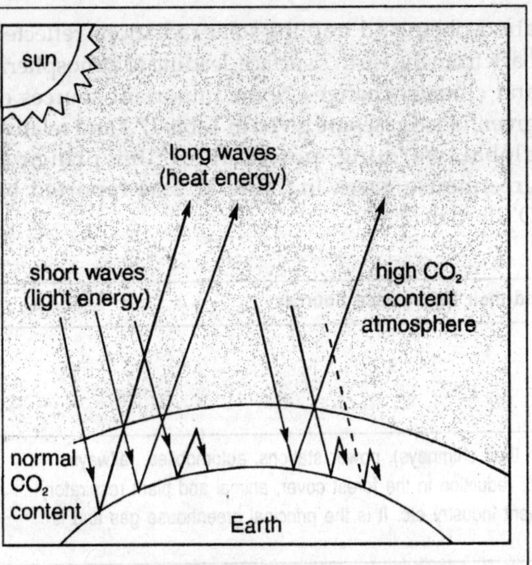

Fig. 12.4 : Greenhouse effect

place to mention here that global background of CO_2 in the atmosphere was around 210 ppm, before the Industrial Revolution (started about 1760). After about 200 years, in 1981-82, CO_2 in the atmosphere was about 340 ppm — which might double up in another 50 years. It may change the climatic conditions, especially by raising the global temperature. This effect caused by CO_2 is chiefly due to the absorption of the long-wave radiation of the sun and retention of the back-radiation from the earth. It is similar to the role of glass panels in a greenhouse (greenhouse is a structure built with glass walls and roof, inside which experimental studies with plants are carried out by scientists (Fig. 12.4), which reduces the convective cooling by the outside air. Similarly, CO_2 accumulating in the atmosphere environment prevents the heat escape and increases the temperature of the earth surface that is within its coverage. Such a phenomenon is called Greenhouse Effect. Such effect would result in the warming up of the earth. Mentioned above O_3 and SO_2 also act as serious pollutants in causing global warming. The Greenhouse gases such as CH_4 and CFCs contribute about 18% and 14%, respectively, to the global warming.

One molecule of CH_4 absorbs 20 times more heat than CO_2 molecule and one molecule of CFC absorbs 10,000 - 20,000 times more heat than CO_2 molecule.

All the greenhouse gases are increasing at a rapid rate. For example, CH_4 has approached at present a level of approximately 1.65 ppm from a pre-industrial value of 0.7 ppm. N_2O is increasing at the rate of 0.25%, while CFCs at the rate of 5%. It is expected that in a period of about 25 years, i.e. around the year 2030, the non-CO_2 effects are going to be one and half times longer than those of CO_2 itself!

Principle of Greenhouse Effect

The principle of the phenomenon of Greenhouse Effect is based upon the principle of infrared absorption characteristics of gases. Higher the concentration of CO_2, greater would be the absorption of thermal radiation, which means that more infrared radiation (heat energy) are reemitted back to the earth's surface, resulting in *'heat trap'* — increasing mean global temperature (Fig. 12.5).

Illustration of Greenhouse Effect

Greenhouse gases, for example, thick CO_2 layer, acts like a glass panel of greenhouse or the window glass of a closed car, allowing the sun rays to filter through but preventing the heat from being escaping in the outer space; thereby warming the troposphere of the atmosphere. This phenomenon may well be understood when one returns to the car after a while whose window glasses are closed and have trapped the heat inside on a warm day. This is exactly what happens in a greenhouse. Greenhouse gases have no effect on the incoming short wave radiations.

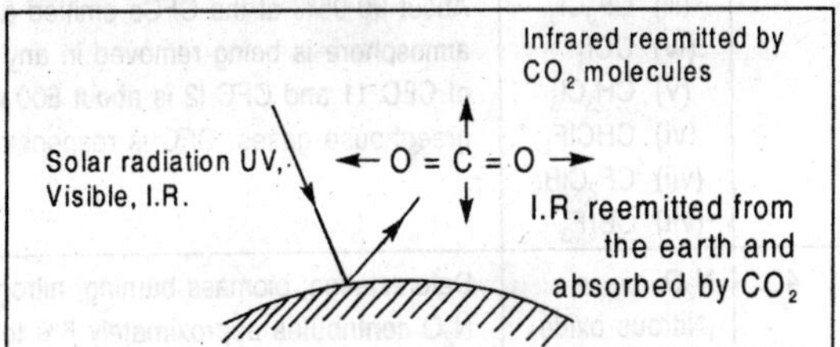

Fig. 12.5 : An illustration of greenhouse effect, I.R. radiation absorbed by CO_2 molecules is reemitted in all directions deflecting some back to the earth's surface.

Major sources of Greenhouse Gases

W.M.O. (World Meteorological Organization) has suggested that a rise in global air temperature could occur in coming 40 - 50 years and that will exceed any climatic change ever experienced in human history. The earth's temperature has not varied by more than 1 or 2°C during the past 10,000 years : but with the increasing concentration of CO_2 in the atmosphere, the earth may become warmer by about 1.5 - 4.5°C by the year 2025-2050 A.D.

Various industrial as well as agricultural operations generate and emit waste gases into the atmosphere. For instance, from burning of fossil fuels alone, more than 2.5×10^{13} tons of CO_2 is being emitted into the atmosphere each year. Growing paddy or live-stock releases CH_4. The application of aerosols as coolants in refrigerators and air-conditioning devices or sprays release CFCs into the atmosphere. These gases create a canopy in then atmosphere and trap the solar radiations reflected back from the earth's surface, leading to atmospheric and climatic changes. Some important sources of greenhouse gases are given in Table 17.1 and relative global warming potential of the principal greenhouse gases in Table 12.2 as reported by O'Riordan (2000).

Table 12.1 : Greenhouse Gases and their Use/release Sources

Sl. No.	*Name of Greenhouse gases*	*Use/Release* sources
1.	CO_2	Burning of fossil fuels, factories (through their chimneys), power stations, automobiles, railways, aircraft, burning of firewood, deforestation, reduction in the forest cover, animal and plant respiratory end product, petroleum by-products, cement industry etc. It is the principal greenhouse gas and is responsible for 50% of global warming.
2.	CH_4	Forest fires are the principle sources of CH_4 and emit one unit of CH_4 for every 100 units of CO_2. Other sources are anerobic decomposition in biological systems, termites, biomass burning, landfills and wetlands. The main sources are paddy rice farming and cattle. Methane is responsible for approximately 18% of global warming.
3.	Man-made CFCs as (i) $CFCl_3$ (ii) CF_2Cl_2 (iii) CF_3Cl_3 (iv) CCl_4 (v) CH_3Cl_3 (vi) $CHClF_2$ (vii) CF_2ClBr (vii) $CBrF_3$	CFCs are released to the atmosphere during the operations and maintenance of appliances and equipments using these molecules as coolants and propellants. CFCs are inert, non-toxic and easily liquefied. CFCs are being frequently used in refrigerators, air-conditioners, in packaging (Styrofoam cups, egg cartons), insulation, in aerosol propellants, and as a solvent for cleaning microchips and electronic circuit boards in computers, automobile air-conditioning etc. About 90-95% of the CFCs emitted are still accumulating in the atmosphere; only 1% of CFCs in the atmosphere is being removed in any one year; annual increases are about 10-11%. Recent global use of CFC 11 and CFC l2 is about 800 million kg. CFCs are considered to be the most potent greenhouse gases. CFC is responsible for approximately 14% of global warming.
4.	N_2O (Nitrous oxide)	Deforestation, biomass burning, nitrogenous fertilizers, and combustion of fossil fuels, N_2O contributes approximately 6% to global warming
5.	SO_2	Combustion of fossil fuels, industries, transportation etc.
6.	Ozone	At ground level, O_3 is used as a disinfectant for air and water bleaches, textiles and oils. It is produced as a result of chemical manufacturing processes, electrical discharges, occurring with lightning during thunderstorms or with the sparking of electrical appliances. It contributes approximately 12% to global warming.

Table 12.2 : Global warming potential of selected principal greenhouse gases

Greenhouse gas	Chemical Formula	Lifetime (Years)	Global warming potential (years) 20	100	500
Carbon dioxide	CO_2	140	1	1	1
Methane	CH_4	12	56	21	6.5
Nitrous oxide	N_2O	120	280	310	170
HFC group	C-H-F	1.5 50	5,000	3,000	500
Sulphur hexafluorides	SF_6	3,200	16,300	24,900	35,900
Perflurocarbons	C-F	3 - 10,000	6,000	8,000	14,000

Possible Consequence of Greenhouse Effects

1. Temperature changes

If the present rate of CO_2 emission continues unchanged, the temperature of the earth would increase by 22°C (40°F) in another 50 years. Considering the effects of other greenhouse gases (CH_4, O_3, N_2O, CFCs) this effect could be more pronounced.

2. Melting of polar ice cap

Scientists feared that an increase in the average global surface temperature by 4.5°F - 10°F could lead to increased melting of the polar ice sheets and glacial ice. The melting glacier may raise the sea level so much that many major coastal cities like New York, Los Angeles, London, Venice and Shanghai will be flooded. Kolkata and Mumbai in India may similarly be affected.

3. Climatic changes

An average rise in the global temperature by only 3°C could mean increase of more than 10°C at high latitudes in some seasons (U.N.E.P, 1987). The "Greenhouse Effect" will make winters shorter and wetter, and summer longer and drier in the Northern latitudes. Subtropical regions might become even drier than they are at present and tropical regions would become wetter. This is because the rainfall pattern may also be affected. With increasing temperature and thus evaporation, a corresponding rise (11 - 15%) a year in rainfall has been estimated. This could mean that cropping pattern and regions might be affected as well.

4. Crop production and nutrient depletion

The promising effect of climatic change due to greenhouse effect will be on agriculture. A warmer climate is likely to move the areas suitable for growing specific crops such as wheat towards the poles. There will also be widespread changes in natural ecosystem with grasslands and desert areas expanding and forests shrinking and moving pole-wards.

Because CO_2 is a natural fertilizer, most plants would be high yielding and faster in a high CO_2 world, but would demand more fertilizers since the soil could be improvised faster, as it supports an increased crop cover.

If the CO_2 levels double as expected by 2050 the yields of many crops and weeds could increase by an average of 30%; yields of some crops could even double while that of others change very little. Response actually depends on how a plant photosynthesizes. Plants that produce intermediary chemicals with 3 carbon atoms (C3 plants) would respond well to an increase in the atmosphere levels of CO_2 but those plants that produce metabolic intermediaries with 4 carbon atoms (C4 plants) would do much less so.

Of the world's 20 major food crops, maize, sorghum, millet and sugarcane are C4 plants, whereas the other sixteen crops (e.g., soyabeans, rice, wheat etc.) are C3 plants whose yields would increase in a CO_2 enriched atmosphere. Undoubtedly this would alter the economic viability of farming, agricultural employment, commodity prices and patterns of world trade.

Miscellaneous Effects

I. Now due to population explosion, forest destruction have a serious effect on the level of atmospheric CO_2. If the present trend of deforestation continues, then the amount of CO_2 in the atmosphere might reach up to approximately 0.0355% by volume.

At higher altitudes in the atmosphere, CO_2 undergoes photochemical reactions producing CO, which is drastically dangerous :

$$CO_2 + h\gamma \rightarrow CO + O$$

II. Because of increased concentration of CO_2 and due to much warmer tropical oceans, there may occur more cyclones and hurricanes and early snow melt in mountains may cause more floods during monsoon.

According to UNEP, within next 25 years or so, there will be a rise in sea level by 1.5 to 3.5 metres. In Bangladesh alone 15 million people will have to shift or drown.

III. It is further notable that if CO_2 level continues to increase, it would accumulate and may inhibit the cooling effect of aerosols and particulate contaminants — consequently temperature of the atmosphere may rise again.

IV. The cloud cover blanket may rise above 50 - 60%.

V. Overall equilibrium of the ecosystem will be disbalanced.

Remedial Measures of Greenhouse Effect

UNEP has chosen the slogan "*Global Warming*" : "*Global Warning*" to alert the public on World Environment Day, June 5, 1989.

Slight increases in global temperature can adversely affect world food production. Biological productivity also decreases due to warming of surface layer. It is, therefore, absolutely necessary that the forest cover should be maintained at least to one-third of the total land area of the world.

The greenhouse effect can be controlled to certain degree by taking the following important measures :

(i) Reducing the consumption of fossil fuels such as coal and petroleum. This can be achieved by depending more on non-conventional renewable sources of energy such as wind, solar, nuclear and biogas energies.

(ii) Disposing of the greenhouse gases as they are formed elsewhere than in the atmosphere.

(iii) Recovering greenhouse gases present already in the atmosphere and disposing them off elsewhere.

(iv) Learn to adapt and accept the changing climate.

(v) International cooperation operating for attempting the reduction of greenhouse gases.

(vi) Enriching oceans with iron can help in absorbing greenhouse gases such as CO_2. Actually the theory is that many parts of the ocean lack iron, which stunts the growth of algae. Enriched with enough iron, the algae will bloom luxuriantly, soaking the atmospheric CO_2, the *principal culprit behind global warming*.

(vii) Enhancing forestation will certainly reduce the CO_2 level, thereby decreasing the "Greenhouse Effect". As trees are the big natural 'sink' of CO_2, they can utilize CO_2 during photosynthesis. According to an estimate, if the forest-cleared lands are reforested in India, Brazil, Indonesia, Africa, Pakistan and Columbia etc., about 50 - 60% of the total released CO_2 may be effectively reduced. Thus there should be a legal ban on deforestation which is increasing alarmingly, particularly in the developing countries.

(viii) There should be a restriction on the emission of dangerous CO_2 and CFCs. from the factories and automobiles.

(ix) '*No Driving*' day once a week should be imposed on all the countries — especially in developed countries like U.S.A.

3. SMOG

Term coined by — Voeux, H.A. Des (1905).

Introduction

The word smog is derived from an elision of smoke and fog. The oldest smog, a mixture of coal smoke and fog, has plagued human beings since 14th century. Urban smog, also known as smoky fog (occurs normally over industrial areas), happened in Dec. 1930 in Meuse Valley of Belgium, which killed about 600 people. A classical example of smog (smoke + fog) is London smog of Dec. 1952, which killed approximately 5,000 people in a single year and thousands became ill. Another example of smog is Donora Pennsylvania Smog (USA) which killed 20 people and made hundreds ill. This smog persisted for 5 days and was preceded by the formation of thick fog due to strong inversion temperature on Dec. 26, 1948.

The other type of predominant smog is called Photochemical or Los Angeles or Killer Smog or Oxidizing Smog. Its first clue appeared in 1944 in Los Angeles and hence the name Los Angeles Smog. It was mainly a mixture of oxidizing pollutants, hence it is known as *oxidizing smog*, while London Smog consisted mainly of a mixture of reducing pollutants and so it has been called *reducing smog*. Photochemical smog problems also occur in several other countries like Tokyo (Japan), Mexico, Sydney, Melbourne etc. This problem is also experienced in some cities of India, such as Mumbai, Chennai, Delhi, Kolkata, Kanpur, Bangalore and Ahmedabad. During 1987, Mumbai experienced heavy smog for about 10 days.

Components of London Smog or Reducing Smog or Classical Smog

(i) SO_x (ii) Particulates, such as soot, $(NH_4)_2SO_4$ etc. (iii) humidity or water from fog. It involves no photochemistry and is commonly known as sulphurous smog.

Components of Photochemical Smog or Los Angeles Smog or Oxidizing Smog

1. Inorganic gases such as O_3, nitrogen oxides (NO_x), H_2O_2 and CO.
2. Organic peroxides, peroxyacetyl nitrate (PAN), peroxybenzoyl nitrate (PBN), Peroxy propionyl nitrate (PPN), peracetic acid (CH_3COOOH), acetylperoxide [CH_3 (CO) OO (CO) CH_3], ethylhydroperoxide etc.

The components of typical photochemical smog, along with their optimum level, are given in the Table 12.3 :

Table 12.3 : Components of photochemical smog and their levels

Sl. No.	Components	Optimum level (ppm)
1.	Carbon monoxide	2.0
2.	Nitric oxide	0.15
3.	Unsaturated hydrocarbons	0.05
4.	Aromatic compounds	0.20
5.	Nitrogen oxide	0.20
6.	Ozone	0.10
7.	Aldehydes	0.20
8.	Organic peroxides	0.03

London Smog was characterized by bronchitis and irritation whereas Los Angeles Smog was characterized by reduced visibility and eye irritation (lacrimation).

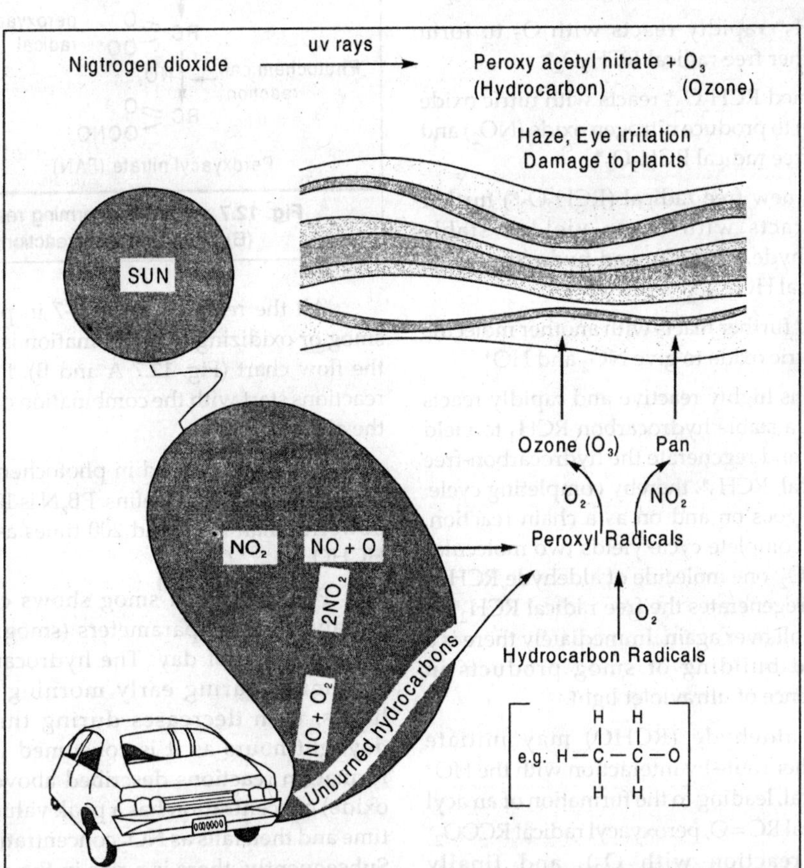

Fig. 12.6 : Some main chemical reactions in photochemical smog

Smog is formed of fog and smoke whereas photochemical smog is formed by a complex series of chemical reactions, when oxides of nitrogen (NO_x) and hydrocarbons of the exhausted gases from factories and engines of automobiles (Fig. 12.6) are exposed to ultraviolet (UV) rays of sunlight.

Photochemical smog is characterized by brown-hazy fumes, which irritate the eyes and lungs, lead to the cracking of rubber and extensive damage to plant life.

Mechanism of Photochemical Smog (Oxidizing Smog) Formation

The most probable mechanism of photochemical smog forming reactions are :

1. Reactive hydrocarbons (C = C group) from automobile exhaust interact with O_3 to form a hydrocarbon-free radical RCH_2*.
2. RCH_2* rapidly reacts with O_2 to form another free radical RCH_2O_2*.
3. Formed RCH_2O_2* reacts with nitric oxide (NO) to produce nitrogen oxide (NO_2) and the free radical RCH_2O_2*.
4. This new free radical (RCH_2O_2*) further interacts with O_2 to yield a stable aldehyde, (RCHO), and hydroxyperoxyl radical HO_2*.
5. HO_2* further reacts with another molecule of nitric oxide to give NO_2 and HO*.
6. HO* is highly reactive and rapidly reacts with a stable hydrocarbon RCH_3 to yield H_2O and regenerate the hydrocarbon-free radical, RCH_2*, thereby completing cycle. This goes on and on as a chain reaction. One complete cycle yields two molecules of NO_2, one molecule of aldehyde RCHO, and regenerates the free radical RCH_2* to start all over again. Immediately there is a rapid building of smog products in presence of ultraviolet light.
7. The aldehyde (RCHO) may initiate another route by interaction with the HO* radical, leading to the formation of an acyl radical RC = O, peroxyacyl radical $RCOO_2$* (by reaction with O_2), and finally peroxyacyl nitrate PAN (by reaction with NO_2). PAN is one of the most potent eye irritants found in smog.

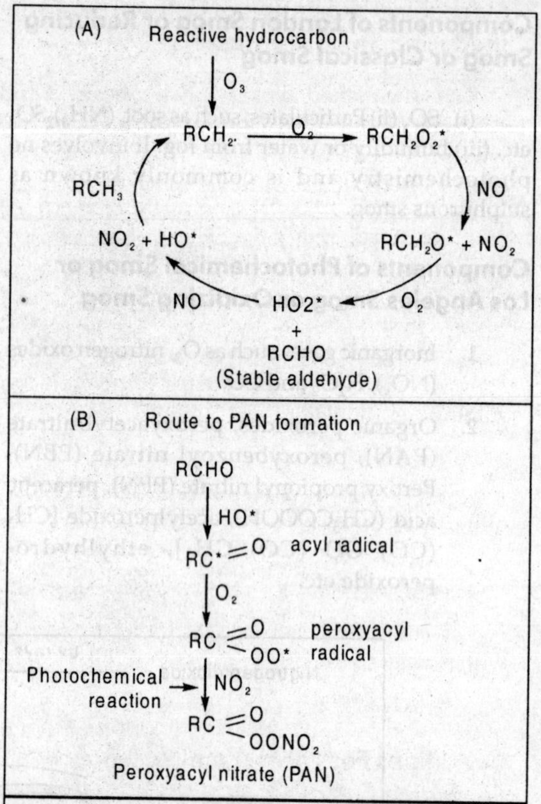

Fig. 12.7 : (A) Smog-forming reactions (B) PAN formation reactions

All the reactions from 1-7 in photochemical smog or oxidizing smog formation is illustrated in the flow chart (Fig. 12.7 A and B). Photochemical reactions start with the combination of N_2 and O_2 in the air.

PB_zN is produced in photochemical smog in presence of NO_2 and olefins. PB_zN is 100 times more powerful than PAN and 200 times as powerful as HCHO.

Photochemical smog shows characteristic variations of the parameters (smog ingredients) with the time of day. The hydrocarbon level is maximum during early morning traffic rush hours, then decreases during the remaining daylight hours as it is consumed in the smog-formation reactions described above. NO (nitric oxide) concentration has a peak value at the same time and then falls as NO_2 concentration increases. Subsequently, there is a rise in the concentration of oxidants (aldehyde, PAN), which are active oxidizing agents and contain the irritating ingredients of smog.

Mechanism of London or Sulphurous Smog (Reducing Smog) Formation

As mentioned above, "smog" means an odd combination of smoke and fog, which in Dec. 1952 was prevalent in London. This is, however, chemically reducing with high level of SO_2 and is called reducing smog. It is also called *Sulphurous Smog*. The steps in the formation of London smog or sulphurous smog or reducing smog are shown in the form of following reactions :

$$S + O_2 \rightarrow SO_2 \quad (1)$$

SO_2 combines with moisture from fog to form sulphurous acid (H_2SO_3).

$$SO_2 + H_2O \rightarrow H_2SO_3 \quad (2)$$

The oxidation of sulphurous acid (H_2SO_3) is quite rapid, more so in the presence of metal salts :

$$H_2SO_3 + ½\,O_2 \xrightarrow{\text{metal salts}} H_2SO_4 \quad (3)$$

It is also notable here that H_2SO_4 may combine with NH_3 in moist air forming $(NH_4)_2SO_4$:

$$H_2SO_4 + 2NH_3 \rightarrow (NH_4)_2SO_4$$

Conditions For Photochemical Smog Formation

Automobile exhaust in the presence of NO_2 and UV radiation produce photochemical smog. Also, when organic substances like alcohol, olefins, paraffins, and carbonyl compounds are irradiated by UV radiation in the presence of NO_2, the photochemical smog may be formed. Hot climate (>24°C) with low humidity is favourable for the formation of photochemical smog. It does not rise high above in the air, instead, it circulates in the lower atmosphere of the earth.

Effects of Photochemical Oxidants on Humans

1. O_3 and PAN (peroxy acyl nitrate) chiefly cause irritation of the eyes creating lacrimation and affect severely the respiratory tract of human beings.

   Lower concentration of O_3 irritates the throat and nose while its higher concentration causes cough, headache, throat dryness, chest pain, and impairment in breathing. Exposure to 1.0 - 3.0 ppm of ozone produces fatigue and lack of coordination in central nervous system.
2. Exposure to 50 ppm of ozone for several hours may lead to mortality due to pulmonary edema i.e., storage of blood in the lungs.
3. NO_2 produces nose and eye irritation and also pulmonary discomfort.
4. NO_x also cause several chronic diseases of heart, lungs and eyes.
5. PBN (peroxy benzoyl nitrate), a secondary photochemical pollutant, occurs in polluted atmosphere involving aromatic hydrocarbons, NO_x and O_3. It acts as a powerful eye irritant and lachrymator.

Effects of Photochemical Oxidants on Plants

1. Smog having O_3, PAN and other photochemical oxidants is regarded to produce early maturity or senescence.
2. PAN causes injury in beets, spinach, lettuce, alfalfa, and coniferous trees. It also causes silvering of leaves.
3. NO_x and PAN cause death of forest trees.
4. PAN inhibits "hill reaction" of photosynthesis.
5. O_3 promotes excessive transpiration from the leaves of plants causing dehydration.
6. O_3 is also harmful for vegetables like radish, carrot and tobacco. It alone causes 50% crop loss.
7. O_3, together with PAN, form small drops in air forming smog, thus blanketing the sunshine, which indirectly inhibits the rate of photosynthesis in plants.
8. Photochemical oxidants (products) like O_3, PAN (peroxy acetyl nitrate), PBzN (peroxy benzoyl nitrate) are phytotoxic and kill microflora and fauna of soil and water.
9. O_3 causes leave tip burn, a disease of white pines.
10. Some sulphates and nitrates which are formed during smog formation due to the oxidation of sulphur-containing components (SO_2 and H_2S) and NO_x (N_2O_3, N_2O_5, NO_2), nitric acid and some nitrates are important toxicants of smog. They adversely affect plant growth and damage crops.

Ecological Effects of Photochemical Smog

Photochemical smog produces regional ecological effects such as

1. Disturbs thermal equilibrium
2. Reduces soil fertility
3. Affects freshwater lake and pond ecosystem.

Effects of Photochemical Oxidants on Materials

Scientists now believe that degradation of materials commonly attributed to "weathering" is actually the result of attack by photochemical pollutants. The smoke containing fog, dust, mist, soot etc. in the smog causes corrosion of metals, stones, building materials, textile, paper, rubber, leather and painted surface.

Organic polymers including rubbers, natural and synthetic textiles are subjected to chemical alteration upon exposure to varying quantities of ozone.

Long exposure of ozone to rubber in a relaxed state brings about no characteristic crack. However, exposure to a concentration of 0.01 - 0.02 ppm ozone causes cracks in rubber.

Secondary photochemical pollutants also attack cellulose in textile fabrics. The harmful effects of ozone exposure on textiles increase in the order for fabrics made of cotton, nylon, terrycot and polyester.

Control of Photochemical pollutants

The control of primary precursors such as hydrocarbons and NO_x may ultimately control O_3 and PAN formation, which are secondary pollutants. In present days some techniques have been devised to control hydrocarbon emissions from stationary sources. Important techniques are :

1. Incineration method
2. Adsorption method
3. Absorption method, and
4. Condensation method.

1. Incineration method

This method is based on the fact that hydrocarbon removal efficiencies are greater in flame after burner in which flame is applied to complete the oxidation of hydrocarbon into CO_2 and HOH.

2. Adsorption method

In this method, exhaust gases are released through a bed of granulated adsorber, which consists of activated carbon. Hydrocarbon vapours are adsorbed over the surface of activated carbon, which remain there until they are periodically removed by passing steam through the system. The hydrocarbons are then condensed to liquids and can be used for further purpose.

3. Absorption method

Liquid containing hydrocarbons is brought to intimate contact of exhaust gases. The scrubbed exhaust then passes on, leaving the hydrocarbons trapped in the scrubbing liquid. The contact between absorbing liquid and exhaust gases is generally carried in tall scrubbing towers.

4. Condensation method

Condensation of hydrocarbons requires low temperature. It causes gaseous hydrocarbon, i.e., smog forming pollutants, to condense to liquids, which are collected and reused. However, the control of hydrocarbon emission from automobiles is more complicated. For this, application of oxidizing catalyst may convert hydrocarbons and CO to CO_2 and HOH :

1. Hydrocarbons $\xrightarrow{\text{combustion}}$ $CO_2 + H_2O$
2. CO $\xrightarrow{\text{combustion}}$ CO_2

Researches are being conducted to utilize smog-forming gases and hydrocarbons.

Table 12.4 : Characteristics of "Classical" and "Photochemical" Smog

Sl. No.	*Characteristics*	*"Classical"*	*"Photochemical"*
1.	First occurrence noted	London (U.K.)	Los Angeles (U.S.A.)
2.	Principal pollutants	Sulphur oxides, particulate matters	Ozone, nitrogen oxides, hydrocarbons, carbon monoxide, free radicals
3.	Principal sources	Industrial and household fuel combustion (coal, petroleum)	Motor vehicle fuel combustion (petroleum)
4.	Effects on human	Lung and throat irritation	Eye irritation
5.	Effects on compounds	Reducing	Oxidizing
6.	Time of occurrence	Winter months (especially in morning)	Around midday of summer months
7.	Popular worst episodes	London Smog	Los Angeles Smog

4. OZONE DEPLETION

Introduction

Ozone (O_3), a natural constituent of the atmosphere, is composed of three oxygen atoms held together in their configuration by electrical attraction. This electrovalent bond is much weaker than the covalent bond that holds oxygen (O_2), so the molecule is always susceptible to having one of its oxygen atoms stripped by an 'oxygen-seeking' molecule. The reactivity of the chemical is in large part determined by the strength of the bonds between atoms. The weak bonds in the ozone molecule make it an even more highly reactive gas than oxygen. The word 'ozone' (in Greek, *ozein* means 'to smell') was first introduced by Shoenbein in 1840 to characterize a chemical species which had a pronounced odour. This pale-blue, stinky and explosive gas is formed when an oxygen molecule (O_2) dissociates thermally. Each atomic oxygen (O) combines with a molecule of oxygen (O_2) to form an ozone molecule (O_3). Ozone is unstable and destroyed mainly in upper stratosphere through photochemical reactions with UV waves. However, nature maintains equilibrium as shown below :

O_2 + Ultraviolet light in the wavelengths $<24^2$ nm $\rightarrow O + O$

The free oxygen then joins an O_2 molecule to form ozone :

$$O + O_2 \rightarrow O_3$$
$$2O_3 \rightarrow 3O_2$$
$$O_3 \rightarrow O_2 + O$$
$$O + O \rightarrow O_2$$
$$O_2 + O \rightarrow O_3$$
$$O + O_3 \rightarrow O_2 + O_2$$

Ozone is highly unstable relative to oxygen. Normally, about 90% of the ozone is present in the upper part of the stratosphere extending from 18 to 40 km average height (22 - 33 km) from the sea level. This upper layer of the atmosphere enveloped by ozone is commonly known as *ozonosphere* or *ozone layer,* or *stratospheric ozone layer* or *protective layer* or *ozone umbrella.* The major subdivisions of the atmosphere and the location of the ozone layer are shown in Fig. 12.8.

Actually, composition of air in the stratosphere remains fairly uniform throughout the region. However, O_3 concentration varies by about 10 ppm in the stratosphere compared to 0.02 ppm in winter to 0.05 ppm in summer in the troposphere. Normally, it has been present in the atmosphere at about 0.75 ppm at sea level. The presence of ozone layer in the stratosphere is of vital significance for

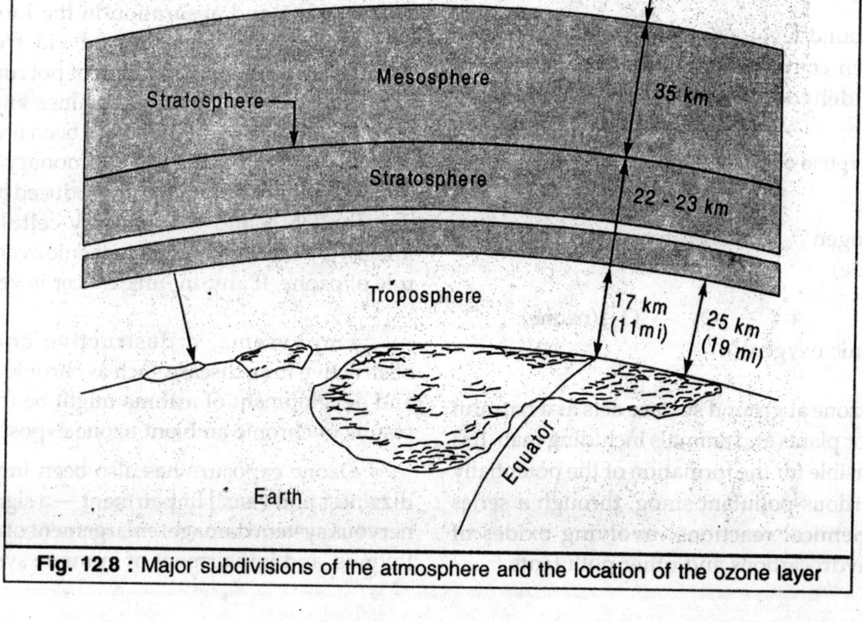

Fig. 12.8 : Major subdivisions of the atmosphere and the location of the ozone layer

all biota, because the harmful solar radiation (220-230 nm), such as ultraviolet rays, which are lethal to life on the earth are not allowed to enter the earth's atmosphere by ozone layer or ozone umbrella. In absence of this layer, all the ultraviolet rays of the sun will reach the earth's surface and, consequently, the temperature of the lower atmosphere (troposphere) may rise to such an extent that the "biological furnace" of the biosphere may turn into a "blast furnace". It functions as a mechanical barrier for not permitting short wave ionising ultraviolet rays and, therefore, protects the life on earth from severe radiation damage. It shall not be out of place to mention that low-altitude (tropospheric) ozone is toxic, and is produced by photochemical smog resulting from fossil fuel emissions.

Formation of Ozone

Natural process of formation of ozone is from oxygen in the stratosphere. In the lower mesosphere, the atmospheric oxygen absorbs UV radiation < 242 nm and photo-dissociates into two oxygen atoms. Theses atoms subsequently combine with molecular oxygen of upper stratosphere producing ozone:

$$O_2 + h\nu\ (242\ nm) \rightarrow O + O$$

$$O + O_2\cdot \rightarrow O_3$$

Or

$$O_2 + h\nu\ (242\ nm) \rightarrow 2O$$

$$2O + 2O_2 \rightarrow 2O_3$$

At ground level, ozone may also be formed from oxygen containing pollutants such as NO_2, SO_2 and aldehyde on absorption of ultraviolet radiation:

absorption of solar UV radiation

NO_2 (Nitrogen dioxide) $\rightarrow$ NO_2 (excited form) $\rightarrow$ $O^* + NO$ (atomic form)

O^* (Atomic oxygen) + $O_2 \rightarrow O_3$ (ozone)

Ozone at ground surface acts as a harmful pollutant for plants and animals including man. It is also responsible for the formation of the potentially more hazardous pollutant smog, through a series of photochemical reactions involving oxides of nitrogen, hydrocarbons and other pollutants.

Sources of Ozone at Ground Level

Exposure to ozone at ground levels (tropospheric) is because of its use as a disinfectant for air and water bleaches in textiles, oil wax and inorganic synthetic industries. Ozone is a key component of oxidant smog and is produced as a result of electrical discharges, occurring with lightning during thunderstorms or with the sparking of electrical appliances. Other sources of ozone in the environment are gamma radiation used in food preservation, commercial UV lamps used for sterilization, high voltage electric equipment, and dermatological phototherapy equipments and even in photocopying machines.

Toxicity of tropospheric ozone

Tropospheric ozone is an extremely reactive molecule. When ozone occurs in concentration above normal, it creates pollution and may damage human health as shown in Table 12.5.

Table 12.5 : Effects of Ozone on Human Health

Sl. No.	*O_3 Level (ppm)*	*Observed effects*
1.	0.2	No ill-effect
2.	0.3	Nose and throat irritation
3.	0.8 (long exposure)	Genetic alteration in the lung within 5 days
4.	1.0-3.0	Extreme fatigue after two hours
5.	9.0	Severe pulmonary edema

The toxic effects of ozone are manifested upon its inhalation and absorption in the lungs. Ozone toxicity is, in part, produced via free radical mechanisms. The per oxidation of polyunsaturated fatty acids and oxidation of amines and proteins which produce free radicals have been implicated in ozone-induced damage in pulmonary and extra pulmonary sites. Free radicals produced by oxidants damage DNA and thus modify cellular genetic integrity, which lead to carcinogenic events. But the role of ozone, if any, in lung cancer is very rare.

Emphysema, a destructive and chronic obstructive lung disease such as chronic bronchitis, and development of asthma might be the ultimate results of chronic ambient ozone exposure.

Ozone exposure has also been implicated in dizziness and visual impairment — a sign of central nervous system damage, enlargement of spleen and thymus, and impairment of immune system.

Photochemical smog is the major cause of ozone exposure causing urban air pollution posing a threat to human health.

Toxic Effects of tropospheric O_3 on Human Beings

As mentioned in above section.

Toxic effects of tropospheric O_3 on Biotic Community

1. Ozone has been known to be highly toxic to fishes in the concentration ranging from 0.1 to 1.0 ppm.
2. Directly or indirectly, ozone increases the mortality rate of larvae of zooplanktons.

Toxic effects of tropospheric O_3 on plants

1. Exposure to air having O_3 results in the lesions to plants usually confined to the upper surface of leaves. The uniformly distributed white or brown flecks characterize their lesions and stipples(dots) in irregularly distributed blotches.
2. Ozone flecking is especially observed with the plants of grape, citrus and tobacco.
3. In pine seedlings, ozone causes tip burns.
4. 0.02 ppm of ozone has been observed to damage pea, pine, tomato and other plants.
5. In plants, ozone enters through stomata. It causes visible damage to leaves, thereby reducing their photosynthetic rate and consequently decrease in the yield of certain food crops.
6. Ozone is now known to be the most toxic photochemical product causing injury to plants.
7. Tobacco plants, when exposed for 5.5 hours at 0.10 ppm concentration of O_3, showed 50% reduction in pollen germination and pollen tube growth.
8. In californ fruits and vegetables, yields have been reduced due to ozone pollution. Grapes are no longer produced at places chiefly because of ozone pollution.
9. Ozone, along with other environmental pollutants like SO_2 and NO_x, is affecting crop losses of over 50% in European countries. In Denmark, ozone affects spinach, potato, clover, alfalfa etc.
10. In Netherlands, ozone level was high enough to reduce yields of beans, potato and poplars. In UK alone, ozone concentration exceeded 400 μg/m^3 in 1976 due to industrial pollution.

Bombawale (1986) presented evidence that a serious leaf spot disease of potato, which appeared in India during each year since 1978, was primarily due to ozone.

From the above discussion it may be concluded that ozone in the stratosphere protects us but it proves to be lethal when it comes in direct contact with us and plants at earth's surface.

Unit of Ozone Thickness

The thickness of ozone layer is measured in Dobson Unit (DU), where 1 DU = 0.01 mm of the compressed gas at 0°C and 760 mm Hg pressure.

Depletion of Stratospheric Ozone Layer (Earth's Protective Umbrella)

Whereas an increase in the concentration of ozone at the ground level is undesirable because it could induce toxicity, a decrease in the stratospheric ozone is also dangerous to human beings, their lives and ecosystem. This is because the stratospheric ozone acts as a protective layer or sheath, which absorbs UV radiations from the sun.

Ultraviolet light arriving in the atmosphere may be divided into three categories according to wavelength. UV-A has the longest wavelength, and UV-C the shortest. Ozone absorbs more than 99% of the UV-C wavelengths, about half of the lower energy UV-B and little of the relatively harmless UV-A. If the equilibrium of the reactions that absorb UV radiation is perturbed, a change in the influx of UV radiation to the surface of earth will occur.

In simple words, an umbrella of ozone in the stratosphere (about 25 km above the surface of the earth) is able to protect us from the harmful solar radiations. In spite of being in a small proportion, it plays a vital role in the climatology and biology of the earth. It filters out all solar radiations below 320 nm, which are biologically harmful, and thereby controlling the thermal budget of the earth. The amount of ozone, therefore, must have been intimately connected with the life-sustaining process. Any large-scale depletion of ozone content, therefore, will exert catastrophic influence on biotic systems including plants.

In the last years of 20th century, there has been much hue and cry about the destruction of stratospheric ozone and the issue has now assumed *global* dimensions. The problem of ozone depletion and its adverse consequences have threatened the existence of life on the planet. IOTP (International Ozone Trend Panel) concluded that the depletion of ozone layer is a global affair and not a seasonal phenomenon confined only to the atmosphere lying over Antarctica.

World Ozone Day— September 16

Causes of Stratospheric Ozone Depletion

Two principal causes may be attributed for ozone depletion — (1) The Natural processes, and (2) The Anthropogenic (man-made) processes.

1. The Natural Processes

The mechanism of ozone depletion by natural processes is not well-understood. However, it is believed that ozone is depleted by reactions with atomic oxygen, reactive hydroxyl radicals and mainly by NO as shown in reaction numbers 1, 2 and 3, respectively :

Reaction 1

$$O_3 + O \rightarrow O_2 + O_2$$

Reaction 2

$$O_3 + HO^* \rightarrow O_2 + HOO^*$$

$$HOO^* + O \rightarrow HO^* + O_2$$

**Indicates the excited atom.*

Reaction 3

$$O_3 + NO \rightarrow NO_2 + O_2$$

$$NO_2 + O \rightarrow NO + O_2$$

It may be noted that nitric oxide (NO) is produced in the stratosphere below 30 km by the reaction of N_2O (nitrous oxide) with excited oxygen atoms, and above 30 km by ionising radiation on nitrogen :

$$N_2O + O^* \rightarrow 2NO$$

$$N_2 + h\nu \rightarrow N + N$$

$$O_2 + N \rightarrow NO + O$$

It may further be noted here that NO_x is also anthropogenic.

2. The Anthropogenic Processes

Anthropogenic processes have been known to cause 50-70% damage to stratospheric ozone. The important anthropogenic processes include the CFCs and NO_x release.

Role of Chloroflurocarbons (CFCs) in O_3 Depletion

Since 1950 it has become evident that the ozone concentration of the stratosphere is gradually decreasing largely due to increasing use of a group of chemicals called chlorofluorocarbons (CFCs) and halons (Table 12.6) :

Table 12.6 : O_3-Depleting Chemicals (Anthropogenic)

Sl. No.	*Compounds*	*Chemical Compositions*	*Chemical names*
1.	CFC-11	$CFCl_3$	Trichloro fluro methane
2.	CFC-12	CF_2Cl_2	Dichloro difluro methane
3.	CFC-13	CF_3Cl_3 CCl_4 CH_3Cl_3	Trichloro trifluro methane Carbon tetrachloride Methyl chloroform
4.	CFC-22	$CHClF_2$	Dichloro fluro methane
5.	Halon-1211	CF_2Cl Br	Di fluro chloro bromo methane
6.	Halon-1301	$CBrF_3$	Tri fluro bromomethane

CFCs also include chloroflurомethane (CFM) or freon are the result of a deliberate search for an ammonia substitute for refrigeration uses. These are chemicals that are inert, non-toxic and easily liquefied. The CFCs and halons are entirely man-made being widely used in air-conditioners, refrigerators, electric and metal clearing, foam blowing, aerosol propellants and as a solvent for cleaning microchips and electronic circuit boards in computers.

The major use of CFC-11 is in foam blowing and CFC-12 is in automobile air conditioning; almost all of CFC-113 is used as a solvent. About 90-95% of the CFCs emitted are still accumulating in the atmosphere; only 1% of CFCs in the atmosphere is being removed in any one year; annual increases are approximately 3-5%. To date, about 15-20 billion kg of CFCs have been produced and used up; these are seriously depleting the earth's protective shield. Earlier models predicted a 6.5% ozone depletion by the year 2030, but it could well be as much as 16%.

Unlike other compounds, CFCs are neither destroyed nor removed in the lower atmosphere by precipitation; sun rays, however — where their chlorine atoms are released into the atmosphere under the effects of UV radiation — encounter and destroy ozone. Also CFCs and halons have 10,000 times more heat absorption power than O_2. At current emission rates the heating potential of CFCs may rise from 15% to the equivalent of CO_2 by the year 2025. CFCs and halons remain inactive in the troposphere and it takes about 20-40 years for these chemicals to travel to reach the stratosphere, but after that their intermediate product (chlorine atom) remains active for 75-110 years. Once these CFCs and halons reach to stratosphere, the chlorine and bromine atoms present in these chemicals are released as a result of interaction with UV radiations in the stratosphere. The ultraviolet light splits up the tight chemical bonds of CFCs and halons and chlorine or bromine atom is set free. These chlorine or bromine atoms act as catalyst in splitting up (breaking down) ozone into oxygen molecule and chlorine monoxide (ClO). The chlorine monoxide molecule is evenly split up by interaction with another ozone molecule to give up two oxygen molecules and a freed chlorine atom. The reactions are :

$$\text{CFCs} \xrightarrow{\text{UV radiation}} Cl + F + C$$

$$Cl + O_3 \rightarrow ClO + O_2$$

$$ClO + O_3 \rightarrow Cl + 2O_2$$

Or

$$ClO + h\nu \rightarrow Cl + O_2 \text{ (photodissociation)}$$

Or

$$ClO + O \rightarrow Cl + O_2$$

This chlorine atom again repeats the cycle of destruction of ozone in the stratosphere.

Similar pattern of reaction takes place for halons :

$$\text{Halons} \xrightarrow{\text{UV radiation}} Br$$

The chlorine or bromine atoms remain active for 75-110 years and are capable of breaking thousands of ozone molecules before the released chlorine gets converted into dilute HCl and come down in the form of acid rain. It is important to note that each atom of chlorine chain reacts with more than 100,000 molecules of ozone converting ozone into oxygen.

Wide range of CFCs are known as shown in Table 12.6, but widely used are CFC-11, CFC-12 and CFC-13. Almost all the CFC-13 is used as solvent. The developed countries release more CFCs into the air than the developing ones. USA, having 5% of global population, releases 70% of CFCs whereas India and China (having 17% and 20% of world population, respectively) release only 1.8% of CFCs.

Role of NO_x in ozone depletion

The rate of destruction of ozone is further enhanced by the oxides of nitrogen (NO_x) released from the exhaust of large fleet of supersonic aircraft and also from fertilizers. The nuclear explosions directly inject superheated NO_x into the stratosphere. The SS aircraft generally fly at ozonospheric level and their exhaust gases directly provide water vapours and NO_x into the stratosphere.

Curtsey (1971) and his collaborators pointed out that SS aircraft fleet could add significant quantities of NO_x leading to about 40% reduction in the O_3 concentration. Following reactions between O_3 and NO_x are known to exist :

$$NO + O_3 \rightarrow NO_2 + O_2$$

$$NO_2 + O_3 \rightarrow NO_3 + O_2$$

Further, OH species have been known to cause ozone depletion in the stratosphere :

Photodissociation

$$HOH \rightarrow OH + H$$

$$OH + O_3 \rightarrow HOO + O_2$$

$$H + O_3 \rightarrow OH + O_2$$

Also, the process of decomposition of ozone may be enhanced in the presence of some greenhouse gases such as CO_2 and CH_4.

It is due to continuous breakdown of ozone molecules that ozone layer in stratosphere is getting depleted and at certain locations holes have been created in the ozone layer.

Ozone Hole

Hole actually represents depletion of ozone concentration in the stratosphere, not an empty space. Besides CFCs, NO_x, OH species etc., methyl bromide and hydrochloroflurocarbons also deplete ozone layer. Depletion of ozone concentration at Antarctica (South Pole) is more than Arctic (North Pole).

There has been gradual decrease in the ozone concentration since 1970 to 1993 :

1970	306 DU
1971	245 DU
1993 (October)	240 DU

Ozone hole was first noticed in 1979 in Antarctica. If hole develops in Arctic (North Pole), then consequences would be more dangerous. However, at Arctic, a layer of ClO cover over the ozone has been noticed in 1987. Chlorine monoxide (ClO) is highly active and encounters ozone molecule.

$$ClO + O_3 \rightarrow ClO_2 + O_2$$
(chlorine monoxide) (chlorine dioxide)

The ClO_2 is soon broken down by UV light to leave a free chlorine atom and an oxygen molecule :

$$ClO_2 + \text{UV light} \rightarrow Cl^- + O_2$$

It is estimated that a single chlorine atom can break down 100,000 ozone molecules.

WMO (World Meteorological Organization) in 1995 reported record increase in the size of the ozone hole at South Pole, which was 20 million km^2, twice the size of Europe. At present the ozone hole covers the entire Antarctica. Globally, it appears that about 4-5% of the earth's ozone has been degraded since the early 1980s. If the trend of ozone loss continues, it is predicted that the ozone shield will be depleted by a further 10% by 2050. It is conjectured that this may lead to an additional 300 million cases of skin cancer in human.

Effects of Ozone Depletion

It is universal fact that ozone layer in the stratosphere protects humans from the deleterious UV radiations coming from the sun. Though ozone accounts for only 3 parts in ten million of the earth's atmosphere, it plays various crucial roles in the radiation balance of the planet. The depletion of ozone-layer by chemical pollution created by man's activities have imparted serious implications by letting "ozone-eaters". The following would be the consequences of ozone - depletion :

The consequences of ozone-depletion :

A. Effect on Human beings

1. Eye damages including tumors of the conjunctiva and cornea, cataracts and retinal damage may be enhanced.
2. Three kinds of skin cancer — *Basal cell carcinoma, Squamous cell carcinoma* and *Melanoma*— are rapidly climbing the list of human diseases caused by UV rays.
3. Suppression of certain immune responses in human occurs due to UV rays.
4. UV radiations cause blood vessels near the skin's surface to carry more blood, making the skin hot, swollen or red causing *sun-burns.*
5. Long exposure to UV radiations caused by O_3 depletion may cause leukemia and breast cancer. Studies reveal that a 10% decrease in stratosphere ozone may lead to 20-30% increase in cancer. Nearly 7,000 people die of such diseases in USA each year. Such cases have increased by 10% in Australia and New Zealand.
6. Photochemical changes occurring in such biological molecules as nucleic acids, proteins, lipids, steroids and melanins may cause damage to the cell nucleus, cytoplasm, organelles and membranes.

B. Effects on Biotic Community

1. UV solar radiation causes heavy mortality in phytoplanktons.
2. The marked reduction in the productivity of phytoplanktons would, in turn, adversely affect zooplanktons.
3. Enhanced UV radiation also impairs fish productivity.
4. Researches conducted on microbes reveal that photo irreversible and reversible types of injury are caused.

C. Effects on Plants

1. Important crops (wheat, corn, rice and soyabean) are sensitive to UV-B rays. Exposure to UV-B radiation may cause reduction in the crop yield.
2. UV-B radiation reaching the earth through the depleted ozone layer in the stratosphere also reduces the effectiveness of photosynthesis and multiplication of phytoplanktons.
3. Due to ozone reduction, intense UV radiation causes greater evaporation of surface water through the stomata of the leaves and decreases the soil moisture content.

D. Climatic effects of Ozone Depletion

Ozone depletion changes spectral composition of solar electromagnetic radiation. The increased solar UV radiation causes greenhouse effect changing the global energy and radiation balance. Needless to mention that every molecule of either of two common CFCs — CCl_3F and CCl_2F_2 — has the same global warming effect as 10,000 CO_2 molecules.

E. Ecological Disturbances due to O_3 Depletion

Scientists are of opinion that the depletion of ozone, if not controlled, would enormously affect the ecosystem productivity, ecological stability and overall environmental equilibrium. It would also trigger several changes in the biospheric ecosystem.

The resultant climatic alterations would cause certain physiological changes in man and animals. Change in energy balance and radiation would affect the survival and stability of living biota. It appears important to mention here that UV radiation disrupts DNA replication to the point of reproductive failure or death.

The changes in thermal conditions of biosphere would affect type, density and stability of vegetation, which, in turn, would affect several biogeochemical cycles occurring in nature.

Control of Ozone Depletion

Since CFCs cause maximum damage to ozone layer, hence its application should be completely banned. No safe substitute of CFCs has yet been searched out. In cooling system, natural products should be used but not the CFCs. Indonesia is using a natural product (Kapok fibre) in foam mats in cooling system. Much is yet needed to be done to stop further ozone depletion.

Several of the EEC (European Economic Community) nations have banned CFC production altogether, a protocol for the protection of the ozone layer was adopted in Vienna in March 1985 by 21 countries and the EEC. The convention pledged to protect human health and the environment from the effects of ozone-depletion.

V. EUTROPHICATION

"Eutrophication" is a natural process, derived from the Greek word *'eutrophos'* meaning well-nourished or enriched. It means eutrophication is a natural process by which lakes are enriched naturally or, more usually, by human activity. As a result, an *oligotrophic lake ("underfed"* or "nutrient poor") may become *eutrophic* ("well-fed" or "nutrient rich") in due course of time. The key nutrients responsible for eutrophication are *nitrogen, phosphorus* and *calcium*. The enrichment leads to other slow processes referred to as *"natural ageing of lakes"*.

Sources of Nutrients : Eutrophication escalates rapidly, however, when abnormally high amounts of nutrients from fertilizers, industrial and domestic wastes, urban drainage, animal wastes, detergents and sediments enter water streams.

Types of Eutrophication

Eutrophication is chiefly of two types — (1) Natural eutrophication, and (2) Cultural eutrophication.

1. Natural Eutrophication

The process of lake-aging characterized by nutrient enrichment is called *natural* eutrophication. This process gradually converts oligotrophic lake into a eutrophic lake. Due to increase in natural fertility, it pursuits the production of phytoplanktons, algal blooms and aquatic vegetations including water hyacinth, water fern, water lettuce and aquatic weeds which, in turn, provide ample food for zooplanktons, herbivores, and fishes.

2. Cultural Eutrophication

When the process of eutrophication is speeded up by human activities, it is called *cultural* eutrophication. Approximately 80% of the nitrogen and 75% of the phosphorus added to lakes and streams in developing and developed countries has its source through human activities. *Lake Mendota* and *Lake Washington* have undergone rapid eutrophication due to human activities. In India, recreational value of Kashmir lakes is diminished while Nainital Lake is undergoing a rapid eutrophication as a result of sewage, domestic waste and detergent addition.

When the average concentration of soluble inorganic nitrogen exceeds 0.30 ppm and the soluble inorganic phosphorus content exceeds 0.01 ppm, algal blooms may appear in any oligotrophic lake.

Effects of Eutrophication

Eutrophication causes several physical, chemical and biological changes, which considerably deteriorate the water quality. Eutrophication may otherwise be called *"nutrient pollution"* of lake etc. Their adverse effects are :

1. Eutrophication reduces the recreational values of a lake.
2. During eutrophication, algal blooms release deleterious (toxic) chemicals which kill fishes, birds, and other aquatic fauna — causing water to stink.
3. The algal blooms impair water quality by giving it a bad taste and odours. If the lake is a source of drinking water, considerable expense may be involved in improving its quality.

4. Decomposition of algal bloom leads to O_2 depletion in water. Thus, with a high CO_2 level and poor O_2 supply, aquatic biota begin to die and the clean water turns into a stinking drain.
5. Toxic gas (H_2S) that emanates (originates) from rotting algae has foul smell. It indirectly impairs the physiology, metabolism and behaviour of aquatic biota.
6. Dense algal blooms at the lake surface reduce penetration of sunlight to the lake bottom. As a result, the deeper waters contain less amount of DO (dissolved oxygen), which is further reduced by decomposing algae and organic matter. accumulated at lake bottom. The O_2 depletion is often responsible to the *winterkill* of fishes in northern lakes of India.
7. Many pathogenic microbes, protozoa, bacteria, viruses etc. grow on sewage products under anaerobic conditions. It results into spread of fatal water-borne diseases such as polio, dysentery, diarrhea, typhoid and viral hepatitis.
8. Macrophytes, particularly *Hydrilla, Potamogeton, Ceratophyllum* and *Myriophylium,* assume high population densities making near-shore and shallow regions unsuited for any purpose.
9. The lake undergoing eutrophication may become O_2 - deficient, destroying fish habitats, leading to the elimination of several desirable aquatic species in water.
10. Prolonged eutrophic conditions lead to "dystrophic state". The lake receiving huge amounts of organic matter from allochthonous sources are called dystrophic. These lakes contain bog flora and high amounts of humic acid while planktonic productivity is very low.
11. The rooted weeds gradually reduce the lake size and also interfere with fishing, navigation, irrigation, and the production of hydroelectric power.
12. As a result of sediment pollution, the lake basin is gradually filled in, rendering the lake shallow, and, in due course of time, the lake becomes "extinct" as a result of eutrophication.

Control of Eutrophication

The role of various nutrients, especially nitrogen and phosphorus, in eutrophication is so overwhelming that an international resolution was ratified at the 9th International Congress of Theoretical and Applied Limnology in 1974. Several technical devices along with prevention of further inflow of effluents have been used to control eutrophication. For example :

1. The waste water must be treated before its discharge into water streams to limit the nutrient value.
2. Recycling of nutrients can be checked through harvest.
3. Eutrophication can be reduced by removing nitrogen and phosphorus at the source and by dilution.
4. Algal blooms should be removed upon their death and decomposition.
5. Algal foodweb should be disrupted to stimulate bacterial multiplication.
6. Algal growth can be controlled by limiting the dissolved nutrients. The most suitable, feasible and effective method involves the use of chemicals to precipitate additional phosphorus. Such precipitants include alum, lime, iron and sodium aluminate.
7. Physicochemical procedures can be applied to remove dissolved nutrients. For example, phosphorus can be removed by precipitation and nitrogen by nitrification or denitrification, electrodialysis, reverse osmosis and ion exchange methods.
8. To minimize algal population, copper sulphate may be applied.

Ecological Classification of Fresh water (Lentic) Biota

A. Pond Flora : The floral community of the pond constitutes the producers (autotrophs).

Pond flora are of four types :

1. **Phytoplanktons** : These are plant plankton having chlorophyll. They are capable of synthesizing their own food by the process of photosynthesis and, hence, are called producers.

   Important phytoplanktons of pond (lentic habitat) are :

Volvox, Euglena, Anabaena, Chalamydomonas, Ceratium, Fragilaria, Microcystis, Eudorina, Closterium, Navicula, Asterionella, Richterilla etc.

2. **Floating Plants** : These includes plants which float freely on the water surface and also rooted plants with floating leaves. Examples — *Wolfia, Lemna, Nymphaea, Pistia* etc.
3. **Submerged Plants** : This comprises plants which remain submerged in water. Examples — *Hydrilla, Vallisneria, Ceratophyllum, Utricularia* etc.
4. **Emergent Plants** : This comprises rooted but emergent plants. Examples — *Sagittaria, Typha, Nymphaea, Marsilea, Nelambo* etc. Few important (emergent plants) flora of a pond have been shown in Fig. 11.4

B. Pond Fauna : The fauna of the pond (lentic habitat) constitute the consumers and decomposers. The consumers are divided into five principal groups : (1) Benthos, (2) Periphyton, (3) Zooplankton, (4) Nekton, and (5) Neuston.

1. **Benthos** : Animals which live on the bottom are called benthos. Examples — *Pila, Lamellidens, Earthworms, Turtles* etc.
2. **Periphytons** : Animals attached or clinging to aquatic plants are called periphyton. These are also called Aufwuchs. Examples — *Hydra, Planaria.*

Revision Questions

A. Long-Answer Questions

1. Name the major environmental effects. Describe any one of them well-studied by you.
2. Define "acid rain" and its causes. Give effects of acid rain on :
   (i) Aquatic ecosystem
   (ii) Forest ecosystem
   (iii) Terrestrial ecosystem
   (iv) Human beings
   (v) Buildings.
3. Write an essay on Eutrophication.
4. Define Greenhouse effect. Name the greenhouse gases and their release sources.
5. Describe the principle and consequences of greenhouse effect.
6. Write an essay on London and Los Angeles Smog.
7. Give the causes of ozone depletion and its mechanism.
8. Justify the statement "Ozone is a protector of human on one hand and killer on the other."

B. Short-Answer Questions

1. *Briefly describe :*
   (i) Wet and dry acid rain
   (ii) Role of wind in acid rain
   (iii) Taj Mahal and acid rain.
2. *Write short notes on :*
   (i) Principle of greenhouse effect
   (ii) Possible consequences of greenhouse effect
   (iii) Remedial measures of greenhouse effect.
3. *Briefly describe :*
   (i) Photochemical Smog
   (ii) Conditions for Photochemical Smog formation
   (iii) Mechanism of Photochemical Smog (Los Angeles Smog) formation
   (iv) Mechanism of London Smog formation
   (v) Control of photochemical pollutants.
4. *Write short notes on :*
   (i) Formation of ozone
   (ii) Toxicity of ozone
   (iii) Depletion of ozone layer
   (iv) Effects of ozone depletion
   (v) Eutrophication.

13

Toxic Metals

SYNOPSIS

Introduction, Toxicology of Arsenic, Lead, Mercury, Cadmium, Chromium, Copper, Zinc, Aluminium and Manganese.

Introduction

Some 65 of the known 92 elements are metallic in nature. The term 'metal' designates an element that is a good conductor of electricity whose electrical resistance is directly proportion to absolute ^{0}T. In addition, metals share some distinctive characteristics such as high thermal conductivity and high density. Metallic elements are intrinsic components of the environment. Erosion of surface deposits of minerals, forest fires and volcanic activity are some of the natural causes of environmental pollution with the toxic metals. The processes involved in the extraction of metals from the ores and their extensive use resulting from improved technology are also responsible for their dispersion in the biosphere. With the increasing use of a wide variety of metals in industry and in our daily life, problems arising from toxic metal pollution of the environment have assumed serious dimensions. Further, their use without appropriate recycling has also precipitated the prospect of exhausting these non-renewable resources.

Trace elements generally found in coal fly ash are Ba, Be, Cd, Cr, Co, Cu, Fe, Ga, Ge, La, Pb, Li, Mn, Mo, Hg, Ni, Sc, Se, Sr, Sn, V, Yb, Zn and Zr. These trace elements are transferred during combustion to particles that range in size from less than 0.1 μm to greater than 100 μm. Some elements also present in traces are As, Bi, Ce, and Ti.

Toxic metals, to a large extent, are dispersed in the biosphere through industrial effluents, organic wastes, refuse burning, transport and power generation (Fig. 13.1). They can be carried to places many kilometers away from the source of origin by the winds depending on their physical state, viz., whether they are in the gaseous form or as particulates. Metallic pollutants are ultimately washed out of the air by rain onto land or the surface of the waterways. Thus, air becomes a major route for the contamination of the rest of the living environment. Industrial smoke containing arsenic, antimony, copper, managanese, nickel, etc. is a potential source of contamination of air. The inhalation of these elements at enhanced levels can produce adverse physiological effects. Metals from the air are accumulated by mosses and lichens since these plants depend on the atmosphere for their mineral supply. The levels of copper and zinc in plants have also been found to be high near the copper smelter plants.

National Institute of Occupational Health, Ahmedabad (Gujarat) recently has collected baseline data on metals in blood and urine of Indian population. Twentysix occupationally unexposed human volunteers from Ahmedabad city were investigated for lead, manganese, zinc, cobalt, nickel, copper, chromium and cadmium contents in blood and urine. The contents of most of these metals were almost similar to values reported from abroad;

however, the nickel content in the blood was found to be on the higher side. This relatively higher content of nickel was assumed to be due to the consumption of vegetable ghee or due to use of stainless steel vessels in cooking.

Metal containing industrial effluents constitute a major source of metallic pollution of the hydrosphere. Metal containing industrial effluents when discharged into coastal waters are, of course, eventually dispersed and diluted in the ocean. In polluted coastal waters there is a possibility of concentration of toxic elements within the food chain in the marine ecosystems. This may seriously affect life of birds and mammals which are dependent on sea for their food.

Mercury-containing effluents are being discharged into waterways by some of the industries in India which can possibly create problems similar to Minamata disease in Japan. Reports on fish from coastal waters of Mumbai reveal that all varieties of fish, except salmon, contained about 100 ng/gm of mercury on fresh weight basis. The greater part of the metal load emitted into the environment is transported by water, lead being an exception.

such as smelting and mining of ores, releasing industrial wastes, use of metallic compounds in pesticides, fertilizers, textile, paper, cosmetics, medicines, detergents, war weapons and in the production of vehicles and aeroplanes have released a large number of toxic metals and their salts in to the environment.

General Principle of Metal Toxicity

The toxicity of metal increases with the increase in atomic number and electropositivity. Among metal salts the toxicity rises from Nitrates < Chlorides < Bromides < Acetates < Iodides < Perchlorates < Sulphates < Phosphates < Carbonates < Fluorides < Hydroxides < Oxides.

A number of metals and metalloids, for example Arsenic, Chromium, and Nickel, and their compounds are carcinogenic in animals. Some of them are considered to be carcinogens in human also. Arsenic was thought to be an exception in that it is carcinogenic in humans but not in animals. Nickel is a biologically active metal which can produce allergy and lung cancer (Sunderman, 1973). Exposure to Ni has been reported to produce nasal

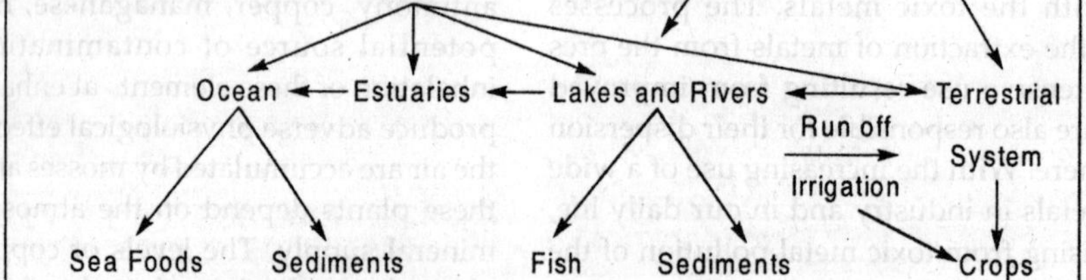

Fig. 13.1 : Dispersion of Metals in Environment

Actually toxic metals are present in small quantities in the earth's crust; hence these are called trace metals. Trace metals by definition are those that occur at less than 1,000 ppm levels in the earth's crust. These are further classified as heavy and light metals based on their densities being over or under 5 g/cm³.

The toxic heavy metals are present at < 0.1% levels in the earth's crust but anthropogenic activities

sinus cancer in refinery workers and in experimental animals. The most important hazard in the nickel-using industry arises as a result of inhalation exposure to nickel carbonyl even in extremely low concentrations. Two phases of acute poisoning are described. Headache, dizziness, nausea and vomiting appear in the first phase and chest pain and tightness, dry cough, dyspnea, cyanosis and extreme weakness in the second phase.

Important Toxic Metals

The following are the important toxic metals of global concern :

1. Arsenic
2. Lead
3. Mercury
4. Chromium
5. Cadmium
6. Copper
7. Zinc
8. Aluminium
9. Manganese.

Manganese is of special interest to India because of the position occupied by the country as one of the leading exporters of manganese ore. The demand for manganese by our steel and chemical industries has also risen.

1. Arsenic

Arsenic is particularly difficult to characterize as a single element because its chemistry is so complex and there are many different arsenic compounds. It may be trivalent or pentavalent and is widely distributed in nature. The most common inorganic trivalent arsenic compounds are arsenic trioxide, sodium arsenite, and arsenic trichloride. Pentavalent inorganic compounds are arsenic pentoxide, arsenic acid, and arsenates, such as lead arsenate and calcium arsenate. Organic compounds may also be trivalent or pentavalent, such as arsanilic acid, or may even occur in methylated forms as a consequence of biomethylation by organisms in soil, fresh water, and sea water. Arsenic trioxide (As_2O_3) is known as white arsenic. It constitutes about 97% of all arsenic produced and used in end-product manufacturing. Arsenic has been known to mankind from ancient times. Already in the 8th century its compounds were known to Arabian chemists, while in the 9th century Avicenna described arsenous anhydride in his works. T. Paracelsus (1493-1541) was the first to use arsenic compounds for medical purposes. Arsenic and its compounds in small doses are valuable drugs having a tonic effect on an organism. They improve the state of an organism and increase metabolism. A solution of potassium arsenite, KA_sO_2, called Fowler's Solution, is used in medicine.

Uses and Pollutional Sources

Natural Sources : Mine wastes and land erosion.

Anthropogenic Sources : Inorganic arsenic is released into the environment from a number of anthropogenic sources which include primary copper, zinc, and lead smelters, glass manufacturers that add arsenic to raw materials, and chemical manufacturers.

The arsenical compounds are used as :

(i) *Pesticides* : Lead arsenate; sodium arsenite; calcium arsenite etc.

(ii) *Herbicides* : $NaAsO_2$.

(iii) *Wood Preservatives* : Monosodium arsenite; Dimethyl arsenic compounds like fluorochrome arsenate phenol (FCAP) and chromated copper arsenate (CCA).

(iv) *Chemotherapic Agents* : Methylated Arsenic acid ($CH_3AsO(OH)_2$ and cocydylic acid $(CH_3)_2AsOOH$.

(v) Lead and copper based alloys to increase hardness and heat resistance.

(vi) *Warfare* : Lewistios ($CH_3CH = CH\,As\,Cl_2$).

Researches (Pandey and Shukla, 1982) reported release of arsenic as its trioxide from HPC Tower of Urea manufacturing plants.

Environmental Fate of Arsenic

Sequential methylation of arsenic by certain soil microbes (fungi and bacteria) releasing dimethyl and trimethyl arsine has been reported.

Toxicokinetics of Arsenic

Actually arsenic is a metalloid and not a metal. The toxicity of arsenic in trivalent state is greater than pentavalent state. It is because the arsenites bind strongly with the -SH-groups of proteins resulting in enzyme inhibition, whereas the arsenates neither bind to-SH-groups nor inhibit enzyme systems. However arsenates inhibit ATP synthesis by oxidative uncoupling of certain reaction. To sum up, As (III) is most potent, followed by As (v), monomethylarsenate, and dimethylarsenate (Kreppal et al, 1993). In fact, ingested arsenite gets oxidized to arsenate and the latter is reduced to arsenite again. Both these inorganic forms are partly methylated to form monomethyl arsenic acid and dimethylarsenic acid (also called cacodylic acid). The formation of methylated forms is regarded as a detoxication mechanism, as far as arsenic toxicity is concerned.

Excretion of absorbed arsenic is mainly via urine. The biological half-life of ingested inorganic arsenic is about 10 hours and 50 to 80% is excreted in about 3 days. The biological half-time to methylated arsenic is about 30 hours.

When low chronic doses of arsenic are ingested, it tends to accumulate in lipid-rich tissues. It also concentrates in nails, hair and skin. Arsenic in nails produces Mee's lines (transverse white bands across finger nails). Human milk contains about $3\mu g$/liter of arsenic. Inhaled arsenic may retain in the lung tissue for relatively long periods of time.

Arsenic crosses placental membranes and is a known animal teratogen. It induces skin lesions and may lead to skin cancer. Inorganic arsenicals are known lung carcinogens in humans.

Trivalent arsenic exerts its toxic action by attacking -SH-groups of an enzyme, thereby inhibiting enzyme action :

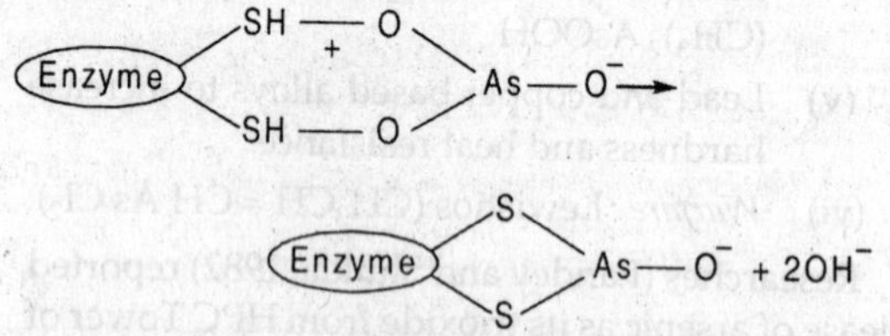

The enzymes which generate cellular energy in the TCA cycle are adversely affected. The inhibitory action is based on bioactivation of pyruvate dehydrogenase by complexation with As (III), whereby the generation of ATP is presented.

As (III) compounds at high concentrations coagulate proteins, possibly by attacking the sulphur bonds, maintaining the secondary and tertiary structures of proteins.

Arsenic affects mitochondrial enzymes and impairs tissue respiration (Brown et al, 1976), which seems to be related to the cellular toxicity of arsenic.

Arsenic compounds are inducers of metallothionein in vivo. Potency depends on the chemical form of arsenic.

Toxic Effects of Arsenic on Humans

Large doses of arsenic (70-180 mg) ingestion may be fatal. Symptoms of acute illness possibly leading to death consist of fever, hepatomegaly, melanosis, and cardiac arrhythmia. Other features include peripheral neuropathy, gastrointestinal, cardio-vascular and hematopoietic effects. Sensory loss in the peripheral nervous system is the most common neurological effect.

Chronic exposure to inorganic arsenic compounds may lead to loss of appetite and weight, diarrhea and neurotoxicity of both the peripheral and central nervous system. Neurotoxicity usually begins with sensory changes, paresthesia, and muscle tenderness followed by weakness. Liver injury is also a characteristic of chronic exposure to arsenic.

Arsenic may induce skin, liver, blood, nasopharyngeal, kidney and bladder cancers in human only but not in any experimental animal. Presence of 0.6-0.8 ppm concentration of arsenic in drinking water in Latin (U.S.A.) caused endemic poisoning and the *Black-foot disease* (WHO, 1972).

Toxic Effects on other Animals

Chronic exposure reduces growth of fingerlings of freshwater fishes (Pandey and Shukla, 1982). Shukla and Pandey (1984, 86, 87) reported impairment in spermatogenesis, oogenesis, nucleic acids (DNA-RNA), and protein metabolism in nature tropical fresh water fishes. Fish liver dysfunctioning at 1-20 mg/1 cone has also been observed. Chronic effects in mammals produce skin damage, liver damage and other pathological effects.

2. Lead

Lead is the most ancient industrial toxicant known to man. Earth contains about 0.00002% of lead by weight. It occurs in nature as its sulphide (galena), cerrusite (lead carbonate), and anglesite (lead sulphate). The average lead content of mined ores ranges from 3-8%. Lead smelting and refining are probably the most hazardous operation in regard to exposure to lead with recorded mean concentration of lead in air of 80-4,000 $\mu g/m^3$.

Chemical Forms

Lead occurs in two valence states i.e., +2 and +4. Lead in inorganic formulations normally exists in the divalent (+2) state; but in the +4 form occurs in lead acetate and tetra alkyl lead (R_4Pb) compounds.

Uses and Pollutional Sources

Lead is extensively used in printing, manufacture of paints, water pipes, storage, battery manufacture, pottery and soldering operations, etc. Besides, it is used as an antiknock agent in gasoline. Lead in gasoline, accounting for 20% of lead used by makind, is responsible for about 96-98% of the pollution problems caused by lead. Lead arsenate is used as pesticide.

Toxicokinetics

Air-borne lead of small particle size is readily absorbed through the lungs. About 50% of inhaled particle lead (size < 0.9 μm) is absorbed; as much as 90% may be absorbed when the size is still smaller (< 0.1 μm). About 90% of lead particles in ambient air that are deposited in the lungs are small enough to be retained. Absorption of retained lead through alveoli is relatively efficient and complete.

Net absorption of lead in the alimentary tract is very low (5-15%) because of its low solubility. More then 90% of the lead in blood is in R.B.C. Lead in bone may contribute as much as 50% of blood lead. Mobilization of lead from maternal bone is of particular concern during pregnancy and lactation and may be mobilized in later years in persons with osteoporosis (Silbergeld et al, 1988). The total lifetime accumulation of lead may be as much as 200 mg to over 500 mg for a worker with heavy occupational exposure. Most of the lead intake by a typical city dweller is from diet (about 200-300 μg/day), air and water adding a further 10-15 μg/day each. Of this total intake, 200 μg of lead is excreted while 25 μg is stored in the bones each day. Renal excretion of lead is usually with glomerular filtrate with some renal tubular resorption. Lead also crosses the placenta. Possibility of lead poisoning in Indian women who apply *sindoor* (red lead) to the scalp (hair parting) and forehead has also been reported.

Toxic Effects of Lead on Human

The major toxicological effects of lead is its interference with heme synthesis, which leads to haematological damage. Lead inhibits several of the

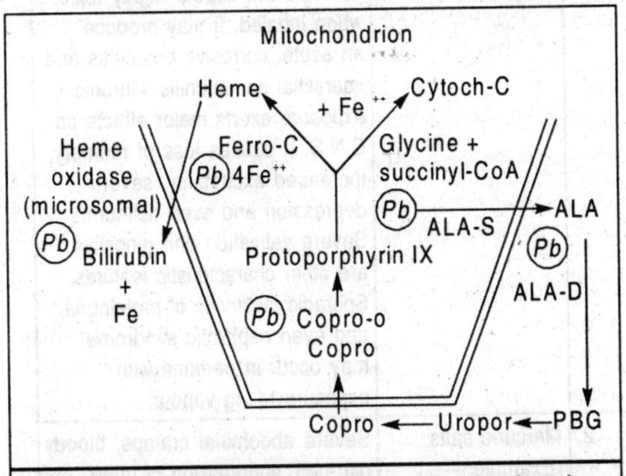

Fig. 13.2 : Scheme of heme synthesis showing sites where lead has an effect

CoA. Coenzyme A : ALA-S aminolevulinic acid synthetase : ALA-δ-aminolevulinic acid : ALA-D. aminolevulinic acid dehydratase; PBG. Porphobilinogen; Uropor, Uroporphyrinogen : Copro, coproporphyrinogen : Copro-O, copro-porphyrinogen oxidase; Ferro-c, ferrochelatase; Cytoch-C, cytochrome c; Pb, site for lead effect.

key enzymes involved in the overall process of heme synthesis whereby the metabolic intermediates accumulate as shown in Fig. 13.2.

Since lead interferes with the synthesis of heme, anemia appears to be the first symptom of its chronic poisoning in animals in general and humans in particular.

Chronic exposure to lead causes weight loss, constipation, and loss of teeth.

Lead also damages liver, kidney and central and peripheral nervous system.

Kidney dysfunction occurs due to the impairment of energy metabolism, leading to the expression of *Fanconi Syndrome* (characterized by an increased loss of amino acids, glucose and phosphate in the urine due to the inability of the damaged tubular cells to resorb these substances). Lead is a renal carcinogen in rodents but whether it is carcinogenic to human kidney is unclear.

An increase in blood pressure is probably the most sensitive adverse health effect from lead exposure occurring in the adult human population.

Lead is also a carcinogen (IARC, 1987). The most common tumours found in people working in lead industry is of the respiratory and digestive systems, not the kidney.

Lead is now a well-known teratogen leading to embryo toxicity. Due to lead intoxication, the rate of abortions goes up and it may also lead to low birth-rate, brain damage and low reproductive ability. In higher concentration, it tends to accumulate in foetal bone and liver and leads to deformity and physiological malfunctioning. Adverse effects of lead intoxication enhances due to iron deficiency. Cases of lead poisoning have also been noticed in India after the consumption of acidic beverages stored in glazed pottery.

Toxic Effects on Other Animals

In fishes lead inhibits liver enzymes, causes muscle tremors, reduces haemoglobin production. Chronic exposure to lead produces renal carcinoma in mammals.

3. Mercury

Mercury is a well-known toxic metal which came to the limelight after the incidence of "Minamata Disease" in 1953-1960 in Japan (For Minamata Disease refer Chapter 5). No other metal better illustrates the diversity of effects caused by various chemical species than does mercury. On the basis of chemical speciation, there are three forms of mercury (i) elemental (ii) inorganic, and (iii) organic compounds, each of which has characteristic toxicokinetics and health effects as shown in Table 13.1.

Uses and Pollutional Sources

In the environment, mercury occurs as metallic mercury and as HgS and $HgCl_2$ in the earth's crust. Anthropogenic sources consist of mining, smelting, paper pulp, paints, batteries, lamps, switches, caustic soda, medicine and instruments. Twenty five percent of the total production of mercury is consumed by the chlor-alkali plants, rest is used in electrical equipments, paints, in measurement and control like thermometers, sphygmomanometers, in dental practice and in agriculture. Methyl and ethyl mercury have been extensively used in seed dressings. Mercury is released into the environment during its production and by human activities like combustion of fossil fuels, waste disposal and industrial activities. Annual production of mercury in the world is estimated to be about 9,000 tons. About 50% of which is estimated to be lost into the environment. As per WHO (1971) standards, the maximum permissible limits of mercury in drinking water should not be more then 0.001 ppm.

Environmental Fate of Mercury

Inorganic mercury released into the environment is converted into more toxic methyl mercury compounds by the action of certain anaerobic bacteria present in the sediments and bottom muds of waterways. The major source of exposure to methyl mercury for people in general population is from the consumption of fish.

Toxicokinetics

Metallic or elemental mercury volatizes to mercury vapour at ambient air temperatures, and most human exposure is by inhalation. Mercury vapour readily diffuses across the alveolar membrane and is fat soluble so that it has an affinity for R.B.C. and the central nervous system. Metallic mercury is very slowly absorbed by the alimentary tract (0.01%).

Inorganic mercury salts may be divalent (mercuric) or monovalent (mercurous). Kidneys contain the maximum concentrations of mercury following exposure to inorganic salts of mercury and mercury vapour, whereas organic mercury has greater affinity for the brain, especially the posterior cortex. All forms of mercury cross the placenta to the fetus in experimental animals.

Elemental or metallic mercury is oxidized to divalent mercury after absorption which is mediated by catalases. Also inhaled mercury vapour absorbed into R.B.C. is transformed into divalent mercury. Methyl mercury may undergo biotransformation to divalent mercury compounds in tissues by cleavage of the CHg bond. Formation of any organic form of mercury, however, in mammalian tissues has not yet been reported.

Within cells, mercury may bind to a variety of enzyme system producing non-specific cell injury or cell death.

Toxic Effects of Different Species of Mercury on Humans

As mentioned above, the toxicity of mercury depends upon its chemical species which may be represented in tabular form :

Table 13.1 : Toxicity of Different Species of Mercury

Sl. No.	*Species of Hg*	*Toxic Effects*
1.	Elemental Mercury (Hg vapour)	Elemental Hg is relatively inert and nontoxic but vapour highly toxic when inhaled. It may produce an acute, corrosive bronchitis and interstitial pneumonitis. Chronic exposure exerts major effects on C.N.S. It causes loss of memory, increased excitability, severe depression and even delirium. Severe salivation and gingivitis are other characteristic features. Sporadic instances of proteinuria and even nephrotic syndrome may occur in persons with exposure to Hg vapour.
2.	Mercuric salts (Example — Mercuric chloride)	Severe abdominal cramps, bloody diarrhea, suppression of urine, circulation collapse, renal failure etc. HgS is trapped in soil.

Contd...

3.	Mercurous Salts	Less toxic than mercuric salts because of less solubility. In excess it may cause acrodynia or Pink Disease in children. It causes fever, pink coloured rash, swelling of the spleen, lymph nodes and fingers in children.
4.	Methyl Mercury	Highly toxic, causes severe neurotoxic effects viz., paresthesia; ataxia; neurasthenia; vision and hearing loss and finally coma and death in adults. Also causes genotoxic effects resulting in chromosomal aberrations. **Special** : Methyl mercury interacts with DNA and RNA and binds with sulfhydryl groups, resulting in changes of the secondary structure of DNA and RNA synthesis.

Toxic Effects on other Animals

Inhibits enzymes and protein synthesis in fishes. In mammals it reduces DNA synthesis.

4. Chromium

Chromium is generally an abundant element at about 100 ppm levels in the earth's crust and occurs in oxidation states ranging from Cr^{2+} to Cr^{6+} as shown in Table 13.2, but only trivalent and hexavalent forms are of biological significance.

Table 13.2 : Valency states of chromium compounds

Valency State	Examples
0	Metallic chromium
+ 2	Chromous chloride ($CrCl_2$) Chromous sulphate ($Cr\,SO_4$)
+ 3	Chromic oxide ($Cr_2\,O_3$) Chromic chloride ($Cr\,Cl_3$) Chromic sulphate ($Cr_2(SO_4)_3$)
+ 6	Chromium trioxide (CrO_3) Dichromate salts ($Na_2\,Cr_2\,O_7$)

Uses and Pollutional Sources

Chromium is extensively used in electroplating, polishing, paint-pigment industry. Chromium in ambient air originates from industrial sources, particularly ferrochrome production, ore refining, chemical and refractory processing, and combustion of fossil fuels. Cement-producing plants are another potential sources of atmospheric chromium. The potential sources of chromium in aquatic environment are effluents from tanneries and textile mills.

Tannery wastes contain 10-50 ppm of chromium. Its permissible limit in drinking water is 0.05 ppm.

Toxicokinetics

Chromium (+ 3) is the most common form found in nature and in biological components. Now report is available that trivalent chromium is converted into hexavalent forms in biological systems. However, hexavalent chromium readily crosses cell membranes and is reduced intracellularly to trivalent chromium. Trivalent chromium compounds are considerably less toxic than the hexavalent chromium compounds.

The excretion of chromium increases with the duration of exposure.

Toxic Effects on Human

Chromium (III) is considered an essential trace nutrient serving as a component of the "glucose tolerance factor". No conclusive evidence exists on the toxic effects of chromium in its trivalent state. The adverse effects of chromium are attributed to the hexavalent forms which are highly toxic to humans. Hexavalent chromium is corrosive and causes chronic ulceration and perforation of the nasal septum. It is because chromium (VI) is a powerful oxidant which can easily penetrate the cell membrane and irritate cells. Chromium (III) is neither irritating nor corrosive. Chromium (VI) is a teratogen and also carcinogenic (inducing bronchogenic cancer).

Studies of chronic exposure on animals to Cr (VI) have not revealed evidence of toxicity.

Toxic Effects on other Animals

Fishes as well as mammals are far less susceptible to the toxic effects of chromium.

5. Cadmium

Cadmium is a toxic heavy metal belonging to the same family as zinc and mercury from toxicology view point. It is relatively rare but toxic to all systems and functions of animals and humans at high levels of exposure.

The occurrence of free metallic cadmium in nature is rare but it exists in close association with other metallic ores. It occurs mainly with zinc (in the ratio of 1 Cd : 200 Zn in zinc sulphide ores) and is obtained as a byproduct from the refining of zinc, copper and lead.

Uses and Pollutional Sources

Cadmium has been used in a number of industrial processes, as in electroplating, in the manufacture of pigments and paints, as a stabilizer in plastics and in welding electrodes. As copper-cadmium alloy, it is used in automobile radiators and cadmium may serve as an electrode component in alkaline accumulators.

It is of interest to note that an extremely wide range of cadmium concentration have been reported in foodstuffs from many parts of the world. In water, cadmium is found mainly in the bottom sediments and suspended particles. Pollution of drinking water with cadmium can occur as a result of leaching from solders containing the metal in fittings of water heaters, coolers and taps. Cadmium in the ambient air occurs mostly in the particulate form. Smoking tobacco may be an important route of exposure for the general population. Both water-borne and air-borne cadmium can cause an increased concentration of cadmium in soil. The use of cadmium containing sewage sludge and superphosphate fertilizers lead to contamination of the soil. Cadmium present in the soil is taken up by plants. It has been observed that more then 50 percent of the soil samples from Punjab and Haryana were sandy with low organic matter, low pH and were deficient in zinc; these conditions are favourable to promote cadmium uptake by the plants. Fertilizers like superphosphate, rock phosphate, diammonium phosphate etc. used in Punjab contained cadmium. Sea water from Bhavnagar, Gogna and Mandapam used for preparation of salt contained cadmium more than the recommended permissible limits. Studies carried out on flyash produced by casal burning power plants during electricity generation shows that the dust contains high contents of cadmium (Hayes et al, 1980).

Respiratory absorption of cadmium is about 15 to 30 percent. Workplace exposure to cadmium is particularly hazardous where there are cadmium fumes or airborne cadmium. Most airborne cadmium is respirable. A major nonoccupational source of respirable cadmium is cigarettes. One cigarette contains 1 to 2 μg cadmium, and 10 percent of the cadmium in a cigarette is inhaled (0.1 to 0.2 μg). Smoking one or more packs of cigarettes a day may double the daily absorbed burden of cadmium.

Toxicokinetics

Of various heavy metals present in the aquatic environment, cadmium pollution is especially a problem because it is not only highly toxic to the organisms but its toxicity is also cumulative.

Alimental absorption is less than respiratory absorption and is about 5 to 8 percent. Absorption is actually enhanced by dietary deficiencies of calcium and iron and by low protein diets. Low dietary calcium stimulates synthesis of calcium-binding protein, which enhances cadmium absorption. Women with low serum ferritin levels have been shown to have twice the normal absorption of cadmium (Flanagan et al, 1978). Zinc decreases cadmium absorption, probably by stimulating production of metallotheionein.

Cadmium is transported in blood by binding to erythrocytes i.e., red blood cells and large-molecular-weight proteins in plasma, especially albumin. A small fraction of blood cadmium may be transported by metallotheionein. Blood cadmium levels in adults without excessive exposure is usually less then 1 $\mu g/dL$. Newborns have a low body content of cadmium, usually less than 1 mg total body burden. The placenta synthesizes metallotheionein and may serve as a barrier to maternal cadmium, but the fetus may be exposed with increased maternal exposure. Human breast and cow's milk are low in cadmium, with less than 1 $\mu g/kg$ of milk. About 50 to 75 percent of the body burden of cadmium is in the liver and kidneys; its half-life in the body is not exactly known, but it is many years and may be as long as 30 years. With continued retention, there is progressive accumulation in the soft tissues, particularly in the kidneys, through ages 50 to 60, when the cadmium burden in soft tissues begins to decline slowly. Because of the potential for accumulation in the kidneys, there is considerable concern about the levels of dietary cadmium intake for the general population. However, presence of higher concentrations of zinc in water may reduce cadmium toxicity.

Principle of Cd Toxicity

Once absorbed into the body, cadmium has a strong affinity for the imidazole moiety of histidine, phosphates, sulfhydryl groups and inhibits a number of enzymes. About a third of the cadmium absorbed is stored in the kidney, which constitutes the principal target organ. Also lung is the major target organ effected by air-borne Cd. The symptoms are the development of severe tracheobronchitis, pneumonitis, emphysema and pulmonary edema (Goyer, 1986). Sampson et al (1984) reported development of pulmonary fibrosis. Cadmium is proved immunosuppressive as indicated by IgG mediated rosette formation (Hadley et al, 1977).

Toxic Effects on Human

The symptoms of acute poisoning in man are vomiting, abdominal cramps and headaches. After inhalation, shortness of breath, general weakness and respiratory insufficiency occurs. Chronic exposure to cadmium results in respiratory disorders, kidney damage, osteoporosis, i.e., decalcification of the skeleton, liver dysfunction, anaemia and hypertension. It has also been found to be highly toxic to the gonads of experimental animals.

Cadmium has recently been accepted as a pulmonary carcinogen (IARC — 1994) in human and rat.

Since the kidney is the principal target organ for cadmium toxicity, the following disorders have been recorded in kidney of human :

1. Aminoaciduria	Excretion of amino acids in the urine.
2. Glucosuria	Excretion of blood sugar in the urine.
3. Hypercalciuria	Excretion of excess calcium in the urine.
4. Proteinuria	Excretion of proteins in the urine.

Itai-Itai Episode

Rice, as in India, is the staple food in most of the regions of Japan. Normally, people living in uncontaminated parts of the globe ingest about 25-60 μg of cadmium daily. However, the reports reveal that the people living along the banks of *Juntsu river* in northern Japan were ingesting 100-1,000 μg of cadmium every day. It was because of the fact that the rice they consumed were grown with such irrigation water supplies (effluent from zinc smelter) that contained 0.35-3.5 ppm of cadmium. It caused about one hundred deaths until the end of 1965. As a result of the analyses carried out by Jun Kobayashi at Okayama University, it became apparent that cadmium intake was the causative factor. The victims accumulated 500 mg of cadmium in their body over the lifetime. This disorder was termed 'Itai-Itai' (literally meaning "ouch-ouch") characterized by renal malfunction (reduced ability to resorb water), drop in the phosphate level of blood serum, loss of minerals from the bone and osteomalacia, a ricket-like condition characterized by pathological bone fracture and intense pain. Actually the onset of symptoms occurs only several years after the consumption of cadmium. Nogawa (1987) reported that serum 1, 25 $(OH)_2$ Vit. D levels were lower in Itai-Itai diseased patients.

Toxic Effects on other Animals

In fishes, Cd accumulates in muscle, liver, gills and bones where it produces toxicity. Cadmium chloride can produce acute but reversible effects on ovarian tissue in sexually immature rats. The ovarian stimulation induced by $CdCl_2$ was manifested as follicular maturation and corpora lutea formation but did not seem to lead to the production of sex hormones. A single intraovarian injection of this salt (1 to 3 mg/kg) sterilized adult goats, all the germinal elements and corpora lutea were destroyed. The stromal tissue with vasculatures was also damaged. Similar changes have also been observed in other species of animals. The change commenced with damage to the vascular structure indicating the primary role of vascular injury in producing acute degeneration of the tissue.

6. Copper

Copper (Cu) is an essential element and is widely distributed in nature. Substantial copper deposits in India are known to occur in Rajasthan. It occurs primarily as its oxide or sulphide ores. Copper deficiency is characterized by hypochromic microcytic anemia resulting from defective hemoglobin synthesis. Oxidative enzymes viz., peroxidase, catalase, cytochrome oxidase and others also require copper.

Uses and Pollutional Sources

Next to iron and aluminium, copper is the third most important metal used in the industries. It is second to silver in its thermal and electrical conductivities, hence it finds use as a conductor in many alloys. It is also used in paints and ceramics. Copper sulphate is used medicinally as an emetic and also as an antihelmintic. Copper sulphate mixed with lime has been used as a fungicide.

The environmental contamination with copper primarily results from its discharge from the industrial plants producing non-ferrous metals. Additional sources comprise wood combustion and steel production. Copper mining and metallurgical operations contribute to contaminations of aquatic environment. Some of the copper salts are used as algicides and fungicides. Apart from these, copper is also used as antifouling agents in paints. Most of the paints contain 100-200 g copper oxide/liter. Fertilizer production and disposal of industrial/ municipal sewage wastes represent minor sources of copper in the environment.

Toxicokinetics

Alimental absorption of copper is generally regulated by body stores (Sarkar et al, 1983). It is transported in serum bound initially to albumin and later more firmly bound to ceruloplasmin, where it is exchanged in the cupric form. The normal serum level of copper is 120 to 145 μg per liter. The bile is the normal excretory pathway and plays an initial role in copper homeostasis. Most copper is stored in liver and bone marrow where it may be bound to metallothionein. The amount of copper in milk is not enough to maintain adequate copper levels in the liver, lungs, and spleen of the newborn. Tissue levels gradually decline up to about ten years of age, remaining relatively constant thereafter. Brain levels, on the other hand, tend to almost double from infancy to adulthood. The ratio of newborn to adult liver copper levels shows considerable species difference : human, 15 : 4; rat, 6 : 4, and rabbit, 1 : 6. Since urinary copper levels may be increased by soft water, under these conditions concentrations of approximately 60 μg per liter are not uncommon.

Copper is an essential component for various enzymes, incluing tyrosinase, involved in the formation of melanin pigments, cytochrome oxidase, superoxide dismutase, amine oxidase (involved in the formation of two proteins — elastin and collagens) and uricase. It is essential for the utilization of iron. Iron deficiency anemia in infancy is sometimes accompanied by copper deficiency. Molybdenum also influences tissue levels of copper. The recommended daily intake of copper ranges from 2-3 mg/day. The impairment of the ability to absorb copper resulting in its deficiency is called Menke's Disease whereas Wilson's Disease is its opposite (i.e., excessive accumulation of copper).

Toxic Effects on Human

Acute poisoning resulting from ingestion of excessive amounts of oral copper salts may produce death. The ingestion of 15-75 mg of copper causes gastrointestinal disturbances. Further increase in amount may cause hemolysis, hepatotoxic and nephrotoxic effects.

Inhalation of airborn copper causes irritation of the respiratory tract and metal fume fever and ultimately lung cancer after chronic exposure.

Toxic Effects on other Animals

Reduces growth, survival and rate of reproduction. It also causes necrosis of kidney cells and brain haemorrhage.

7. Zinc

Zinc is a nutritionally essential trace metal to all organisms, as it is necessary for the normal functioning of various enzymes. More than 200 metalloenzymes require zinc as a cofactor.

Zinc is ubiquitous in the environment so that it is present in most foodstuffs, water and air. The important ores of zinc are : Willemite (Zn_2SiO_4), Zincite (ZnO), Zinc Blends (ZnS), Smithsmite ($ZnCO_3$), etc. Zinc forms about 40 mg/kg of the earth's crust. Atmospheric zinc levels are higher in industrial areas.

Uses and Pollutional Sources

Zinc is used in dry batteries, construction materials, pigments and printing processes. It is also used for protective coatings on iron, steel, brass and alloys.

Smelting of ores contributes appreciably to the atmospheric levels of zinc. Municipal refuse and automobiles serve as additional pollutional sources. Agricultural use of $ZnSO_4$ — containing pesticides and fungicides, for example — Manozeb (16% Mn and 2% Zn), Zineb and Ziram (1-18% Zn) may be yet another source of Zinc in the environment.

Toxicokinetics

Approximately 20-30% of ingested zinc is absorbed. Its absorption is generally influenced by prostaglandins E_2 and F_2 and is chelated by tryptophan derivatives like picolinic acid. Zinc concentration in tissues varies widely. The liver receives up to about 40% zinc. It is bound there to metallotheionin. The greatest concentration of zinc in the body is in the prostate, probably related to the rich content of zinc-containing enzyme and phosphatase.

It is interesting to note that Zinc is necessary for the normal functioning of the cell, viz., protein synthesis, carbohydrate metabolism, cell division and growth. Recommended daily dietary allowances of zinc are at 15 mg for adults and 10 mg for children over a year old. The average dietary intake of zinc in India is about 12-15 mg, mostly from food.

Toxic Effects on Human

Zinc at concentration over 15 mg/m³ in air may produce metal fume fever. Large doses of zinc salts (220-440 mg of zinc as $ZnSO_4$) cause gastrointestinal disorders including vomiting and diarrhea. There is no evidence of zinc being teratogenic or mutagenic. Zn shows no hematological, hepatic or renal toxicity.

Toxic Effect on other Animals

In fish zinc causes decrease in O_2 consumption. It also damages gills, decreases growth and fecundity.

8. Aluminium

Aluminium (Al) is one of the most ubiquitous elements in the environment. It is the third most abundant element in the earth's crust (about 8-15%). In industrial societies it is the second most important metal. Bauxite, the principal ore source of aluminium, is widely distributed in India. The most important deposits are found in Bihar, M.P. and Maharashtra.

Uses and Pollutional Sources

Aluminium is a non-ferrous metal, chiefly extracted from bauxite. It is extensively used for canning, food packaging, and as foil for covering and preserving foodstuff. Beverage cans and cooking utensils constructed of Al are extensively used around the world.

• Contamination of the total environment with Al may result from the indiscriminate disposal of Al-containing products and wastes. Workers in the mining and manufacturing industries may be exposed to aluminium particles. Human exposure to Al comes from foods and drinking water as well as pharmaceuticals.

Toxicokinetics

Average daily intake of Al by human has been reported to be 18-20 mg/day through foods and drinking water. In plasma 80-90% of Al binds to transferrin, an iron-transport protein for which there are receptors in many body tissues.

Aluminium at higher levels of exposure interferes with phosphate metabolism. It reduces the bioavailability of ATP and casein, and inhibits certain active enzymes by complexing them.

Bone and lung have the highest concentration of aluminium. Actually bone may be a "sink" for aluminium. It does not normally accumulate in blood to any great extent.

Aluminium compounds may affect absorption of other elements in the alimentary tract and alter its motility by inhibition of acetylcholine-induced contractions. Aluminium has been found to inhibit fluoride absorption and may decrease the absorption of calcium and iron compounds. Al may cross the placental barrier. Al in its dissolved form in water is much more toxic than suspended form.

Toxic Effects on Human

Generally, human beings are least affected by aluminium; however, inhalations with aerosol produce pneumonia. In general, Al binds with bone and causes osteomalacia. Al competes with or alters calcium metabolism in several organ systems including the brain. Calcium rises in brain tissue following Al exposure and disrupts neurotubular integrity and transport.

Toxic Effects on other Animals

In fishes, Al causes gill hyperplasia and mortality.

9. Manganese

Manganese (Mn) is an essential trace element and is a cofactor for a number of enzymatic reaction, particularly those involved in phosphorylation, fatty acid and cholesterol synthesis. It is toxic at higher concentrations. The most important commercial source of manganese is Pyrolusite (MnO_2) containing 60-63% manganese. India ranks third in the production of manganese. About 75% of Mn is produced in Orissa, Karnataka, M.P. and Maharashtra.

Uses and Pollutional Sources

Besides export, indigenous demand for this metal is constantly increasing with the growth of the steel industry. Manganese is also used for the manufacture of ferromanganese alloys and in non-ferrous industries. MnO_2 is utilized in dry cell batteries as a polarizer; metallic manganese is coated in the electrodes in welding rods and fluxes for iron and steel. Manganese is also used in the production of zinc by electrolysis, manufacturing pigments, paints, ceramics, photographic material, as a wood-preservative and in fertilizers. Organic manganese compounds are used as additives in gasoline, fuel oil and diesel.

Burning of fossil fuels (coal, oil) is the principal source of Mn in the environment. Industrial processes of battery, glass and steel manufacture provide additional input of Mn into the environment. The use of Mn in some fertilizers contributes further to air and water pollution.

Chemical Forms

Manganese can exist in seven possible oxidation states : O, +1, +2, +3, +4, +6 and +7. Of these, only two valence states, +2 and +4, are encountered commonly from toxicological view-point.

Toxicokinetics

Mn is present in all living organisms. Vegetables, the germinal portions of grains, fruits, nuts, tea, and some spices are rich in manganese. A daily allowance of 2.5-9.0 mg of Mn is recommended for humans. But when exposed to higher levels of Mn, approximately 1-4% of it is absorbed, retained, and accumulated — chiefly in kidney, liver, and bones. Gastrointestinal absorption is less than 3-4%. Mn concentrates in mitochondria so that tissues rich in these organelles have the maximum concentration of Mn, including the pancreas, liver, kidneys and intestines. Biological half-life of the manganese in the body is 37 days. It readily crosses the blood-brain barrier and its half-time in the brain is longer than in the whole body. The principal route of its excretion is with feces.

Toxic Effects on Human

Evidence of occupational Mn poisoning was reported for the first time as early as 1837 in France in a few workers engaged in the grinding of manganese ores.

A definite relationship was established only in 1919. Since then several cases of chronic manganese poisoning have been detected in miners, steel plant and foundry workers, welders, workers employed in dry cell battery manufacturing industries and in families consuming water with high manganese contents. Industrial toxicity from inhalation exposure, generally to manganese dioxide, in mining or manufacturing, is of two types : The first, manganese pneumonitis, is the result of acute exposure. Men working in plants with high concentrations of manganese dust show an incidence of respiratory disease 30 times greater than normal. Pathological changes include epithelial necrosis followed by mononuclear proliferation.

The second and more serious type of disease resulting from chronic inhalation exposure to manganese dioxide, generally over a period of more than 2 years, involved the central nervous system. Chronic manganese poisoning (*manganism*) produced a neuropsychiatric disorder characterized by irritability, difficulty in walking, speech disturbances, and compulsive behaviour that may include running, fighting and singing. If the condition persists a masklike face, retropulsion or propulsion, and a Parkinson's Disease-like like syndrome develops. To sum up, the symptoms of neurological syndrome of manganese poisoning are usually insidious and progressive. An initial phase of psychomotor excitement is followed by asthenia, somnolence, imbalance while walking, slurred speech, difficulty in fine movements, impotence, etc. Established cases show tremors, paresthesia, memory loss, difficulty in walking and swallowing. Bradykinesia, impaired postural stability, diminished associated movements, gait disorder described as "Cock-walk", retropulsion and masked faces are also observed.

Victims of chronic manganese poisoning tend to recover slowly, even when removed from the excessive exposure. The oral absorption of Mn is increased by iron deficiency.

Toxic Effects on other Animals

Mn is neuro- and pancreotoxic to fishes and mammals.

Revision Questions

A. Long-Answer Questions

1. Describe the toxicology of Arsenic.
2. Describe the toxicology of Mercury.
3. Describe the toxicology of Cadmium.
4. Describe the toxicology of Lead.
5. Describe the toxicology of Manganese.

B. Short-Answer Questions

1. *Write short notes on :*
   (i) Heavy metals
   (ii) Dispersion of metals in environment
   (iii) Principle of metal toxicity.
2. *Briefly describe :*
   (i) Minamata Disease
   (ii) Itai-Itai episode
   (iii) Toxicity of different species of mercury
   (iv) Zinc toxicity.
3. *Write short notes on :*
   (i) Toxicokinetics of Arsenic
   (ii) Toxicokinetics of Manganese
   (iii) Toxicokinetics of Mercury
   (iv) Toxicokinetics of Cadmium.

14

Xenobiotics Translocation [Absorption; Distribution; Biotransformation and Excretion]

SYNOPSIS

Introduction, Translocation, Mechanism of xenobiotics transfer, Distribution of xenobiotics, Storage depots, Biotransformation, Excretion of xenobiotics, Deposition of xenobiotics, Membranous barriers.

Introduction

Xenobiotics are the environmental chemicals. It is a term derived from the Greek word — *xenon* = a stranger + *bios* = life i.e., stranger to life. The salient xenobiotics of concern to human health and environment are listed in Table 14.1 :

Table 14.1 : Xenobiotics of Concern to Human Health and Environment

Products	*Examples*	
Synthetic compounds of use	Foods additives, Biocides, Pesticides, Dyes, Explosives, Detergents, Polymers, Intermediates etc.	
Inadvertently formed waste and byproducts	Industrial waste water, automobile exhaust, products of incinerations, chlorination drinking water	CO, SO_x, NO_x metals, PAH, NH_3, Chloroform, Hydrocarbons, Chlorophenols, Chloroform, Particulate matter
Environmental transformation products of xenobiotic (Industrial) compounds	DPE, photomirex, 2, 4-dichlorophenol, 3,4 — dichloroaniline, benzo (a) pyrene, quinines	

The metabolism of xenobiotics provides a basic knowledge to the understanding of pharmacology, toxicology, cancer research, drug addiction. All these areas involve exposure to xenobiotics.

Translocation

The biological system may be exposed to the high amounts/concentration of xenobiotics, but if the concentration at the specific sites remain low due to any reason viz., defensive mechanism, there appears little or no effect. In order to produce a measurable effect, a xenobiotic must be transported from the site of exposure to the specific sites of action. The process of *transport of xenobiotics* (toxicants) from the site of their application to the specific site of action or more diffused sites of action is designated as translocation of xenobiotics. In other words, dynamics of movement of xenobiotics in the living system from its penetration into the blood to its final elimination from the body is termed *translocation.* The factors which bring translocation are referred to as translocation factors including absorption, distribution, binding and excretion.

The process of translocation involves four principal steps :

1. Absorption from the site of action via skin or GIT or lungs.
2. Transport of xenobiotics by blood to —
   (a) The biotransformation sites, (b) Other tissues.
3. Accumulation of xenobiotics and their release into the blood.
4. Elimination from the body mainly through :
   (a) G.I.T. (b) Kidney

The entire process of translocation is represented in the Fig. 14.1. The major routes by which xenobiotics enter the body are skin, G.I.T. and lungs. Xenobiotics thus absorbed are eliminated from the circulation by metabolism and circulation. The quantification and determination of the time course of absorption, distribution, biotransformation and excretion of xenobiotics are referred to as toxicokinetics.

deleterious effects at one site or several sites in the body. Exceptions are caustic and corrosive agents (acids, bases, salts, oxidizers), which act topically. A toxicant absorbed into the blood stream through any barrier (described ahead) is distributed, at least to some extent, throughout the body, including the site where it produces damage. This site is often called the target organ or target tissue.

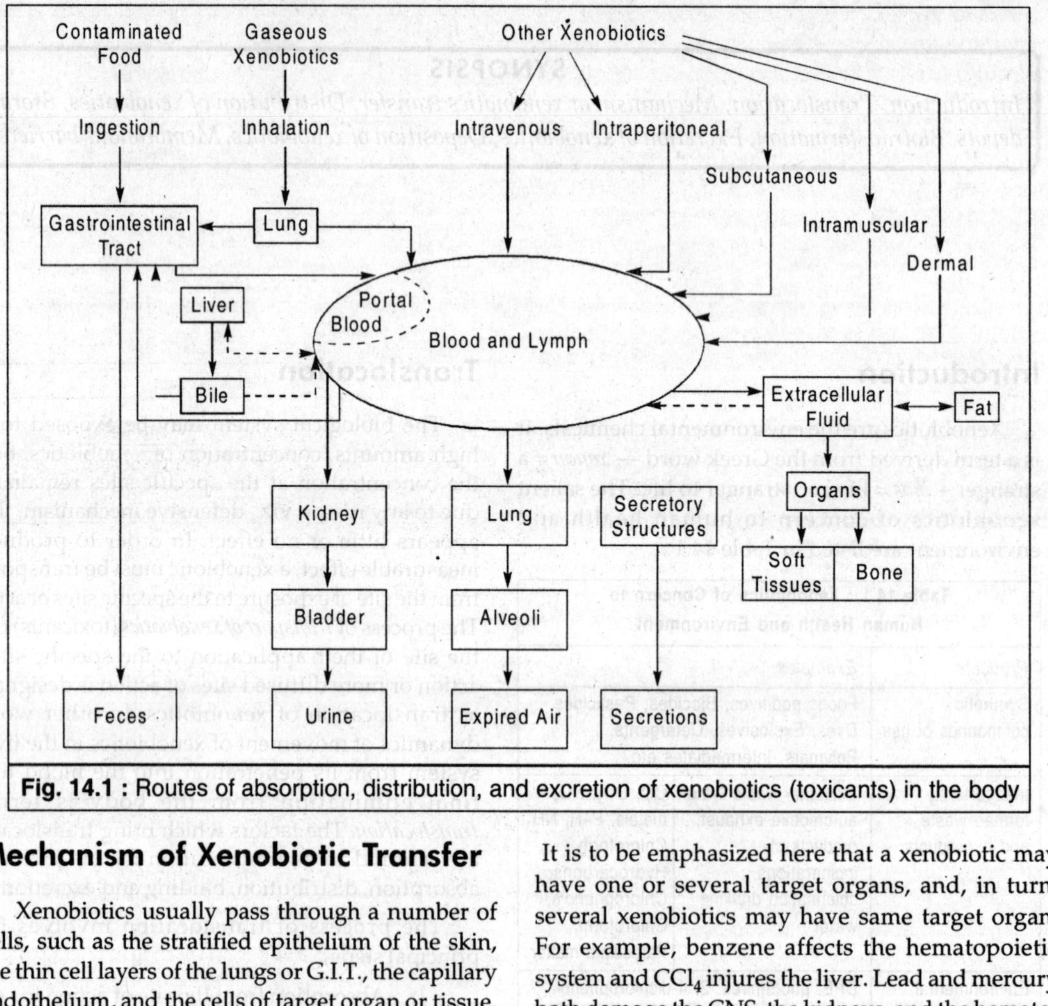

Fig. 14.1 : Routes of absorption, distribution, and excretion of xenobiotics (toxicants) in the body

Mechanism of Xenobiotic Transfer

Xenobiotics usually pass through a number of cells, such as the stratified epithelium of the skin, the thin cell layers of the lungs or G.I.T., the capillary endothelium, and the cells of target organ or tissue. The cell membrane (plasma membrane) surrounding all these cells are remarkably similar. It is approximately 100 A^0 thick with pores of size ranging from 4 to 40 A^0. The lipid layer of the plasma membrane is permeable to and readily penetrated by lipid soluble substances. In order to enter the animal and reach some target organ, any xenobiotic (toxicant) must penetrate one or more living membrane(s). The membranes form barriers through which toxicants have to cross to exert their

It is to be emphasized here that a xenobiotic may have one or several target organs, and, in turn, several xenobiotics may have same target organ. For example, benzene affects the hematopoietic system and CCl_4 injures the liver. Lead and mercury both damage the CNS, the kidneys, and the hematopoeitic system. Toxic agents in different ways may modify the permeability of plasma membrane. For instance, size, shape, degree of ionization, and lipid solubility of ionized and non-ionized forms of a xenobiotic (toxicant) are its important physio-chemical properties that influence its passage across the plasma membrane.

A toxicant may pass through plasma membrane by one of the two general processes :

1. Passive transport (diffusion according to Fick's law) in which the cell expends no energy, and
2. Specialized transport, in which the cell provides energy to translocate the xenobiotic (toxicant) across its membrane.

1. Passive Transport

Simple Diffusion : Most xenobiotics cross cell membranes by simple diffusion. Small hydrophilic molecules (up to a molecular weight of about 600) presumably permeate membranes through aqueous pores (Benz et al, 1980), whereas hydrophobic molecules diffuse across the lipid domains of membranes. The smaller a hydrophilic molecule is, the more readily it traverses membranes by simple diffusion through aqueous pores. Consequently, ethanol is absorbed rapidly from the stomach and intestine and is distributed equally rapidly throughout the body by simple diffusion from blood into all tissues.

The rate of passage is related directly to the concentration gradient across the membrane and to the lipid solubility. For example, mannitol is hardly absorbed (< 2%); acetylsalicylic acid is fairly well absorbed (21%); and the thiopental is even more readily absorbed (67%).

Many toxicants are ionizable. The ionized form is often unable to penetrate the cell membrane because of its low lipid solubility.

The degree of ionization of a chemical depends on its p*Ka* and on the pH of the solution. The relationship between p*Ka* and pH is derived by the Henderson-Hasselbalch equation :

$$\text{For acids: pKa} - \text{pH} = \log \frac{[\text{Nonionized}]}{[\text{Ionized}]}$$

$$\text{For bases: pKa} - \text{pH} = \log \frac{[\text{Ionized}]}{[\text{Nonionized}]}$$

Some examples are :

For Weak Acids

$$\text{pKa} - \text{pH} = \log \frac{[\text{nonionized}]}{[\text{ionized}]}$$

Benzoic acid pKa ≈ 4

Stomach pH ≈ 2

$$4 - 2 = \log \frac{[\text{nonionized}]}{[\text{ionized}]}$$

$$2 = \log \frac{[\text{nonionized}]}{[\text{ionized}]}$$

$$10^2 = \log \frac{[\text{nonionized}]}{[\text{ionized}]}$$

$$100 = \log \frac{[\text{nonionized}]}{[\text{ionized}]}$$

Ratio favours absorption

Intestine pH ≈ 6

$$4 - 6 = \log \frac{[\text{nonionized}]}{[\text{ionized}]}$$

$$-2 = \log \frac{[\text{nonionized}]}{[\text{ionized}]}$$

$$10^{-2} = \frac{[\text{nonionized}]}{[\text{ionized}]}$$

$$\frac{1}{100} = \frac{[\text{nonionized}]}{[\text{ionized}]}$$

For Weak Bases

$$\text{pKa} - \text{pH} = \log \frac{[\text{ionized}]}{[\text{nonionized}]}$$

Aniline pKa ≈ 5

Stomach pH ≈ 2

$$5 - 2 = \log \frac{[\text{ionized}]}{[\text{nonionized}]}$$

$$3 = \log \frac{[\text{ionized}]}{[\text{nonionized}]}$$

$$10^3 = \log \frac{[\text{ionized}]}{[\text{nonionized}]}$$

$$1{,}000 = \log \frac{[\text{ionized}]}{[\text{nonionized}]}$$

Intestine pH ≈ 6

$$5 - 6 = \log \frac{[\text{ionized}]}{[\text{nonionized}]}$$

$$-1 = \log \frac{[\text{ionized}]}{[\text{nonionized}]}$$

$$10^{-1} = \frac{[\text{ionized}]}{[\text{nonionized}]}$$

$$\frac{1}{10} = \frac{[\text{ionized}]}{[\text{nonionized}]}$$

Ratio favours absorption

Filtration : The membranes of the capillaries and the glomeruli have relatively large pores (about 40 A^0) and allow molecules smaller than albumin (mol. wt. 60,000) to pass through. Bulk flow of water through these pores results from hydrostatic and/ or osmotic pressure and can act as carrier of toxicants. The pores in most cells, however, are relatively small (about 4 A^0) and allow chemicals only up to a mol. wt. of 100 - 200 to pass through. Chemical of larger molecules, therefore, can filter into and out of the capillaries. They can, therefore, establish equilibrium between the concentrations in the plasma and in the extracellular fluid, but they cannot do so by filtration between the extracellular and intracellular fluids.

2. Specialized Transport

There are numerous compounds whose movements across membranes cannot be explained by simple diffusion or filtration. Some compounds are too large to pass through aqueous pores or too insoluble in lipid domains of membranes. Nevertheless, they often are transported very rapidly across membranes, even against concentration gradients. To explain these phenomena, the existence of specialized transport systems has been postulated. These systems are responsible for the transport across cell membranes of many nutrients, such as sugars, amino acids, nucleic acids, and also that of some foreign compounds.

Active Transport : The following properties denote an active transport system :

1. chemicals are moved against electro-chemical or concentration gradients.
2. the transport system is saturated at high substrate concentrations and thus exhibits a transport maximum (Tm).
3. the transport system is selective for certain structural features of chemicals and has the potential for competitive inhibition between compounds that are transported by the same transporter, and
4. the system requires expenditure of energy so that metabolic inhibitors block the transport process.

Substances actively transported across plasma membranes presumably form a complex with membrane-bound macromolecular carrier on one side of the membrane. The complex subsequently traverses to the other side of the membrane, where the substance is released. Afterward, the carrier returns to the original surface to repeat the transport cycle.

Active transport is especially important in regard to eliminating xenobiotics from an animal. The central nervous system (CNS) has two transport systems at the choroid plexus to transport compounds out of the cerebrospinal fluid (CSF) — one for organic acids and one for organic bases. The kidney also has two active transport systems, whereas the liver has at least four, two of which transport organic acids, one organic base and one neutral organic compound.

Facilitated Diffusion : Facilitated diffusion applies to carrier-mediated transport that exhibits the properties of active transport, except that the substrate is not moved against an electrochemical or concentration gradient and the transport process does not require the input of energy; that is, metabolic poisons do not interfere with this transport. The transport of glucose from the gastrointestinal tract (GIT) across the basolateral membrane of the intestinal epithelium, from plasma into red blood cells and from blood into the CNS, occurs by facilitated diffusion.

Additional Transport Processes : Other forms of specialized transport have been proposed, but their overall importance is not as well-established as that of active transport and facilitated diffusion. Phagocytosis (cell eating) and pinocytosis (cell drinking) are proposed mechanisms for cell membranes flowing around and engulfing particles. This type of transfer has been shown to be important for the removal of particulate matter from the alveoli by phagocytes and from blood by the reticu-loendothelial system of the liver and spleen.

ABSORPTION

The process by which xenobiotics cross body membranes and enter the blood stream is referred to as absorption. There are no specific systems or pathways for the sole purpose of absorbing toxicants. Xenobiotics penetrate membranes during absorption by the same processes, as do biologically essential substances such as oxygen, foodstuff and other nutrients. The main sites of absorption are the GIT, the lungs, and the skin. However, absorption may also occur from other sites, such as the subcutis, peritoneum, or muscle, if a chemical is administered by special routes. Experimentalists often distinguish between parenteral and enteral administration of drugs and other xenobiotics. It shall not be out of place to mention that enteral administration includes all routes pertaining to the alimentary canal (sublingual, oral, and rectal), whereas parenteral administration involves all other routes (intravenous, intraperitoneal, intramuscular, subcutaneous) etc.

Overall, the absorption of xenobiotics may take place through 4 ways :

A. Absorption by the skin (Dermal absorption)

B. Absorption by the lungs (Pulmonary absorption)

C. Absorption by gastro-intestinal tract (Gastro-intestinal absorption)

D. Absorption through special routes.

A. **Absorption by the Skin** : The body's primary barrier to the toxic chemicals is the skin, which has a surface area of about 18,000 cm^2 in humans and constitutes about 10% of body weight. Thus, dermal absorption is very important in view of the surface area of skin. A xenobiotic entering through the skin would normally have to penetrate several layers of intact cells to reach the systemic blood circulation. Lipid-soluble molecules such as tetraethyl lead and benzene are more easily absorbed via the skin than water-soluble polar molecules and ions (e.g., inorganic leads, H_2SO_4). Solvents such as CCl_4, highly lipid soluble, absorbed percutaneously exert systemic toxic effect following absorption by this route. Small hydrophilic xenobiotics such as hydrazine is sufficiently absorbed through skin, causes local as well as systemic toxic effect.

B. **Absorption by the Lungs** : Humans breathe 22,000 times a day taking in 16 kg of oxygen. Thus pulmonary absorption can be the principal route of entry for gaseous and particulate air pollutants and volatile organic chemicals. The large surface area of lungs makes it particularly vulnerable to the toxic injuries caused by inhaled gases and particulates. The most frequent cause of death from poisoning by CO and occupational disease *silicosis* are the consequences and the results of absorption of air-borne poisons by the lungs. Bhopal Gas Tragedy may be considered as an extreme example of absorption of MIC gas through the lungs.

The air-borne particulate matters deposited in the lungs are absorbed by processes known as 'pinocytosis' (a special type of transport, in which cells engulf small droplets of extracellular fluid) and 'phagocytosis' (a specialized transport process that involves the engulfing of a particle or colloidal material by the cells) and moved by the action of cilia into the pharynx from where it may be swallowed in. But the soluble chemical molecules could penetrate the epithelium and enter the blood circulation directly. Pinocytosis and phogocytosis may also be referred to as cell-drinking and cell-eating, respectively.

C. **Absorption by the Gastro-intestinal Tract**: The oral uptake is the most important pathway at least for the water-borne and food-borne pollutants. The absorption of xenobiotic molecules along the gastro-intestinal (G.I.) tract (consisting of stomach, small intestine and colon) is characteristically due to the epithelial wall of the G. I. Tract, which exhibits the properties of the lipoid membrane. Hence, un-ionized polar toxicants are more readily absorbed, the absorption rates for these chemicals being influenced by their relative octanol-water partition coefficients. The substances ingested by oral route do not produce systemic effects unless they are absorbed. The stomach having a pH 1 to 3 represents the most significant site of absorption for xenobiotics that are actually acidic or near neutral. Along the small intestines and colon (whose pH lies in the range of 6 to 8), weak acids and bases are more efficiently absorbed, the rate of absorption being inversely related to their water solubilities.

D. **Absorption through special routes** : Generally the xenobiotics enter the blood stream following absorption through skin, lungs and gastrointestinal tract. But, sometimes, the toxicants are administered in the body of experimental animals through special routes such as (1) intra-peritoneal (2) subcutaneous (3) intramuscular, and (4) intravenous. The toxicants are directly administered into the blood stream through intravenous route. Administration of toxicants through intraperitoneal route results in rapid absorption of toxicants primarily through the portal circulation because of the large surface area and rich blood supply to the peritoneal cavity. Toxicants administered subcutaneously and intramuscularly are absorbed at slow rate. This can be altered by changing the blood flow to the area and the medium in which the toxicant is administered. Most of the toxicant administered intra-peritoneally enters the liver via portal circulation before it reaches the general circulation of the animals. Sometimes the xenobiotics may be completely biotransformed and eliminated into the bile and consequently it never reaches the target sites.

DISTRIBUTION

A xenobiotic after entering the blood by absorption or intravenous administration is ready for distribution throughout the body. Distribution usually occurs rapidly. The rate of distribution to organs or tissues is determined primarily by blood flow and the rate of diffusion out of the capillary bed into the cells of a particular organ or tissues. The final distribution depends largely on the affinity of a xenobiotic for various tissues. In general, the initial phase of distribution is dominated by the blood flow, whereas the eventual distribution is determined largely by affinity.

Some toxicants do not readily cross plasma membranes and, therefore, have restricted distribution; whereas other xenobiotics rapidly pass through plasma membranes and are distributed throughout the body. During the course of distribution, the xenobiotics may get concentrated in a particular tissue or organ and such tissues or sites are termed *storage depots*. The site of accumulation of toxicant may be its site of major toxic action, but more often it is not. Actually xenobiotics (toxicant) accumulate at a site other than the target organ or tissue. The accumulation may be viewed as a protective process. Xenobiotics enter the systemic blood circulation as a result of the absorption processes. Since the blood-stream passes through all the body tissues, it can distribute the chemical throughout the body. Among the body fluids, plasma water is relatively more important than the interstitial or intra-cellular water. In humans, the blood plasma accounts for 4% of the total body weight and for about 53% of the total blood volume.

The concentration of a chemical (xenobiotic) in blood plasma, is very important since it determines the amount of the chemical (1) that can reach the site of toxic action (i.e., target organ), (2) that might be directed to a storage depot (e.g., bone, adipose tissue) and (3) that is transported to the liver where the process of biotransformation to activate or inactivate an organic pollutant occurs. The concentration of a toxicant in blood is primarily dependent on its volume of distribution. The volume of distribution can be quantified by the following formula :

$$VD = \frac{\text{Dose (mg)}}{\text{Plasma concentration (mg/l)}}$$

The volume of distribution sometimes indicates that the xenobiotic compound is localized in a particular tissue or is mainly confined to the plasma.

Various factors influence the distribution of toxicants from the blood stream; of prime importance is the binding of chemical molecules in blood to the plasmatic proteins, particularly, the serum albumins. The chemical molecules that are bound to the plasmatic proteins are rendered unavailable for distribution or for immediate toxic action (until dissociated, or competitively displaced by another chemical molecule). Only the free and unbound chemical forms are susceptible for metabolic alterations and are responsible for the elimination of toxic effects.

Several organophosphorus insecticides and also the organo-chlorine insecticides, dieldrin, for example, are known to be firmly bound to plasmatic proteins. Problems could emerge when a second substance is introduced into the blood-stream, especially if the latter replaces the previously bound substance(s) from the binding sites on the plasma proteins. This can result in unpredictable increase of the unbound fraction of the first substance in the plasma, leading to serious toxicological consequences.

Storage-depots or store — house (Storage of toxicants in certain tissues)

Xenobiotics often are concentrated in a specific tissue. Some xenobiotics attain their highest concentrations at the site of toxic action and in fact such substances prove to be highly deleterious. For example, CO, which keeps a high degree of affinity for hemoglobin and paraquet shows high affinity for the lungs, hence it accumulates in the lung. Other xenobiotics get concentrated at sites other than the target organ. For example, lead is accumulated in bone, but its poisoning appears in soft tissues.

Likewise, organochlorines, which are well known neurotoxicants, accumulate in tissue having high lipid content. The compartments where xenobiotics get concentrated or accumulated are denoted as *storage depots* or *store house* of toxicants. As mentioned earlier, the storage may be viewed as a protective device, which prevents the deposition or accumulation or storage of xenobiotics (toxicants) at the site of their action, and consequently minimize their toxic actions. Elimination of the toxicants of storage depots takes place through biotransformation and excretion.

The principal storage sites for xenobiotics are :

1. Plasma proteins as Storage Depot

Several plasma proteins — for example albumin, transferring, ceruloplasmin, α & β lipoprotein etc.— may bind some xenobiotics (toxicants). The extent of plasma protein binding varies considerably among xenobiotics. For example, antipyrine is not bound. Others such as secobaribital are bound to about 50% and some, like warfarin, are about 99% bound. Plasma proteins are capable of binding various acidic compounds such as phenylbutazone, basic compounds such as imipramine, and netural compounds such as digitoxin. Most xenobiotics that are bound to plasma proteins bind to albumin. The principal metal binding protein in plasma is ceruloplasmin. The binding of xenobiotics with plasma protein is of special significance in toxicology. Because of their binding with high molecular weight plasma protein, they cannot cross capillary walls and other plasma membranes. Consequently, the bound form of toxicant may not enter the target site to produce injury. However, addition of another toxic agent can displace the first one from the plasma proteins, thus making it available in free form. For example, if a sulphonamide drug is provided to a patient already taking any diabetic drug, the former replaces the latter and induces hypoglycemic coma.

2. Body Fat as Storage Depot

Tissues and areas in the body rich in fat tend to store some xenobiotics that are highly soluble in lipids. These chemicals tend to localize in adipose tissues by partitioning between intracellular lipids and body water. Such accumulation in adipose tissue has been demonstrated for a number of xenobiotics, including chlordane, DDT, and polychlorinated and polybrominated biphenyls. Storage lowers the concentration of the toxicants in the target organ and serves as protective mechanism. Following starvation, the fat mobilization leads to a sharp rise in the plasma concentration of the stored xenobiotics.

Age, sex and race play an important role in the storage (accumulation) of lipid soluble xenobiotics. For example, males accumulate more of organochlorie insecticides than females, older people accumulate more OCIs than young people and Afro-Americans accumulate more of these insecticides than white men.

3. Liver as Storage Depot

Actually liver may be regarded as a principal storage organ. Since it performs variety of functions viz., storage of glycogen, deamination of amino acids, production of bile, synthesis of blood proteins, synthesis of urea through ornithine cycle and biotransformation, it receives all kinds of molecules from GIT through portal circulation and houses a large amount of all types of xenobiotics as well as proteins, bile and glycogen. The xenobiotics, entering liver for the process of biotransformation, bind with different constituents or groups and with other molecules present in the liver, according to their solubility and affinity. Thus liver acts as storage depot for various xenobiotics.

A protein in the cytoplasm of liver cells called ligandin, a γ protein, has been identified which shows a high affinity for many organic acids. Researches reveal that this protein may play a key role in the transfer of organic anions from plasma into liver. This protein also binds azo dye carcinogens and corticosteroids. Metalothionin, another binding protein of liver, binds to the cadmium. Liver is capable of binding lead 50 times greater than plasma within 30 minutes after single administration.

4. Kidney as Storage Depot

Like liver, kidney has a high capacity for binding multitude of xenobiotics. Since kidney is the principal excretory organ, all kinds of xenobiotics reach kidney glomeruli for filtration. The glomerular basement membrane restricts excretion of protein and lipid molecules, stores various metalloprotein molecules and chlorinated hydrocarbon contaminated fat molecules. These go on accumulating at this site and cause extensive damage to the kidney glomerular basement membrane. It shall not be out of place to mention that liver and kidney probably concentrate more xenobiotics than do all the other organs combined.

5. Bone as Storage Depot

Bone is known to serve as a major storage site for some xenobiotics viz., lead, fluoride, and radioactive strontium. Of these xenobiotics, accumulation of lead in skeletal tissues is more pronounced, as it comprises of about 90% of the total lead in the body. Actually, skeletal uptake of xenobiotics is essentially a surface chemistry phenomenon, with exchange taking place between the bone surface and the fluid in contact with it. Deposition and storage of toxicants in bone may or may not be detrimental. For instance, lead is not toxic to the bones whereas the accumulation of fluoride for long term exposure causes skeletal flurosis. Likewise, long term storage of radioactive strontium produces osteosarcoma and other neoplasma. Xenobiotics may be released from the bone by ionic exchange at the crystal surface and dissolution of bone crystals through osteoclastic activity.

6. Brain and Nervous Tissues as Storage Depots

Some xenobiotics have high affinity for the brain tissues. For example, CNS drugs like thiopentone, barbiturates and psychotropic drugs, chlordiazepoxide and chlorpromazine enter the brain tissue rapidly after intravenous administration in experimental animals.

The myelin sheath of axon also accumulates lipid-soluble xenobiotics and acts as storage depot. The axon sheath accumulates a large amount of lipophilic chlorinated hydrocarbons and retains them for long time. Likewise, mercury also accumulates in myelin sheath and produces *Hatter's Shake.*

7. Erythrocytes as Storage Depot

Pharmacological agents, viz. nitroalamine, becomes localized in the R. B. C. by binding with hemoglobin. Similarly, other xenobiotics, viz. triethyltin (TET) has been reported to bind rapidly with rat hemoglobin. Arsenic, cadmium and lead also localize with erythrocytes for short time.

8. Gonads as Storage Depot

Researches reveal that epididymal fat of testes and of prostate gland accumulate and store lipid-soluble xenobiotics, example — DDT.

9. Respiratory Tract as Storage Depot

Xenobiotic molecules may get deposited on the mucosal lining of the respiratory tract and remain trapped there for short time.

10. G.I.T. as Storage Depot

Various lipid soluble xenobiotics and metals like mercury get deposited and remain trapped in the G.I.T. and accumulate. Also intestinal mucosal lining absorbs lipid soluble xenobiotics and retains them.

11. Gall Bladder as Storage Depot

Gall bladder acts as storage organ of bile. It also receives and accumulates all kinds of fat-soluble xenobiotic molecules that come from liver and finally pours them into the intestine. In this way, gall bladder also acts as a temporary storage depot.

12. Spleen as Storage Depot

Spleen being the principal organ for the removal of dead RBCs, also collects large number of RBCs from blood; and along with them, it collects protein binding as well as –SH group binding toxic molecules and stores them for considerable time.

13. Fetus as Storage Depot

Various lipophilic and hydrophilic xenobiotic molecules cross the placental barrier and reach the fetus. The lipophilic xenobiotics get absorbed in the fat content and accumulate, while hydrophilic molecules get absorbed in the fluid compartments of the fetus and are released slowly.

Effect of Accumulation and Retention of Xenobiotics

Some xenobiotics (toxicants) do not readily cross plasma membrane and, therefore, have restricted distribution. Whereas, other toxicants rapidly pass through plasma membrane and are distributed throughout the body. Some toxicants accumulate in certain parts of the body as a result of protein binding, active transport, or high solubility in fat. As described earlier in the preceding section, the site of accumulation of a toxicant also may be its site of major toxic action but more often it is not. If a toxicant accumulates at a site other than the target organ or tissue, the accumulation may be viewed as a protective process in that plasma levels and, consequently, the concentration of a toxicant at the site of action, are diminished. In this case, it is assumed that the xenobiotic in the storage depot is toxicologically inactive. However, because any chemical in a storage depot is in equilibrium with the free fraction of toxicant in plasma, it is released into the circulation as the unbound fraction of toxicant is eliminated, for example, by biotransformation.

The following are the toxic effects due to accumulation and retention of xenobiotics :

1. Accumulation of high concentration of various xenobiotics viz., 1,000 ppm of organochlorinated hydrocarbons in liver may cause cytoplasmic vacuolation, swelling, necrosis and fatty degeneration of liver.
2. In lungs aldrin, dieldrin etc. causes congestion.
3. Lindane and BHC cause fatty degeneration of kidney, congestion of the bladder and also nephritis. These xenobiotics also cause hemorrhage of GIT.
4. Metallic xenobiotics, after accumulating in the glomerular basement membrane of kidney produce nephritis.
5. Chlorinated hydrocarbons, after accumulating in the axon sheath, disrupt impulse transmission.
6. Xenobiotics, after crossing placental barrier, accumulate in the fetus and produce deleterious effects. For example — thalidomide causes phocomelia i.e., absence of limbs in fetus.

BIOTRANSFORMATION

Biocatalytical conversion of xenobiotics (toxicants) into hydrophilic forms to facilitate their excretion from the body is termed biotransformation. In other words, biochemical modification of xenobiotics (toxicants) into non-toxic form may be termed as biotransformation. It is also called metabolic transformation. In principle, biotransformation may be defined as a process which converts the lipophilic compounds to more hydrophillic metabolites. In simple words, the conversion of toxic xenobiotics into less toxic metabolites and then forming conjugates may be termed biotransformation.

The principal site of biotransformation is the liver, the other being the lungs, stomach, intestine, skin and kidneys. Actually almost all parts of the body possess some activity against xenobiotic substances, but the major enzyme systems are found mainly in the liver. A detailed account of biotransformation of xenobiotics has been described ahead in a separate chapter (21).

Excretion

Actually elimination of the xenobiotics from the body system is a bi-phasic process. First the xenobiotics reach blood from various routes of entry, then they are distributed by blood to different organs of the body, from where the toxic molecules re-enter blood for redistribution to the organs of elimination or deposition. The rate of elimination of xenobiotics from blood stream, therefore, depends on the rate of their distribution and redistribution by the blood and the organs of elimination.

The principal organs of the elimination (excretion) of xenobiotics and their metabolites are :

1. Kidneys (urinary or renal excretion)
2. GIT
3. Liver (Biliary excretion)
4. Lungs
5. Sweat glands on skin
6. Mammary glands
7. Vagina (Vaginal secretion)
8. Salivary glands.

1. Kidneys

Kidneys are the organs that seem to be especially designed for the excretion of xenobiotics and their metabolites. Toxicants and their metabolites are steadily removed from the blood as they pass through the kidneys. Blood, amounting to one-fourth of the cardiac output, actually flows via the kidneys, which filter an estimated 180 liters of plasma every day. The filtration of toxic substances (of molecule weight < 70,000) from the blood is carried out at the glomeruli; the remaining filtrate passes into proximal tubules where re-absorption of lipophilic substances takes place. Toxicants may also be excreted into the urine by a process called active secretion. The urinary excretion of xenobiotics is influenced primarily by the pH of urine; the bases are more readily eliminated if the urine is acidic, and vice versa. Conversely, acidic urines may exhibit a high capacity of re-absorption for organic acids. The various conditions that affect the kidney would eventually influence the urinary excretion of metabolites.

To sum up, kidney excretes xenobiotics/their metabolites by the following three mechanisms :

1. Passive glomerular filtration
2. Active tubular secretion
3. Passive diffusion across tubules.

All ionized and polar forms of xenobiotics, example — organophosphates, alcohol etc., and also weak acids and weak bases are completely excreted by glomerular filtration. Organic bases and weak electrolytes are secreted into tubular lumen. Polar and lipophilic substances are not excreted at all from kidney.

2. GIT

Many insoluble molecules and molecules in colloidal form are eliminated through fecal content. The excretion may be due to the agent/compound not absorbed after oral ingestion. Some ionized xenobiotics may be excreted into the gut, if concentration gradient is favourable. Heavy metal ions in the form of metallo-protein are eliminated through feces.

3. Liver (Biliary Excretion)

Generally, the free and conjugated forms of toxicants or their metabolites of molecular weight < 300 are conveniently excreted through the kidneys. But the biliary excretion represents a relatively more important route for the elimination of compounds with higher molecular mass (300 - 700). This route is indeed a prominent route of elimination. This route provides way for the elimination of metabolites formed in the liver, directly into the bile without entering the systemic blood circulation (Fig. 14.2). Some xenobiotics may undergo entero-hepatic circulation, the mechanism by which a compound excreted in the bile is re-absorbed from the intestine and returned to the liver. Recirculation of non-polar toxicants in this manner might lead to the cumulative poisoning of the liver cells.

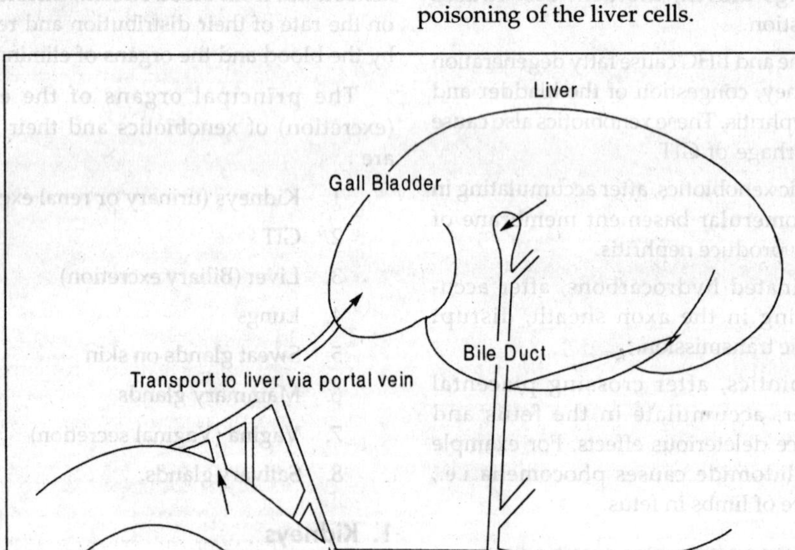

Fig. 14.2 : Biliary excretion route for xenobiotic compounds

The biliary route of excretion is important in elimination of anions, cations and non-ionized molecules. Liver also secretes organic cations. Non-ionized molecules are also secreted in bile and perhaps liver has also a transport system for the excretion of metallic xenobiotics.

Biliary excretion of xenobiotics varies with species, and is generally maximum in dogs and rats. Hepatic excretion system is not fully developed in infants and, therefore, some xenobiotics are more toxic to infants than adults.

4. Lungs

Many volatile compounds like alcohol and paraldehydes are eliminated through expiration from lungs. Lungs also excrete Flurobenzene and CO. Gases with low blood/gas solubility, such as benzene and nitrous oxide, are excreted at rapid rates.

5. Sweat Glands

Many water-soluble toxic molecules, entering sweat glands, are excreted with the sweat. Generally, sweat glands excrete very small amounts of xenobiotics. However, drugs used for leprosy, viz., diethyl dithioliosphthalate, are largely excreted through sweat.

6. Mammary Glands.

Certain xenobiotics are partly excreted into milk of lactating mothers. Several xenobiotics as DDT, thiouracil, tetracycline, morphine, heroin, sedatives like barbiturates, tranquilizers like diazepam and phenothiazine, oral anticoagulants and purgatives etc. have been reported to be excreted in milk.

7. Vagina (Vaginal Secretion)

Though the role of vagina is not much understood, hippuric and glycine conjugates of benzoic acid have been reported in the estrus secretions of cow.

8. Salivary Glands

Water-soluble toxic molecules enter salivary glands from the GIT and are eliminated through saliva. Example — iodides, fluorides, sulphonamides. Drugs like pentobarbitone are excreted in higher concentration in the saliva of ruminants.

Organs of Deposition

Many xenobiotics get deposited in the dead tissues of the body and are eliminated.

For example :

1. Metallic ions viz, arsenic get deposited in nails and hair follicles.
2. Toxic molecules that are highly lipophilic, for example, dieldrin, get deposited in the carcass of ruminates.
3. Some metals like lead and cadmium get deposited in bones.

Concept of Membranous Barrier

When a xenobiotic (toxicant or drug) is administrated to an animal, its action is obstructed by certain barriers. Chiefly these are :

1. Blood-Brain Barrier
2. Blood-Nerve Barrier
3. Placental Barrier
4. Blood-Renal Barrier
5. Others.

1. **Blood-Brain Barrier (BBB)** : It protects the CNS from certain neurotoxins. This site is less permeable than other areas of the body. Blood-brain barrier is located at the capillary wall. Since the capillary endothelial cells are tightly joined, leaving few or no pores between these cells (Bradbury, 1984), the toxicant has to pass through the capillary endothelium itself. Lack of vesicles in these cells further reduces their transportability. Finally, the protein concentration of the interstitial fluid in the brain is low, in contrast to that in other organs. Therefore, protein-binding does not serve as a mechanism for the transfer of toxicants from the blood to the brain. Due to these reasons, the penetration of toxicants into the brain is dependent on their lipid solubility. An outstanding example is methyl mercury, which enters the brain readily and its main toxicity is observed on central nervous system. In contrast, inorganic mercury compounds, being insoluble in lipid, do not enter the brain readily nor adversely affect it. However, the kidney is not spared.

   The BBB is absent where the cells produce hormones or act as hormonal or chemoreceptors. Glutamate and a number of related compounds, such as aspartate, have been shown to affect certain areas in the brain unprotected by the BBB, e.g., the arcuata nucleus of the hypothalamus.

The BBB is effective in excluding many neurotoxins, such as diphtheria, staphylococus, and tetanus toxins. This is also true with doxorubicin, which affects the dorsal root ganglia but not the CNS. Mercury chloride has a small molecule but is hydrophilic and mainly in ionic forms. Its concentration in the brain is minimal and so are its CNS effects. The BBB is not completely developed at birth and, because of this reason, some toxicants prove to be more deleterious in newborns than in adults.

2. **Blood-Nerve Barrier (BNB)** : Peripheral nerves are covered by the connective tissue sheaths, the perineurium and epineurium interlaced with the endoneurium. The BNB is provided with the blood vessels in the endoneurium and supplemented by the lamellated cells of the perineurium sheath. It is not as effective as the BBB; therefore, the dorsal root ganglia are generally more susceptible to neurotoxins than the neurons in the CNS (Jacob, 1980).

3. **The Placental Barrier** : A number of toxicants may pass through the placenta by diffusion. The placental barrier differs anatomically as well as histologically among various animal species. There are six layer of cells between fetal and maternal blood in some species, whereas in others there is only one layer. Furthermore, the number of layers may change as the gestation progresses. Although the relationship of the number of layers of the placenta to its permeability needs quantitative determination, the placental barrier does impede the transfer of toxicants to the fetus, which is, therefore, protected to some extent. However, the concentration of a toxicant such as methyl mercury may be higher in certain fetal organs, such as the brain, because of the less effective fetal blood-brain barrier. On the other hand, the foetal concentration of the food colouring amarnath is only 0.03-0.06% of that of the mother (Munro and Willer, 1978). Placenta has capacity of metabolizing the xenobiotics (toxicant/drugs) which may prevent the access of some toxicants to the fetus (Jachan, 1972).

4. **Blood-Renal Barrier** : Actually the kidney glomeruli act as membranous barrier for all protein buond xenobiotics. This barrier slows down the rate of transport of protein bound toxicants as well as lipophilic toxicants.

5. **Other Barriers** : Other barriers are also present in such organs as the eyes and testicles. In addition, the erythrocyte plays an interesting role in the distribution of certain toxicants. For example, its membrane acts as barrier against the penetration of inorganic mercury compounds but not to alkyl mercury. Furthermore, there is affinity of the erythrocyte cytoplasm for alkyl mercury compounds. Because of these factors, the concentration of inorganic mercury compounds in the erythrocyte is only about half than that in the plasma, whereas that of methyl mercury in the erythrocyte is about ten times to that in the plasma. Also, lipid layer situated in between two protein layers in plasma membrane functions as barrier to the action of bioactivated toxicant.

Overall process of translocation of xenobiotics (toxicants) viz., Penetration, Biotransformation, Storage, Elimination (Excretion) and Deposition is shown in Fig. 14.3.

Responses To Xenobiotics

When the xenobiotic is a *drug*, Phase I reactions — as describe in Chapter 21 — may produce its active form or terminate its action if it is pharmacologically active in the body without prior metabolism. Here it is important to mention that drugs act through biochemical mechanisms.

Certain xenobiotics are very *toxic* (e.g., cyanide) and some xenobiotics are not so much toxic — even sufficient amounts of these are administered. The toxic effects of xenobiotics cover the wide spectrum. Three general types of effects are mentioned here and shown in Fig. 14.4 :

1. The first one is *cell injury*, which can result in cell death. The macromolecular targets include DNA, RNA and protein.

2. The reactive species of a xenobiotic may bind to a protein. The xenobiotic is said to act as a hapten. The resulting antibody can damage the cell by several immunologic mechanisms.

3. Reactions of activated species of chemical carcinogens with DNA are of great importance in *chemical carcinogenesis.* Some chemicals require activation by mono-oxygenases in the ER to become carcino-

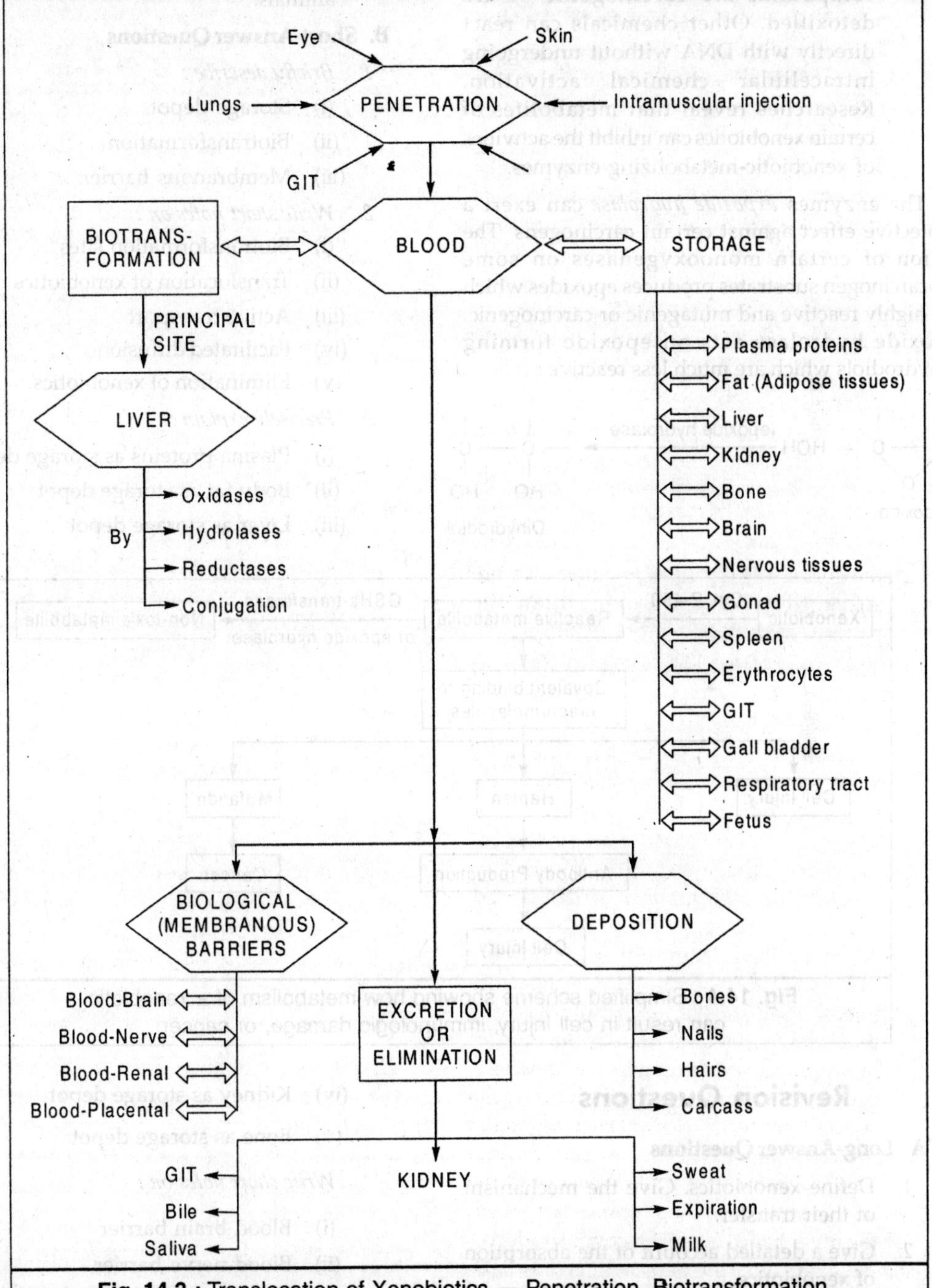

Fig. 14.3 : Translocation of Xenobiotics — Penetration, Biotransformation, Storage, Excretion and Deposition

genic. The activities of the monooxygenases known as cytochrome P-450 species and of other xenobiotic-metabolizing enzymes present in the ER help to determine whether such compounds are carcinogenic or are detoxified. Other chemicals can react directly with DNA without undergoing intracellular chemical activation. Researches reveal that metabolites of certain xenobiotics can inhibit the activities of xenobiotic-metabolizing enzymes.

The enzymes *exposide hydrolase* can exert a protective effect against certain carcinogens. The action of certain monooxygenases on some procarcinogen substrates produces epoxides which are highly reactive and mutagenic or carcinogenic. Epoxide hydrolase acts on epoxide forming dihydrodiols which are much less reactive :

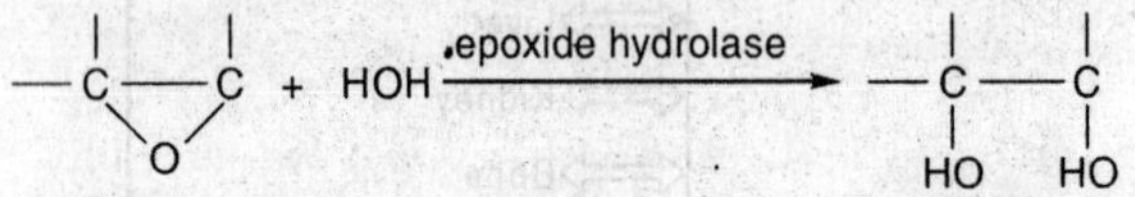

Epoxide Dihydrodiol

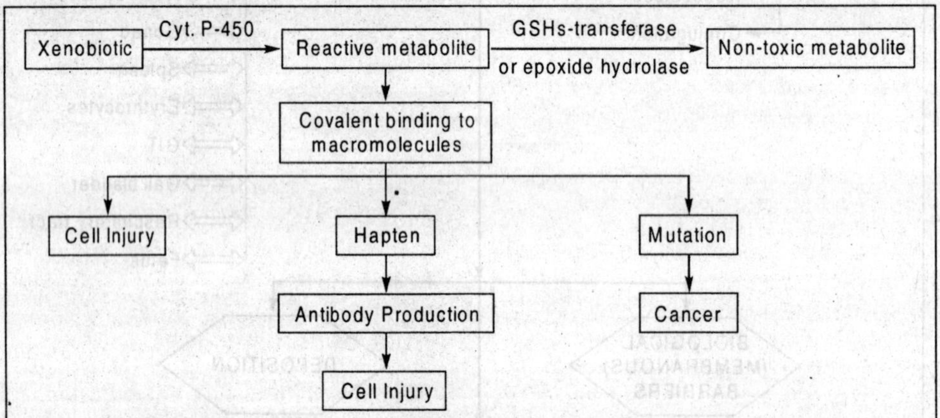

Fig. 14.4 : Simplified scheme showing how metabolism of a xenobiotic can result in cell injury, immunologic damage, or cancer

Revision Questions

A. Long-Answer Questions

1. Define xenobiotics. Give the mechanism of their transfer.
2. Give a detailed account of the absorption of xenobiotics.
3. Define storage depots and their types found in mammals.
4. Describe the salient method of absorption of the xenobiotics.
5. Give the various methods for elimination of the xenobiotics.
6. Describe various barrier systems found in animals.

B. Short-Answer Questions

1. *Briefly describe :*
   (i) Storage depots
   (ii) Biotransformation
   (iii) Membranous barrier.
2. *Write short notes on :*
   (i) Biotransformation sites
   (ii) Translocation of xenobiotics
   (iii) Active transport
   (iv) Facilitated diffusion
   (v) Elimination of xenobiotics.
3. *Precisely explain :*
   (i) Plasma proteins as storage depot
   (ii) Body fat as storage depot
   (iii) Liver as storage depot
   (iv) Kidney as storage depot
   (v) Bone as storage depot.

Write short notes on :

(i) Blood-brain barrier
(ii) Blood-nerve barrier
(iii) Blood-renal barrier
(iv) Placental barrier.

15

Toxicity Tests

SYNOPSIS

Introduction, Types of toxicity tests, Acute toxicity test, Bioassay of LD_{50} and LC_{50}, Sub-acute toxicity tests, Chronic toxicity tests, Toxic effects and their categories.

Introduction

Barnes (1963) defined the toxicity of a chemical/ xenobiotic/product as its capacity to cause injury while the hazard attributable to a xenobiotic/ chemical/product represents the probability that it will do so. Barners's definition is considered to be fundamentally correct and is accepted by the National Academy of Sciences, U.S.A. (Oser, 1971). However, hazard ranking is rather difficult process in which acute toxicity is but one factor (Jones, 1978).

Toxicity tests are experiments or trials designed to assess or evaluate the doses/concentration of xenobiotics (toxicants drugs) and the duration of exposure required to produce a criterion effect. The criteria of the effect may be mortality or death of the target species or any other parameter such as histological, physiological, biochemical, behavioural, immunological, hematological parameters etc. In other words, toxicity tests are those which are performed to evaluate the degree of response of a target species or a group of organisms produced by a specific level of stimulus. These tests provide a database that may be applied to assess the risk associated with a situation in which the selected toxicant, the species and the exposure conditions are well-profiled.

The overall effects produced by a xenobiotic (drug or toxicant) in experimental animals may be used to predict the possible and probable effects in humans. On the basis of dose per unit area of body surface, toxic effects in human beings are generally in the same range as those in the experimental animals.

The aquatic toxicity tests are frequently referred to as bioassays. It may be defined as a test which is designed to assess (evaluate) the relative potency of a xenobiotic (toxicant) by comparing its effect on a living organism with that of a standard preparation. Sprague (1973) defined bioassay as a test in which the quantity or strength of a xenobiotic (toxicant) is determined by the reaction of living organisms to it. Cairns (1980) defined bioassay as merely a dose-response evaluation.

Types of Toxicity Test

The toxicity tests may be conducted at various levels in accordance with the need.

(A) **Test types based on number and conditions of species** : On the basis of test organisms and conditions, the test may be of the following types :

(1) Single species test

(2) Multispecies test, and

(3) Ecosystem test

1. **Single species test** : As the name indicates this test is generally performed in the laboratory with single species. These tests are conducted with individual species that

are considered representative of broad classes of organisms (For example, invertebrates, fish, mammals etc.) The results of these tests provide information on the toxicity of specific type of xenobiotic in different organisms under defined conditions.

Such tests are easy to conduct and may be standardized and replicated. These tests under controlled laboratory conditions are helpful in establishing the cause-effect relationship. But, these tests cannot be used to know the adaptive ability of natural populations.

2. **Multispecies test** : As the name indicates these tests are conducted with several species of organisms and can be performed in laboratory. In this test, effects on several species of organisms can be identified which may provide information more directly related to ecological consequences of the chemical. Multispecies laboratory tests are useful in understanding the cause-effect relationship that prevails in natural systems.
3. **Ecosystem test** : These tests may also be conducted in laboratory. These studies involve model ecosystem consisting of some common autotrophs and heterotrophs of various grades. The laboratory ecosystem tests are more close to the natural conditions. Hence, the data obtained from ecosystem tests can be easily relied upon.

(B) **Test types based on exposure of toxicants** : On the basis of exposure of toxicants or xenobiotics, the toxicity tests may be categorized into two groups :

(a) Single dose test, and

(b) Multiple dose test.

(a) **Single dose test** : In this test, the toxicant is applied only once and the desired effect is recorded after certain duration.

(b) **Multiple dose test** : The information obtained from single dose test is used for the identification of multiple dose. The upper dose for multiple exposure should not be more than the dose expected to cause 5-10% mortality. The multiple dose test is usually performed in two species, one being non-rodent with exposure by the route of intended use. A minimum three dose levels and, more commonly, five dose levels are employed using 10-15 animals per sex per dose level. The multiple dose study not only characterizes the dose-response relationship of a test substance but also provides data for more reasonably predicting the maximum tolerance levels for the species during potential life-time exposures.

(C) **Types of Test based on length of exposure** : In order to examine the various effects associated with various length of exposure, the ideal and most widely accepted testing framework includes the following three types of tests :

1. Acute toxicity test
2. Subacute toxicity test
3. Long-term (chronic) toxicity test.

Actually, the acute and subacute toxicity tests are time-bound experiments, whereas time is not taken into consideration in chronic toxicity tests.

1. ACUTE TOXICITY TEST

Definition : The most relevant definition (meaning) of 'acute' given in the Oxford English Dictionary is 'coming sharply to a crisis' opposite to 'chronic'; however, it is linked to disease processes. To the toxicologist the term 'acute' can be used to define either the exposure or the response within short duration exposure.

The most common use of the term 'acute' in toxicology relates to exposure on one occasion only (i.e., single exposure). However, the term 'acute' is sometimes considered acceptable for multiple exposure during a short period of time (Hagan, 1959). The period set for multiple exposure to be defined as acute is not clearly defined but restricted in 24 hours is mostly acceptable; however, it may be up to 96 hours.

Occasionally, the term 'acute' is also used to define the response and not the exposure. From the above description a precise definition of 'acute toxicity' may be cited as severe effects experienced by organisms during short-term (24-96 hours) exposure to toxicants.

The results of acute toxicity tests are generally represented in terms of LD_{50} (median lethal dose) or LC_{50} (median lethal concentration) or EC_{50} (median effective concentration).

LC_{50} Value : The LC_{50} value represents the ppm or mg/1 or % per liter concentration of the toxicant in the media, that would kill 50% of population of a particular aquatic species, viz. fishes and insects.

LD_{50} Value : The LD_{50} value represents the amount of toxicant, expressed as mg/kg body weight of the test animals, that would kill 50% of the population of a particular terrestrial vertebrate species.

EC_{50} Value : In case of some invertebrates (for example, Daphnids, midge larvae, etc.) death is not easily determined and the criterion of effect is immobilization of 50% test species — termed EC_{50}.

Objectives of Acute Toxicity Test

The objectives of the acute toxicity test in the experimental toxicology of various toxicants (xenobiotics) are :

1. To predict hazard to non-target species.
2. To assess toxicity in target species when the pest is a vertebrate animal.
3. To provide information on the mechanism of toxic action.
4. To provide data on which user risk-benefit relationship may be assessed.
5. To aid the establishment of exposure levels in studies designed to assess long-term effect in experimental toxicology.
6. To observe the behavioural responses of the experimental animals.
7. To assess $LD_{50}/LC_{50}/ED_{50}$ and toxicity index of various toxicants.

Most of the acute toxicity studies are designed to determine the median lethal dose/concentration (LD_{50}/LC_{50}) of the toxicants. The LD_{50} or LC_{50} has been defined as a statistically derived expression of a single dose of a toxicant that can be expected to kill 50% of the test animals in short duration i.e., from 24-96 hours. LD is applied for non-aquatic animals and LC is applied for aquatic animals such as fishes etc.

When the route of exposure is inhalation, the endpoint is then either the median lethal concentration (LC_{50}) with a given duration of exposure or the median lethal time (LT_{50}) with a given concentration of the chemical in the air.

Experimental Design for Acute Toxicity Tests

1. Selection of animal species (Test animals)

Choice of test species for predictive toxicity testing is often affected by economic consideration, by the limitation of laboratory facilities or by the problems associated with the real target species.

Mammals : Two rodent species are most often used for the study of acute toxicity, the laboratory rat *(Rattus norvezicus)* and the laboratory mouse *(Mus musculus)*. These two species are mostly used in determining the LD_{50}. Their preference is due to the fact that they are economical, readily available, and easy to handle. Further, there are more toxicological data on these species of rodents, a fact that facilitates comparisons of toxicants with other chemicals.

In addition, a non-rodent species is desirable. This is true specially when the LD_{50} values in rats and mice are markedly different or when the pattern or rate of biotransformation in humans is known to be significantly different from rats and mice.

The LD_{50} determination should be done in animals of both sexes, and also on adult and young animals, because of their differences in susceptibility.

Birds : Many species of birds are vulnerable to intoxication by pesticides. However, there are some species of birds that may be considered as pests and their susceptibility to the avicidal properties of chemicals may be considered as an advantage. As an example, for carbophenothion, a pesticide, toxicity to some bird species has been tabulated in Table 15.1 :

Table 15.1 : The acute oral toxicity of carbophenothion to some bird species in 24 hours

Species	*LD_{50} value (mg/kg)*	*Source*
Starlings	5.6	Schafer (1972)
Redwing	7.5	Schafer (1972)
Pigeons	34.8 (31.1 - 38.9)	Jenning et al (1975)
Japanese quail	56.8 (50.8 - 63.6)	Jenning et al (1975)
Mallards	121.0 (95.9 - 152.0)	Tucker and Crabtree (1970)
Domestic fowl	316.0	Jenning et al (1975)

Fish : Fish may be exposed to (1) pesticides because of the deliberate addition of pesticides to the water e.g., herbicides for aquatic weed control, insecticides for malaria eradication and other means, (2) to metallic pollutants which get their way into the aquatic media through effluents etc.

The pisicidal properties of pesticides and metallic toxicants can be examined in the laboratory. Most of these studies are made with freshwater species (Cope, 1961; Muirhead-Thomson, 1971; Pandey and Shukla, 1979; Shukla and Pandey, 1985, 1987 and 1988), less often with estuarine fish (Butler, 1965) and even less frequently with sea fish (Holden, 1973).

Ideally, fish acute toxicity tests are carried out under natural conditions but most predictive acute toxicity tests have to be carried out in aquaria.

Following fishes have been frequently used for acute toxicity tests in different part of the world :

1. Brown trout, *Salmo trutta*; 2. Rainbow trout, *salmo gairdneri*; 3. Fathead minnow, *Pimephales*; 4. Bluegill, *Lepomis macrochirus;* 5. Goldfish, *Carassius auratus;* 6. Flag fish, *Jordanella floridane;* 7. Common guppy, *Poecilia reticulata;* 8. Tropical fish, *Tilapia mossambica;* 9. Mosquito fish, *Gambusia affinis,* and 10. Sunfish, *Lepomis gibbosus.*

In India, the following fish species are more commonly used for acute toxicity studies :

1. Puntius sps *(sophore, ticto);* 2. Mystus sps. *(cavasius, vittatus);* 3. *Heteropneustes fossilis;* 4. *Clarius batrachus;* 5. Chanda sps. (*nama* and *ranga*); 6. Channa sps. *(striatus; punctatus* and *gachua);* 7. *Colisa fasciatus.*

2. Route of Administration

Generally, the toxicant should be administered through the route by which humans would be exposed. The oral route is most commonly used. When a xenobiotic (toxicant) is to be administered orally, it should be given by gavage (force-feeding by tube). Mixing a chemical in the diet has the disadvantage of imprecise dosage and generally reduce the toxicity of chemical/toxicant.

A vehicle (solvents, carriers, surfactants etc.) is usually required in dissolving or suspending the toxicant to facilitate its administration. Even when the toxicant is liquid, it may need a diluent. The vehicle should have little or no toxic effect, and it should not react with the toxicant. Common vehicles are water, saline, vegetable gums and cellulose derivatives.

The volume of the toxicant in solution or suspension can affect the toxicity. Excessively large volumes of a liquid can cause untoward effects to the animals. On the other hand, if the volume is reduced the concentration will be higher, a fact that can increase the toxicity (Griffith, 1964). Therefore, when a large amount of a toxicant is to be administered to an animal, it may be advisable to give it in divided doses.

The dermal and inhalation routes are increasingly used, not only for chemicals that are intended for human use but also for chemicals with health hazards to personnel handling.

3. Dosage and Number of Animals

The term 'dose' is defined as the quantity of toxicant/xenobiotics administered to, received by the animals at one time or in a given period of time. 'Dosage' refers to the administration of a toxicant in doses.

The aim of the LD_{50} test is to determine the dose that would kill 50% of the animals and to determine the slope of the dose-response curve. It is, therefore, necessary to try to select a dose that would kill about half of the selected animals, another that would kill more than half (preferably less then 90%), and a third dose that would kill less than half (preferably more than 10%) of the animals. Four or more doses are often used with the hope that at least three of them would fall in the right range.

In general, the precession of the LD_{50} is improved by increasing the number of animals per dose and by decreasing the ratio between successive doses. Many investigators use about 50 animals/dose. However, others (Weil, 1952) proposed the use of four animals/dose. According to the results achieved by Weil, the LD_{50} values obtained with this simplified method are similar to those obtained with ten animals/dose. However, the fiducial ranges are much wider. It is, therefore, obvious that for routine LD_{50} tests, the procedures proposed by Weil might be adequate.

On the other hand, there are occasions when a more precise determination of the LD_{50} is desired. A more precise LD_{50} can be obtained with a smaller ratio between successive doses.

In determining the LD_{50} in large animals, e.g., dogs, generally a much smaller number of them is used.

4. Observations and Examinations

After administering the toxicant to the animals, they should be examined not only for the number and time of death but also for central, autonomic, and behavioural effects, including their onset, intensity, and duration. The frequency of these non-lethal effects should also be noted for each dosage group so that the LD_{50} dose for these effects might be estimated.

Gross autopsies should be performed on all animals that have died as well as at least some of the survivors, especially those that are morbid at the termination of the experiment.

Autopsy can provide useful information on the target organ, especially when death do not occur. Shortly after the dosing, histopathological examination of selected organs viz., liver, kidney gonads, intestine etc. may also be indicated.

Evaluation (Bioassay) of LD_{50} and LC_{50}

Sprague (1973) defined bioassay as 'a test in which the quantity or strength of material (xenobiotic or toxicant) is determined by the reaction of a living animal to it'. According to Cairns (1980) 'bioassay is merely a dose-response evaluation', i.e., living organisms are used to determine their response to a series of different toxicant/chemical concentration (i.e., dose). The exposure may be to a beneficial substance such as Vit. B 12 or a potentially toxic material such as cyanide. These are usually given in a graded series of concentrations in which the increased dose is accompanied by an increased response. The response may be crude or brude or brutal (death) or more subtle such as changes in growth rate, cell division, behaviour, respiration etc. Without such tests it is difficult to decide about the concentrations that are essential to life for the beneficial materials or those detrimental to life for the toxic substances.

Object of Bioassay

The object of a short-term bioassay of a pollutant or toxicant or contaminant or waste is usually to answer one or more of the following questions :

(i) Is it deleterious?

(ii) How much deleterious?

(iii) Does it vary in toxicity?

(iv) How effective are methods in reducing toxicity?

(v) Which fraction of the compound (waste) is most toxic?

The bioassay are particularly useful in evaluating such substances as industrial wastes, which are composed of a mixture of several chemical compounds rather than pure chemicals frequently used in the more academic tests.

Terminologies : Following terms are used to describe the type of experiment (bioassay) which is being conducted :

1. *Lethal* : Causing death.
2. *Sublethal* : Below the level which directly causes death.
3. *Cumulative* : Brought about, or increased in strength, by successive additions at different times or in different ways.
4. *Delayed effects* : Toxic symptoms not appearing until an appreciable time after the exposure.
5. *Static bioassay* : It is the bioassay in a tank of standing test water, which may or may not be changed during the period of test.
6. *Continuous flow bioassay* : In continuous flow or flow-through bioassay the test solution is renewed continuously or by frequent periodic additions.
7. *Lethal Concentration (LC)* : LC is used to express the result of bioassays having lethality as the criterian of toxicity. A numerical is used with it. For example, LC_0 — a concentration of toxicant at which mortality is zero percent of the test animals; LC_{25} — a concentration of toxicant at which mortality is 25% of the total population of the test animals; LC_{50} — a concentration of the toxicant at which mortality is 50% of the total population of the test animals; LC_{100} — a concentration of the toxicant at which mortality is 100 percent of the test animals. This value is expressed for fixed exposure period viz., 24, 48, 72 and 96 hours.
8. *Median Tolerance Limit (TLm)* : The symbol TLm or TL_{50} have been used to express the result of bioassays. In dealing with lethality, TL_{50} is the same as LC_{50}, but TL_{10} is equivalent to LC_{90} and TL_{90} to LC_{10}. Likewise, TL_{25} is equal to LC_{75} and TL_{75} is equal to LC_{25}.

9. *Median Effective Concentration (EC_{50})* : The symbol EC_{50} is used when some effect other than lethality is being examined. The EC_{50} is the concentration of the toxicant producing a specific effect or response such as loss of equilibrium or deformity, etc. in 50% of the test animals. A time period is also recorded, for example 96 hours EC_{50}.

10. *Median Lethal Dose (LD_{50})* : Described earlier in this chapter.

11. *Median Lethal Concentration (LC_{50})* : Described earlier in this chapter.

Considerations in selection of test animals

The most important considerations in the selection of test animals for bioassays are :

(i) Their sensitivity to the material or environmental factors under consideration.

(ii) Their geographical distribution, abundance and availability throughout the year.

(iii) Their economic, recreational and ecological importance.

(iv) Knowledge of their environmental requirements and culture methods for their rearing in the laboratory.

(v) Their general physical conditions and freedom from parasites and diseases, and

(vi) Their suitability for bioassay tests.

MEDIAN LETHAL DOSE (LD_{50})

LD_{50} has been defined as "statistically derived expression of a single dose of a material (toxicant/chemical) that can be expected to kill 50% of the test animals", or LD_{50} means the amount of material (toxicant/chemical) which is actually received inside the body, for example, by injection or by ingestion; and which is lethal to 50% of the total population of the test animals. In simple form, the amount of toxicant/drug which proves lethal to 50% of the total population of the test animals is known as LD_{50}, or the amount of toxicant/drug which brings 50% mortality in the total population of the test animals under specific period is known as Median Lethal Dose (LD_{50}).

LD_{50} is generally utilized in case of non-aquatic animals where the drug/toxicant can be administered conveniently, through various methods. Acute toxicity studies are in fact designed to determine the Median Lethal Dose (LD_{50}).

Experimental design (Selection of species of animals; Route of administration of toxicant; Dose and number of animals and experimental factors affecting the LD_{50}) has been described ahead, in section LC_{50}.

Uses of LD_{50} values

These values are useful in various ways :

1. Classification of chemicals (xenobiotics) according to their relative toxicity.

   A common classification in per kg body weight is :

Category	*LD_{50}*
Super toxic	5 mg/kg or less
Extremely toxic	5 - 50 mg/kg
Highly toxic	50 - 500 mg/kg
Moderately toxic	0.5 - 5.0 g/kg
Slightly toxic	5.0 - 15.0 g/kg
Practically non-toxic >	15.0 g/kg

2. Evaluation of the hazard from accidental overdose.

3. Planning sub-acute and chronic toxicity studies in animals.

4. Providing information about :

   (a) The mechanism of toxicity.

   (b) The influence of age, sex and environmental factors.

   (c) The variations in response among different animal species and strains.

5. Providing information about the reactivity of a particular animal population.

6. Contributing to the overall information required in planning therapeutic trials of drugs in humans.

7. Quality control of chemicals, to detect toxic impurities and physico-chemical changes affecting bioavailability.

· Acute LD_{50} values for a variety of chemical agents (xenobiotics) have been illustrated in Table 15.2 :

Table 15.2 : Acute LD_{50} values for a variety of chemical agents (Loomis, 1978)

Name of chemical agent	*Species*	*LD_{50} (mg/kg body wt.)*
Ethanol	Mouse	10,000
NaCl	Mouse	4,000
Ferrous sulphate	Rat	1,500
Morphine sulphate	Rat	900
DDT	Rat	100
Picrotoxin	Rat	5
Strychnine sulphate	Rat	2
Nicotine	Rat	1
α - Tubocurarine	Rat	0.5
Tetrodotoxin	Rat	0.1
Dioxin (TC DD)	Guinea pig	0.001
Botulinum toxin	Rat	0.00001

MEDIAN LETHAL CONCENTRATION (LC_{50})

The median lethal concentration is the usual method of reporting acute toxicity results. The recent trend is to use the symbol LC_{50}. For example, the 96 hour LC_{50} is the concentration of a toxicant/pollutant/drug that is lethal to 50% of the test animals in 96 hours. Lethal concentration is applied mostly in aquatic media. However, in the terrestrial animals, when the route of exposure is inhalation, the endpoint is either the median lethal concentration (LC_{50}) within a given duration of exposure or the median lethal time (LT_{50}) with a given concentration of the chemical in the air.

Actually, LC_{50} is merely a convenient reference point for expressing the acute lethal toxicity of a given toxicant to the test animals. Obviously, the safe concentration which permits successful reproduction, growth and all other normal life processes in the animal's natural habitat is usually much lower than the LC_{50}.

Toxicant Bioassay (LC_{50}) Using Fish

1. *Selection of Fish Species :* It has been stated earlier in this chapter.
2. *Size of Fish and other considerations :* Fish should be small, hardy and preferably air-breather. Fish should be free from diseases and mortality should be negligible in laboratory conditions Acclimatization of the fish to the laboratory conditions required at least 8-14 days. The fish should not be overcrowded. It is desirable to have 10-20 fishes for each test concentration. Fish should not be fed during usual acute bioassay, nor for at least 24 hours before testing. Fecal matter should be removed daily from the test tank.
3. *Test Tanks :* The tanks should be of non-toxic material. Glass is the best.
4. *Characteristic of Test Water :* Before and during bioassay some important physico-chemical characteristic of the test water viz., temperature, CO_2, pH, total hardness, D.O. and electrical conductivity should be measured.
5. *Range Finding or Exploratory Test :* The object of an exploratory test is to have a general idea of the range of concentrations suitable for the full-scale tests. For example, one might put 3 or 4 fishes into a ten liter jar of test water and check mortality over one or two days. The important thing is to use a wide range of concentrations in order to find out the concentrations of the toxicants/pollutants for the final test.
6. *Calculations of Results to Estimate LC_{50} :* It is desirable to have 10 or 20 fishes at each concentration and tests should be continued for 96 hours. A control test must run for each series. All the test tanks should have identical conditions, except for the toxicant. Concentrations of toxicants/pollutants must be chosen to span a range from that causing zero mortality to that causing complete or absolute mortality. Geometric or logarithmic series of concentrations are efficient to use. Mortality should be carefully recorded. A fish should be considered dead when there is no respiratory or other movement and no response to gentle padding. Dead fish should be immediately removed, and total length or wet weight or both should be recorded. Mortality in control set should be absent in acute tests, or it should not be more than 10% representing the occasional very weak fish.

- LC_{50} (also LD_{50}) values may be calculated by either of the two methods :

A. Graphical interpolation method, and

B. Statistical method.

A. Graphical Interpolation Method

Estimation of the LC_{50} by interpolation involves plotting the data on semilogarithmic coordinate paper with concentrations of chemical substance on the logarithmic and percentage of test animals (fish, etc.) died on the arithmetic scale. Thus, mortality observed in each tank is plotted against the concentration in the same tank. A straight line is drawn between the two points representing the percentage dead at the two successive concentrations that were lethal to more than half and to less than half of the test animals. The concentrations at which this line crosses the 50% lethality line is the essential LC_{50} value. Fig. 15.1 illustrates this procedure, which is referred to as straight-line graphical interpolation :

B. Statistical Method

Based on the data obtained from acute toxicity test, time-dependent LC_{50} or LD_{50} values and their 95% confidence limits can be calculated by any of a variety of statistical methods (Litchfield and Wilcoxon, 1949; Goulden, 1959; Finney, 1971). But, most widely used methods are probit analysis method (Finney, 1971) and Litchfield and Wilcoxon's (1949) method. In probit analysis method, the logarithmic values are read for each concentration/ dose exposed and probit values are read for the percent mortality. Then a graph is plotted between logarithmic doses/concentrations and the probit mortality. The curve so obtained is termed *probit-mortality curve* (Fig. 15.2).

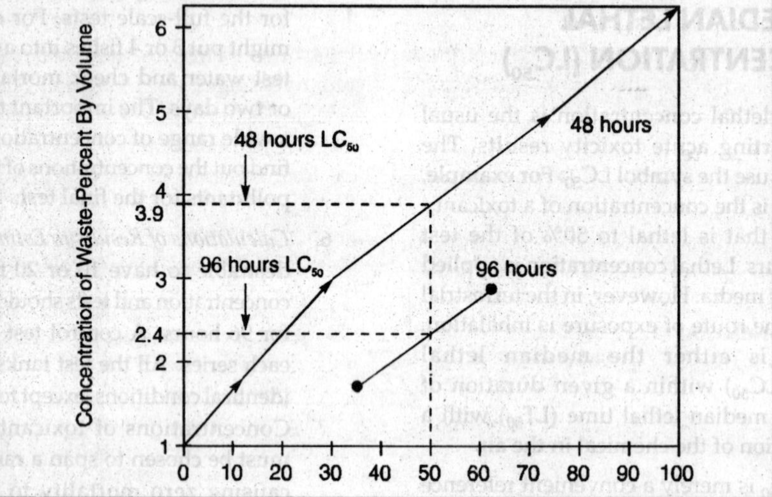

Fig. 15.1 : Estimation of LC_{50} by straight-line graphical interpolation

Hypothetical results used in developing the graph have been given in Table 15.3. The LC_{50} values for 48 and 96 hours are 3.9 and 2.4 mg/l, respectively.

Table 15.3 : Hypothetical data used for Fig. 15.1

No. of test animals (fish)	*Toxicants concentration in mg/l*	*Mortality in test animals (fish)* 48 hrs. No.	48 hrs. %	96 hrs. No.	96 hrs. %
20	1.0	0	0	2	10
20	1.8	4	20	7	35
20	3.2	7	40	13	65
20	5.6	14	70	19	95
20	10.0	20	100	20	100

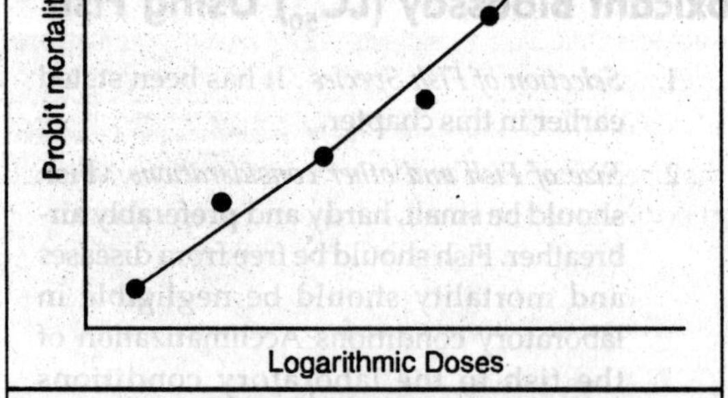

Fig. 15.2 : Probit-mortality curve

From this curve the values of expected probits are read and then corrected probits are calculated by the formula given by Finney (1971) to finally calculate LC_{50} or LD_{50} values.

In case of mortality in control groups, the data are corrected by Abbott's formula (Abbott, 1925) :

$$\text{Corrected mortality} = \frac{\%\text{ test mortality} - \%\text{ control mortality}}{(100 - \%\text{ control mortality})} \times 100$$

Significance of acute toxicity data

Although laboratory toxicity tests are conducted under unnatural conditions, the results are generally accepted as conservative estimates (slightly higher) of potential effects of test chemicals in the field. Therefore, much emphasis is given to perform more realistic tests in the laboratory, because the ultimate goal is to conduct tests in the field to validate the results of laboratory tests.

The data of acute toxicity tests enable to determine no observed effect dose/concentration which is maximum dose/concentration of test chemical causing no harmful effect on test animals. It also provides data on lowest observed effect. Besides providing LC_{50} or LD_{50} value, the acute toxicity data are also important and beneficial in the fixation of sublethal doses/sublethal concentrations for chronic toxicity tests. It shall not be out of place to mention that LC_{50} and LD_{50} values are estimated in similar manner.

Experimental Factors Affecting Acute Toxicity Tests

1. **Status of the test animal** : The response of any animal, including man, to a xenobiotic (toxicant) is associated with the biological and physical status.

The results of acute toxicity tests are generally expressed in terms of either body weight or body surface-area, and, of these two, body weight is more commonly used.

Within reasonable limits, the body surface-area may be considered to be a function of body weight and this function is expressed by the equation :

$A = k.W^x$

where, A = body surface-area (cm^2)

W = body weight (g)

k and x = variables for different species.

Values for k and x have been published for many species (Spector, 1956; Holliday et al, 1967). However, there are many other physiological functions (F) that might be related to body weight by the general formula :

$F = K_s W^y$

F = specified function

K_s = 1st constant for specified function

y = 2nd constant for specified function

W = body weight (g)

2. **Age** : Age variation may give rise to differences in susceptibility of acute intoxication by different toxicants and there is no simple rule for relating age to toxic response (Goldenthal, 1971). For a great majority of toxicants, the young are 1.5 - 10 times more susceptible than adults. K. Pandey and J. P. Shukla (1984) reported that fries and fingerlings of tropical freshwater fishes are most sensitive stages for toxicants. However, not all xenobiotics (toxicants) are more toxic to the young. Certain chemicals, notably CNS stimulants, are much less toxic to neonates. Lu et al (1965) reported that the LD_{50} of DDT was more than 20 times greater in newborn rats than in adults, in sharp contrast to the effect of malathion on age (Table 15.4) :

Table 15.4 : Effect of age on acute toxicity of DDT and Malathion in rats as reported by Lu et al (1965)

Pesticide	*Age*	*LD_{50} (mg/kg)*
DDT	Newborn	> 4,000.00
	Pre-weaning	437.8
	Weaning	355.2
	Adult	194.5
Malathion	Newborn	134.4
	Pre-weaning	925.5
	Adult	3,697.0

3. **Sex** : Male and female animals of the same strain and species usually react to toxicants similarly. There are, however, notable quantitative differences in their susceptibility, especially in rats. As a result of a survey of the acute oral and percutaneous toxicities of 98 environmental toxicants to rats, Games (1969) concluded that when tested by the oral route, the majority were more toxic to females than to males.

4. **Nutritional status** : Boyd (1972) reported that nutritional status of animals can affect the acute response to toxic chemicals.

The biotransformation of toxicants is mainly catalyzed by the microsomal mixed-function oxidases (MMFO) system. A deficiency of essential fatty acids, proteins, vitamins A, C and E generally depresses MMFO activities and makes the experimental animal for acute toxicity bioassay susceptible.

5. **Temperature** : Temperature of the environment can alter the toxic effects. For example, the toxicity of strychnine, nicotine, atropine, malathion and sarin is increased in animals exposed to cold temperature. Likewise, many xenobiotics show increase in toxicity in animals exposed to hot temperature.

6. **Humidity** : A higher relative humidity may increase the acute toxicity, resulting in a lower LD_{50} dose.

7. **Caging** : Caging can affect the LD_{50} of a chemical in several ways. For example, the LD_{50} of isoproterenol was less than 50 mg/kg in rats caged individually, whereas, it was often about 800 mg/kg in rats caged in groups (Balazs, 1972). However, the LD_{50} value of most of the chemicals is only slightly affected by this factor. The type of cage (mesh versus solid) and the type of material can also affect the reaction of the animals to toxicants.

Because of the potential of these factors on the LD_{50}, the precise conditions under which the test is conducted should be noted.

2. SUB-ACUTE TOXICITY TESTS

Sub-acute toxicity is also known as short-term toxicity. It is also known as sub-chronic toxicity. This study involves a stimulus which is less severe than the acute stimulus and produces a response in longer time, and may become chronic. This study involves repeated administrations, usually on a daily or 5 times per week basis, over a period of about 10% of the span, namely 3 months in rats, and 1 or 2 years in dogs. However, shorter durations such as 14 and 28-days' or 15 and 30 days' treatments have also been used by some investigators (Shukla and Pandey, 1982, 84).

Objective of the Sub-Acute Toxicity Tests

Human beings are more often exposed to chemicals at levels much lower than those that are acutely fatal, but they are exposed over a longer period of time. To assess the nature of the toxic effects under these more realistic situations, sub-acute and long-term (chronic) toxicity tests are conducted.

Experimental Design

1. **Animal species selection and Number** : Generally, for sub-acute toxicity experiments, two or more species of animals are used. Ideally, the animals chosen should biotransform the chemical in a manner essentially identical to humans. Since this is often unattainable, the rats and the dogs are usually selected. This preference is based on their appropriate size, availability and the greater preponderance of toxicological information on chemicals in these animals.

Equal numbers of both sex animals should be used. Generally 10-30 rats are used in each dose group as well as in the control group. However, smaller numbers of dogs (4 - 8 dose/group) are used because of the greater number of examinations that can be made and because of their size and expense involved.

2. **Route of Administration** : The toxicant/chemical should be administered by the route of the intended use or exposure in humans. For most of the chemicals/toxicants/drugs, the common route is oral. The preferred procedure is to incorporate the chemical/xenobiotic in the diet, although the drinking water is sometimes used as the vehicle. The latter is advisable when the chemical may react with a component in the diet. The chemical/toxicant may be administered by gavage or in gelatin capsules, a procedure more often used in dogs. The daily dose of the chemical may also be incorporated in a bolus of canned dog food.

Dermal application, exposure by inhalation, and parenteral route are used for special purposes, e.g., industrial and agricultural toxicants and drugs.

3. **Dosage and Duration** : Since the aim of these studies are to determine the nature and site of the toxic effects as well as the 'no effect level', it is advisable to select three doses :

(i) a dose that is high enough to elicit definite signs of toxicity but not high enough to kill many of the animals; (ii) a low dose that is expected to induce no toxic effect, and (iii) an intermediate dose. Sometimes one or more additional doses are included to ensure that the above objectives are achieved. As mentioned earlier, a control group should also be included. Control group animals will not receive the chemical/toxicant under test but must be given the vehicle used in experimental group.

The doses are generally selected on the basis of the information obtained in the acute toxicity studies, using both the LD_{50} and the slope of the dose-response curve. Any information on related chemicals/toxicants on its metabolism, especially the presence or absence of bioaccumulation, is also taken into account.

A range-finding study has also been proposed for seven days. The criteria for toxic effects include mortality, body weight gain, relative liver and kidney weights (hepatic and renal index), and food consumption.

In experiments with rat, the dose levels may be kept at constant concentrations and expressed in mg/kg or constant dosage and expressed in mg/kg body weight of the animals. As the animal grows, there are changes in the body weight as well as in the rate of food consumption.

The duration of such studies in rats is generally 90 days. In the dog, the duration is often extended to six months or even 1 or 2 years.

4. **Observations**

(i) *General Observations :* This includes appearance, behaviour, and any morphological abnormality. Dead and moribund animals should be removed from the cages for gross and possible microscopic examinations. Frequent observations is necessary to minimize cannibalism.

(ii) *Body Weight and Food Consumption :* It should be determined weekly. Decreasing body weight is a simple yet sensitive index of toxic effects. Food consumption is also useful indicator.

(iii) *Laboratory Tests :* Hematological examinations usually include hemoglobin percentage, RBC count, total WBC count, and differential leukocyte count. Special tests such as reticulocyte count, platelet count, methemoglobin, glucose-6-phosphate dehydrogenase and ATPase estimation may also be conducted.

Clinical laboratory tests usually include blood glucose, serum glutamic oxaloacetic acid transaminase (SGOT), serum glutamic pyruvic acid transaminase (SGPT), alkaline phosphatase, total protein, albumin, globulin, blood urea nitrogen (BUN), and such elements as Na, K, Ca and P.

Urine analysis usually includes colour, specific gravity, pH, protein, glucose, ketones, formed elements (RBC etc.), and any crystalline and amorphous substance.

Postmortem Examinations

Whenever possible, all animals that are found dead or dying should be subjected to a gross histopathological examination. Additionally, the weights of a number of organs — either in absolute values or in terms of the body weights — should also be determined as they serve as useful indicators of the toxicity. The organs that are usually weighed are liver, kidneys, adrenals, heart, brain, thyroid and gonads.

Those that are histologically examined are : all gross lesions, brain, spinal cord, eye and optic nerve, salivary gland, thymus, thyroid, heart, aorta, lung, pancreas, liver, small and large intestine, adrenal gland, spleen, kidney, gall bladder — if present, pituitary gland, urinary bladder, gonads, skeletal muscle, bone and its marrow.

5. **Evaluation** : A comprehensive sub-acute (short-term) toxicity study, using various parameters and examinations described above, usually may yield information on the toxicity of the chemical/toxicant under test, with respect to the target organs, the effect on these organs, and the dose-effect and dose-response relationships.

Such information often provides indication on the additional specific types of studies that should be conducted. The quantitative data from the short-term and other studies have been suggested for use in the determination of the "no effect level" (NEL). However, this suggestion has not been widely accepted by various workers including Gehring and Rao, 1979; Shukla and Pandey, 1985.

For certain types of chemicals/toxicants, the NEL from sub-acute (short-term) studies, along with data on their acute toxicity and metabolism as well as information from genetic, reproductive, and any other studies, may be used in determining their 'acceptable intakes' for humans.

3. CHRONIC TOXICITY TEST

Chronic toxicity is also known as long-term but definite toxicity, which involves repeated administrations over a considerable period of the lifespan of the test animals or at least a major fraction of it.

Objectives of chronic toxicity test : The basic objectives of chronic toxicity tests is :

(1) To study adverse effects of chemicals on the structure and function of organs, tissues and cells after prolonged period of exposure.

(2) To study the secondary effects of toxicants such as carcinogenicity, teratogenicity and mutagenicity.

(3) To evaluate no effect levels or safe levels of toxicants.

Experimental design for chronic toxicity test

1. *Selections of animals; species and number :* Usually one or more species of animals are used in chronic toxicity experiments. Unless otherwise indicated the rat is used; dogs and nonhuman primates are next in preference. Because of small size, mice are not suitable for general chronic (long-term) studies, although they are often used in carcinogenesis studies.

   An equal number of male and female animals are used. Generally, 40 - 100 rats are placed in each dose group and the same number in control group. Much smaller numbers of dogs and nonhuman primates can also be used.

2. *Route of administration, dose and duration :* Route of administration and criteria for the selection of doses is the same as in short-term (sub-acute) studies. But due to a longer span of time and expense involved in conducting such studies, greater care should be taken in selecting doses.

   The duration of such studies in rats is usually 24 months or even more. However, duration should not exceed 30 months. Dogs and monkeys are kept for 6 years or more.

3. *Observations :* Usually it includes body weight, food consumption, general observations, laboratory tests, and postmortem examination, as illustrated under sub-acute (short term) studies.

4. *Evaluation :* The purpose of the chronic toxicity study is to define the nature of the toxicity of a particular chemical (toxicant) and to determine the no effect level (NEL).

If the nature of the toxicity is not very serious, and has been observed nil, an acceptable intake (AI) can be extrapolated from the animal data. This is generally done by the application of a safety factor. This factor allows partly for any difference in sensitivity between the animal species and human, partly for the wide variations in sensitivity among the human population and partly for the fact that the number of animals tested is small compared to the size of the human population that may be exposed.

Generally three categories of tests are frequently used to predict the chronic effects of toxicants on aquatic organisms :

(1) The toxicants are exposed for the entire life-cycle of test animals to measure the effects of chronic exposure to a xenobiotic on survival, growth, reproduction and other parameters over one or more generation of a population of test animals.

(2) Tests on most sensitive life-stages are conducted to observe the effects of chronic exposure to a xenobiotic on survival and growth of toxicologically most sensitive life-stage of a species.

(3) Effects of toxicants on various physiological parameters of test animals are studied in functional tests.

The data from all these types of tests are used to estimate chronic toxicity thresholds :

(i) *Life-cycle test :* In these tests, groups of test animals are exposed to a series of concentration of xenobiotics. The test is initiated either with eggs, larvae or juveniles and continued until the test animals attain maturity to reproduce. It may continue for several generations as per requirements or objectives of the test. The concentration of toxicants may range from those having significant adverse effects on survival, growth and reproduction to at least one causing significant adverse effects on these parameters. The test species are limited to those who can complete their life-cycle under laboratory conditions.

(ii) *Test with sensitive life-stages :* The life-cycle toxicity test involves considerable time and expense, especially in case of vertebrates. Therefore, the test procedures have been developed to conduct experiments on most sensitive life-stages of test animals to predict chronic toxicity thresholds.

Generally, the early life-stages of fish are most sensitive (Shukla and Pandey, 1982) to test chemicals, hence these stages are selected for this type of test. Apart from saving time and expense, these tests can be conducted with a large number of species than life-cycle tests. Thus, the estimate of these tests can be applied for a much wider variety of species from wide range of habitats and trophic levels as compared to life-cycle toxicity tests. The estimates of such tests do not differ significantly from those estimated from life-cycle toxicity tests (Mc Kim, 1985). Thus these tests are reliable too.

(iii) *Functional test* : The animals respond to sublethal levels of toxicants physiologically and behaviourally. These organismic responses may be ascribed to :

(a) the functional responses occur as a result of action of toxicants, and/or

(b) the functional responses may occur as adaptive responses of organisms to the toxicants exposed.

These functional responses (sublethal effects) may be studied in the laboratory employing a variety of analytical procedures. The functional responses of animals observed following exposure to sublethal levels of toxicants can generally be divided into following four main classes :

(i) Behavioural responses
(ii) Physiological responses
(iii) Biochemical responses, and
(iv) Histopathological responses.

The behaviour actually represents an integrated response of test animals consequent upon complex biochemical and physiological changes. Toxicant-induced behavioural changes or responses indicate their effects on internal homeostasis. The behavioural endpoints may thus be used as sensitive indicators of sublethal effects of toxicants. The common behavioural responses of aquatic organisms receiving considerable attention are locomotion and swimming, attraction avoidance, predator-prey relationship, aggression, territoriality, opercular functioning, mucus secretion, and learning.

The commonly studied physiological and biochemical responses are respiration and oxygen consumption, studies on the activity of enzymes, clinical chemistry and hematology. The histological changes may significantly modify the function of tissues and organs. Thus, histological studies following chronic exposures to toxicants are also useful. It may, therefore, be concluded that the functional tests employing behavioural, physiological, biochemical and histopathological parameters are, on one hand, useful for evaluating the environmental hazards of xenobiotics (toxicants) and, on the other hand, they may provide valuable information on the mode of action of xenobiotics (toxicants/days).

MAXIMUM ACCEPTABLE TOXICANT CONCENTRATION (MATC)

The chronic toxicity tests are also useful in determining the maximum acceptable toxicant concentration (MATC) and presumable safe concentrations.

MATC, as defined by Mount and Stephan (1967), is threshold concentration that produces statistically significant toxic effects but for the death of the animals exposed. It is a hypothetical concentration and is bounded by a range at lower end by the highest concentration in the chronic test which produces no effect (no observed effect concentration, NOEC) and at higher end by the lowest concentration that produces statistically significant effect (lowest observed effect concentration, LOEC).

MATC may thus be represented as :

NOEC < MATC < LOEC

Keeping in view the importance of toxicity tests the following terminologies are of great importance :

1. **No Effect Level** : This is the maximum dose level of toxicant/chemical that has not induced any sign of toxicity in the most susceptible species of animals tested using the most sensitive indicator of toxicity. However, certain signs of toxicity, e.g. delayed neurotoxicity, cataract etc. are demonstrable in short-term (sub-acute) studies. The NEL is not necessarily an absolutely no-effect level; rather it should be treated as 'no observed effect level', because the use of a more sensitive indicator of toxicity or a more susceptible

animal species may reveal a lower NEL. In addition, an effect might well be demonstrable if a sufficiently large number of animals are used in the test. However, the general consensus is that there is an absolute NEL — no matter how large the dose groups are.

On the other hand, certain effects are generally considered as physiologic adaptations, or otherwise 'non-toxic'. These effects are, therefore, excluded in establishing the NEL. For example, liver enlargement may result from stimulation in the activity of hepatic mixed-function oxidases. A decrease of body weight may follow reduced food consumption, which, in turn, may be a result of unpalatability of the chemical/toxicant. A reduction in liver glycogen or alteration in WBC count may not be indicative of a toxic effect but an adaptation to a stress situation. However, before disregarding these effects in evaluating the toxicity of a toxicant/chemical, care must be taken to ensure that these are not manifestations of toxicity.

2. **Safety** : Safety can be expressed as the inverse function of the product of toxicity and bioavailability :

$$\text{Safety} \propto \frac{1}{\text{Toxicity} \times \text{Bioavailability}}$$

when bioavailability is the characteristic attributable to the toxicant that defines its ability to be assimilated by the organism.

Since it can be argued that hazard is an inverse function of safety, it may be simply stated that :

Toxic Hazard = (Toxicity × Bioavailability)

To exatrapolate from the NEL in animals to an acceptable intake in humans, a safety factor of 100, originally proposed by Lehman and Fitzhugh (1954), is generally used. This factor is intended to allow for differences in sensitivity of the animal species and humans, to allow for wide variations in susceptibility among the human population, and to allow for the fact that the number of animals tested is small as compared to the size of the human population that may be exposed (WHO, 1974).

While the factor 100 is often used, the WHO Expert Committees have used figures that ranged from 10 - 2,000. The size of the safety factor is determined according to the nature of the toxicity.

Biochemical data relating to the absorption, distribution, and excretion of the toxicant and its biotransformation (detoxication and bioactivation) in various species of animals and in humans are often useful in determining the size of the safety factors.

3. **Risks** : The term 'Risk' was first adopted by the International Commission on Radiological Protection (ICRP, 1970) in evaluation of the health hazards related to ionizing radiation. The use of this term stems from the realization that often a clear-cut 'safe' or 'unsafe' decision cannot be made. Thus the risk may be defined as the expected frequency of undesirable effects arising from exposure to a pollutant. It may be expressed in absolute terms as the risk due to exposure to a specific pollutant. It may also be expressed as a relative risk, which is the ratio of risk among the exposed population to that among the unexposed (WHO, 1978).

4. **Acceptable Daily Intake (ADI)** : The term acceptable daily intake (ADI) was coined by the Joint FAO/WHO Expert Committee of Food Additives in 1961. It has been adopted by the joint FAO/WHO meeting of experts on Pesticides Résidues (WHO, 1962).

ADI is defined as 'the daily intake of a chemical, which, during an entire lifetime, appears to be without appreciable risk on the basis of the known facts at the time'. It is expressed in mg of the chemical/kg of body weight.

5. **Virtual Safety** : The concept of virtual safety was first introduced by Mantel and Bryan (1961). The term was defined as a probability of carcinogenicity of less than 1/100 million (10^{-8}) at a statistical assurance level of 99%. The U.S. Food and Drug Administration, however, found that the doses associated with such a low risk level were too small to be enforceable in most acute situations and thus adopted a risk level of 10^{-6} (FDA, 1977). These levels of

risk are so low that the doses associated with them are referred to as Virtual Safe Doses (VSD). It is also worth mentioning that estimated VSDs for the same chemical often differ by a factor of several orders of magnitude.

TOXIC EFFECTS

In fact, toxic effects are greatly variable in nature, target organ and mechanism of action. A better understanding of their characteristics can improve assessment of the associated health hazards. It can also facilitate the development of rational preventive and therapeutic measures.

Actually, all toxic effects result from biochemical interactions between the toxicants (and/or their metabolites) and certain structures of the organism. The structure may be non-specific, involving a particular sub-cellular structure. A variety of structures, including the 'receptors', may also be affected. The nature of effects may vary from organ to organ.

Categories of Toxic Effects

The varieties of toxic effects can be grouped in different ways.

Some important toxic effects are :

1. Local and Systemic Effects

Certain chemicals (toxicants) can cause injuries at the site of first contact with an animal. Such effects are termed *local effects.* These local effects can be induced by caustic substances on the gastrointestinal tract, by corrosive substances on the skin, and by irritant gases and vapours on the respiratory tract. Respiratory frequency is a sensitive indicator of local irritation and is often concentration-related. Certain gases, e.g., Ozone and NO_2, increase the frequency, whereas others, e.g. SO_2 and HCHO, decrease it. Such local effects represent generalized destruction of living cells. The local effects which are produced only at the application spot are incapable of affecting the distant organs.

Examples of Local Effects

1. Ingestion of caustic soda and various other irritants.
2. Irritation and burning produced by the industrial toxicants on the mucus membranes of eye, nose, throat etc. of the workers.

Actually, local effects depend upon the route of administration of the toxicants, e.g., skin, gastro-intestinal tract and respiratory tract. NH_3 and Cl_2 are classic examples of irritant gases. They cause bronchial constriction and edema, which result in dyspnea, but chronic effects are rare. Arsenicals induce irritation on acute exposure.

Systemic effects result only after the toxicant has been absorbed and distributed to other parts of the toxicant in the system, and, thereafter, their distribution in the various parts of the body of animal are called systemic effects. Most toxicants exert their main effects on one or few organs. These organs are referred to as the 'target organs'. A target organ does not necessarily have the highest concentration of the toxicants in the animal. For example, the target organ of methyl mercury is the central nervous system, but its concentration is much higher in the liver and kidney. Likewise, the target organ of DDT is also the central nervous system, but it is concentrated in the adipose tissues. A toxicant may exert systemic effects after its absorption from the respiratory tract, and distribution to other tissues, or it may induce local effects on the respiratory tract, or both. A toxicant may also affect the respiratory tract after exposure from other routes. Many chemicals can be absorbed from the inspired air. After absorption, they are carried by the circulating blood to various parts of the body and exert their effects, such as general anesthesia.

Certain organic solvents, such as perchloro-ethyene and xylene, are rapidly absorbed after inhalation and distributed to various parts of the body including the liver, which is the major site of biotransformation. Part of the solvents enters the lung through circulation and may form reactive metabolites, leading to covalent binding with macromolecules there. This process, in turn, produces cellular damage and edema showing systemic effects.

Decreasing order of toxicity on various parts of the body may be shown as :

CNS → CIRCULATORY SYSTEM →

VISCERA → BONE → MUSCULAR SYSTEM

2. Reversible and Irreversible Effects

Reversible effects of toxicants are those that disappear following cessation of exposure to them. Irreversible effect, in contrast, will persist or even progress after exposure is discontinued. Certain effects are obviously irreversible. These include carcinomas, mutations, damages to the neurons, and hepatic cirrhosis.

Certain effects are considered irreversible even though they disappear some time after cessation of exposure. For example, the 'irreversible' cholinesterase-inhibiting insecticides inhibit the activity of this enzyme for a period of time that approximates to the required time for the synthesis and replacement of the enzyme.

The effect produced by a toxicant may be reversible if the animal is exposed at a low concentration and/or for a short duration, whereas irreversible effects may be produced at higher concentrations and/or for longer durations of exposure.

3. Immediate and Delayed Effects

Various toxicants produce immediate toxic effects, which develop shortly after a single exposure, a notable example being cyanide poisoning. Delayed effects occur after a lapse of some time. Carcinogenic effects are generally manifested 10 - 20 years after the initial exposure in humans, even in rodents a lapse of many months is required. To determine these and other delayed effects of toxicants, long-term (chronic) studies are essential.

4. Allergic and Idiosyncratic Reactions

In allergy the xenobiotics (toxicants) actually combine with an endogenous protein to form an antigen, which, in turn, elicits the formation of antibodies. A subsequent exposure to the chemical may result in an antigen-antibody interaction, which provokes the typical manifestations of allergy. Thus this reaction is different from the usual toxic effects, firstly because a previous exposure is required, and secondly because a typical sigmoid dose-response curve is usually not demonstrable with allergic reactions (Loomis, 1978).

Generally, an idiosyncratic reaction is a genetically determined abnormal reactivity to a xenobiotic (chemical). Some patients exhibit prolonged muscular reaction and apnea following a standard dose of succinylcholine. These patients have a deficiency of serum cholinesterase, which normally degrades the muscle relaxant rapidly.

5. Application Factor (AF)

AF concept was proposed by Mount and Stephan (1967). It is regarded as a concentration range, which can be derived by using LC_{96h} obtained from acute toxicity test and MATC derived from chronic toxicity test

$$AF = \frac{MATC}{LC_{50}\ 96h}$$

In other form, $MATC = AF \times LC_{50}$ (96h)

The AF is actually specific. Its value determined for one species may be applied to other species as well. The AF may be used to derive estimates of chronic toxicity without performing chronic toxicity tests. This saves the time and expenses of chronic toxicity tests.

Revision Questions

A. Long-Answer Questions

1. Define acute toxicity and describe evaluation (bioassay) of LD_{50} or LC_{50}. Add a note on the factors affecting acute toxicity.
2. Give a detailed account of sub-acute toxicity.
3. Give a detailed account of chronic toxicity.
4. Give a detailed account of different types of toxic effects.

B. Short-Answer Questions

1. *Explain :*
   (i) Toxicity test type based on number of species
   (ii) Toxicity test type based on exposure of toxicants
   (iii) LC_{50}, LD_{50} and ED_{50} values
   (iv) Objective of acute toxicity tests.
2. *Write short notes on :*
   (i) Lethal concentration
   (ii) TLm
   (iii) EC_{50}
   (iv) Uses of LD_{50} values
   (v) Significance of acute toxicity data.
3. *Describe briefly :*
   (i) No effect level
   (ii) Safety
   (iii) Risks
   (iv) MATC
   (v) Application factor.
4. *Explain in brief :*
   (i) Local and systemic effects
   (ii) Reversible and irreversible effects
   (iii) Immediate and delayed effects
   (iv) Allergic and Idiosyncratic reactions.

16

Outlines of Toxicological Testing Methods

SYNOPSIS

Behavioural responses, Behavioural tests, Toxicological importance of behavioural test, Analytic tests, Functional tests, Potentiation and synergism, Pathological and histological examinations, Other toxicity tests.

After exposing the animals to a toxicant, their number and time of mortality (death) should not be only considered, but also the central, autonomic and behaviour effects, including their onset, intensity and duration must be examined. The frequency of the non-lethal effects should also be noted for each dosage group so that the LD_{50} dose for these effects might be estimated. In this regard, the following aspects have been considered :

1. Behavioured responses (tests)
2. Analytic tests
3. Functional tests
4. Potentiation
5. Pathological and histological examinations.
6. Other toxicity tests
   A — Teratogenic tests
   B — Reproductive tests
   C — Carcinogenic tests
   D — Skin toxicity tests
   E — Eye toxicity tests.

1. BEHAVIOURAL RESPONSES (TESTS)

The pharmacological response/toxicological response to a drug/toxicant, which manifests itself through changes in the behaviour patterns of the test animal, is known as *behavioural response.* Mc Namara (1976) compiled a list of toxic signs and the systems and organs affected as a guide to facilitate toxicological observations in animals (Table 16.1) :

Table 16.1 : Relationship Between Behavioural Toxic Signs and Body Organ/Systems in Animals

S. No.	Systems	Toxic signs
1.	Behavioural	Sedation, restlessness, sitting position-headup, staring straight ahead, severe depression, excessive preening, gnawing paws, panting, irritability, aggressive and defensive hostility, fear, confusion, bizarre activity in non-aquatic mammals. Increased mucus secretion, irregular movement, restlessness and gulping of more air in aquatic animals, for example fishes (Shukla and Pandey 1985-1986).
2.	Autonomic	Relaxed nictitating membrane, exophthalmos, nasal discharge, salivation, diarrhea, urination, piloerection.
3.	Sensory	Sensitivity to pain; corneal labyrinth; placing and hind limb reflex; sensitivity to sound and touch; nystagmus; phonation.
4.	Neuromuscular	Decreased and increased activity; fasciculation; tremors; convulsion,

Contd...

S. No.	Systems	Toxic signs
		ataxia, prostration; straub tail, hindlimb weakness; pain and hindlimb reflexes (absent or diminished); opisthotonos; muscle tone; death.
5.	Cardiovascular	Increased and decreased heart rate, cyanosis, vasoconstriction, vasodilation, hemorrhage.
6.	Respiratory	Hyponea; dyspnea, gasping; apnea.
7.	Ocular	Mydriasis; lacrimation; ptosis; nystagmus; cycloplegia; pupillary light reflex.
8.	Gastrointestinal; Gastrourinary	Salivation; retching; diarrhea; blood in stool and urine; constipation, rhinorrhea, emesis, involuntary urination and defecation.
9.	Cutaneous	Alopecia, piloerection, wet dog shakes, erythema, edema, necrosis, swelling.

The observation period should be sufficiently long so that delayed effects, including death, would not be missed. The period is usually 4-14 days but may be even longer.

Gross autopsies should be performed on all animals that have died as well as at least some of the survivors, especially those that are morbid at the time of termination of the experiment. Autopsy can provide useful information on the target organ, especially when death does not occur shortly after the dosing. Histopathological examination of selected organs and tissue may also be indicated. Different drugs produce different types of behavioural and psychological alterations in the subject when administered e.g. :

1. Reserpine produces suicidal tendencies.
2. Amphetamine produces confusion and inability to concentrate.
3. Glucocorticoids in certain amount may cause euphoria (feeling of well-being), restlessness and psychological upsettings.
4. Heroin causes a fixed intention — whatsoever a person thinks about.

There is a large body of information on behavioural toxicology, resulting from a widespread feeling that behaviour is a subtle and sensitive indicator of toxicity. However, this view has been questioned, for example by Norton (1980), who stated : "Scientific data supporting this view are not only scanty but the available evidences often flatly contradict this assumption. It is hoped that the improved testing procedures would increase the sensitivity and utility of this approach in neurotoxicology and this branch will undoubtedly grow."

Behavioural Tests

In simple words, behaviour may be defined as the whole complex of observable, measurable or recordable activities of an animal. Most of the behavioural responses are adaptive overt movements in relation to physical, chemical and biological variables of the environment. These adaptive responses enable the animal to overcome environmental stresses and thereby help in minimizing the effects of environmental variables.

The behavioural tests have been used frequently in pharmacological evaluations of drugs for several years. These tests have only recently been recognized as an essential part of toxicity screening because of belief that behavioural responses of animals have been selected as a measure of toxicity of xenobiotics because of the following reasons :

(i) The behaviour of the animals is the end-point of functional integration of nervous system encompassing sensory, motor and cognitive aspects.

(ii) The functional capacity of central nervous systems cannot be determined by histological and even physiological studies independent of behavioural analyses.

(iii) The morphological and chemical alterations cannot be detected at time when behaviour is profoundly disturbed.

(iv) The behavioural tests may prove to be more sensitive than other tests in tracing central nervous system toxicosis.

(v) Neurotoxicants may be used as a tool to understand the behavioural phenomena.

Criteria of Behavioural Tests : The main target of behavioural toxicity screening is to obtain information that can be used to protect useful life and humanhealth.

The following criteria must be considered while selecting a behavioural response for toxicity testing :

(i) The selected behavioural pattern should be amenable to laboratory or field controlled investigations. In other words, much of the background on the behaviour should be already known.

(ii) The behavioural patterns that depend on diverse sensory and/or motor mechanisms should be examined.

(iii) The behavioural response chosen should be ecologically relevant.

(iv) The behavioural pattern should be closely examined in order to identify the effects of major biotic and abiotic variables.

(v) The behavioural test procedure should be easy to conduct in a routine way. It should be sensitive to measure even subtle changes in behaviour. The technique should be able to measure such behavioural changes in a wide range of animals in relation to a variety of toxicants.

U.S. National Academy of Science, National Research Council and Environmental Protection Agency in 1975, have jointly prepared a report for behavioural tests. As per their report, the behavioural protocol consists of two types of behavioural tests :

(i) Elementary behavioural tests, and
(ii) Sensitive behavioural tests.

(i) ***Elementary Behavioural tests*** : The elementary behavioural tests for acute toxicity studies are :

(a) **Activity responses** : Locomotion of a standard distance in standard time, gross movements, excitement or depression in spontaneous exploratory locomotion, abnormal locomotion such as circling etc.

(b) **Objective signs** : Tremors, lacrimation, defecation, convulsions, abnormal postural changes etc.

(c) **Reflex responses** : Hypo or hyper reflexia such as corneal and pinna righting, placing etc.

(d) **Elicited responses** : Grasping response to nose or tail pinch, handling or noise induced convulsion, orientation etc.

(e) **Body weight changes** : Immediate and brief weight losses and failure to gain weight as compared to normal growth curves.

(ii) ***Sensitive behavioural tests*** : The sensitive behavioural tests include :

(a) **Sensory function** : Vision, audition and olfaction.

(b) **Motor control** : Rotarod, tremor and force transducer.

(c) **Complex processes** : Operant behaviour.

(d) **Phylogenetically determined performances** : Sex or parental behaviour, aggressive behaviour, thermoregulatory behaviour, circadian rhythms and ingestive behaviour.

(e) Tolerance and dependence.

However, Tilson et al (1979) rectified aforementioned behavioural screening proposals and cited following measures in the primary test protocols :

(a) **Motor Tests** : Motor tests include spontaneous motor activity, forelimb and hindlimb grip strength and tremor.

(b) **Sensory tests** : These tests include the behavioural responses such as visual orientation/localization, auditory localization, tactile stimulation, orientation, negative geotaxis and pain.

(c) **Tests for arousal deficits** : These tests include emergence and startle response such as, air-puff, auditory cue etc.

(d) **Tests for cognitive functions** : These tests cover rapid escape/avoidance, conditioning test, retention and passive avoidance.

(e) **Tests for physiological and consumatory responses** : These tests include body-weight, rectal temperature, autonomic signs and respiration.

Some common behaviours of aquatic organisms which are used as a measure of toxicity of toxicants in aquatic systems are :

(a) Feeding behaviour,
(b) Locomotory behaviour including directed and undirected behaviour,
(c) Predatory behaviour — such as predator-prey interaction,
(d) Reproductive behaviour,
(e) Social interactions — such as territoriality, dominance and aggregation.
(f) Breathing mode.

Behavioural Testing Procedures

Direct visual observations of behaviour — both in the laboratory and field — are tedious, time-consuming, often difficult to evaluate and quantify and sometimes too subjective. Recent technological advances — such as acoustic and radiotelemetry, advanced photography, sonar and closed-circuit television, mini-computers, microprocessors and other electronic tools have greatly enhanced our ability to collect and analyse behavioural toxicity data.

(a) **Methods of exposure** : If the goal of behavioural test is to collect information concerning the protection of human health, the selection of route of adminstration is important. The route of entry of toxicants should be similar to that encountered in the environment. The toxicants show maximum effect when administered via the intravenous route. The most common routes of exposure in the environment for human beings are by inhalation, topical application and oral ingestion. Inhalation exposure is often not practical because of expensive and complex facilities required to control and monitor the exposure. Therefore, oral route is often used as an substitute.

(b) **Test animals** : Keeping in view the moral and ethical considerations, human being cannot be used as a suitable subject for behavioural toxicity screening. Therefore, animal models are generally used for behavioural toxicity studies and the data obtained from such studies are employed to extrapolate for human beings. The behaviour is generally species specific. The primates are generally the subjects of choice. Generally, rats and mice are used as animal models. Apart from rats and mice, other animals such as rabbits, cats, dogs, squirrels and monkeys are generally used.

In aquatic toxicology, aquatic species such as fathead, minnows, perches, daphnids, mysids, sheep-head minnows etc. have been used for acute early stage or chronic toxicity studies. But, there is little background data on their general behaviour and biology, so that their use in behavioural tests are limited. For these reasons, a wealth of information on behavioural effects of drugs/ toxicants is available on fishes like *Betta spendens, Colisa fasciatus, Clarius batrachus, Carassius auratus, Cyprinus carpio, Mystus species, Channa species* etc.

(c) **Test design** : Becuase of many unanswered questions in behavioural toxicology, it is difficult to describe a specific test design for a behavioural study. A simple generalized design for beha-vioural study is given here :

1. Test animals after 15 days acclimatization may be first exposed to toxicant and then monitored in experimental situation to determine the behavioural responses of exposure.
2. The behavioural responses should be studied following exposure to a range of 3-5 sublethal doses/concentrations of toxicants.
3. Concurrent control groups (untreated animals) should also be tested much in the same way as toxicant exposed groups for comparative considerations.
4. The duration of exposure is also important in behavioural testing because certain cyclic phenomena (e.g., daily, tidal and lunar) may make the animals more susceptible to environmental stimuli including toxicants.
5. The duration of exposure is dependent on test animals and nature of toxicants. For example, for animals having short life-cycles, animals selected from sea-shores and toxicants having short half-life (e.g., organophosphate insecticides), exposure should be of short duration.
6. Varieties of tests with different behavioural parameters should be included. The tests should be conducted in stepwise manner. The simple responses should be studied first and complex behaviours later on.
7. Samples of test animals from overall population should be selected randomlyfor study. The size of each group must be large enough to apply statistical procedures viz., student 't' test to determine the significance of data.

The behavioural responses are especially susceptible to endogenous and environmental variations. The responses of animals are highly variable — both with the same species and within same animal — at different times depending upon the biological status and environmental conditions. It is, therefore, important to adhere to proper experimental procedures such as large number of animals, controlled experimental environment and statistical analysis of results.

Toxicological Importance of Behavioural Tests

Behavioural tests have been extensively used in assessing the neurotoxicity of solvents, heavy metals (especially lead, cadmium and mercury); metalloids viz., arsenic; pesticides, and certain CNS drugs like alcohol and amphetamines.

The behavioural effects are especially susceptible to endogenous and environmental variations. For example, Norton (1980) reported data to indicate a large variability of results both between animals of the same species and within the variations in the same animal at different times. It is, therefore, important to adhere to proper experimental procedures, e.g., sufficiently large number of animals rigorously controlled experimental environment and statistical analysis of results.

To facilitate proper interpretation of behavioural effects of neurotoxic agents, Tilson et al (1979) recommended validation with agents that are known to produce specific effects. For example, they recommended the use of triethyltin, methyl mercury and inorganic lead in studies involving mixed central and peripheral neuropathies.

The significance of a neurotoxic effect depends on its reversibility. In general, irreversible effects are more serious than reversible ones. The site of the effect also plays an important role. There are areas in the nervous system more critical to physiologic function than others. Flarov (1974) pointed out that the toxicological nature of the surrounding can be assessed with the study of behavioural responses of the animals.

2. ANALYTIC TESTS

Analytic test includes the analysis of various toxicants. At present, several toxicants are in use and the number is increasing steadily every year. For some, analytic data are meagre while for others more than one analytic methods are available. Rapid advances in analytic chemistry, instrumentation, and techniques are being speedily evolved for identification of toxicants, their metabolites and degradation products from food and biological materials. In addition, modifications are often made in analytic method for their improvements.

There are various analytic methods frequently used in chemical analyses of environmental toxicants :

1. Neutron Activation Analysis (NAA)
2. Anodic Stripping Voltametry (ASV)
3. Atomic Absorption Spectrophotometry (AAS)
4. Optical and Emission Spectroscopy (OES)
5. Classical Polarography
6. Potentiometry with Ion-Selective electrodes
7. Gas Chromatography
8. Gas Chromatography/Mass Spectrometry (GC/MS)
9. High Performance (Pressure) Liquid Chromatography (HPLC)
10. X-ray fluorescence
11. Fourier Transform Infrared Spectrometry
12. Immuno Assays
13. Chemiluminescence.

3. FUNCTIONAL TESTS

In general, three basic types of test may be recommended for the precise assessment of toxic effects :

1. Morphological tests — adopted to evaluate morphological changes;

2. Biochemical tests — adopted to evaluate biochemical changes;

3. Functional tests — adopted to evaluate functional changes.

Morphological changes refer to gross and microscopic alterations in the morphology of the tissues. Some of these changes, e.g., necrosis and neoplasis are, however, irreversible and serious while functional changes usually represent reversible changes in the functions of target organs. Kidney and liver function tests (e.g., rate of excretion of dyes) are commonly utilized in toxicological studies.

Several investigators have attempted to show that functional changes might be detected earlier in animals exposed to lower dose than those with morphologic changes. The evidence collected in certain hepatic and renal tests, however, does not support this assumption (Grice, 1972 and Zbinden, 1976). Nevertheless, more recent findings indicate that functional tests are more sensitive and are also valuable in following the progress of effects on target organs in long-term studies in animals including humans.

Biochemical effects usually refer to those without apparent morphological changes. An example of such effects is the elevation of serum transminases, which is associated usually with toxic effects on the liver but occasionally with myocardial infarction and muscle injury. Certain biochemical effects lead to changes in functions like the toxic manifestations associated with cholinesterase inhibition following exposure to organophosphate and carbamate insecticides.

From the functional test viewpoint, toxicants do not affect all organs to the same extent. The reason is, however, not always apparent. For example, methoxyflurane is metabolized to release F^-. This takes place in the liver, but the target of toxicity are kidneys. Presumably, for some reasons, the kidney is more sensitive than the liver (Mazze and Cousins, 1973).

Important Functional Tests

(A) Respiratory Functional Tests

Tests on these functions are often carried out because of their greater sensitivity than morphological changes and also because of their ability to detect reversible effects. Consequently, these tests are more useful in human studies to provide more valid comparison between humans and the experimental animals with respect to their relative susceptibility to the toxic effects of the substances being tested.

Respiratory frequency is a sensitive indicator of local irritation and is often related to concentration. Certain gases, e.g., ozone and NO_2, increase the frequency whereas others, e.g., SO_2 and HCHO, decrease it.

Respiratory efficiency can be estimated by measuring O_2 and CO_2 in the blood or by measuring the rate at which inhaled carbon monoxide is taken into the blood (Cobb, 1981).

(B) Kidney Functional Tests

Nephrotoxins can exert adverse effects on various parts of the kidney (Fig. 16.1A and 16.1B) resulting in alterations of different functions. Luwe (1981) reported that urinary concentrating ability, and kidney weight were the most sensitive and consistent indicators of nephrotoxicity. Goldstein et al (1981) observed that urine osmolarity was another most sensitive indicator of the nephrotoxicity of a platinum complex, whereas GFR and ERPF were affected only later and at higher doses. Hence functional examinations of the kidney are routinely carried out as an integral part of short and long-term toxicity studies and may be done in a number of ways :

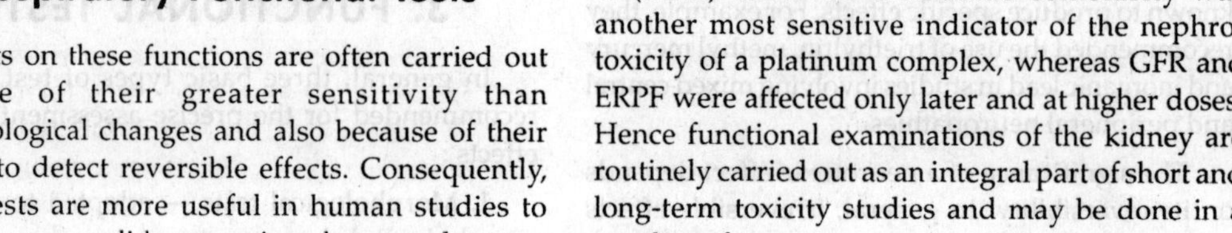

Fig. 16.1 A : Structure of a human kidney

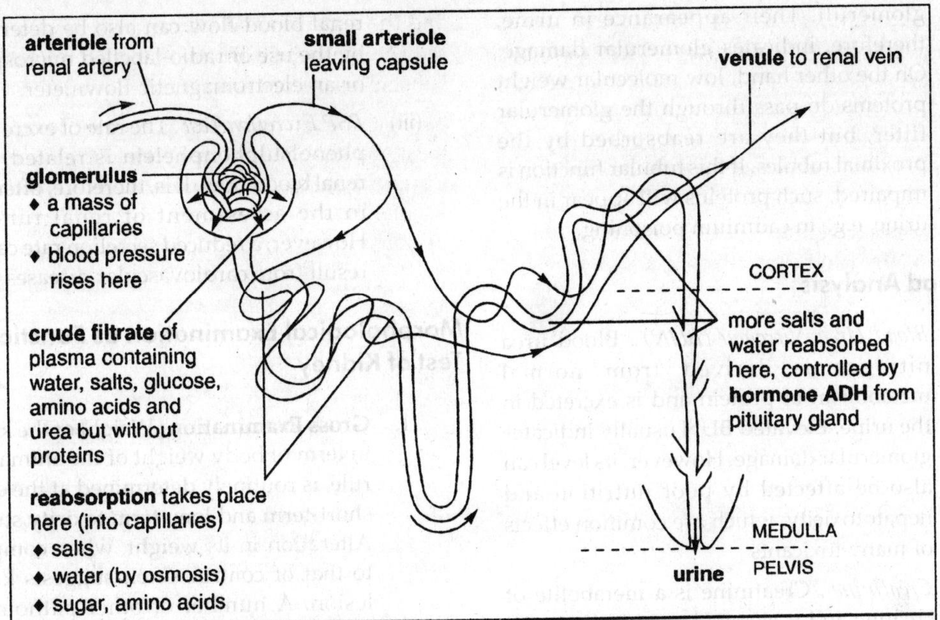

Fig. 16.1 B : One Bowman's capsule and tubule

(a) Urine Analysis

(i) *Proteinuria* : Because of the size of their molecules, a very small amount of proteins of low molecular weight pass through the glomerular filter. The low molecular weight protein are readily reabsorbed by the proximal tubules. The occurrence of large amounts of protein in the urine is thus an indication of a loss of integrity of glomeruli. It is to be noted that normal rat urine may contain some protein. A critical comparison of the treated animals with that of the control is, therefore, important.

(ii) *Glycosuria* : Glucose in the glomerular filtrate is completely reabsorbed in the tubules, provided the amount of glucose to be reabsorbed does not exceed the transport maximum (Tm.). Glycosuria in the absence of hyperglycemia thus indicates tubular dysfunction.

(iii) *Urine Volume and Osmolarity* : These two values are usually inversely related and are useful indicators of renal function in concentration test, wherein water is withheld from the animal, and also in dilution test, wherein a large amount of water is given to the animal. The osmolarity can be estimated from the specific gravity, but the freezing point of urine provides a more accurate measurement. A toxicant may cause high output renal failure or, on the other hand, oliguria or even anuria, resulting from tubular injury with concomitant interstitial edema and intraluminal pigment or debris.

(iv) *Acidifying Capacity* : This can be assessed from the pH of urine, titratable acids, and NH_4^+. The capacity is reduced when there is distal tubular dysfunction.

(v) *Enzymes* : Enzymes such as *maltase* in urine may indicate destruction of proximal tubules. *Lysozyme* level in urine is greatly increased following intoxication with chromium, but only moderately so after mercury poisoning. Urine *alkaline phosphatase*, on the other hand, may be renal or hepatic in origin. Plummer (1981) suggests that the levels of urinary enzymes not only are useful indicators of renal damage but also indicate the subcellular site of origin e.g., *alkaline phosphatase* is located in endoplasmic reticulum, *glutamate dehydrogenase* in mitochondria, and *lactate dehydrogenase* in cytoplasm.

(vi) *Proteins* : High molecular weight proteins are normally not filtered through the glomeruli. Their appearance in urine, therefore, indicates glomerular damage. On the other hand, low molecular weight proteins do pass through the glomerular filter, but they are reabsorbed by the proximal tubules. If this tubular function is impaired, such proteins will appear in the urine, e.g., in cadmium poisoning.

(b) Blood Analysis

(i) *Blood Urea Nitrogen (BUN)* : Blood urea nitrogen is derived from normal metabolism of protein and is excreted in the urine. Elevated BUN usually indicates glomerular damage. However, its level can also be affected by poor nutrition and hepatotoxicity, which are common effects of many toxicants.

(ii) *Creatinine* : Creatinine is a metabolite of creatine and is excreted completely in the urine via glomerular filtration. An elevation of its level in the blood is thus an indication of impaired function. Furthermore, data on its level in blood and its amount in urine can be used to estimate the glomerular filtration rate; however, one drawback of this method is that some creatinine is secreted by the tubules.

Special Renal Functional Tests

(i) *Glomerular Filtration Rate (GFR)* : The glomerular filtration rate can be more accurately determined by the clearance of inulin, a polysaccharide. It is diffused into the glomerular filtrate and is neither absorbed nor secreted by the tubules.

(ii) *Renal Clearance* : This is the volume of plasma that is completely cleared of a substance in a unit of time. The renal clearance of p-aminohippuric acid (PAH) exceeds that of inulin because it is not only filtered through the glomeruli but also secreted by the tubules. A reduction of PAH elimination without a concomitant decrease of GFR indicates tubular dysfunction. PAH is nearly completely (up to 90%) removed from the blood in one passage. The rate of its clearance is, therefore, useful in determining the effective renal plasma flow (ERPF). The renal blood flow can also be determined by the use of radio-labelled microspheres or an electromagnetic flowmeter.

(iii) *PSP Excretion Test* : The rate of excretion of phenolsulphonphelein is related to the renal blood flow. It is, therefore, often used in the assessment of renal function. However, a reduced secretion rate can also result from cardiovascular diseases.

Morphological Examination as Functional Test of Kidney

1. **Gross Examination** : Weight of the kidney in term of body weight of the animal, as a rule, is routinely determined at the end of short-term and long term toxicity studies. Alteration in its weight, when compared to that of control, often suggests kidney lesion. A number of other pathological lesions can also be detected on gross examination.

2. **Light Microscopy** : Histopathological examinations can reveal site, extent and morphological nature of renal lesions. Sharrat and Frazer (1963) found that the histopathological examinations were more sensitive than the functional test used in assessing acute and chronic glomerular or tubular injuries. However, some of the newer functional tests (in vitro studies) appear more sensitive.

3. **Electron Microscopy** : From the functional test viewpoint, this technique is useful in assessing ultrastructural change in the cells, such as mitochondria, other organelles, basement membranes and brush border. For example, prolonged exposure to methyl mercury increased the volume density of mitochondria and lysosomes.

In Vitro Studies

Kidney Slices : Tissue slices are an useful tool for the study of renal tubular function.

In assessing the effects of a toxicant on kidney, external factors that might affect the blood volume or blood pressure should be taken into consideration since they may indirectly impair renal functions. Furthermore, kidney diseases, such as those associated with ageing, may be prevalent and should also be considered (Cotchin and Roe, 1967).

(C) Assessment of Liver by Liver Functional Test

Following specific functional tests are carried to evaluate the liver function :

Serum Bilirubin and van den Bergh Reaction : This test is done for the differential diagnosis of prehepatic, hepatic, and post-hepatic jaundice.

Determination of total serum bilirubin is a good index of determining the severity of jaundice. Its level in the normal adults may vary from 0.4-0.8 mg/100 ml. In acute, infective and toxic hepatitis, the values may increase upto 10 mg/100 ml. In haemolytic jaundice, the values may range from 2-3 mg/100 ml.

van den Bergh Test : When freshly prepared diazotised sulphanilic acid reagent is added to serum, the coagulated bilirubin, if present, gives a reddish-violet colour within a minute. This bilirubin is called direct reacting bilirubin (Direct van den Bergh reaction). The unconjugated bilirubin does not give any colour within one minute. If alcohol is added to the mixture, indirect bilirubin becomes soluble and colour is developed. This fraction of bilirubin is called indirect bilirubin, and the test is known as indirect van den Bergh reaction. Formation of a faint pink colour after one minute and deepening of the colour in 2-3 minutes occurs when serum contains both conjugated as well as unconjugated bilirubin. This is called biphasic or delayed van den Bergh reaction.

A positive direct van den Bergh reaction with raised serum bilirubin indicates that the liver is functionally normal but the biliary tract is obstructed. Positive indirect van den Bergh test indicates hemolytic as well as hepatic jaundice. In hepatic jaundice both conjugated as well as unconjugated bilirubins are found to be raised, hence delayed van den Bergh reaction is often observed.

Functional Tests Based on Carbohydrate Metabolism

Galactose and fructose are converted into glycogen in the liver alone, and, hence, if the liver is damaged due to any hepatic toxicant, these sugars are found in the circulation in high amounts.

(i) *Galactose Tolerance Test :* Fasted individual is given 40 g galactose dissolved in 300 ml water. Blood is drawn at half an hour intervals for 2 hrs after administration of galactose. Galactose content of the four samples is determined after removing glucose by fermenting with yeast. Then galactose index is calculated which is the sum of the galactose values of the four samples. McLagan used a figure of 160 mg as the upper limit of normal galactose index. In infective and toxic hepatitis and in cirrhosis of liver, values up to 500 mg were obtained.

(ii) *Fructose Tolerance Test :* To the fasted individual, 50 g fructose dissolved in 300 ml water is administered. Fasted blood sample is taken after every half an hour up to 2 hrs. The total sugar content of blood (glucose and fructose) is determined. In normal subjects, the highest blood sugar should not exceed the fasting level by more than 30 mg per 100 ml. Higher blood sugar level is obtained in the patients with infective hepatitis.

Test Based on Detoxicating Function of the Liver

The most sensitive and best test is the detoxication of benzoic acid by liver to hippuric acid by conjugating it with glycine. It is known as the *hippuric acid test.*

Test : Three hours after light breakfast and emptying the urinary bladder, the patient is given to drink 6 g sodium benzoate dissolved in 200 ml. water. Subsequently, Urine is collected for 4 hours. The urine should have 4.5 g hippuric acid. Smaller quantities are excreted in acute and chronic liver diseases.

Test Based on Excretory Function of the Liver

Liver can remove a dye BSP (Bromsulphthalein) from its combination with plasma albumin and excrete it in the bile. When the dye is injected, it immediately combines with the plasma albumin and circulates in blood. Liver removes this dye and excretes it in the bile.

Test : 5 mg/kg. body weight of dye is injected. The blood retains less than 10% of the dye in 30 minutes and 7% in 45 minutes. At 60 minutes, no dye is found to be retained. This is considered to be the most sensitive and reliable liver function test.

If the liver function is impaired due to any toxicant/drug, the dye is excreted very slowly. About 50% dye is retained at the end of 45 minutes. The test loses its value in obstructive jaundice. The test is found to be very useful in the diagnosis of liver cell damage without showing clinical jaundice in chronic hepatitis and in cirrhosis of liver.

Serum Enzymes in Determination of Liver Functions

The level of SGPT (Serum Glutamic Pyruvic Transaminase), SGOT (Serum Glutamic-Oxaloacetic Transaminase), Isocitrate dehydrogenase and lactic dehydrogenase increases in circulation in the damage of liver cells due to toxicants.

Example : In normal case SGOT and SGPT activity remain 4-20 (I.U./liter) whereas in toxic liver necrosis this value may be 15-100 (I.U./liter) times of normal value.

Blood Ammonia as an Index of Liver Function

Actually, liver is the only organ which converts ammonia into urea. In advanced liver damage or in hepatic coma itself, due to damage of the parenchymal cells of the liver, it is not able to convert ammonia into urea. Thus, the level of ammonia rises from the normal range of 40-70 μg/100 ml to even up to 250 μ/100 ml (measured in terms of nitrogen).

4. POTENTIATION AND SYNERGISM

Potentiation in toxicology is used to describe the situation in which the toxicity of a chemical substance on an organ is markedly increased by another chemical substance that alone has no toxic effect on that organ. For example — isopropanol has no effect on the liver, but it can increase considerably the hepatotoxicity of carbon tetrachloride (CCl_4). Another known example of potentiation is administration of scopolamine and morphine. These drugs are sedative in action. The simultaneous administration of these two drugs produce more sedation than the sum of sedation produced by the two drugs, if given at different times.

Thus, the administration of two drugs simultaneously to exert a greater response than the sum of their independent responses may be termed '*potentiation*'. In simple words, when one toxicant potentiates the toxicity of the other toxicant, the phenomenon is referred to as potentiation and can be mathematically represented as :

Mathematical Representation of Potentiation

	Drug used	Degree of response
1st Case	X	A
2nd Case	Y	B
3rd Case	(X + Y)	C

According to the definition of potentiation

$C > A + B$

Hewitt et al (1982) studied that the potentiation is medicated through a variety of routes e.g., increasing the bioactivation of CCl_4, favouring the formation of phosgene over CCl_3 depicting hepatic glutathione and enhancing the susceptibility of subcellular organelles.

There are other chemicals, such as anti-metabolites, that exert their toxicity through interference with metabolic pathways. Phalloidin acts via attachment to membrane receptors.

A number of chemically related compounds, e.g. chloroform, tetrachloroethane, and carbon tetrabromide, as well as phosphorus, appear to act in similar ways.

Potentiation Test : The potentiation tests may be performed very easily. Certain fraction of lethal doses of two toxicants having similar mode of action may be exposed to test animals and a parameter may be selected on the basis of their mode of action as a measure of toxic effect.

Similarly, the sum fraction of both toxicants may be applied together and the same parameter may be again measured to quantify the toxic effect produced by those toxicants.

Examples of Potentiation

1. Application of DDT with C_6H_6 (benzene) potentiates its toxic action on insect pest

DDT + Benzene → More toxicity

2. Use of alcohol with barbiturate potentiates its toxic action

C_2H_5OH + Barbiturate → More toxicity

3. Use of morphine with heroin potentiates its toxic action

Morphine + Heroin → More toxicity

SYNERGISM

The chemical compounds that produce greater effects in combination than when they act individually are termed synergists and the phenomenon as synergism.

Example :

1. Many drugs produce greater effect in combination viz. (i) use of filixmas with castor oil raises action of the former against tapeworm. (ii) Adrenaline, when applied with Procaine, becomes more effective.

2. Most widely used synergist of pesticides is pyrethrin with piperonylbutoxide. The latter increases pyrethrin toxicity ten-fold.

5. PATHOLOGICAL AND HISTOLOGICAL EXAMINATIONS

After the administration of a toxicant or exposure to environmental toxicants, the affected organs show various pathological and histological alterations. Histological and pathological examinations in combined form constitute histopathological examination.

For example, first of all, the gross morphology of the liver of an animal viz., rat, is used to study the effect of any chemical/toxicant. The colour and appearance of target tissue by toxicant in comparison to control group often indicate the nature of toxicity, such as fatty liver or cirrhosis. The net weight is often a very sensitive indicator of toxic effect on the liver.

Histopathological examinations can reveal the site, extent and morphological nature of specific tissue lesions viz. renal lesions, hepatic lesions etc.

Sharrat and Frazer (1963) reported that histopathological examinations in certain cases, viz. kidney toxicology, prove to be more sensitive than the functional tests.

Histological examinations may be of two types :

1. Light Microscopy; and
2. Electron Microscopy.

Light microscopy of any target organ, viz. liver, can reveal a variety of histological abnormalities e.g.,

(i) Fatty changes

(ii) Necrosis

(iii) Cirrhosis

(iv) Hyperplastic nodules

(v) Neoplasia

(vi) Sinusoidal congestion

(vii) Cellular vacuolations etc.

Electron microscopy proves to be much sensitive than light microscopy. It is useful in assessing ultrastructural changes in the cells, such as mitochondria, lysosomes and other organelles. So much so, it is even useful in the detection of density of various cell organelles like mitochondria and lysosomes.

Observations of subcellular changes, along with the biochemical findings, are often useful in depicting the mechanism of action of toxicants.

Certain histopathological studies on specific tissues of animals made by various authors under different toxicants have been summarized in Table 16.2 (P. 250).

Significance of Pathotoxicological Data

Although pathological and histological examination in toxicology have been found useful in assessing the extent of damage in various tissues, yet a variety of pathotoxicological data may also be useful indicators of immune functions. For example, thymic atrophy appears to be a very sensitive indicator of immunotoxicity. Paucity of lymphoid follicles and germinal centers in the spleen is indicative of B-cell deficiency, whereas T-cell deficiency is characterized by lymphoid hypoplasia in the paracortical areas. Important pathotoxicological assessments are :

1. *Haematology Profile*—hemoglobin, erythrocyte count, leukocyte count, differential cell count.
2. *Clinical Chemistry* — creatine phosphokinase (CPK), creatinine acid and alkaline phosphatase, lactic dehydrogenase (LDH), cholinestersase,
3. *Serum Proteins* — albumin, globulin, albumin/globulin ratio.
4. *Weights* — body, spleen, thymus, liver, kidney, and
5. *Histology* — liver, thymus, lung, kidney, heart, spleen.

Table 16.2 : Histopathological Examinations of Different Tissues of Animals Exposed to Different Toxicants

Tissue	*Toxicant*	*Animal used*	*Histopathological observations*	*Reference*
Testis	Mixture of Ethyl parathion, methylparathion and DDT	Rat	(1) Seminiferous tubles degeneration with atrophy. (2) Seminiferous tubules filled with edematous fluid and debris matter. (3) Hemorrhage in the intertubular area. (4) Necrosis in the tubules. (5) Enlargement of interstitum.	Datta, K. K. and Dikshith, T.S.S. (1973) (Exp. Path. Bd.) 8, 5-363-370
Liver	Mixture of Ethyl Parathion, methylparathion and DDT	Rat	(1) Degenerative changes of the hepatocytes. (2) Cellular vacuolation and foamy appearance of the cytoplasm. (3) Nuclei of the hepatocytes showed karyorhexis karyolysis and pyknosis. (4) Inflammation of the bile duct. (5) Irregular necrosis of the hepatocytes.	Datta, K. K. and Dikshith, T.S.S. (1973) (Exp. Path Bd.) 8, 5-363-370.
Testis	Manganese	Rat	(1) Disorganized tubular pattern. (2) Most of the tubules collapsed and lined by a single layer of epithelium. (3) Lumen contained eosinophilic material with few entangled spermatids. (4) Tunica albuginea thickened and fair degree of fibrosis observed in the interstitium. (5) Seminiferous tubules showed degeneration. (6) Interstitial tissue showed edema and slight fibrocytic reaction revealed.	Seth, P. K. et al. Env. Physiol. Biochem. (1973) 3, 263-267.
Skin	Lindane	Albino rats	(1) Mild dermatitis. (2) Hyperkeratinization. (3) Epidermis with inflammatory cells. (4) Formation of abscess, filled in with Polymorphonuclear cells. (5) Mild acanthosis. (6) Cells of epidermis showed necrosis.	Dikshith, T. S. S. et al (1973) 29, 684-685 experientia
Liver	NH_3	Fish (Cutthroat) trout fry	(1) Gill epithelium detached from supportive cells and edematous. (2) Lamellae showed hypertrophy and scattered necrosis.	Thurston, R. V. et al (1978) 107, 12, 363-368 trans. Am. Fish. Soc.
Testes	Arsenic	Fish, *Colisa fasciatus*	(1) Edamatous lobular wall. (2) Distroted labules. (3) Breaking and dissolution of lobular wall. (4) Crenation of the blood cells. (5) Pyknosis and clumping in Leydig cells.	Shukla, J. P. and Pandey, K. (1984) 30 (3), 227-231 Cell. Mol. Biol.
Ovary	Arsenic	Fish, *Colisa fasciatus*	(1) Prominent follicular spaces and decrease in the development of 2nd stage of oocyte. (2) Decreased Cortical Yolk vesicle. (3) Seized development of 3rd stage of oocyte during mature phase.	Shukla, J. P. and Pandey, K. (1984) Bull. Inst. Zool. Academia Sinica 23(1) 69-74.

6. OTHER TOXICITY TESTS

A. TERATOGENIC TESTS

Teratogenesis is the formation of congenital defects (Greek *teras* = monster). This type of illness has been known for decades and is an important cause of morbidity and mortality among newborns. Out of 320 human pregnancies reviewed by Murphy (1928), 14 cases of children with small head circumference and mental retardation have been noticed. Their mothers were, however, exposed to therapeutic radiation in the first trimester.

A new era in teratology was initiated as a result of the clinical use of thalidomide, a sedative hypnotic.

Numerous drugs, food additives, pesticides, environmental contaminants and radiation etc. cause potential teratogenicity.

Teratogenic agents : The toxicants which cause teratogenesis are known as teratogenic agents. Teratogenic agents have been listed in Table 1.6 of Chapter 1.

Causes of Teratogenecity : A gestating-embryo exhibits great dynamicity of the living cells. The embryonic cells multiply and differentiate at a tremendous rate making the embryo more susceptible to the drugs.

Stage Sensitivity for Teratogenicity

Pre-differentiation stage : During this stage the embryo is not susceptible to teratogenic agents. These agents either cause death to the embryo by killing all or most of the cells, or have no apparent effect on the embryo. Even when some widely harmful effects have been produced, the surviving cells can compensate and form a normal embryo. This resistant stage varies from 5-9 days depending on the species.

Embryonic stage : In fact this is the period when the cells undergo intensive differentiation, mobilization and organization. It is during this period that most of the organogenesis takes place. As a result, the embryo becomes most susceptible to the effects of various teratogens. This period generally ends sometimes from the 10th-14th day in rodents and in the 14th week of the gestation period in humans. All organs are, however, not susceptible in the same period of the pregnancy. Rat embryo is most susceptible between days 8 and 12 for most organs, but the palate and urinogenital organs are more susceptible at a later stage for teratogens.

J. G. Wilson (1965) observed teratogenic treatment on the 10th day of gestation which resulted in the following incidences of malformations in rat :

Brain defects	35%
Eye defects	33%
Heart defects	24%
Skeletal defects	18%
Urinogential defects	6%

Fetal stage : This stage is characterized by growth and functional maturation. Teratogens are thus unlikely to cause morphological defects during this stage, but they may induce functional abnormalities. Whereas, morphologic defects are, in general, readily detected at birth or shortly thereafter functional abnormalities, viz., CNS impairment, may not be diagnoised for some time even after birth.

Mode of action of Teratogens : Various mechanisms are involved in teratogenic effects :

1. **Interference with Nucleic Acids** : Various teratogenic agents interfere with nucleic acid replication, transcription, or RNA translation. These include alkylating agents, antimetabolites, intercalating agents and amino acid antagonists.
2. **Inhibition of Enzymes** : Inhibitors of enzymes, e.g. 5-flourouracil, may induce malformation through interference with differentiation or growth by inhibiting *thymidylate synthatase*. Other examples include 6-aminonicotinamide, which inhibits *glucose-6-phosphate dehydrogenase*, and folate antagonists which inhibit *dihydrofolate reductase*.
3. **Deficiency of Energy supply and Osmolarity** : Certain teratogens can affect the energy supply for the metabolism by restricting the availability of substrates either directly (e.g., dietary deficiencies) or through the presence of analogs for antagonists of vitamins, essential amino acids, and others. In addition, hypoxia and agents i.e., CO and CO_2, can be teratogenic by depriving the metabolic process of the required O_2 and probably also by the production of osmolar imbalances. These can induce edema, which, in turn, cause mechanical distortion and tissue ischemia. Physical agents that can cause malfor-

mations include radiation, hypothermia, hyperthermia and mechanical trauma.

t shall not be out of place to mention that the mode of action of many teratogens is yet uncertain. Furthermore, a potential teratogen may or may not exert teratogenic effects depending on such factors as bioactivating mechanism, stability and detoxifying capability of the embryonic tissues. Appropriate experimental testing for the teratogenicity of toxicants is, therefore, essential.

Testing Procedures

Animals : For teratogenic tests, the animals should be young, mature and healthy. Usually, *Prima gravida* females are preferred. Rats, rabbits and hamsters are the commonly used animals, because of their ready availability, easy handling, little size and short gestational period. Pigs, are sometimes also used because they are phylogenetically more similar to humans. WHO (1967) suggested the use of nonhuman primates because of their phylogenetic proximity to humans. Other animals such as dogs and cats have also been used by some investigators.

With rats and rabbits, at least 20 and 12 females, respectively, are placed in each dose group of teratogenic agent (EPA 1982). Smaller numbers of large animals such as dogs and nonhuman primates are also used.

Administration of Teratogenic Agent

Dosage : At least three dosages are usually used. The lowest dosage should be approximately interspersed between the two extremes.

In addition, two control groups are included. One of these is given the vehicle or physiologic saline and the other receives a substance of known teratogenic activity. These groups provide information on the incidence of spontaneous malformation and the sensitivity of the specific lot of animals under the existing experimental conditions. In addition to these contemporary controls, data from historical controls are also useful.

Route and Timing : The test compounds should be administered through route that stimulates the human exposure situations. For food additives and contaminants, the chemical is preferably incorporated in the animal feeds. Oral drugs are generally administered by gastric gavage.

The timing of administering the substance is of great importance. For routine teratologic studies, it is customary to administer the substance during the entire period of organogenesis when the embryo is most susceptible. This period varies from one species to another.

Observations

The Pregnant Animals : The animals should be examined daily for gross signs of toxicity and many females that show signs of impending abortion or premature delivery (e.g., vaginal bleeding) should be examined.

The Fetuses : Fetuses are usually surgically removed from the mother about one day prior to the expected delivery. This procedure is intended to avoid cannibalism and permit counting of resorption sites and dead fetuses.

Following observations are to be made and recorded :

1. Number of corpora lutea
2. Number and position of implantations
3. Number and position of resorptions
4. Number and position of dead fetuses
5. Number and position of live fetuses
6. Sex of each live fetus
7. Weight of each live fetus
8. Length of each live fetus, and
9. Abnormalities of each fetus.

Detailed Examinations

To determine the different types of abnormalities, each fetus is examined for external defects. In addition, about 2/3rd of random sampled fetuses are closely examined for skeletal abnormalities after staining with Alizarin Red (Dawson, 1926). The remaining one-third of the fetuses are examined for visceral defects after fixations in Bowin's fluid and sectioned by microtome (Wilson, 1965). With larger animals, e.g., dogs, pigs, and non-human primates, the skeletal structure is generally examined with X-ray instead of staining.

Delayed Effects

With toxicants that are suspected of having effects on the central nervous system or genitourinary system, a sufficient number of pregnant females are allowed to deliver their pups. These pups are nursed either by their biological mothers — thus possibly being exposed to the toxicants via the milk — or by foster mothers. In the latter case, the potential effects of postnatal exposure are eliminated.

Neuromotor and behavioural tests may be applied to detect CNS effects. These include posture, mother activity, coordination, endurance, vision, hearing, learning ability, response to foreign environment, mating behaviour and maternal behaviour.

EVALUATION OF TERATOGENIC EFFECTS

Categories and Relative Significance

Aberrations : As mentioned earlier, in addition to functional abnormalities, morphologic defects may involve external/or internal structure. Not all types of aberrations have the same significance. For example, supernumerary ribs decrease, or abnormal sternal ossification might have little or no visible effect on external morphology, functional activity, or survival of the fetus. These have been considered as deviations.

Malformations of doubtful significance include curly tail, straight legs, malrotated limbs and paws, wristdrop, protruding tongue, enlarged atria and/or ventricles, abnormal renal pelvic development, and translucent skin. In general, these have been characterized as minor anomalies. There are, at the other extreme, major malformations that are incompatible with survival, growth, development, fertility, and longevity, e.g., *Spina bifda, hydrocephalus.*

In practice, the distinction between these categories is not always clear cut. It is then necessary to take other factors into consideration :

(i) **Resorption** : This is a manifestation of death of the conceptus. Although the site of resorption can be readily identified with a close examination of the uterus; the number of resorptions is more reliably determined by subtracting the total near term offsprings from the total implantations, as indicated by the number of corpora lutea. If there is an appreciable increase in the number of resorptions in the treated group, it may be necessary to alter the testing procedure to differentiated embryotoxicity from teratogenicity, e.g. by lowering the dose used to reduce the toxicity or shorten the exposure period.

(ii) **Fetal toxicity** : This may manifest as reduced body weight on non-viable foetus. This type of data is often useful in assessing the teratogenicity of the toxicant in question. With rabbits, the viability of fetus, if in question, may be determined by incubating it for 24 hours.

Sources of Error

1. The animals used may exhibit an excessive number of spontaneous malformation or may be resistant to teratogenic effects. These errors can usually be assessed by the response of the animals to the negative and positive control agents.
2. Poor animal husbandry and mishandling of the animals may also result in an increased incidence of malformations.
3. The food consumption can be affected by the toxicant used. This fact may then alter the body weight of the mothers and indirectly affect the fetuses.
4. Excessively large doses can result in many resorptions but few or no malformations. On the other hand, if the doses are too small, there may not be any evidence of teratogenicity.

Analysis of the Results

In comparing the treated and control groups, the proper experimental unit is the litter rather than the individual fetuses. The number of litters with malformed fetuses, resorptions, or dead fetuses are the parameters to be used in statistical analysis. However, an increase in the average number of fetuses with defects per litter may provide evidence of teratogenicity.

If the results indicate a relationship between the doses and the response (incidence of malformation), it is generally justifiable to conclude that the agent is teratogenic under the specific experimental conditions.

When the incidence of malformation does not provide a definite conclusion, an analysis of the data from the historical controls may be valuable. Furthermore, a close examination of the data on other parameters of the fetus and on the mother is sometimes useful.

Extrapolation to Humans

The results obtained in teratogenesis studies in animals cannot be readily extrapolated to humans. The lack of a suitable animal model is evidenced by the fact that the most potent human teratogen, thalidomide, which is effective at a dose of 0.5 - 1.0 mg/kg, has no teratogenic effect in rats and mice at 4,000 mg/kg. Only moderate embryopathy is noted in rabbits. On the other hand, acetylsalicylic acid has a long history of safe use in human pregnancy but is a potent teratogen in rat, mice and hamsters.

The mechanism of teratogenesis and the differences in response among various species of animals are poorly understood. The cause of spontaneous congenital malformations in humans are unknown. More basic animal studies and prospective epidemiologic studies are essentially required.

Nevertheless, since all chemicals that are teratogenic in humans were found to be active in certain animals as well, it is, therefore, prudent to carry out appropriate animal tests on all chemicals to which females of child-bearing age are usually exposed. If positive results are achieved with a substance — especially this is so in more than one species of animal — exposure of females of childbearing age to this substance should be avoided, if possible. In assessing the teratogenic effects of a chemical, not only the incidence but also the severity of the aberrations should be taken into account (Khera, 1981).

In Vitro Tests

Actually, these tests are not in routine use as yet, though they may show the mode of actions of teratogens. Some of these tests are — cell culture, organ culture, etc.

B. REPRODUCTIVE TESTS

The process of gamete formation is termed gametogenesis — development in males is spermatogenesis, while in females it is called oogenesis. The entire process of gametogenesis and fertilization has been shown in Fig. 16.2.

Chemobiokinetical studies reveal that toxicants may interfere with the various events of the reproductive cycle of test animal. Toxicants may act directly on the reproductive system or indirectly via certain endocrine glands. Before toxicants act directly they must reach the target organs in sufficiently high concentration. This concentration may be higher or lower than that in the blood. For example, with DDT, the concentration is almost 80 times higher in the ovary than in the plasma. Fabro (1978) observed that a number of other substances may penetrate the oocyte, oviduct, uterine fluid, and the blastocyst.

Toxicity of xenobiotics on the testis is influenced by the blood-testis barrier (BTB) and by the activities

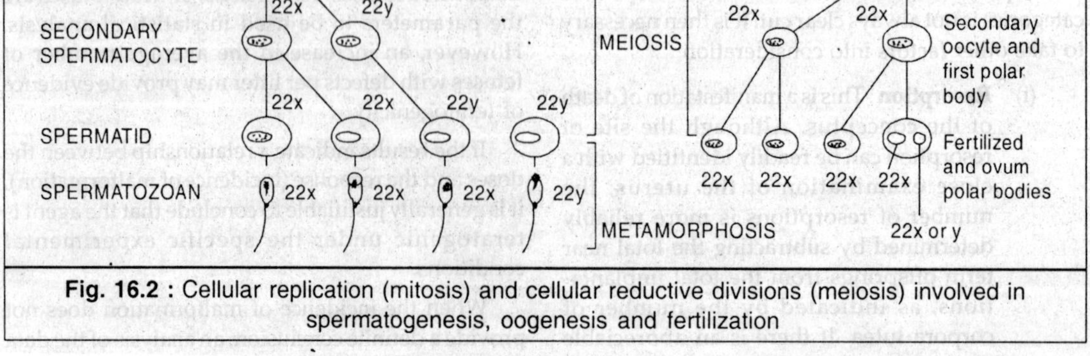

Fig. 16.2 : Cellular replication (mitosis) and cellular reductive divisions (meiosis) involved in spermatogenesis, oogenesis and fertilization

of certain enzymes (Lee and Dixon, 1978). BTB is a complex multicellular system composed of both myoid cells surrounding the seminiferous tubules and several barriers. The penetration rate of chemicals (toxicants) into the testis is proportional to their molecular weights and their partition coefficients.

Testis contains both activating and detoxicating enzyme systems. These two enzyme systems are capable of, respectively, increasing and decreasing the toxicities of toxicants. Furthermore, there is an efficient DNA repair system in the premeiotic spermatogenic cells, but there is none in spermatids nor in spermatozoa. Mutations can, therefore, be induced by electrophilic substances :

Reproductive Toxic Effects

Various toxicants have been reported to adversely affect the spermatogenesis and cause testicular atrophy as shown in Table 16.3 :

Table 16.3 : Toxicants Reported to Affect Male Reproductive System

Group of toxicant	*Name of toxicant*
Metals and *trace metals*	Arsenic, aluminium, boron, cadmium, cobalt, lead, mercury, methyl-mercury, molybdenum, nickel, silver.
Pesticides: Insecticides, Herbicides, Fungicides & Fumigants	Hexachlorocyclohexane, lindane, aldrin, chlorodane, dieldrin, DDT, methoxychlor, DDVP, kepone, carbaryl, 2, 4-D, 2, 4, 5-T diquat, paraquat.
Pesticides: Rodenticides	Apholate, capran, cineb, maneb, carbon disulphide, Dibromochloropropane (DBCP), ethylene dibromide, Fluroacetate (fluoroacetamide).
Food additives and *Food contaminants*	Aflatoxins, cyclamate, diethlystilbesterol, dimethyl nitrosamine, gossypol, metanil yellow, monosodium, glutamate.
Industrial chemicals	Polychlorinated biphenyls, vinyl chloride, benzene, carbon disulphide, hexane, thiophene, toluene, xylene, plasticizers-phthalate esters.

Hypotensive drugs such as guanethidine (like thalidomide) can induce non-pregnancy by interfering with seminal emission rather than by any direct effect on the fetus.

A variety of chemicals may be toxic to female reproductive functions (Dixon, 1980). Certain toxicants affect the oocyte, others (e.g., Haloperidol) prevents implantation or affect the development and growth of the conceptus, e.g., nicotine and DDT lower fetal weight in rabbits (Fabro, 1978). Some toxicants, e.g. spironolactone, have various effects on reproductive functions, such as interference with ovulation and implantation and retardation of development of the sex organs of the offspring (Nagi and Vigro, 1982).

Shukla J. P. and Pandey K. (1983, 1984, 1985) have made extensive studies of certain metallic toxicants like arsenic on the reproductive cycling (testicular and ovarian cycle) of freshwater teleost and observed impaired reproductive development in testis and ovary during different phases.

Reproductive Routine Testing

Multigeneration Reproduction Studies

Procedure

In general, rats or mice are the animals of choice for reproductive tests. A minimum of 20 females and 10 males are placed in each of three dose groups along with a control group. The doses are so selected that the high dose will produce some minimal toxic signs but will not result in a mortality greater than 10%, i.e., LD_{10} for a specified period. The low dose will not induce any observable effect.

Prior to breeding, the males are dosed throughout the period of spermatogenesis (60-80 days) and the females are dosed throughout the development of the ova (16 days). Dosing of the females is continued through the gestation and lactation periods. The test chemical (toxicant) should be administered by a route that most closely resembles the human exposure condition.

To determine the potential effect of the chemical on the reproductive function of the offsprings (F_1), a second generation offsprings (F_2) is bred and reared to reproductive immunity.

Observations and Examinations

The adult animals are observed for the body weight, food consumption, general appearance, oestrus cycle, and mating behaviour. Also, the fertility and nesting and nursing behaviours are noted. Half of the maternal rats are killed on day 13 of gestation for examination of corpora lutea, implantations, and resorptions (Jones, 1981).

The other half of the maternal rats are allowed to deliver their pups. The pups are examined for litter size, number of the stillborn, sex distribution, and congenital anomalies. In addition, the viability and pup weight are recorded at birth, on day 4, and at weaning, and preferable also on day 12 or 14. Auditory and visual functions and behaviour are also examined for subtle congenital defects.

An appropriate number of males and females are randomly selected from all generations and examined histopathologically, especially with respect to the reproductive organs.

Evaluation

From the data described (EPA, 1978), the following reproductive indices can be derived :

1. Fertility index : The percentage of matings resulting in pregnancy.
2. Gestation index : The percentage of pregnancies resulting in the birth of live litters.
3. Viability index : The percentage of live pups that survive 4 days or longer.
4. Lactations index : The percentage of pups alive on day 4, that survive 21 days, or longer.

Other indications include

5. Mating index (the percentage of oestrus cycles that had matings).
6. Male fertility index (the percentage of males exposed to fertile nonpregnant females that resulted in pregnancies).
7. Female fertility index (the percentage of females exposed to fertile males that resulted in pregnancies), and
8. 12 or 14 days' survival indices.

Multigeneration reproductive study can reveal a variety of deleterious effects on the reproductive functions. However, non-specific response (e.g., non-pregnancy) is common. It may follow unusual routes of administering the chemical, such as inhalation, topical application to the eye or nose, and parental administration. Excessive handling may also disturb the normal reproductive function.

The pathological examination may reveal suppression of spermatogenesis, advanced stages of follicle atresia, and ovulatory or meiotic failure.

Adjunct Test

In order to save time and expenses, the males of the parent generation (F_0) used in the multigeneration study can be utilized in a dominant lethal test, and a second litter of the first generation offsprings (F_{1b}) may be bred to study teratogenicity.

Other Reproductive Tests

The effects revealed in multigenerational reproduction studies outlined above may result from paternal, maternal, or fetal exposure. Thus additional tests are often conducted to establish the cause of the effect. A few such reproductive tests are described :

1. **Prenatal and Postnatal Studies** : Such studies are often conducted to determine the effects on late development of the offspring. The procedure is essentially the same as that of the multigeneration study, except that only the females are treated and their treatment covers only the last third of gestation and throughout, and lactation. The litter data, as described above, are noticed and evaluated (Palmer, 1976).
2. **Analysis of the Semen** : It is useful in epidemiologic and animal studies. The sperm count and their motility, survival and morphology often provide information on deleterious effects on testis (Eliasson, 1978).
3. **The Profused Male Reproductive Tracts Tests** : The profused male reproductive tracts have been found as an useful model in studying the effects of toxicants on the secretion and accumulation of androgens (Bardin et al., 1978).
4. **In Vitro Study** : To study the deleterious effects of toxicants, Brackett (1978) described a procedure using *in vitro* study of deleterious effects by observing the development of fertilized ova.

C. CARCINOGENESIS AND CARCINOGENIC TESTS

(Refer Chapter 11)

D. SKIN TOXICITY AND TESTS

Introduction

The body of all animals, including humans, is almost entirely covered by a layer termed skin which is usually exposed to a variety of physical and chemical exposures such as cosmetics, household products, topical medication, industrial as well as environmental pollutants, radiations etc. Such an exposure of the epidermis and the dermis of the skin rests over the subcutaneous tissue (Fig. 16.3).

The epidermis of humans is relatively thin, averaging 0.1-0.2 mm in thickness, whereas the dermis is about 2 mm thick. These two layers are separated by a basement membrane. The epidermis in turn consists of a basal cell layer (stratum germinativum), which provides new cells to other layers. These new cells become prickle cells (stratum spinosum) and, later, the granular cells (stratum granulosum). The nuclei in these cells disintegrate and dissolve. In addition, these cells produce keratohydrin, which later becomes keratin in the outermost stratum corneum. This layer is gradually shed. The epidermis also contains melanocytes, which produce pigments, the Langerhans cells, which act as macrophages, and lymphocytes. The latter two types of cells are involved in immune responses. The epidermis thus forms protective covering of the body.

The dermis is primarily composed of collagen and elastin which are important structures to support the skin. In this layer, there are different types of cells, the most abundant being the fibroblasts, which participate in the biosynthesis of fibrous proteins and ground substances such as hyaluronic acid, chondroitin sulphates, and mucopolysaccharides. The other types of cells include fat cells, macrophages, histiocytes, and mast cells.

There are, in addition, various other structures. These include the hair follicles, sweat glands, sebaceous glands, small blood vessels, and neural elements, including Meissner's corpuscles.

Types of Toxic Effects

Different types of effects may result from dermal exposure to toxicants. Most of the effects involve the skin itself, but some of them affect its appendanges — hair, sebaceous glands, and sweat glands.

Primary Irritation

Actually, irritation is a local reaction of the skin to such chemicals as strong alkalis, acids, solvents, soaps, detergents and environmental contaminants. It ranges in severity from hypermia, edema, and vesiculation to ulceration (corrosion). Primary irritation occurs at the site of contact and, in general, on the first contact. It is thus different from sensitization.

Sensitization Reaction

The skin may show little or no reaction on contact with chemical. However, a reaction occurs after a subsequent exposure to the chemical. This delayed reaction, with a latent period ranging from a few days to years, apparently involves a complex immune mechanism. After penetration through the epidermis, the toxicant forms a covalent bond with certain proteins.

A sensitization reaction often results from topical medications such as antibiotics, antiseptics, local anesthetics, antihistamines and stabilizers (e.g., ethylenediamine). Other substances include plants (e.g., poison ivy), metal compounds, dyes, cosmetics and industrial pollutants.

Phototoxicity and Photoallergy

These skin reactions are almost similar in that both are light-induced and may follow either systemic administration or topical application of the offending chemical. However, phototoxicity involves no immune reactions, whereas photoallergy does.

Phototoxicity is much more common than photoallergy. The very commonly phototoxic chemicals in humans are chlorothiazides, phenothiazines, coal tar, pyridine, acridine, sulfonamides and phenanthrene. The skin reaction consists of a delayed erythema, followed by hyper-pigmentation and desquamation.

The well-known photoallergic chemicals are multifungin, promethazine, cyclamates, thiazides, chlordiazepoxide etc. Many compounds are, therefore, photoallergic and phototoxic.

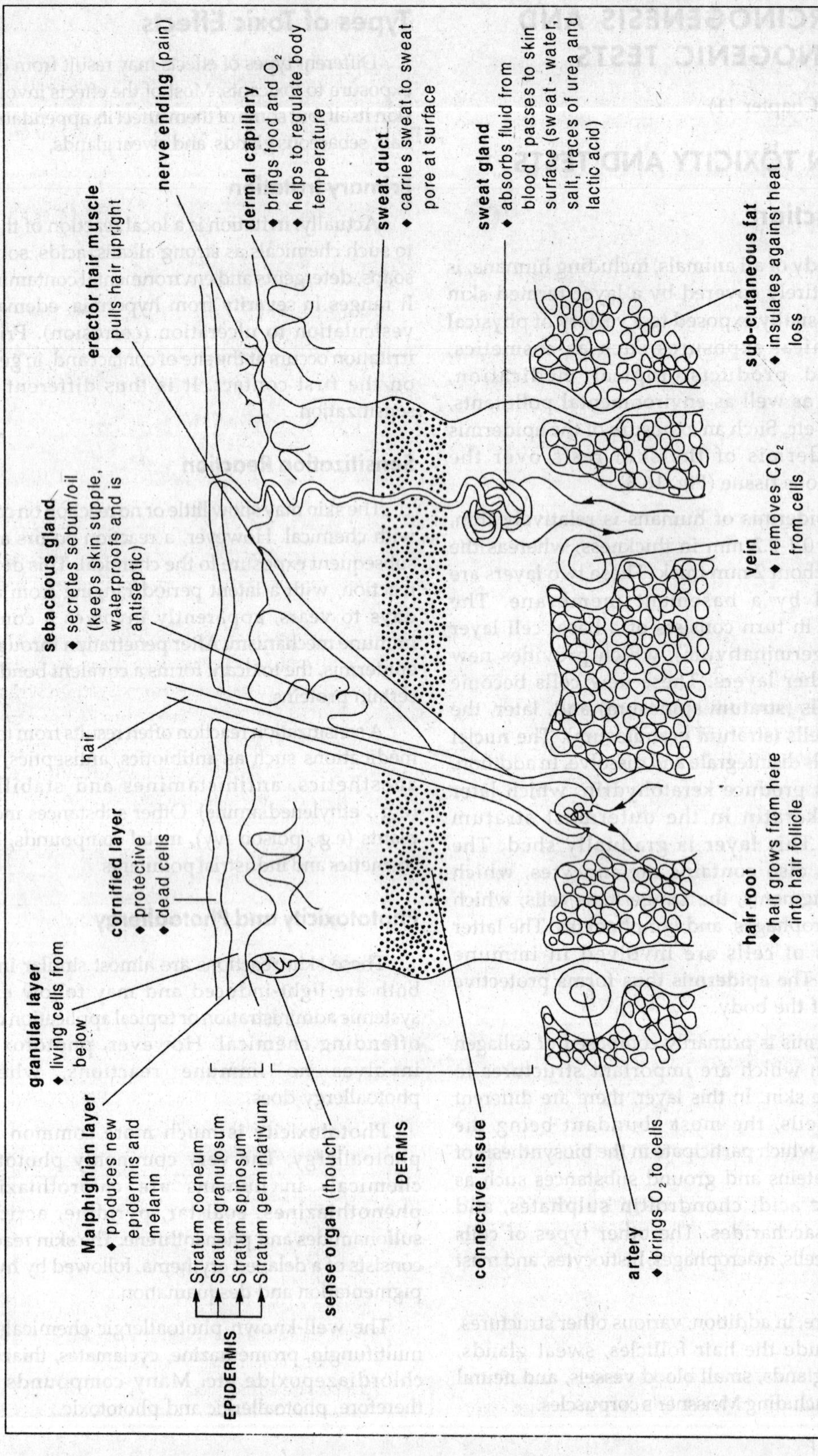

Fig. 16.3 : Skin Showing Structure and Function

Contact Urticaria

A special type of reaction, involving mainly the skin in contact with the offending physical or chemical agents, but sometimes there are also systemic effects. The systemic effects include generalized urticarial and anaphylactoid reactions.

Cutaneous Cancer

It has been known for over two centuries that soots may cause skin cancer. Recent studies confirm that soots and related substances, e.g., creosote oil, coal tars, shale oils, etc., cause cancers of the skin and other sites in animals and humans (International Agency for Research on Cancer, 1979).

Various polycylic aromatic hydrocarbons (e.g., benzo (a) pyrene) and heterocyclic compounds (e.g., benz (c) acridine) are known to induce skin cancer after topical application on animals (IARC, 1973).

Effects On Epidermal Adnexa

I. Hair

Various antimitotic agents used in cancer chemotherapy may induce hair loss. These agents hit at the anagen phase of hair growth. The affected hair sheds after about two weeks of therapy, but hair growth takes about two months after the withdrawal of therapy.

Various other medications are known to cause loss of hair by altering hair follicles in anagen phase to telogen phase. In such cases, hair shedding normally starts 2-4 months after therapy. The medications involved in this type of hair loss include oral contraceptives, anticoagulants, propranol and tripanol.

II. Sebaceous Glands

These glands secrete fat through the expulsion of their lipid laden cells and are, therefore, known as *holocrine*. Their activity is hormone-dependent.

Various chlorinated aromatic hydrocarbons can cause various skin lesions, including chloracne. The severity of acne varies, but it has been confined chiefly to occupational workers.

III. Sweat Glands

Sweating serves useful physiologic functions such as regulation of body temperature. Blockage of the sweat ducts, a disorder known as *miliaria*, may occur after topical application of 95% chloroform and phenol (Shelley and Horvath, 1950). Systematically, atabrine may cause anhidrosis. A number of antimitotic agents can cause necrosis of the sweat glands.

SKIN TESTING METHODS

Selection of Animals

To study the effect of toxicants on the skin, albino rabbits are usually preferred although albino guinea-pigs, white mice and others may also be used.

Test for Primary Irritation

To measure the primary irritation of the skin a patch-test technique is adopted on the hair-free abraded and intact skin of the albino rabbit. This method was outlined by Draize, 1959. The method consists of introducing 0.5 ml or 0.5 gm of the test substance, under an one square-inch (6.25 sq. cm.) patch. The animals are immobilized in an animal holder with patches secured in place by adhesive tape. The entire trunk of the animal is then wrapped with rubberized cloth for the entire 24 hour period of exposure. This latter procedure aids in maintaining the test patches in position and, in addition, retards the evaporation of volatile chemicals. After 24 hrs of exposure the patches are removed and the resulting reactions are evaluated on the basis of scores as cited in Table 16.4 :

Table 16.4 : Evaluation of Skin Reaction

1.	**Erythema and Eschar Formation**	
	No erythema	0
	Very slight erythema (barely perceptible)	1
	Well-defined erythema	2
	Moderate to severe erythema	3
	Severe erythema (Beet redness) to slight eschar formation (injuries in depth)	4
	Total possible erythema score	4
2.	**Edema Formation**	
	No edema	0
	Very slight edema (barely perceptible)	1
	Slight edema	2
	Moderate edema (raised approximately 1 mm)	3
	Severe edema (raised more than 1 mm and extending beyond area of exposure)	4
	Total possible edema score	4

Readings are also made after 72 hrs and the final score represents an average of 24 and 72 hr readings. An equal number of exposures are made on areas of skin which have been abraded earlier. The abrasions are minor incisions through the stratum corneum, but not sufficiently deep to disturb the derma (i.e., not sufficiently deep to produce bleeding).

The total erythema and edema scores are added in both the 24 and 72 hr readings and the mean of the scores for intact and abraded skin are combined. This combined average is referred to as the *primary irritation* index.

Chemical compounds producing combined averages (primary irritation index) of 2 or less are only mildly irritants, whereas with indices from 2 to 5 are moderate irritants, and those with scores above 6 are considered severe irritants.

Test for Sensitization Reaction

The procedure for sensitization reaction was described by Draize in 1959. In this procedure guinea-pigs are given 10 repeated intradermal injections of the chemical on the flank and a challenging dose on the other flank after a 10 to 14 day resting period. A greater reaction after the challenging dose, in comparison with that after the sensitizing doses, indicates sensitization.

However, Draize test is considered insufficiently sensitive to identify allergic potential. So much so, the intradermal injection is regarded inappropriate to assess the safety of chemical to be applied on the surface of the skin.

Various modifications have thus been proposed :

Freund Complete Adjuvant Test

This test involves the use of Freund adjuvant in order to enhance the sensitivity. Furthermore, in challenging the treated animals, four concentrations of the chemical are given, thus providing a basis for quantitative estimation. However, as in Draize test, the test material is given by intradermal injection.

Epicutaneous Methods

One of these methods (Buehler test) stipulates that the test material, after dilution, be applied topically to the skin and held in place for 6 hours by means of an occlusive patch for the induction and challenge. In the other, the open epicutaneous test, the undiluted compound and/or its progressive dilutions are applied to the skin, which remains uncovered.

Test For Phototoxicity and Photoallergy

For the dectection of phototoxicity tests, hairless mouse, rabbit and guinea-pigs are used as test animals. The substance to be tested may be administered topically or by a systemic route. The reaction of the skin to nonerythrogenic light (wavelength greater than 320 nm) is then determined. Significant erythema, compared to controls, indicates phototoxicity.

For the examination of photoallergy, albino guinea pigs are particularly useful. The procedure involves, in principle, an induction of photosensitization by repeatedly applying a small amount of the chemical on a shaved and debilated area of the skin and exposing that area to appropriate ultraviolet rays. After 3 week interval, the guinea-pigs are exposed to the chemical and the UV rays to elicit photoallergy.

Test For Contact Urticaria

Hunziker (1977) performed the patch test on the flank and nipples of guinea-pigs and described the gross and histologic changes produced by various chemicals viz., dinitrochlorobenzene. The open patch test can be applied to human patients suspected of being susceptible to the chemical.

Any immunologic involvement can be shown by the passive transfer test, in which 0.1 ml fresh serum from the patient is injected intradermally into the forearm of a volunteer and challenged 24 hours later by applying the suspect chemical to the injection site.

Test For Cutaneous Cancer

For the detection of cutaneous cancer, the mouse is the commonly used animal.

The procedure involves topical application of the chemical on a shaved area of the skin. The chemical, if a liquid, is applied directly. Otherwise it is dissolved or suspended in a suitable vehicle. The skin painting is usually done once a week or more frequently. It is advisable to include a vehicle control group as well as a positive control group, which is treated with a known skin carcinogen such as benzo(a)pyrene.

Test For Hair Loss

This is usually performed in guinea-pigs. The chemical is administered by the same route as that to be encountered in human exposure conditions. A vehicle control group is included to assist in the interpretation of the results.

Test For Acne

In this test, rabbit is usually preferred as the test animal. In 1968, Kligman and Katz described a test according to which the test substance is inserted in the external ear canal of rabbits (the test animal). Positive reactions are marked by thickened mounds on the follicles, caused by distention with horny materials.

Evaluation of Skin Tests

The aforementioned testing procedures have been widely used, despite certain drawbacks. Seeking to improve on them, Middleton (1981) has proposed the following new approaches, which are designed to measure more specific biological and biochemical events :

1. Alterations in the normal differentiation pattern of epidermis, which are demonstrable by measuring in different layers, the contents of phospholipids, keratin, and a histidine-rich protein.
2. Alterations in epidermal cell proliferation as shown by the mitotic index among basal cells and the rate of DNA synthesis.
3. Alterations in cellular viability as determined by enzyme leakage.

Presently these approaches still require considerable further investigations to ascertain their predictive value and to work out the optimum procedures.

E. TOXICOLOGY OF THE EYE AND TESTS

The eye, being complex in structure, is a spherical body and is covered chiefly by three coats or layers of tissues : the sclera, the choroid and the retina (Fig. 16.4). The sclerotic coat mainly consist of fibrous tissues. The choroids coat consists of pigments and blood vessels, and the retina is composed of the nerve fibers, cells, and special receptors. They are actually nontransparent, however, light is admitted through the front of the eye, where the three coats are replaced by a number of tissues, notably the cornea and the lens.

The cornea is formed by the modification of the sclera. It consists of a relatively thick stroma and is covered, in front, by an epithelium, consisting of several layers of cells and Bowman's membranes and, behind, by Descemet's membrane and an endothelium. The cornea and the front portion of the sclera as well as the inside of the eyelids are covered by a thin layer of conjunctiva.

The lens is formed of transparent fibers enclosed in the lens capsule. It is suspended by the ciliary zonule to the ciliary body and its curvature is adjustable by the contraction and relaxation of the ciliary muscle (Fig. 16.5).

The space between the lens and the cornea is filled with the aqueous humor. Also, in this space and just in front of the lens, is the iris. The iris is heavily pigmented and rich in blood vessels. The iris has a central opening — the pupil.

The retina is a photosensitive structure that responds to light stimuli. It consists of several layers. The outermost is a pigmented epithelium. Next to it are the retinal rods and cones, which are the light-responsive neural structures. They are linked to the ganglion cells via the bipolar cells. The axons from the ganglion cells converge and exit from the eye at the optic papilla.

Due to the diverse physiological nature and spatial relations, these ocular structures may exhibit a variety of effects as a consequence of exposure to toxicants.

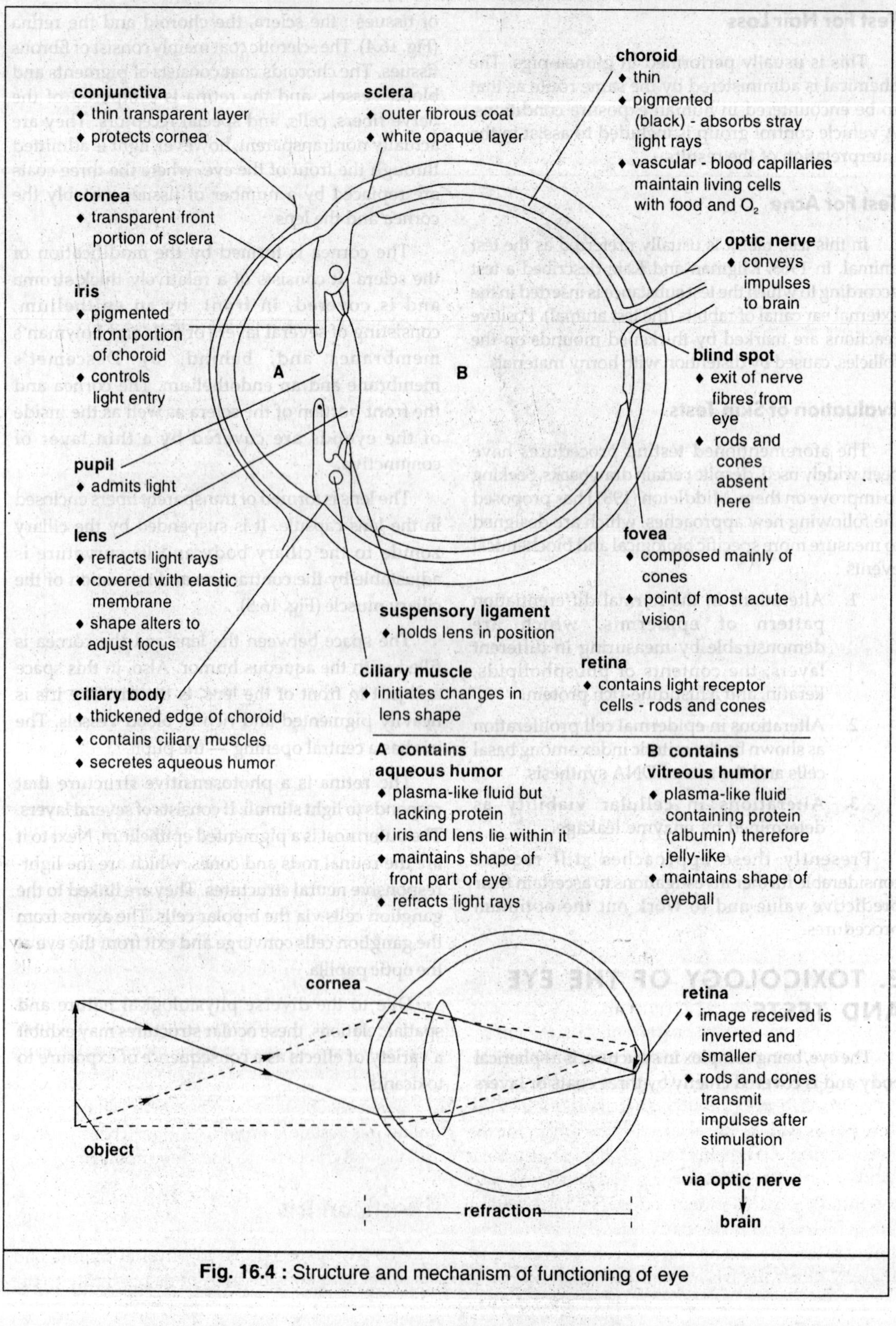

Fig. 16.4 : Structure and mechanism of functioning of eye

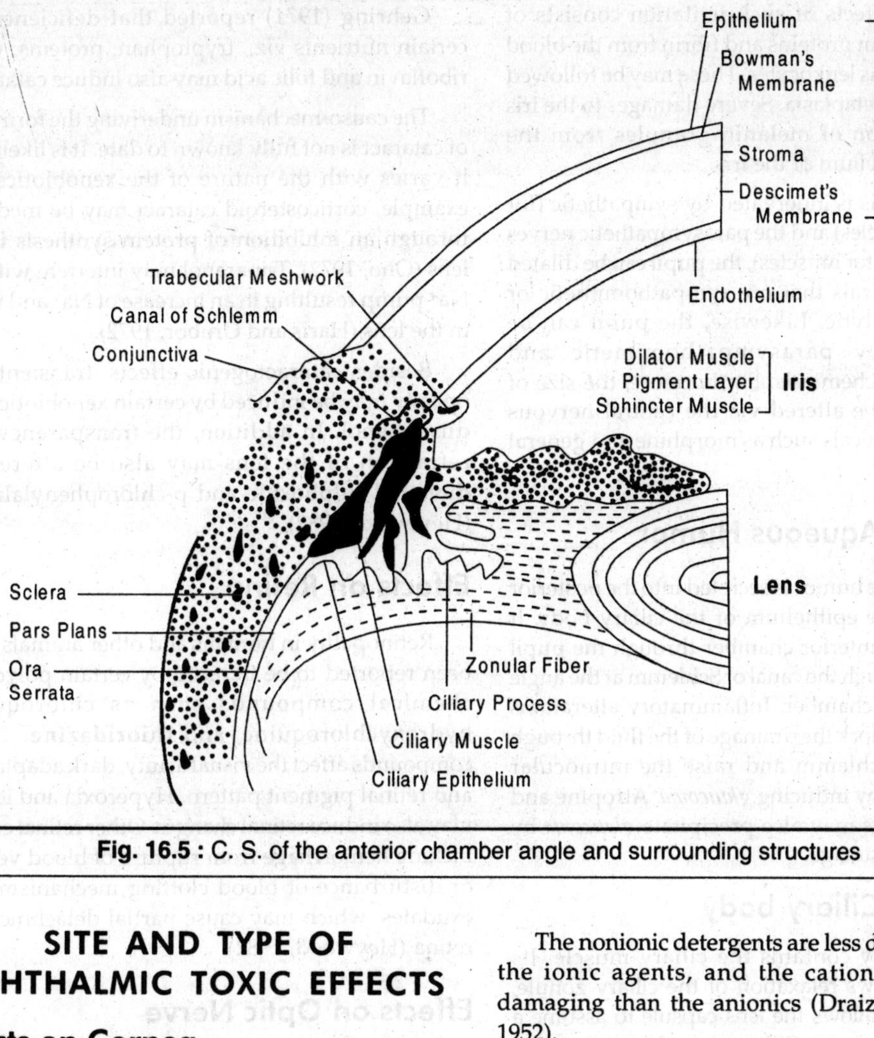

Fig. 16.5 : C. S. of the anterior chamber angle and surrounding structures

SITE AND TYPE OF OPHTHALMIC TOXIC EFFECTS

Effects on Cornea

This delicate structure is subjected to toxic effects of xenobiotics chiefly through external exposures. Xenobiotics that impair the cornea function include acids and alkalies, detergents, organic solvents, and smog. Acids and alkalis can readily damage the cornea. The extent of damage ranges from minor, superficial destruction of the tissue, which heals completely, to opacity of the cornea or even perforation. Acid burns are related to the low pH as well as the affinity of the anion for the corneal tissue. The effects of alkalis usually have slower onset than those caused by acids and essentially pH-dependent. However, ammonium ion, which is present in many household products, penetrates the cornea more readily and can thereby affect the iris.

The nonionic detergents are less damaging than the ionic agents, and the cationics are more damaging than the anionics (Draize and Kelley, 1952).

Organic solvents, viz. acetone, hexane and toluene, may enter the eye as a result of industrial or laboratory accidents. These xenobiotics may dissolve lipid and damage the corneal epithelial cells.

Industrial smoke and fog combine to form smog. It adversely affects the respiratory tract, but, even at low concentrations, it irritates the corneal sensory nerve endings and cause reflex lacrimation.

Other xenobiotics may adversely affect the cornea following systemic administration. These include quinacrine, chloroquine, and chlorpromazine.

Effects on Iris

The iris is susceptible to physical trauma and chemical irritations because of its proximity to the cornea. The effects of such irritation consists of leakage of serum proteins and fibrin from the blood vessels as well as leukocytes. These may be followed by fibroblast metaplasia. Severe damages to the iris cause liberation of melanin granules from the posterior epithelium of the iris.

Since the iris is innervated by sympathetic (for the dilator muscles) and the parasympathetic nerves (for the constrictor muscles), the pupil can be dilated by toxic chemicals that are sympathomimetic or parasympatholytic. Likewise, the pupil can be constricted by parasympathomimetic and sympatholytic chemicals. Furthermore, the size of the pupil can be altered via the central nervous system by chemicals such as morphine and general anesthetics.

Effects on Aqueous Humor

The aqueous humor is secreted into the posterior chamber by the epithelium of the ciliary body. It flows into the anterior chamber through the pupil and drains through the canal of Schlemm at the angle of the anterior chamber. Inflammatory alterations of the iris may block the drainage of the fluid through the canal of Schlemm and raise the intraocular pressure, thereby inducing *glaucoma*. Atropine and other mydriatics may also precipitate *glaucoma* by blocking the drainage.

Effects on Ciliary body

Ciliary body contains the ciliary muscle. Its contraction allows relaxation of the ciliary zonule, which, in turn, allows the lens capsule to assume a more spherical form. Ciliary muscle is parasympathetically innvervated, there *cholinesterase inhibitors* and parasympathetic agents like atropine can cause the lens to be fixed in different state of accommodation.

Effects on Lens

Various xenobiotics are known to alter the lenticular transparency, resulting in the formation of cataract. Examples are 2, 4-dinitrophenol, corticosteroids, triparanol, thallium, busulfan, alkylating agents, naphthalene, iodoacetic acid, DMSO (Dimethyl sulphoxide) etc. Their cataractogenic property has been reported in humans and other animals, such as rat, rabbit and young fowl. The effects usually follow systemic exposure, but with certain chemicals viz., corticosteroids, anticholinesterases, etc. they may occur after topical application (Axelsson, 1968).

Gehring (1971) reported that deficiencies of certain nutrients viz., tryptophan, proteins, vit. E, riboflavin and folic acid may also induce cataract.

The causo-mechanism underlying the formation of cataract is not fully known to date. It is likely that it varies with the nature of the xenobiotics. For example, corticosteroid cataract may be mediated through an inhibition of protein synthesis in the lens (Ono, 1972). Triparanol may interfere with the Na^+ pump resulting in an increase of Na^+ and water in the lens (Haris and Grubor, 1972).

Besides cataractogenic effects, transient lens opacity may be induced by certain xenobiotics like diisophenol. In addition, the transparency and refraction of the lens may also be altered by dimethyl sulphoxide and p-chlorophenylalanine (Heywood, 1982).

Effects on Retina

Retinopathy in humans and other animals have been reported to be induced by certain polycyclic chemical compounds such as chloroquine, hydroxychloroquine, and **thioridazine**. These compounds affect the visual acuity, dark adaptation, and retinal pigment pattern. Hyperoxia and iodate may also induce retinal changes. Other retinal effects include hemorrhage from rupture of blood vessels or disturbance of blood clotting mechanism and exudates, which may cause partial detachment of retina (Heywood, 1982).

Effects on Optic Nerve

Xenobiotics at one time may affect either the ganglion cells or the optic nerve. Damage to one of them results in the degeneration of the other. Certain toxicants affect mainly the central vision. The most notable example is methanol (CH_3OH). Others include CS_2, disulphuran, ethambutol and thallium. On the other hand, quinine, chloroquine, arsenic^{+5}, and CO cause constriction of visual fields by damaging the structures responsible for peripheral vision. However, nitrobenzol affects both the central and peripheral vision (Harrington, 1976).

It would not be out of place to mention that certain organic solvents viz., trio-cresyl phosphate, acrylamide, n-hexane, and methyl n-butyl ketone can induce peripheral neuropathy but spare the visual system (Grant, 1980).

The mode or action of these toxicants is not well understood, however, quinine appears to act through an inhibition of DNA, which, in turn, inhibits RNA transcription and protein synthesis. As^{+5} appears to act on the ganglion cell neurons (Potts and Gonasum, 1980).

Eye Testing Methods

Effects on the eye can be examined after topical application of the toxicants. In certain cases, systemic administration can also result in adverse ocular changes.

Test Animals

Mainly albino rats are used to determine the ocular irritancy of ophthalmic medications and other chemicals that might come in contact with the eyes. Dogs and non-human primates, viz. rhesus monkeys, may also be used as test animal.

Gross Examination

For testing the ocular irritancy, the test described by Draize and Kelley (1952) has been accepted as standard procedure.

It specifies the use of nine rabbits. Into one eye of each rabbit is instilled 0.1 ml of the test material. In three of the nine rabbits, the test material is washed with 20 ml of lukewarm water 2 seconds after the instillation, and in three other rabbits, the washing is done after a 4-second delay. In the remaining three rabbits, the eyes are left unwashed. The ocular reactions are read with the unaided eye or with the aid of a hand slip lamp at 24, 48 and 72 hours and at 4 and 7 days after treatment. The reactions of the conjunctiva (redness, chemosis, and discharge), cornea (the degree and extent of opacity), and iris (congestion, swelling and circumcorneal injection) are scored according to a specified scale. A series of coloured pictures — as a guide for grading eye irritation by the U.S. Food and Drug Administration (1965) — has also been used.

Further, in 1980, Griffith et al recommended following points to be taken into account in conducting eye irritation tests :

1. A 0.01 ml dose, or its weight equivalent of solids and powders, be applied directly to the central corneal surface of at least six eyes without subsequent rinsing or manipulation of the eyelids.
2. Evaluation of the irritancy may be based on the median duration for the eyes to return to normal, instead of using a scoring system based on the type and extent on the deleterious effects produced by toxicants.

In 1983, U.S.C.P.S.C. (United States Consumer Product Safety Commission) reported the need of six albino rabbits for each test chemical substance. For test liquids, 0.1 ml is used, and for solids and pastes, 100 mg. The test substance is instilled into one eye of each rabbit without washing. Ocular examinations are made after 24, 48 and 72 hours. A rabbit is considered as having positive reaction if the eye shows ulceration or opacity of the cornea, inflammation of the iris or swelling of the conjunctiva with partial aversion of the eyelids.

INSTRUMENTAL EXAMINATIONS OF OPHTHALMIC TOXICITY

Ophthalmoscopy

The ophthalmoscope is used in assessing the deleterious effects of xenobiotics on different parts of the retina. The examination is usually intended to trace the existence of edema, hyperaemia, atrophy of the optic disk, pigmentation, or the state of blood vessel. Alterations in the vitreous humor, lens, aqueous humor, iris and cornea can also be traced.

Visual Perimetry

Effect on the visual field can be readily determined in the humans only but not in the laboratory animals, except the non-human primates. It may reveal the loss of peripheral vision resulting from exposure to any xenobiotic.

Other Procedures

Procedures involving instrumentations, viz., electro-oculography, and visually evoked responses are also useful in ophthalmic toxicity experimentation (Grant, 1980).

Histological and Biochemical Examinations

Light microscopy may also reveal the site of action of toxicants on the eye. Electron microscopy may reveal ultrastructural changes and the biochemical studies can elucidate the mechanism of toxic effects.

For example, choroquine has been observed with light microscopy to cause a thickening of the pigment epithelium, followed by migration of the pigment to the outer-nuclear layer, and, finally, total atrophy of the photoreceptors (Meier-Ruge, 1968). Electron microscopy showed mitochondrial swelling and disorganisation of the endoplasmic reticulum in the photoreceptor inner segment (Sotze and McConnell, 1970).

Biochemical studies revealed inhibition of many enzymatic reactions, especially those related to protein metabolism of the pigment epithelium.

Evaluation

Eye irritation tests are widely used to assess the ocular irritancy of the xenobiotics. In general, the albino rabbits are the animals of choice. Despite the variability of the results, the tests in rabbit have proved useful in predicting eye irritation in humans.

A large number of animal experimentations and clinical studies indicate that there is fair correlation between humans and animals in their reactions to xenobiotics with respect to cataract formation and retinopathy (Grant, 1980).

Revision Questions

A. Long-Answer Questions

1. Give a detailed account of the relationship between behavioural toxic signs and body organs/system in animals.
2. Give an account of elementary and sensitive behavioural tests for acute toxicity. Add a note on the behavioural testing procedure.
3. Describe two important functional tests studied by you.
4. Describe various histopathological changes in some tissues of animals under toxicant stress.
5. Define teratogenesis and describe its testing procedure.
6. Define carcinogenesis and describe its testing procedure.
7. Describe the toxicity of the eye and its tests.
8. Describe the toxicity of the skin and its test.
9. Give an account of the reproductive toxic effects and its tests.

B. Short-Answer Questions

1. Briefly describe :
   (i) Toxicological importance of behavioural tests.
   (ii) Potentiation.
   (iii) Respiratory functional test.
   (iv) Special renal functional tests.
   (v) Synergism.
2. Write short notes on :
   (i) Urine analysis.
   (ii) Blood analysis.
   (iii) Blood ammonia as an index of liver function.
   (iv) Functional tests.
   (v) Criteria for behavioural test.
   (vi) Behavioural toxic signs.

17

Dose-Response Relationship

SYNOPSIS

Introduction, Selection of doses, Duration of exposure, Types of human exposure, Categories of toxic effects, Responses, Dose-response relationship, Cumulative responses, Threshold limit, Margin of safety (Therapeutic index), Potency versus Efficacy.

Introduction

The term 'dose' is defined as the quantity of toxicant administered to, or received by, the animal at one time or in given period of time. 'Dosage' refers to the administration of toxicant in doses. The amount is readily ascertained when the toxicant/ chemical/substance is given by a route such as oral, subcutaneous, or intraperitoneal. This definition is not applicable to exposure by inhalation, because the amount that is retained in the animal depends on the concentration of the toxicant in the inhaled air as well as the duration of the exposure (contact reaction between a chemical and biological system).

Selection of Doses

To observe the responses, in acute lethality studies, one concentration should be aimed causing the death of about 50% of the animals. At least two other concentrations are selected, one killing more than 50% but less than 90% and the other killing 10-50% of the animals.

In longer-term studies, at least three concentrations should be selected. The highest concentration is intended to cause some toxic effects and the lowest concentration should have no apparent effects. At present, for long term toxicity studies of any toxicant, 8 or 10% concentration (dose) of LC_{50} for 96 hours is selected.

Duration of Exposure

Overall, based on the duration, the exposure of animals to toxicants may be divided into four categories :

1. **Acute exposure or Short-term exposure**

   It is exposure of animals to a chemical up to 96 hours.

2. **Subacute exposure**

   The repeated exposure of animals to sublethal doses of toxicants for one month or less duration is termed as subacute exposure.

3. **Subchronic exposure**

   The repeated exposure of animals to sublethal doses of toxicants for one to three months time is known as subchronic exposure.

4. **Chronic or Long-term exposure**

   The repeated exposure of animals to sublethal doses of toxicants for more than 3 months time is known as chronic exposure.

Types of Human Exposure

The humans are exposed to toxicants by any of the following means :

(a) **Intentional exposure** : People consume various drugs and food-additives in their daily life. Sometimes, these substances are taken over for long span of time. As an example, many people use alcohol and cigarettes on a long-term basis, which may lead to chronic toxic effects.

(b) **Occupational exposure** : Humans, on account of their occupations, are exposed to a wide variety of xenobiotics. Occupational exposure to toxic substances is mainly chronic and continuous. The routes of exposure are either inhalation or skin contact. Lung diseases and dermatitis are very common diseases among industrial workers. Sometimes, acute exposure may occur in the event of an accident such as an explosion, spillage, leakage or because of bad working practices.

(c) **Environmental exposure** : A wide variety of chemical substances in liquid and gaseous state reach the environment on account of various anthropogenic activities and affect the health of humans. Such exposure is recognized as *environmental exposure i.e.,* exposure through the environment. This form of exposure is usually chronic but sometimes accidents at factories may lead to acute exposure such as, the leakage of methyl isocyanate (MIC) by Union Carbide factory at Bhopal. Chronic exposure to gases such as sulphur dioxide, nitrogen oxide, ammonia and carbon mono oxide occur in industrial areas and in the regions of heavy traffic.

(d) **Accidental poisoning** : Mistaken ingestion of poisonous/toxic substances, cleaning fluids and other household products, drugs and excessive doses of drugs cause accidental poisoning. Such type of exposure is normally acute rather than chronic.

(e) **Intentional poisoning** : Intentional poisoning and intentional consumption of poisonous substances for suicidal purposes come under this category of exposure.

Biological Effect Monitoring

Biological effect monitoring (BEM) may be defined as the continous or repeated measurement and assessment of early biological effects in exposed humans, to evaluate ambient exposure and health risk by comparison with appropriate reference values based on knowledge of the probable relationship between ambient exposure and biological effects. BEM is used in environmental and occupational toxicology — and also sometimes in forensic toxicology to monitor possible problems at an early stage. Generally, BEM is carried out using readily available specimens such as urine, blood, and feces.

When a BEM parameter is exceeded, the agent non-specific signs and symptoms are not necessarily related to exposure to specific agents, nor to conditions of work environment. BEM parameters are sensitive, but non-selective, and thus may serve as a safety net when the exposure is actually not known (e.g., exposure to various different solvents or a mixture of solvents).

Categories of toxic effects

The exposure of animals to certain doses/ concentrations of toxicants brings about certain alterations in the animals. Such alterations are called *responses* or *effects* of toxicants. Toxicant-induced alterations may be at the level of whole animal (death of organisms), at cellular levels (histopathological changes) or at molecular levels (biochemical changes) leading to some physiological responses. When the criterion of effect is lethality or death of organisms, the obtained response is *'all or none'* type.

Following exposure to a toxicant, the animal either dies or survives i.e., either the animal responds fully to toxicants and the response is death of the animal or it does not respond at all. Hence, this type of effect is termed as *'all or none'* type. Categories of toxic effects have been described in Chapter 15.

Responses

Actually responses are the incidences produced by the different doses of different toxicants on the animals. Responses may also be defined as the signals of toxicity in the animals. The ultimate consequences of the toxicity of various toxicants may also be referred to as responses.

The ultimate response, as a result of acute exposure, depends on the amount of toxicant at the critical target sites. In other form, the amount of toxicant at the target site is partly dependent on the dose.

Between exposure to the dose and the response there are four major occurrences. The first three occurrences are absorption, metabolism, and excretion and in this context, metabolism includes both detoxification and toxification processes. The fourth occurrence is distribution within the body, which is critical to the toxicology of xenobiotics.

The nature of toxic effect often varies with the dose. For instance, as the dose of barbiturate increases, the effect progresses through no observable effect to mid-sedation, to loss of righting reflex (sleep), and, finally, death. Furthermore, because of differences of more individual susceptibility, not all animals exposed to the same dose will respond in the same way. Based on the doses, the effects i.e., responses (incidences) may be assessed by simple statistical analyses. It may yield information on the dose-effect and dose-response relationship.

Dose-response Relationship

It may be defined as the graphical relationship between different doses of a toxicant and the corresponding responses produced by them in a test animal.

In simple words, the relationship between the dose and the magnitude of the effect produced by the toxicant may be termed as the dose-response relationship or dose-effect relationship of toxicants.

The exact relationship between the dose of the xenobiotic (toxicant) and the response (effect) depends upon the biological object under observation i.e., the age, size, health, sex and species of the animal and on the nature of chemical (toxicant) used.

Such studies assist medical practitioners to assess the therapeutically effective dose of any drug and the toxicologists to predict the adverse effects of various xenobiotics on biological system.

In toxicology the important concept is the dose. It is also important to recognize the difference between dose and concentration. Dose is actually the concentration multiplied by time.

In the case of essential compounds there is a gradual transition from deficiency to effectiveness depending on the dose given. At low dose, when deficiency occurs, the metabolised or excreted amount of a substance is greater than the uptake, leading to a shortage in the cell or tissue or organ. A well-known example is vitamin deficiency, leading, in the case of vitamin C, to scurvy. An intake of vitamin C will rectify the balance, and the scurvy will disappear. Thus, a certain minimum amount is necessary to prevent deficiency. This is illustrated in Fig. 17.1, where the effect of an essential compound is shown as a function of dose. There is a gradual change of effects from deficiency to an essential level through to fatal toxicity.

Another known example is Zinc. There is a homeostatic control of zinc in blood plasma, but deficiency may lead to skin healing problems which are overcome by zinc treatment at the wound site. In excess, zinc may cause gastrointestinal problems if the dose is high enough (1 g or more).

Actually the dose-effect relationship was originally developed by pharmacologists to study the relation between a drug and the effects or side-effects on the body. These kinds of relationship can also be used in toxicology. They may be applied to compounds as illustrated in Fig. 17.2, which reveals the relationship between the dose and different effects for a non-essential compound. The basic difference between Figs. 17.1 and 17.2 is that, in the case of a non-essential compound, no benefit can be expected and, thus, sometimes such compounds are termed toxic. Figs. 17.1 and 17.2 refer to no effects, early effects, and clinical effects.

The no-effect area is clearly concerned with levels that have no effect. The maximum dose that produces no detectable alteration under defined conditions of exposure is referred to as the no-effect level. It is not easy to determine actually whether an effect is detectable, and what the significance of the effect is. Thus a more practical measure is the *no observed adverse effect level* (NOAEL). An adverse effect is a change in morphology, physiology, growth, development, or life span of an organism that results in impairment of functional capacity or impairment of capacity to compensate for additional stress or increase in suceptibility to the harmful effects of other environmental influences. Early effects are not necessarily adverse effects and a NOAEL, may be above the level causing an early effect but below the level causing a clinical effect.

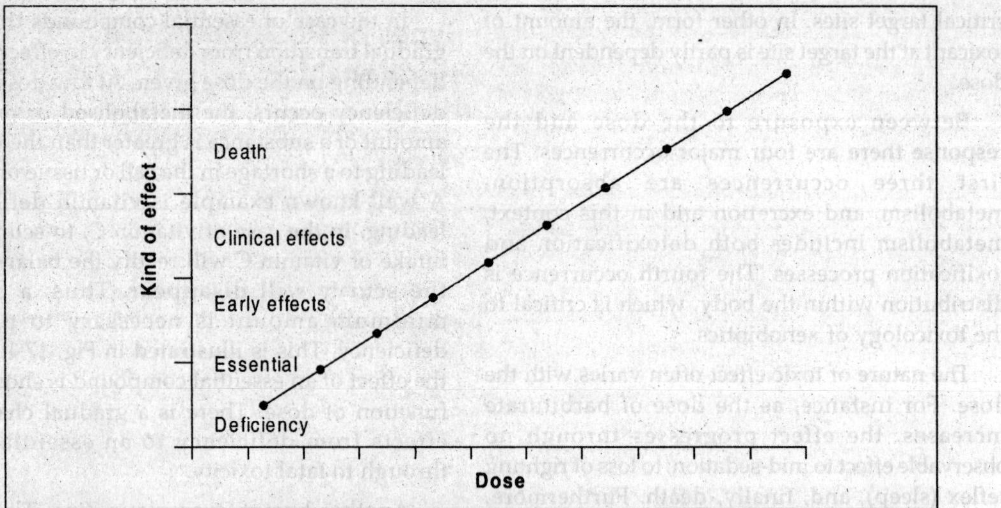

Fig. 17.1 : A simple relationship between the dose and different effects for an essential compound. Note that with enhancing dose the seriousness of the effect increases

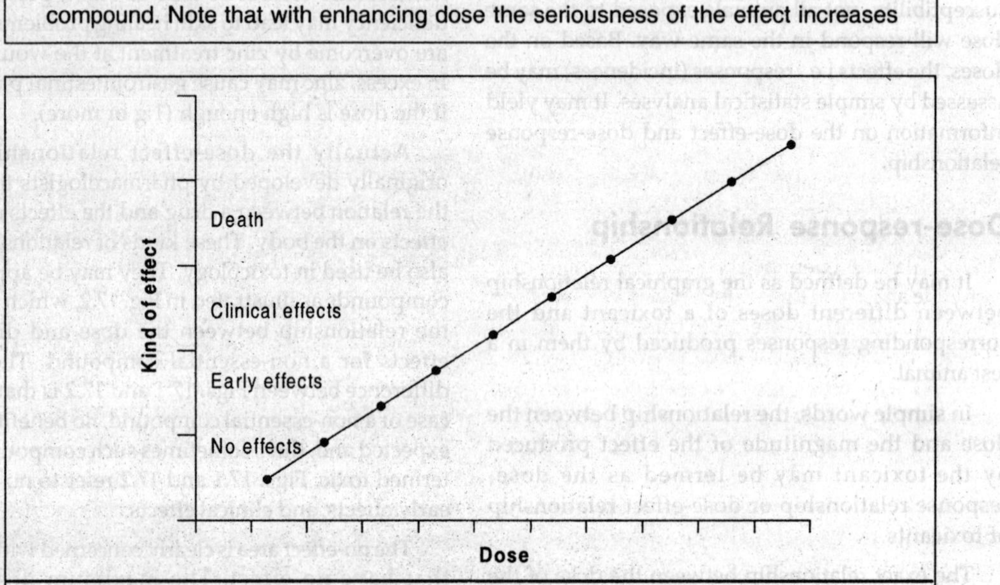

Fig. 17.2 : A simple relationship between the dose and different effects for a non-essential compound

Clinical effects are usually mentioned as adverse effects, e.g., occupational asthma (which may be caused by many compounds). Other examples of clinical effects are disturbances in the peripheral and central nervous system (e.g., tremor, polyneuropathy), and disturbances of liver and renal function. As the body has only a few organs and the number of industrially used chemicals is about 70,000, very few early effects and clinical effects are specific for one compound. Thus, one effect may be caused by different compounds.

The branch of science dealing with the production of effects from the structure of a chemical compound is called quantitative structure-activity relationship (QSAR).

Dose-effect relationships are generally S-shaped (Fig. 17.3). In this example there is an exponential rise above the baseline in urinary β_2 microglobulin, a parameter for assessing renal tubular damage, as a function of the dose of cadmium in urine. Beyond this, enhancing the dose gives only a moderate rise in the effect. The steepness of the dose-effect relationship determines the value at which the NOAEL must be set. The NOAEL has to be set at a lower dose when the curve rises steeply compared with a low rise.

Early effects are those that can be observed before adverse change has occurred. Examples are compounds that smell offensive at a no-effect level. Mercaptans, for example, are used as a warning odour in natural gas. At higher levels these are toxic. Thus, in low concentrations, a certain effect that is not adverse is used for a different purpose. Other examples of early effects are the excretion of certain enzymes and specific proteins into urine as a consequence of low exposure to certain compounds such as metals, solvents, pesticides, and drugs.

A dose-effect relationship exists for nearly all organ systems. But there are two exceptions — cancer and the immune system.

If in Fig. 17.3 the effect is replaced by the response, a dose-response curve is obtained. A response may be the reaction of an animal or part of an animal (such as a muscle) to a stimulus, and, as such, is similar to a dose-effect curve. An alternative definition of response is the proportion of a group of individuals that demonstrate a defined effect in a given time at a given dose rate. These kinds of dose-response relationships are quite different from the dose-effect relationships mentioned earlier. Dose-response (D-R) relationships in this sense can be used to differentiate between groups. If a D-R relationship for one group has a steeper slope than for another, the first group will be more sensitive to the dose than the latter. D-R relationships consequently can be used to detect sensitive groups.

Types of Dose-Response relationships

These are two basic types of dose-response relationship :

1. Quantitative
2. Quantum.

1. Quantitative Dose-Response Relationship

It is also known as graded-dose-response-relationship. Occasionally, such quantitative information can be derived from human data. For example, with increasing intakes of methyl mercury, there were paresthesia, ataxia, dysarthria, deafness, and death, and a greater proportions of individuals manifest these effects (Bakir et al, 1973). Fig. 17.4 shows graphically both the dose-response and dose-effect relationships among individuals exposed to various doses of methyl mercury.

As mentioned above, Quantitative-Dose-Response Relationship is also termed as 'Graded Dose-Response-Relationship'. It is because the degree of response eventually reaches a maximum value corresponding to the subsequent increase in the dose. At this point, further increase in the dose level does not elicit any response at all on behalf of the living subject. This act of cessation with the further increase in the degree of response is termed *"ceiling effect"*. The dose of the toxicant at which the ceiling effect is produced is called the 'ceiling-dose'.

The study of the dose-response relationship proves useful in the evaluation of toxicological data. Because of the variation in individual's sensitivity of

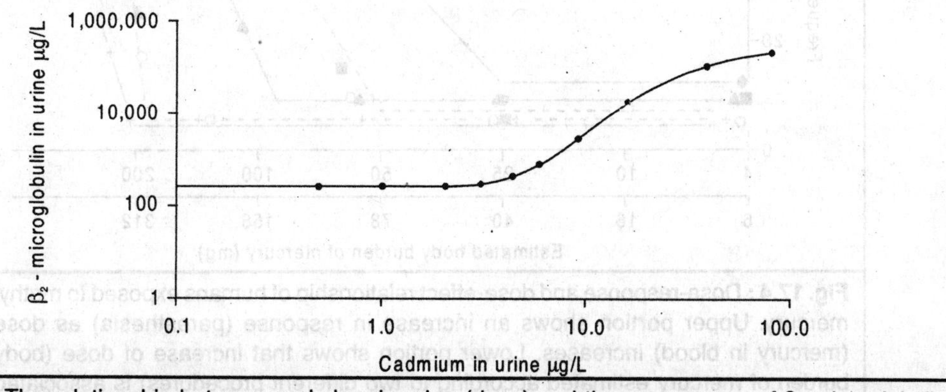

Fig. 17.3 : The relationship between cadmium in urine (dose) and the appearance of β_2-microglobulin in urine (an early measure of renal tubule damage). Note the exponential rise after the horizontal part

any group of organisms, they will not die at the same dose of a chemical. Therefore, the frequency of response, e.g., death, will increase along with the dose. When the mortality or frequency of other effects is plotted against the dose on a logarithmic scale, a S-shaped curve is obtained. The central portion of the curve (between 16% and 84% response) is sufficiently straight for estimating the LD_{50} or ED_{50} dose.

2. Quantum-Dose-Response-Relationship

Such a dose response relationship exhibits quantized alterations in the degree of response elicited by increasing subsequent enhancements of toxicant dose. Hence such a dose-response-relationship is basically 'all or none' type relationship.

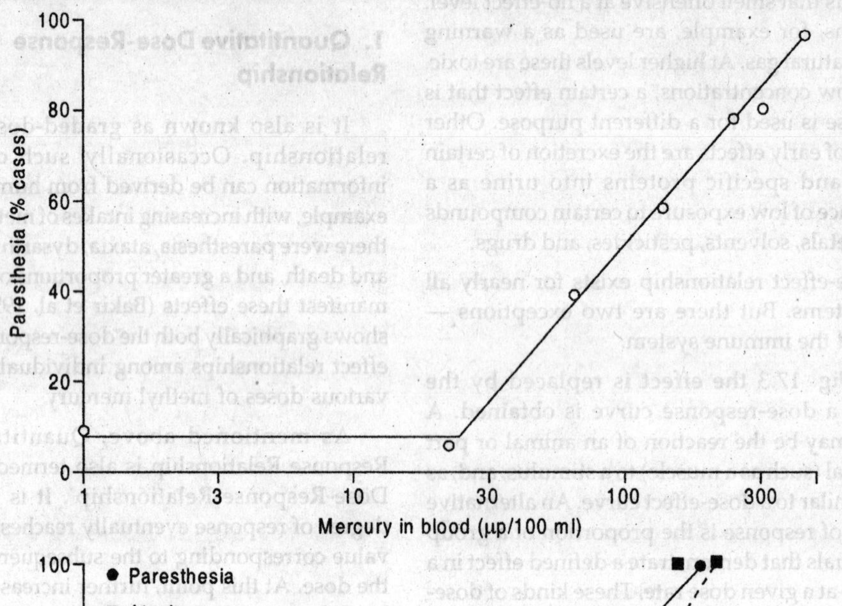

Fig. 17.4 : Dose-response and dose-effect relationship of humans exposed to methyl mercury. Upper portion shows an increase in response (paresthesia) as dose (mercury in blood) increases. Lower portion shows that increase of dose (body burden of mercury estimated according to two different procedures) is associated with not only greater responses but also additional effects (from WHO, 1976)

Alive or dead outcome = Quantal response, but

Alive = Moribund = Dead = Polychotomous Quantal Response.

Quantum-dose-response relationship is observed in case of population studies. In such studies, the percentage of the affected population increases as the dose of the toxicant is raised. The relationship is quantal because a response is stated to be either present (all) or absent (none) in a given individual's population.

Cumulative Response

Soon after the toxicant is absorbed and distributed in the animal, it is excreted slowly or rapidly. They may be excreted as the parent chemicals or their metabolites and/or as their conjugates. When the excretion of a toxicant/drug occurs at a slow rate, the complete-response is a result of the sum of the responses, produced by more than one dose of that toxicant/drug. Such a phenomenon is referred to as cumulation of the toxicant and the response produced is referred to as cumulative response.

Saccharin is rapidly excreted; hence its blood level drops rapidly, even after repeated administration. On the other hand, methyl mercury is excreted very slowly; its gradual accumulation culminates in a near plateau only after 270 days (Munro and Willes, 1978). Due to slow rate of excretion of this toxicant its blood level rises resulting in poisoning.

Certain heavy metals, e.g. lead, may accumulate inside the bones without generating deleterious effects although they may produce certain adverse effects on release into the blood circulation. This incidence is referred to as *passive cumulation.*

Another example of cumulative response is intake of alcohol by human. A single dose of alcohol per day may cause no ill-effect. But if taken repeatedly, the alcohol goes on accumulating slowly in the body because of its slow excretion. The accumulated alcohol damages liver, causes degeneration of intestinal mucosa and, finally, produces cirrhosis of liver, and death. Likewise small doses of X-rays produce slight burns which heal up in a few days. But exposure to these small doses of X-rays over a long period of time causes cumulation which produce deleterious effects, sometimes cancer.

Threshold Limit

Threshold limit is that concentration of the toxicant/drug beyond which any further increase in the toxicant/drug dose causes progressive changes in the degree of response. Below the threshold limit, there is no apparent change in the degree of responses elicitied by an increase in the dose-level. In other words, a dose below which no effect or response is measurable. Such a dose of toxicant is termed as threshold dose. In terms of 50% lethal concentration, it is known as threshold lethal concentration.

The 'threshold lethal concentration' is also known as incipient LC_{50}. Threshold has more theoretical significance than the LC_{50} at an arbitrary time period and is more useful.

A system of classifying different grades of toxicity has been given in a report by an international group (Joint IMCO/FAO/UNESCO, 1969).

The categories given below seem useful for general description of toxicants/pollutants:

Very toxic	— Threshold below 1 mg/l
Toxic	— Threshold 1-100 mg/l
Moderately toxic	— Threshold 100-1,000 mg/l
Slight toxic	— Threshold 1,000-10,000 mg/l
Practically non-toxic	— Threshold above 10,000 mg/l

Loomis (1978) used the concept of complex (chemical-receptor) to explain the existence of threshold that "there must be necessarily be a quantity of chemical agent below which no biological effect would be achieved". The existence of threshold doses for chemicals is important in evaluating safety. Actually safety is the practical certainty that injury will not result from the chemical in question when used in the quantity and in the manner proposed for it. On the other hand, risk denotes the probability (expected frequency) that a chemical will produce undesirable effect under specified conditions.

For workers in industrial environments, standards have been published by the American Conference of Governmental Industrial Hygienists (1974), and their recommendations are given in Table 17.1 for a selected list of pollutants. These data can be used in the design of industrial ventilation system.

Application of Threshold Dose

The concept of threshold dose is of basic importance because it implies that there is no observed adverse effect level (NOAEL). The NOAEL is important for setting exposure limits. Based on NOAEL, the acceptable daily intake (ADI) may be determined. This factor is used to determine the safe intake of food-additives and contaminants such as pesticides and residues of veterinary drugs, hence to establish safe levels in food. The ADI is determined by applying the suitable safety factor which is generally 100 :

$$ADI = \frac{NOAEL}{100}$$

The NOAEL is determined in mg/kg/day and can be used in setting of occupational exposure limits :

Table 17.1 : Threshold-Limit Values (TLVs) for Important Toxic Pollutants

Substance	*Cppm*	*Cmv mg/m³*
Acetaldehyde	100	180
Acetone	1,000	2,400
Acrolein	0.1	0.25
Ammonia	25	18
Arsine	0.05	0.2
Benzene	10	30
CS_2	20	60
CO	50	55
CCl_4	10	65
Cl_2	1	3
Chloroform	25	120
Diphenyl	0.2	1
Ethyl alcohol	1,000	1,900
Ethyl amine	10	18
Ethyl bromide	200	890
Fluorine	1	2
HCN	10	11
H_2S	10	15
Lead fumes and dusts	—	0.15
Naphthalene	10	50
Nitric oxide	25	30
Nitrobenzene	1	5
Phenol	5	19
Phosgene	0.05	0.25
SO_2	5	13

Cppm = Concentration in parts per million

Cmv = Particulate loading of the air — may be defined as total mass of suspended particulate material/ unit volume of mixture

mg/m³ = mg/cubic meter

Therapeutic Index (Margin of Safety)

The concept of the "therapeutic index" was introduced by Paul Ehrlich in 1913.

Therapeutic index indicates the closeness of a lethal-dose to the therapeutically effective dose or therapeutic index is the ratio of LD_{50} and ED_{50} :

$$\text{T.I. (Therapeutic Index)} = \frac{LD_{50}}{ED_{50}}$$

Therapeutic Index is of great importance to a physician and biochemist, as it clearly indicates the 'margin of safety' of a particular drug, by showing that how much extra concentration of a therapeutically effective dose will turn it into a lethal dose.

Thus, the margin of safety is the range of doses of a drug or chemical between a non-effective dose and a lethal dose.

The greater the value (T.I.), the higher the margin of safety.

For example, the therapeutic index of Penicillin is 100 :

$$\text{i.e., T.I.} = 100$$

$$\text{i.e., } \frac{LD_{50}}{ED_{50}} = 100$$

$$\text{i.e., } 100 \times ED_{50} = LD_{50}$$

We can conclude, therefore, that a hundred-fold increase in the ED_{50} dose will exert lethal effects. In simple form, this explains the fact that even large variations from the prescribed dose will not prove deleterious in the animals. In this way penicillin accounts for a very high margin of safety.

To sum up, the concept of therapeutic index is very useful in considering the margin of safety for the practical use of pharmaceuticals. Margin of safety is calculated by a safety factor : 100

i.e., Margin of Safety of any chemical

$$= \frac{\text{Threshold concentration}}{100}$$

Potency versus Efficacy

To compare the toxic effects of two or more chemicals, the dose response to the toxic effects of each chemical must be established. One can then compare the potency and maximal efficacy of the two chemicals to produce a toxic effect. These two important terms can be explained by reference to Fig. 17.5, which depicts dose-response curves to four different chemicals for the frequency of a particular toxic effect, such as the production of tumours. Chemical A is said to be more potent than chemical B because of their relative positions along the dosage axis. Potency thus refers to the range of doses over which a chemical produces increasing responses. Thus, A is more potent than B and C is more potent than D. Maximal efficacy reflects the limit of the dose-response relationship on the response axis to a certain chemical. Chemicals A and B have equal maximal efficacy, whereas the maximal efficacy of C is less than that of D.

B. Short-Answer Questions

1. *Briefly describe :*

   (i) Dose, (ii) Response, (iii) Types of human exposure, (iv) Biological effect monitoring.

2. *Write short notes on :*

   (i) NOAEL,

   (ii) Quantitative-dose-response-relationship

   (iii) Quantum-dose-response-relationship

   (iv) Chronic toxicity

   (v) Therapeutic index.

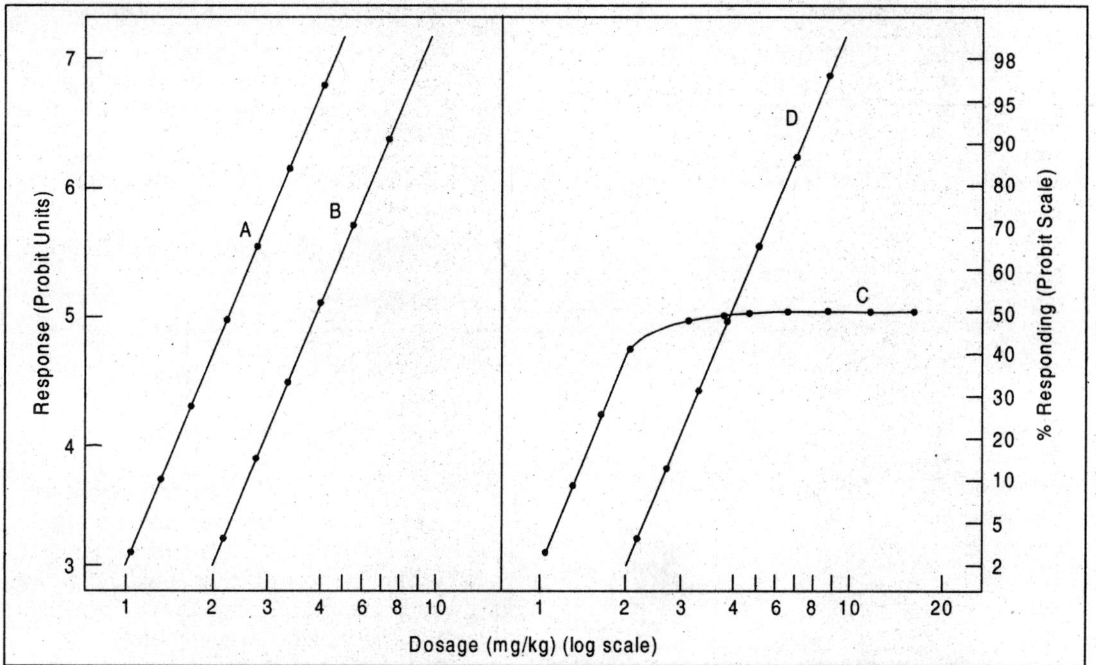

Fig. 17.5 : Schematic representation of the difference in the dose-response curves for four chemicals (A - D), illustrating the difference between potency and efficacy

Revision Questions

A. Long-Answer Questions

1. Illustrate dose-response relationship with suitable examples.
2. Describe types of dose-response relationships with suitable examples.
3. Describe the categories of toxic effects and their objectives.

3. *Briefly describe :*

   (i) Safety and Risk

   (ii) Cumulative responses

   (iii) Threshold limit

   (iv) Margin of safety

   (v) Potency versus efficacy.

compare the potency and maximal efficacy of the two chemicals to produce a toxic effect. These two important terms can be explained by reference to Fig. 17.5, which depicts dose-response curves to four different chemicals for the frequency of a particular toxic effect, such as the production of tumours. Chemical A is said to be more potent than chemical B because of their relative positions along the dosage axis. Potency thus refers to the range of doses over which a chemical produces increasing responses. Thus, A is more potent than B and C is more potent than D. Maximal efficacy reflects the limit of the dose-response relationship on the response axis to a certain chemical. Chemicals A and B have equal maximal efficacy, whereas the maximal efficacy of C is less than that of D.

B. Short-Answer Questions

1. *Briefly describe :*

(i) Dose, (ii) Response, (iii) Types of human exposure, (iv) Biological effect monitoring

2. *Write short notes on :*

(i) NOAEL

(ii) Quantitative-dose-response-relationship

(iii) Quantum-dose-response relationship

(iv) Chronic toxicity

(v) Therapeutic index

Fig. 17.5 : Schematic representation of the difference in the dose-response curves for four chemicals (A - D), illustrating the difference between potency and efficacy

Revision Questions

A. Long-Answer Questions

1. Illustrate dose-response relationship with suitable examples.
2. Describe types of dose-response relationships with suitable examples.
3. Describe the categories of toxic effects and their objectives.

3. *Briefly describe :*

(i) Safety and Risk

(ii) Cumulative responses

(iii) Threshold limit

(iv) Margin of safety

(v) Potency versus efficacy.

18

Mode of Action of Toxicants

SYNOPSIS

Introduction, Receptor concept, Mechanism of action of important toxicants : (1) Heavy Metals (2) Insecticides (organochlorines, organophosphates, carbamates and pyrethroids) (3) Carcinogens (4) Teratogens, and (S) Radioactive pollutants.

Introduction

Toxicants exert their toxic effects by bringing changes at one or more subcellular sites. The potential sites are the nuclei, mitochondria, lysosomes, endoplasmic reticulum, plasma membrane and other cell bodies. The mechanism may also be classified by the chemical nature of the molecular targets. These include proteins, coenzymes, lipids and nucleic acids (DNA and RNA) as described in proceeding section. Carbohydrates, on the other hand, are rarely affected by toxicants. In principal form, a toxicant may affect a specific target organ or tissue because of a lack of the required repair mechanism. For example, N-methyl-N-nitrosourea (MNU) produces tumours in the brain of the rat, occasionally in the kidney, but never in the liver. It is because the liver is fully capable of enzymatically excising the O^6 - alkylguanine induced by MNU from the DNA, whereas the brain is deficient in this respect. The capacity of the kidney falls between those of the liver and brain.

I. PROTEINS

(i) Structural Proteins

Structural proteins like plasma and organelle membranes are usually damaged by the toxicants. Notable examples are hexane and silica. Such damages result in impairment of the structural as well as other functions of the membranes. On the other hand, the structural protein collagen is rarely affected.

(ii) Enzymes

Enzymes are common targets of toxicants. Effects exerted by the enzymes may be specific, such as the case with a number of carbamate insecticides on cholinesterase. The effects may be nonspecific, for example, Hg and Pb are potent inhibitors of a great variety of enzymes. The effects may also be irreversible, for example, diisopropyl flurophosphate (DFP) brings irreversible enzyme inhibition by covalently binding with enzymes.

(iii) Carriers

Carriers, viz., hemoglobin, may be affected by a toxicant through preferential binding. For example, CO can bind hemoglobin at the site where O_2 is generally bound. Because of its higher affinity for hemoglobin, it can inactivate hemoglobin and cause manifestations of O_2 deficiency in tissues.

Certain toxicants, such as nitrites and aromatic amines, may enhance the formation of methemoglobin which is an oxidation product of hemoglobin with no oxygen binding ability.

(iv) Coenzymes

It is an established fact that coenzymes are essential for the normal functioning of enzymes. Their levels in the body may be diminished by the presence of toxicants that inhibit their synthesis. For example, Pyrithiamine can inhibit *thiamine kinase*, which is known to be responsible for the formation of the coenzyme thiamine pyrophosphate (TPP). Likewise, NDAPH can be destroyed in the presence of free radicals, which can be produced by such toxicants as CCl_4.

Metal-dependent enzymes can be inhibited by chelating agents, for example cyanides and dithiocarbamates, through removal of metal coenzymes such as copper and zinc.

(v) Nucleic Acids

Certain toxicants viz., alkylating agents, establish covalent binding with replicating DNA and RNA and consequently cause serious lesion like cancer, teratogenesis etc. Such toxicants may also exert immunosuppressive effects.

Certain antimetabolites such as aminopterin and methotrexate may be incorporated into DNA and RNA and then interfere with their replication.

II. LIPIDS

Peroxidation of polyenoic fatty acids has been suggested as a mechanism of the necrotic action of various toxicants, e.g., CCl_4. The ions of Hg and Cd may complex with phospholipid bases and expand the surface area of the membrane, thereby adversely altering its function. The general anesthetics ether and halothane and also many other lipophilic xenobiotics can accumulate in the plasma membranes and consequently interfere with the transport of O_2 and glucose into the cell. The cells of the CNS are especially susceptible to lowering of O_2 tension and glucose level and are, therefore, among the first to be deleteriously affected by such xenobiotic compounds.

Table 18.1 summarizes the various modes of action, representative xenobiotics, initiating factors and the early molecular target (Bridges et al, 1983). Though the action of a toxicant is highly especialised, its effect may be widespread and disseminated by a variety of secondary processes.

RECEPTOR CONCEPT

The term receptor was first and foremost introduced by Paul Ehrlich (1897). Later on the concept of a 'receptive substance' as the site of action

Table 18.1 : Types of Chemically Induced Toxicity

Reaction's nature	Xenobiotic	Initiating factor	Early molecular target
Enzyme Inhibition	parathion	paraxon	cholinesterase
	fluoroacetate	fluorocitrate	aconitase
	dinitrophenol	dinitrophenol	uncouples oxidative phosphorylation
	rotenone	rotenone	inhibits oxidative phosphorylation
Cofactor depletion	galactosamine	uridine diphosphate galactosamine	uridine tri-phosphate depletion
	ethionine	S-adenosy lethionine	adenine/ATP
	menaquinone	semiquinone	NAD(P)H depletion
Receptor interactions	TCDD	TCDD	binding to cytosolic receptor
Physicochemical changes			
Redox	aniline	phenylhydroxylamine	hemoglobin to methemoglobin
Redox	paracetamol	quinoenimine	oxidatative damage
Solubility	sulfapyridine	acetylsulfapyridine	precipitation in tubules
Solubility	harmol	harmoglucuronide	biliary cholestasis
Physical damage	silica	silica	lysosomal membranes
Solvation	hexane	hexane	membranes

of any Xenobiotic/drug was proposed by Jhon N. Langley in 1905. Langley reflected an idea after his studies on actions of atropine and pilocarpine on the secretion of saliva in the cat. He postulated that at the ending of nerve cells and gland cells there are certain substances which are capable of forming compounds with atropine and pilocarpine.

Definition

Langley defined receptors as chemical groups present on plasma membrane that give a biological response by combining with chemically complementary groups — either of drugs or of chemical messenger. The receptors possess a similar geometry as that of the drug/toxicant molecule and have the capacities to bind and interact with it. The drug receptors may be defined as macromolecules with which drugs interact to exert their characteristic effects. Likewise toxicant receptor may be defined as a macromolecule with which the toxicants interact to produce toxic effects.

The binding of the xenobiotic molecule to the surface of the receptors is facilitated by adsorption or other forces. The receptor-chemical-interaction initiates the characteristic dose-response for the specific chemicals (drug/toxicant).

Location

The receptors are generally localized in the plasma membranes or organelles. Sometimes, they are located in the cytosol of specific cells (e.g., reproductive cells of the ovaries possess steroid receptors). Insulin receptors are located in the membrane of fat cells whereas cholinergic receptors are located in the postsynaptic membranes of neuronal and neuromuscular junctions.

Generally the receptors are named after their drug/toxicants such as morphine receptor. Thus, in view of a large number of drugs, many receptors may be expected. But practically it is not true. It is because the drugs are designed for receptors and not the receptors for drugs.

Chemical Nature of Receptors

The chemical nature of receptor is very complex. Welsh and Taub (1951) firstly suggested that receptors are basically protein macromolecules having high molecular weights and characteristic geometrical surfaces. The response elicited by a drug is a function of the receptor's geometry, rather than any special property, like its solubility etc.

Heilbronn et al (1975) reported that cholinergic receptors (ChR), besides protein, also contain 5.4-6.0% carbohydrates, consisting of mannose, glucose and N-acetylglucosamine. Jacobs and Cuatrecasas (1976) observed that insulin receptors isolated from liver and fat cell membranes are glycoprotein.

β-adrenergic receptors isolated from turkey erythrocyte membrane are entirely composed of proteins.

Categories of Receptors

Although only a few types of receptors viz., cholinergic, insulin and β-adrenergic have been isolated, identified, and characterized. Many others have been identified and characterized in situ, using pharmacological, biochemical and immunological procedures. The results indicate that receptors have a variety of functions. On this ground, the receptors have been categorized into four main types :

1. Neurotransmitter receptors.
2. Hormone receptors.
3. Drug receptors.
4. Silent receptors.

'Neurotransmitter receptors' include the cholinergic (nicotine, located in the ganglia and skeletal muscles : muscarinic in smooth muscles and brain), α and β adrenergic, dopamine, opiates (e.g., endorphin), and histamine (H_1 and H_2) receptors.

'Hormone receptors' are for insulin, ACTH, thyrotropin, estrogen, progesterone, angiotensin, glucagon, prostaglandin and others.

'Drug receptors' include certain chemicals such as antidepressants and antitumor agents bind with specific macromolecules that may be considered as receptors. These receptors may well have endogenous messengers (such as endorphine and enkephalins for opiates) but are as yet undiscovered. Poland and Knutsen (1981) suggested that for certain receptor, a physiologic ligand existed at some point in evolution.

'Silent receptors' are the sites capable of binding with drugs in particular. Silent receptors are also called "inert receptors". The best example of silent-receptors are the plasma proteins which accommodate a host of receptor-sites, but are practically incapable of producing a response, even after binding with the drug molecule.

Mechanism of Action of Receptors

Basically the receptors function as the sites of biological system capable of recognizing substances of specific chemical nature, and, after binding with such substances, they initiate certain biological effects. The principal mechanisms of action are :

I. Involvement of Adenylate (or Guanylate) Cyclase

A common sequence involving this enzyme includes :

1. The substance (ligand) after binding with the receptor undergoes conformational changes.
2. The ligand-receptor complex causes the regulatory protein to bind with guanosine triphosphate (GTP).
3. This protein-GTP complex activates adenylate cyclase and induces a dissociation of the ligand-receptor complex, thus freeing the receptor for further action.
4. The activated adenylate cyclase (or guanylate cyclase) stimulates the conversion of ATP to cyclic adenosine monophosphate (cAMP) or to cyclic guanosine monophosphate (cGMP).
5. Cyclic AMP (or cGMP) acts as second messenger and amplifier by activating certain *protein kinases* that phosphorylate proteins, mainly enzymes.
6. The changed enzyme function triggers the biological effect.

II. Role of Calcium ion Channel

In certain cases, the binding of the ligand to the receptors result in the formation of Ca^{2+} channels, probably through the hydrolysis of phosphatidyl inositol catalyzed by phospholipase. The elevated intracellular Ca^{2+} concentration causes contraction, exocytosis, mitosis, altered membrane permeability and metabolism.

Factors Affecting the Functions of Receptors

Receptor-drug interaction initiates the characteristic dose response for the particular drug. The receptors occur in two forms, the activated and non-activated ones, in dynamic equilibrium. The agonists stabilize the receptors in the activated form, whereas the antagonists stabilize the receptors in the non-activated form (Ariens and De Miranda, 1979). Hence it may be concluded that agonism and antagonism affects the functioning of receptors.

Concept of Agonism and Antagonism

The process of receptor-activation is termed agonism and the substance which agonizes is called 'Agonist'.

When two different drugs are administered, they produce opposing or antagonistic actions on the physiological system. Such a process is termed antagonism and the causative drugs are called 'Antagonists'.

An antagonist blocks the action of an agonist, such as neurotransmitter or a hormone by preventing the binding of the agonist to the receptor.

There are accessory receptor sites for the competitive antagonists and an allosteric interaction exists between agonist and antagonist, each binding to its own receptor site on the receptor molecule. The binding of an agonist excludes the binding of an antagonist to the same receptor, and vice versa.

Role of Receptors in Toxicology

The receptor concept has significantly advanced our understanding pertaining to certain biochemical, physiological and pharmacological effects, as well as facilitating the development of drugs. Although it has hardly played a prominent role in toxicology, however, with the advent of molecular toxicology, the use of the receptor concept may undoubtedly aid in the understanding of the mechanism of action of toxicants. Main (1980) specifically noted that the emergence of method for locating and purifying receptors and the increasing knowledge of the role of receptors in the regulation of body functions are especially useful in this respect.

Poland and Kuntson (1982) reviewed extensive literature on TCDD (2, 3, 7, 8 - tetrachlorodibenzo-p-dioxin) and related halogenated aromatic hydrocarbons with respect to their toxicity, their ability to induce AHH (aryl hydrocarbon hydroxylase) and their affinity for a cytosol receptor. They suggested that the existence of the receptor could satisfactorily explain activity data based on dose-response curves for AHH induction, and receptor binding in vitro.

It is of interest to note that Loomis (1978) used the concept of the complex (chemical-receptor) to explain the existence of thresholds that "there must necessarily be a quantity of chemical agent below which no biological effect would be achieved". The existence of threshold doses for chemicals is important in evaluating their safety. Further it is of interest to mention that a toxicant may induce tolerance to its toxicity by reducing the number of receptors (Costa et al, 1981).

MECHANISM OF ACTION OF COMMONLY USED TOXICANTS

An understanding of the mechanisms of toxicity is of practical and theoretical importance. Such information provides rational basis for interpreting descriptive toxicity data, estimating the probability that a xenobiotic will cause harmful effects, designing drugs and industrial chemicals that are less hazardous, and developing pesticides that are more selectively toxic for their respective target organisms.

As a result of the huge number of potential toxicants and the multitude of biological structures and processes that can be impaired, there are a tremendous number of possible toxic effects. Correspondingly, there are various pathways that may lead to toxicity (Fig. 18.1). The most direct pathway occurs when the chemical causes toxicity by its mere presence at critical sites in the body without interacting with a target molecule (path A). This pathway is followed, for example, when agents precipitate in renal tubules and block urine formation. With this type of toxicity, delivery (step 1) is the most important consideration.

An example of a more complex route (path B) to toxicity is that taken by the fugu fish poison tetrodotoxin. After ingestion, this poison reaches the Na^+ channels of motoneurons (step 1). Interaction of tetrodotoxin with this target (step 2) results in blockade of Na^+ channels, inhibition of the activity of motor neurons (step 3), and ultimately skeletal muscle paralysis. No repair mechanisms can prevent the onset of such toxicity.

The most complex path to toxicity (path C) involves even more steps. First, the toxicant is delivered to its target or targets (step 1), after which the ultimate toxicant interacts with endogenous target molecules (step 2), triggering perturbations in cell function and/or structure (step 3), which initiate repair mechanisms at the molecular, cellular, and/or tissue levels (step 4). When the perturbations induced by the toxicant exceed the repair capacity or when repair becomes malfunctional, toxicity occurs. Tissue necrosis, cancer, and fibrosis are examples of chemically induced toxicities whose development follow this four-step course (Fig. 18.1).

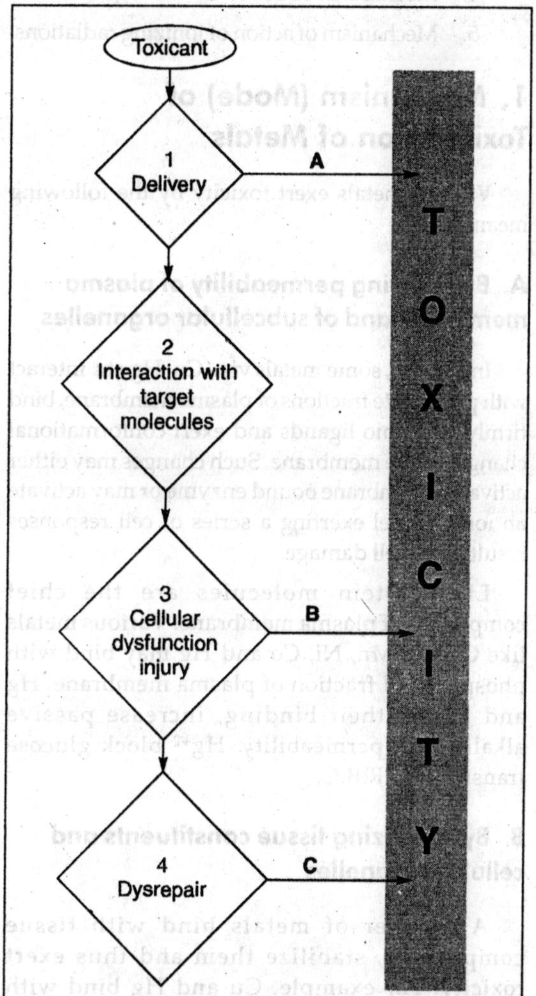

Fig. 18.1 : Generalized scheme showing potential stages in the development of toxicity after exposure to any toxicant

Present section deals with the mechanism (mode) of toxic action of the following commonly used toxicants.

1. Mechanism of action of some metals
2. Pesticides :
   (i) Mechanism of action of organochlorine insecticides.
   (ii) Mechanism of action of organophosphate insecticides.
   (iii) Mechanism of action of carbamates.
   (iv) Mechanism of action of pyrethroids.
3. Mechanism of action of carcinogens.
4. Mechanism of action of teratogen.
5. Mechanism of action of ionizing radiations.

1. Mechanism (Mode) of Toxic Action of Metals

Various metals exert toxicity by the following means :

A. By affecting permeability of plasma membrane and of subcellular organelles

In general, some metals viz., Cu, Hg etc. interact with phosphate fractions of plasma membrane, bind firmly with thio ligands and exert conformational changes in the membrane. Such changes may either activate a membrane bound enzyme or may activate an ion channel exerting a series of cell responses resulting in cell damage.

Lipo-protein molecules are the chief component of plasma membrane. Various metals like Cu, Al, Mn, Ni, Co and Hg may bind with phospholipid fraction of plasma membrane. Hg and Pb, by their binding, increase passive alkaline ion permeability. Hg^{+2} block glucose transfer into R.B.C.

B. By stabilizing tissue constituents and cellular organelles

A number of metals bind with tissue components, stabilize them and thus exert toxicity. For example, Cu and Hg bind with collagen protein; Co, Ni, Cu, and Zn form covalent linking with protein by binding with immidazole nitrogen of histidine and SH groups of cysteine. Hg and Ag bind with Hb irreversibly and form Hg-Hb, and Ag-Hb complexes, respectively, and then obstruct O_2 - carrying function of hemoglobin.

Hg also binds serum albumen and stabilizes it.

Metals show affinity with NH_2 group. These bind with NH_2 groups of protein causing irreversible change. The affinity of metals for NH_2 group is in the following order :

Hg > Cu > Ni > Pb > Zn > Co > Cd > Mn

C. By inactivating enzymes and cell constituents by binding with functional groups

Metal ions bind with functional groups or ligands present in biologically active molecules and form coordinate covalent complexes with nitrogen ligands or other groups such as - SH, -OH, $-NH_2$, -COOH, $-PO4^{-3}$ etc. and exert toxicity.

Examples

1. Lead binds with $-SH_2$, $-NH_2$ inside hemoglobin and causes lysis of porphyrin. It also binds with $PO4^{-3}$ ligands and produces deleterious effects.
2. Cd binds with -SH group of glutamic dehydrogenase and consequently causes irreversible inactivation of the enzyme and inhibits alkaline phosphatase and ATPase activity.
3. Arsenic binds with -SH group of hemoglobin and causes hemolysis.

In general, toxic metals and metallic compounds attack the active sites of enzymes, inhibiting essential enzyme function. Heavy metal ions in particular, e.g., Hg^{2+}, Pb^{2+} and Cd^{2+} act as effective enzyme inhibitors. They have affinity for sulphur containing ligands e.g., $-SCH_3$ and -SH in methionine and cystein amino acids, which are part of the enzyme structure.

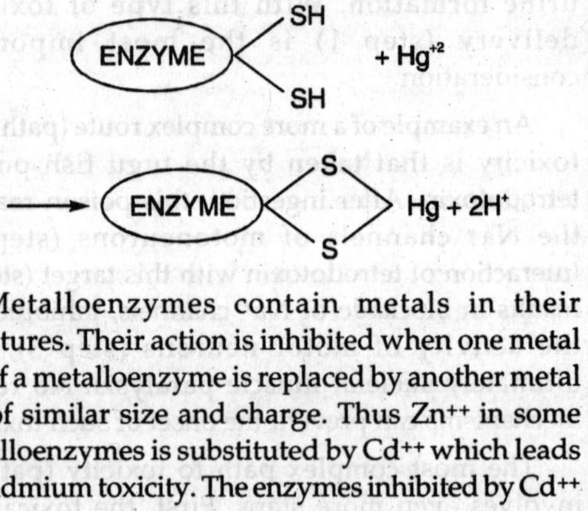

Metalloenzymes contain metals in their structures. Their action is inhibited when one metal ion of a metalloenzyme is replaced by another metal ion of similar size and charge. Thus Zn^{++} in some metalloenzymes is substituted by Cd^{++} which leads to Cadmium toxicity. The enzymes inhibited by Cd^{++}

include adenosine triphosphatase, alcohol dehydrogenase, amylase, carbonic anhydrase and glutamic - oxaloacetic transaminase. Pb^{+2} inhibits acetylcholinesterase, carbonic anhydrase, cytochrome oxidase, alkaline phosphatase, ATPase and some of the key enzymes in the synthesis of heme.

D. By causing irreversible damage to nucleic acids after binding with them

Many metals, for example, Cd, Pb and Al, stabilize RNA. Certain metals like Zn cleaves 5 phosphate bond in RNA, resulting in depolarization. Some metals, for example, Cu unwinds DNA and disrupts it. Few metals like Zn^{+2} and Co^{+2} compete with Mg^{+2} ions necessary for the incorporation of deoxyribonucleotide and ribonucleotides in DNA and RNA synthesis, respectively, causing damage to DNA replication and transcription. Cu and Cd compete with Mg^{+2} of RNA polymerase enzyme system causing damage to RNA synthesis.

2. Mechanism of Action of Pesticides

I. Mechanism of Toxic Action of Organochlorines

The chlorinated hydrocarbons are stimulants of the nervous system. Their mode of action is similar in insects and humans. They affect nerve fibres along the entire length by disturbing the transmission of the nerve impulse. More specifically, the members of this group of insecticides disrupt the Na/K balance that surrounds the nerve fibers. The result of this imbalance is a nerve that sends transmissions continuously rather than in response to stimuli.

Despite the similarity of many of the compounds within each of the three subgroups, the individual toxicities vary greatly. The compounds also vary greatly in their ability to be stored in tissue. For example, the structure of methoxychlor is very similar to DDT, but its toxicity is far lower, as its tendency to accumulate in fatty tissue. Storage in fatty tissue is actually a strategy that the body uses to remove toxic materials from active circulation. Fatty storage prevents the toxic agent from reaching in the circulation and consequently on various organs like liver, kidney etc.

DDT type insecticides interact with the neuronal membrane by altering the membrane permeability (transport) for potassium and sodium and the calcium mediated processes. By inhibiting these functions, the repolarization of the nerves is distributed resulting in hyperexcitability.

Cyclodines and cyclohexane compounds have a central nervous system stimulating mode of action. These compounds antagonize the neurotransmitter, ϒ-aminobutyric acid (GABA), permitting only partial repolarization of the neuron and thus uncoordinated nervous excitation.

Owing to their lipophilicity, organochlorine compounds are partitioned and stored largely in adipose tissue, where they are biologically inactive. There is an equilibrium between body fat and free circulating compounds. Redistribution and mobilization of fat, for example, due to disease, ageing or fasting may result in mobilization of stored organochlorines in quantities that may lead to manifest toxicity. Owing to the different patterns of use of organochlorine insecticides in industrialized and developing countries, the pattern of distribution of the residues in human fat (including breast milk) differs between countries.

Summary of mechanism of action of OCIs :

CHIs act as diffused CNS stimulants. They do not act upon any specific receptors. Their possible mechanism of action may be :

a. They change the electrical activity in the CNS by altering Na^+ and K^+ flux across the nerve cell membranes (O'Brien, 1967).

b. They increase the concentration of free ammonia and glutamine in the brain.

c. They inhibit Na^+, K^+ and Mg^{++} ATPase in the nerve endings of CNS (Matsumara and Patil, 1969).

d. Cyclodienes inhibit GABA receptor mediated Cl^- transport and clamodulin regulated Ca^{++} pump activity in the brain.

II. Mechanism of Toxic Action of OPIs (Organophosphate Insecticides)

The OPIs in general are less persistent than the OCIs, a property due to which they are preferred to the OCIs. In principle, OP compounds react with the active site of AChE (a serine hydroxyl group). In other words, all OP compounds are anticholinesterase agents, i.e., they inhibit the enzyme cholinesterase (ChE) at the neuromuscular junction.

During normal muscle contraction, the cholinergic nerve fibres of autonomic system liberate acetylcholine (ACh) at the neuromuscular junction. The ACh excites muscle fibres and, consequently, muscle contraction takes place. Very soon ACh is splitted by ChE present at the neuromuscular junction into choline and acetyl CoA.

Actually splitting of ACh by ChE prevents its accumulation at the neuromuscular junction and thus assists in muscle relaxation after each contraction :

Reaction

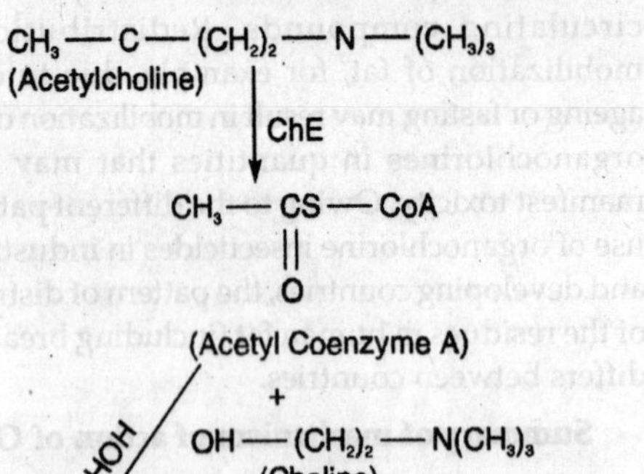

In its usual course of action, the ChE first gets acylated by losing one H^+ and then fastly recovers by using one molecule of metabolic HOH.

Actually the OP compounds are structurally similar to ACh. At the enzyme site, they compete with ACh and bind with ChE, thus inhibiting its normal function. The OP compounds phosphorylate the ChE and form a stable complex. Ultimately the enzyme fails to recover quickly. These events bring about an accumulation and consequently increase of ACh at the neuromuscular junction, which in turn, respectively, excites the muscle fibers and produces repetitive muscle contraction that causes tremors in the body muscles of insect and the insect finally dies of exhaustion.

III. Mechanism of Toxic Action of Carbamates

The primary way that insecticidal carbamates work on both target and non-target species is through the inhibition of the enzyme acetylcholinesterase. Acetylcholine (ACh) is a substance that transmits nerve impulse from a nerve cell to a specific receptor such as another nerve cell or a muscle cell. Acetylcholine, in essence, acts as a chemical switch. When it is present (produced by the nerve cell) it turns the nerve impulse on. When it is absent, the nerve impulse is discontinued. The transmission ends when the enzyme acetylcholinesterase breaks down the ACh into choline and acetic acid. Without the action of this enzyme, ACh builds up at the junction of the nerve cell and the receptor site, and the nerve impulse continues. Carbamate insecticides block (or inhibit) the ability of this enzyme, (acetylcholinesterase) to break down the ACh and end the nerve impulse.

Carbamate inhibition of acetylcholinestrase is reversible process. Estimate of the recovery time in humans range from immediate up to 4 days, depending on the dose, the specific pesticide, and the method of exposure. The breakdown of carbamate compounds within an organism is a complex process and is dependent on the specific pesticide structure.

In nutshell the mode of action of carbamates, like OPIs is the inhibition of acetylcholinesterase. Carbamates are potent reversible inhibitors of cholinesterase. In addition, all the carbamates inhibit aliesterase of insects, but they kill insects and mammals entirely by cholinesterase inhibition. The mechanism of inhibition of AChE and its reactivation is shown in Fig. 18.2 taking an example of carbaryl, a widely used carbamate (Fig. 18.2) :

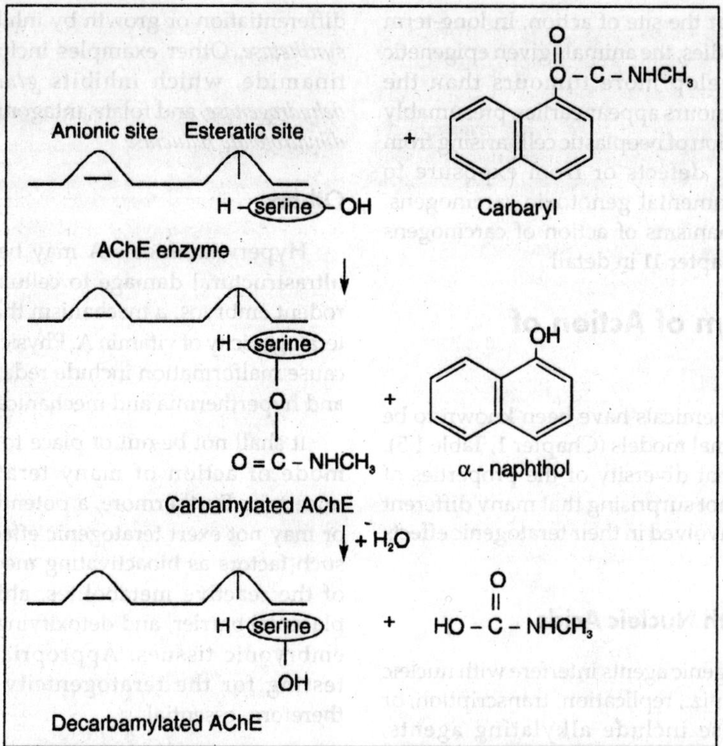

Fig. 18.2 : Inhibition of acetylcholinesterase enzyme by carbaryl

IV. Mechanism of Toxic Action of Pyrethroids

The Type I esters affect the sodium channels in nerve membranes with prolongation of sodium influx causing repetitive neuronal discharge and prolonged afterpotential but no severe membrane depolarization. The type II α cyanoesters lead to greater and more prolonged sodium influx with persistent membrane depolarization and eventually nerve blockade. Whereas the first group exerts its main effects on synaptic transmission causing hyperexcitabilty and tremor, the second group shows its first effects on the sensory nervous system. Other actions include inhibition of Na^+/Mg^{+2} ATPase (adenosine triphosphatase) and alteration of calcium and chloride ion homeostasis.

3. Mechanism of Action of Environmental Carcinogens

Chemical carcinogenesis is a multistage process. Carcinogenic chemicals act either by covalently binding with genetic macromolecules or by promoting the process. The former type of chemicals are known as genotoxic carcinogens because they, or their reactive metabolites, react with the genetic materials.

After the interaction between a genotoxic carcinogen and the DNA, the affected cell either dies or reverts to a normal cell through error-free DNA repair. If neither of these events occurs, the carcinogenic initiation becomes irreversible after the cell undergoes replication. Such an initiated cell may remain dormant for a long period of time before it becomes a tumour through cellular proliferation.

Initiated cells remain dormant probably because of the suppressant influence of the surrounding normal cells. This influence is apparently reduced under varying conditions. These include cell removal (e.g., partial hepatectomy), cell killing (resulting from cytotoxic chemicals, viruses, radiation), growth factors (e.g., hormones), and exogenous factors.

Epigenetic carcinogens increase the tumour yield by promoting the replication of cells initiated by genotoxic carcinogens, or by augmenting the available amount of genotoxic carcinogens and/or their metabolites at the site of action. In long-term carcinogenesis studies, the animals given epigenetic carcinogens develop more tumours than the controls, or the tumours appear earlier, presumably because of promotion of neoplastic cells arising from inherited genetic defects or from exposure to unknown environmental genotoxic carcinogens. The various mechanisms of action of carcinogens are outlined in Chapter 11 in detail.

4. Mechanism of Action of Teratogens

A variety of chemicals have been known to be teratogenic in animal models (Chapter 1, Table 1.5). In view of the great diversity of the properties of these agents, it is not surprising that many different mechanisms are involved in their teratogenic effects as given below :

Interference with Nucleic Acids

Various teratogenic agents interfere with nucleic acids functioning viz., replication, transcription, or translation. These include alkylating agents, antimetabolites, intercalating agents and amino acid antagonists. Some of these chemicals are active; others require bioactivation, such as aflatoxin and thalidomide. Although some chemicals, e.g., carbon tetrachloride and nitrosamines, also yield reactive metabolites, however, their reactive metabolites are too unstable to reach the embryo. Therefore, these toxicants are not potent teratogens.

Deficiency of Energy Supply and Osmolarity

Certain teratogens may affect the energy supply for the metabolism of the organism by restricting the availability of substrates either directly (e.g., dietary deficiencies) or through the presence of analogs or antagonists of vitamins, essential amino acids, and others. In addition, hypoxia and agents inducing hypoxia (CO, CO_2) can be teratogenic by depriving the metabolic process of the required oxygen and probably also by the production of osmolar imbalances. These can induce edema and hematomas, which, in turn, may cause mechanical distortion and tissue ischemia.

Inhibition of Enzymes

Inhibitors of enzymes, e.g., 5-fluorouracil, can induce malformation through interference with differentiation or growth by inhibiting *thymidylate synthetase*. Other examples include 6-aminonicotinamide, which inhibits *glucose-6 phosphate dehydrogenase*, and folate antagonists, which inhibit *dihydrofolate reductase*.

Others

Hypervitaminosis A may be associated with ultrastructural damage to cellular membranes in rodent embryos, a mechanism that can explain the teratogenicity of vitamin A. Physical agents that may cause malformation include redation hypothermia and hyperthermia and mechanical trauma.

It shall not be out of place to mention that the mode of action of many teratogens is as yet uncertain. Furthermore, a potential teratogen may or may not exert teratogenic effects depending on such factors as bioactivating mechanism, stability of the reactive metabolites, ability to cross the placental barrier, and detoxifying capability of the embryonic tissues. Appropriate experimental testing for the teratogenicity of toxicants is, therefore, essential.

5. Mechanism of Action of Ionizing and Non-ionizing Radiations

I. Ionizing Radiations

When ionizing radiations penetrate the living tissue, it wrecks havoc on the atoms and molecules in its path. Actually, when a water molecule in the cell is irradiated, an electron is knocked out of its orbit. The ejected electron may then attach to a normal water molecule creating instability. Such unstable water molecules (HOH) split into hydrogen ions (H^+), hydroxyl ions (OH^-) and the free radicals -H^* and OH^*. Radiation also produces a host of several species e.g. H_2, H_2O^-, H_2O^+, H_2^+, HO_2, H_3O^-, e^-, e^+ and H_2O_2. The free radicals are extremely reactive. They react with protein molecules in the cell, setting a chain of events that can destroy living cells or make them to function abnormally.

The ionic species deactivate the enzymes by dissociating the S-H-S hydrogen bond. Consequently, with enzyme inhibition, cell growth may continue but cell division and multiplication may be stopped. Proteins are mainly the body builders that play an important role in the formation of plasma membranes. Radiation exposure may damage the plasma membrane by making it permeable, while the large doses of ionizing radiations can kill quickly or inflict severe damage. Even the lower doses can initiate cancer throughout the body. Radiation also results in abnormal interchange of materials through an imperfect plasma membrane causing temporary or permanent injury in the body.

Studies of survivors who have received significant doses such as atom bomb survivors, uranium miners and radium watch dial painters etc. showed that damage depends on how the victims were exposed. Experiments revealed that human organs can repair some radiation damage. The sensitivity to radiation damage appears to be directly proportional to the cell's reproductive capacity and inversely proportional to its degree of differentiation.

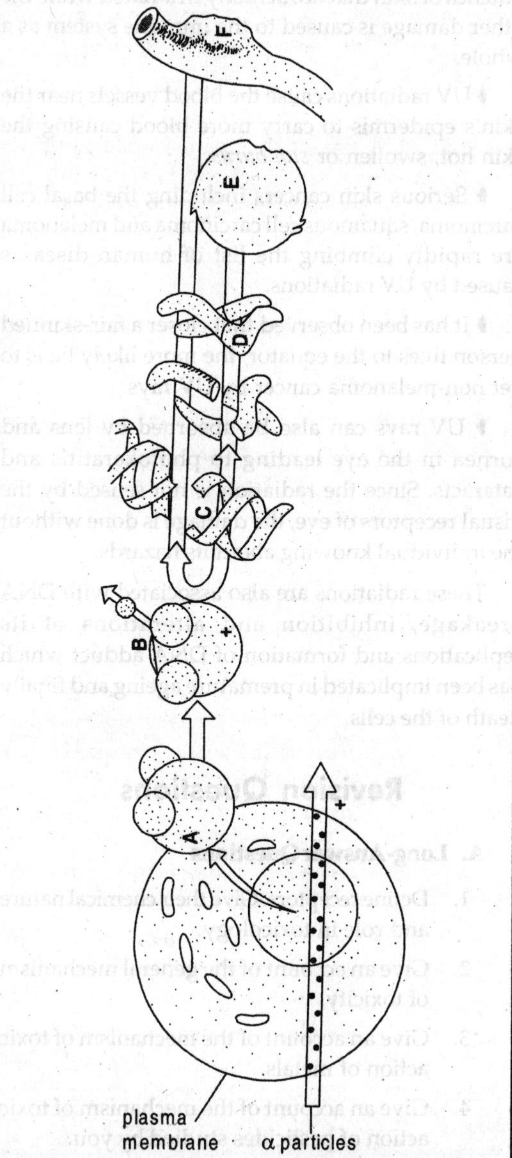

Fig. 18.3 : Tiny paths of destruction

Mechanism of Reaction of Ionizing Radiation

Tiny radiation particles viz., α-particles damage the cell as shown in Fig. 18.3 :

(i) When a beam of high energy positively charged alpha particles penetrates a living cell, it disrupts the atoms and molecules in its path. Water molecule (say A), for example, can be dissociated by the alpha particle's positive charge.

(ii) The alpha radiation's strong positive charge strips the water molecule of an electron (e^-) forming ions (B). The water molecule (H_2O) being neutral in character acquires a positive charge (H_2O^+) and destabilise its relationship with neighbouring molecules.

(iii) A deoxyribonucleic acid (DNA) molecule (C) can also be broken apart or altered by alpha particles or ions. Sometimes the DNA's genetic code is scrambled which then reproduces in later generations.

(iv) In this step damaged strands of DNA can cause chromosomes to break apart and then recombine in an abnormal manner (D). Unless the body's repair system can isolate and treat or overcome the damage, the cell may eventually die (E), often within few hours.

(v) Radiation may also bring chronic alterations in DNA molecules which, if not destroyed completely, may reproduce abnormally for several years, spawning cancerous cells and eventually resulting in acute tumours (F).

How do various particle radiations differ in their hazard?

Radioactive elements emit three types of radiations :

(i) Alpha particles (positively charged).

(ii) Beta particles (negatively charged), and

(iii) Gamma rays without any charge and identical to X-rays.

Actually alpha particles dissipate their energy quickly and hence have less penetrating power and pose minimum hazard. However, when alpha emitters are ingested, they can be very bad since for equal energy they can produce more 'ion-pairs' than either beta or gamma rays.

Beta (β) particles are generally more penetrating than alpha or X-rays. β-emitters external to the body are more dangerous than alpha emitters; since β-rays can penetrate from a few millimeters to a centimeter or so under the skin. When β-emitters are ingested they are more hazardous than they are externally but are comparatively less than the ingestion of alpha emitters because of the lesser specific ionization by beta particles.

Gamma rays have the highest penetrating power and so, like X-rays, are most dangerous. Gamma-rays may destroy tissues and inflict serious burns quite rapidly.

Both gamma and X-rays interact with living matter in three ways :

(i) By ion-pair production (i.e., high hγ on high atomic weight atoms to produce e^- and e^+).

(ii) Photoelectric way (i.e., ejection of e^- by low hγ on high atomic weight atoms)

(iii) Compton (i.e., moderate hγ on atom of any atomic weight to produce e^- and longer λ).

The charged particles so formed then go on to excite and ionize the biological medium. Since gamma-rays and X-rays can penetrate deeply in the body tissue, they constitute a hazard for the entire body. Incidentally, gamma-emission frequently accompanies alpha or beta emissions.

II. Mechanism of Non-Ionizing Radiations

Modern life and radiations seem to increase the risk of radiative pollution. Of all the nonionizing radiations including infrared, radiowaves, micro-waves, radar etc., the action of ultraviolet (UV) radiations has been extensively studied.

In body cells, the protein and nucleic acid are chiefly responsible for the absorption of radiation. In the region of 240 nm to 280 nm wavelength, the absorption by nucleic acid is 10 to 20 times greater than that by proteins of the same weight.

- Normally the pyrimidines (bases found in DNA) Thymine (T) and cytosine (C), undergo photochemical reactions much readily than the purines, adenine (A) and guanine (G), on absorption of UV radiations. During the reaction, all the DNA molecules get charged and acquire the excited state. At this stage, molecules are capable of undergoing further reactions leading to mutations.

- UV radiations are thought to trigger two distinct immunological effects. One is confined to patches of skin that are actually irradiated while the other damage is caused to the immune system as a whole.

- UV radiations cause the blood vessels near the skin's epidermis to carry more blood causing the skin hot, swollen or *sun burns.*

- Serious skin cancers including the basal cell carcinoma, squamous cell carcinoma and melanoma are rapidly climbing the list of human diseases caused by UV radiations.

- It has been observed that closer a fair-skinned person lives to the equator, the more likely he is to get non-melanoma cancer by UV rays.

- UV rays can also be absorbed by lens and cornea in the eye leading to photokeratitis and cataracts. Since the radiation is not sensed by the visual receptors of eye, the damage is done without the individual knowing about its hazards.

These radiations are also associated with DNA breakage, inhibition and alterations of its replications and formation of DNA adduct which has been implicated in premature ageing and finally death of the cells.

Revision Questions

A. Long-Answer Questions

1. Define receptors. Give their chemical nature and role in toxicology.
2. Give an account of the general mechanism of toxicity.
3. Give an account of the mechanism of toxic action of metals.
4. Give an account of the mechanism of toxic action of pesticides studied by you.
5. Illustrate the mechanism of action of environmental carcinogens.

6. Describe the mechanism of action of ionizing and non-ionizing radiations.
7. Illustrate the mechanism of action of teratogens.

B. Short-Answer Questions

1. Briefly describe :
   (i) Types of chemically-induced toxicity
   (ii) Receptors
   (iii) Chemical toxicity
   (iv) Mechanism of action of ionizing radiations
   (v) Mechanism of action of non-ionizing radiations.
2. Write short notes on :
   (i) Mode of action of organochlorines
   (ii) Mode of action of organophosphates
   (iii) Mode of action of carbamates
   (iv) Mode of action of pyrethroids.

19

Modifying Factors of Toxicity of Xenobiotic Chemicals

SYNOPSIS

Introduction, Modifying factors of toxic effect : 1. factors pertaining to xenobiotic chemicals, 2. to exposure, 3. to surrounding medium, and 4. to organisms.

Introduction

While toxicity is an inherent property of a xenobiotic chemical, the nature and extent of the toxic manifestations in an animal that is exposed to the xenobiotic chemical depend upon various factors. The obvious ones are the dose and duration of exposure.

They also include such less obvious host factors as the species, its sex and age and its nutritional and hormonal status. Various environmental factors also play significant role. In addition, the toxic effect of a xenobiotic may be influenced by simultaneous and consecutive exposure to other chemicals.

The toxic effects may be modified in a number of ways : alterations in the absorption, distribution, and excretion of xenobiotic; an increase or decrease of its biotransformation; and changes in the sensitivity of the receptor at the target organ.

For the sake of convenience, the present chapter is divided into four sections :

1. Factors pertaining to xenobiotic chemicals
2. Factors pertaining to exposure
3. Factors pertaining to surrounding medium, and
4. Factors pertaining to organisms.

1. Factors Pertaining to Xenobiotic Chemicals

(i) Physico-chemical properties : Actually the physico-chemical properties of a xenobiotic chemical such as solubility, vapour pressure, ionization, valency state, functional groups etc. affect the pharmacological as well as toxicological properties of any toxicant. For example, the toxicity of arsenic in trivalent (+3) state is greater than pentavalent (+5) state. This is because the former i.e., +3 compounds, bind strongly with the sulph-hydryl (-SH-) groups of proteins resulting in enzyme inhibition, whereas the latter, i.e., +5 compounds, do neither bind to sulph-hydryl groups nor inhibit enzyme systems. However, +5 state compounds (arsenites) inhibit ATP synthesis by oxidative uncoupling of certain reactions. Also the +3 state compounds (arsenates) are easily soluble in water and, hence, are very toxic to animals.

The functional group present in xenobiotic chemical compound appears to be responsible to exert the degree of toxic action. The polar (hydrophilic) chemicals are not easily soluble in lipids. Therefore, these are incapable of crossing the membranous barriers and ultimately cannot easily reach the target sites for appropriate action. However, non-polar or lipophilic substances are easily soluble in lipid and other organic solvents. These can readily penetrate the lipoprotein layers of the membranes, hence, these readily exert their potential toxic effects. It may thus be inferred that the toxicity of a xenobiotic compound is largely dependent on the chemical structure, valency state, type of functional groups, ionic characteristics and lipid solubility.

(ii) **Dose (concentration) of xenobiotic chemical** : Most commonly, the term dose is used to specify the amount of chemical/drug administered, usually expressed per unit body weight. In principle form, the toxic effects exerted by the xenobiotic chemical is directly related to its concentration at the target site. The lower doses of chemical cause less effects whereas high doses may cause pronounced effects. It may, therefore, be inferred that toxicity of xenobiotic chemical is directly proportional to its dose.

(iii) **Translocation of xenobiotic chemical** : Effective translocation is actually a key factory to exert toxic effects, especially for those toxicants which produce systemic effects. During the course of translocation, some of the xenobiotic chemicals interact with macromolecules and may be stored in certain inactive tissues i.e., *storage depots*. The inefficient translocation and storage of toxicants in an inactive form actually diminish their toxic potential, while their effective translocation to the target tissues/organs in an active form enhances their toxicity.

(iv) **Biotransformation and bioactivation of toxicants** : Biochemical modifications of the xenobiotic chemical molecules within the living cells may be termed as biotransformation. It actually reduces the toxic potential of the toxicants [For detail refer Chapter 21]. In biotransformation, the parent toxic compound, after converting into the metabolites, form conjugate and are finally excreted from the body. This ultimately results in the decrease of toxic action of the toxicants. In some instances the metabolites prove more toxic than the parent toxic compound, consequently causing more toxicity and this process is referred to as bioactivation (See Chapter 21).

If the biotransformation leads to conversion of lipophilic forms to hydrophilic forms, then the exertion of xenobiotic is enhanced, consequently toxicant may not proceed at the target sites to exert adverse or deleterious effects. Such processes ultimately reduce the toxicity of xenobiotics.

(v) **Chemical interaction** : The toxicity of xenobiotic in an animal may be increased or decreased by a simultaneous or consecutive exposure to another xenobiotic. If the combined effect is equal to the sum of the effect of each substance given alone, the interaction is considered to be additive, e.g., combinations of most organophosphorus insecticides on cholinesterase activity. If the combined effect is greater than the sum, the interaction is considered to be synergistic and the phenomenon is denoted by synergism. In simple form, synergism is a phenomenon in which the combined effect of two compounds is more than the sum of the individual effects. For example, carbon tetrachloride and ethanol on the liver. The term potentiation is used to describe the situation in which the toxicity of a substance on an organ is markedly increased by another substance that alone has no toxic effect on that organ. For example, Isopropanol has no effect on the liver, but it can increase considerably the hepatoxicity of carbon tetrachloride.

The exposure of an organism to a xenobiotic may reduce the toxicity of another. Chemical antagonism denotes the situation wherein a chemical reaction takes place and produces a less toxic product, e.g., chelation of heavy metals by dimercaprol. Functional antagonism exists when two chemicals produce opposite effects on the same physiologic parameters, e.g., the counteraction between CNS stimulants and depressants. Competitive antagonism exists when the agonist and antagonist act on the same receptor, e.g., the blockade of the effects of nicotine on ganglia by ganglionic blocking agents. Noncompetitive antagonism exists when the toxic effect of a chemical is blocked by another with a non-specific action, namely, not occupying the same receptor, e.g., the reduction of the toxicity of cholinesterase inhibitors by atropine.

MECHANISM OF CHEMICAL INTERACTION

Chemical interactions are achieved through a variety of mechanisms. For example, nitrites and certain amines can react in the stomach to form nitrosamines, the majority of which are potent carcinogens, and thus greatly increase the toxicity. On the other hand, the action of many antidotes (an agent that neutralizes the poison or counteracts its effect) is based on their interaction with the toxicants; for example, thiosulphate is used in cases of cyanide poisoning. Furthermore, a chemical may displace another from its binding site on plasma protein and thereby increases its effective concentration. A chemical may modify the renal excretion of weak acids and weak bases by altering the pH of urine. Competition for the same renal transport system by one chemical can hinder the excretion of another.

An important type of interaction involves the binding of chemicals with their specific receptors. An antagonist (a chemical which opposes or resists the action of other) blocks the action of an agonist (a chemical that initiates a pharmacological response after binding with a receptor) such as a neurotransmitter or a hormone, by preventing the binding of the agonist to the receptor.

The concepts that have been proposed to explain their effects include :

1. There are accessory receptor sites for the competitive antagonists, and an allosteric interaction exists between agonist and antagonist, each binding to its own receptor site on the receptor molecule. The binding of an agonist excludes the binding of an antagonist to the same receptor, and vice versa.
2. The receptor occurs in two forms, the activated and non-activated, which are in dynamic equilibrium. The agonists stabilize the receptor in the activated form, whereas the antagonists stabilize the receptors in non-activated form.

Another important type of interaction results from alterations of the biotransformation of a chemical by another. Some chemicals are inducers of microsomal and nonmicrosomal enzymes. The common inducers include phenobarbital, 3-methylcholanthrene (3-MC), PCB, DDT, etc. The inducers may lower the toxicity of other chemicals by accelerating their detoxication. For example, pretreatment with phenobarbital shortens the sleeping time induced by hexa-barbital and the paralysis induced by zoxazolamine. Such pretreatment also reduces the plasma level of aflatoxins (Wong et al, 1981). In addition, 3-MC pretreatment greatly reduces the liver necrosis produced by bromobenzene, probably by increasing the activity of the epoxide hydrase.

2. Factors Pertaining to Exposure

Toxicity of xenobiotics is greatly affected by various factors pertaining to their exposures. The contact reaction between the organism and the xenobiotic is called exposure.

Salient factors associated to exposures are :

(i) Exposure routes
(ii) Exposure duration, and
(iii) Exposure systems.

(i) Exposure routes

The route of exposure may affect kinetic factors such as absorption, distribution, biotransformation, and excretion; and may ultimately determine the toxicity of a xenobiotic. The xenobiotics may enter the body by dermal, oral, inhalation or by intraperitoneal, intramuscular, subcutaneous and intravenous injections. Actually the route of exposure determines the extent of toxicity of xenobiotics. For example, any xenobiotic produces more rapid and maximum effect when incorporated by intravenous route. It is because through this route the xenobiotic directly reaches to the target sites in active form and thus exerts greatest effects. The rate of penetration of xenobiotics through injection follows in given order :

intravenous injection > intramuscular injection > intracutaneous injection

Overall the exposure routes in descending order (except intravenous which stands first) of toxicity are :

Inhalation > Intraperitoneal > Subcutaneous > Intramuscular > Oral > Topical.

Further, the rate of penetration of xenobiotics may be summarised as :

1. **State of Xenobiotic :**
   Gaseous > Liquid > Solid
2. **Xenobiotic Particle Size :**
   0.2 nm or less > 0.3 nm > more than 0.3 nm
3. **Polarity :**
   Polar > Apolar
4. **Solubility :**
   Hydrophilic > Lipophilic
5. **Surface Area and Vascularity :**
   . Large surface area and vascularity > Less surface area and vascularity
6. **Carrier Solvent :**
   With carrier solvent > without carrier solvent

(ii) Exposure duration

The frequency of exposure may also affect toxicity. Toxic effects can be produced by acute (short-term) or chronic (long-term) exposure to xenobiotics. There are also exposures of intermediate duration, known as subchronic exposures. The acute toxic values in terms of LC_{50} or LD_{50} of xenobiotics decrease with increase in duration of exposure. In other words, the exposure for short-term has less effect in comparison to long-terms exposures. It very simply denotes that the toxicity of xenobiotics increases with increase in exposure duration. Threshold concentration of a xenobiotic is important in toxicity. Xenobiotics elicit an adverse effect only if the organism is exposed to a high enough concentration. At a low concentration, a minimal effect or no effect may result from exposure.

(iii) Exposure Systems

Types of exposure systems affect the toxicity of xenobiotics to certain extent. A variety of exposure systems may be selected for the exposure of toxicants. For instance, in aquatic medium, various exposure systems are used for xenobiotics, such as :

(a) *Renewal System :* Where the toxicant solution is renewed after certain interval.

(b) *Static System :* Where toxicant is mixed in the water and the organisms are exposed to the toxicant in still water.

(c) *Recirculatory System :* Where the toxicant solution is recirculated through certain pumps.

(d) *Flow through System :* Where the toxicant solution flows into and out of test chamber intermittently or continuously.

3. Factors Pertaining to Surrounding Medium (Environmental Factors)

Any factor of the external environment that is involved in the bioavailability of the xenobiotic affects its toxicity in the organism. For instance, DO, pH temperature salinity hardness of water, and dissolved solids in water would affect toxicity of a xenobiotic.

Salient environmental factors that affect toxicity are :

(i) Temperature

Temperature of the environment can alter the toxic effect. For example, the toxicity of strychnine, nicotine, atropine, malathion, and sarin is increased in animals exposed to cold temperature. However, the toxicity of parathion and another organophosphorous insecticides is reduced by hypothermia. Further, colchicine and digitalis are more toxic to the rat than to the frog. But their toxicity to the frog can be increased by raising the environmental temperature. The duration of the response, however, is shorter when the temperature is higher. The effect of environmental temperature on the magnitude and duration of the response is apparently related to the temperature-dependent biochemical reactions responsible for the effect and for the biotransformation of the xenobiotics. In aquatic medium, the temperature is expected to affect greatly the toxicity of xenobiotics. The increased water temperature increases the solubility of many xenobiotic chemicals. Several pesticides have been observed to be more toxic at increased temperatures while some others present stronger lethal action at low temperatures.

Sanders and Cope (1966) reported that DDT was three times more toxic against cladocerans at 10°C than at 27°C.

(ii) Pressure

Interest in the effect of barometric pressure on the toxicity of chemicals stems from human exposure to them in space and is saturation diving vehicles. At high altitudes, the toxicity of digitalis and strychnine decreases whereas that of amphetamine is increased. The influence of changes in barometric pressure on the toxicity of chemicals seems attributable mainly, if not entirely, to the altered oxygen tension rather than to a direct pressure effect.

(iii) Irradiation

Whole body irradiation increases the toxicity of CNS stimulants but decreases that of CNS depressants. However, it has no effect on analgesics such as morphine.

The effects of toxicants often show a diurnal pattern that is mainly related to the light cycle. In the rat and the mouse, the activities of cytochrome P-450 are the greatest at the beginning of the dark phase.

(iv) Caging

Caging can affect the LD_{50} of a chemical in several ways. For example, the LD_{50} of isoproterenol was less than 50 mg/kg in rats caged individually, whereas it was about 800 mg/kg in rats caged in groups of ten (Balazs, 1972). However, the LD_{50} values of most chemicals are only slightly affected, if at all, by this factor. The type of cage (mesh versus solid) and the type of litter material can also affect the reaction of the animals of toxicants.

(v) Humidity

A higher relative humidity may increase the acute toxicity, resulting in a lower LD_{50} dose.

(vi) Oxygen

Oxygen is essentially required for respiration to all biota. Increase in temperature, as an example, decreases the dissolved oxygen content of the water. Reduction in DO content of water imposes stress on the aquatic organisms, which may greatly enhance the toxicity of a xenobiotic.

(vii) pH

Toxicity of certain xenobiotics is greatly affected by pH of the water. For example, cyanide toxicity. Appreciable proportion of ionic cyanide i.e., CN^-, occurs at pH 8.5 whereas the molecular form i.e., HCN, predominates at 5.0 - 60 pH. The undissociated form of cyanide i.e., HCN is twice as toxic as the ionic form. This clearly indicates that changes in pH of water may greatly affect the toxicity of cyanide in particular and certain other xenobiotics viz., NH_3 etc. in general.

(viii) Salinity

In general, salinity does not significantly affect the toxicity of xenobiotics. However, a particular species may be more affected with change in water salinity. For example, toxicity of Zn and NH_4Cl to rainbow trout and Atlantic salmon decreases with an increase in salinity (Herbert et al, 1965).

(ix) Water hardness

. The total hardness of water has little effect on the toxicity of most of the xenobiotics, however, metal toxicity is greatly affected. Many heavy metals are more toxic to fishes and other aquatic organisms in soft water in contrast to the hard water.

4. Factors Pertaining to Organisms

Actually the factors associated to the organisms are of great significance in the study of toxicity of xenobiotics.

The salient toxicity modifying factors pertaining to organisms are :

(i) Test Species

Differences of toxic effect from one species to another have long been recognized. Knowledge in this field has been used to develop, e.g., pesticides, which are more toxic to pests than to humans and other mammals. Among various species of mammals, most effects of toxicants are similar. This fact forms the basis of predicting the toxicity to humans from results obtained in toxicologic studies conducted in other mammals, such as the rat, mouse, dog, rabbit and monkey. There are, however, notable differences in detoxification mechanisms (William, 1974). Even in different individuals of same species, the toxicity of chemical varies because of variation in susceptibility owing to certain genetic factors. It can, therefore, be concluded that certain individuals of a species may be susceptible to a chemical whereas the others may be resistant to the same chemical.

As an example, among various fish species, the toxicity of a xenobiotic may vary. Spear and Pierre (1979) observed that salmonids and minnows are approximately 15 times more susceptible to copper than that of sunfish. Pickering et al (1962) found 6 to 90-fold variations in the sensitivity to organophosphorus pesticides between sensitive (bluegills and guppies) and tolerant species (fathead minnows and gold fish).

(ii) Sex, Hormonal Status and Pregnancy

The toxicity of xenobiotics differ with respect to sexes because the male and females differ in their responses due to hormonal and metabolic differences. Actually male and female individuals of the same strain and species usually react to toxicants similarly. There are, however, notable quantitative differences in their susceptibility, especially in the rat. For example many barbiturates induce more prolonged sleep in female rats than in males. The shorter duration of action of hexobarbital in male rats is related to the higher activity of the liver microsomal enzymes to hydroxylate this chemical. This higher activity may be reduced by castration or pretreatment with estrogen. Similarly, in male rats, demethylate aminopyrine and acetylate sulfanilamide metabolize faster than females, and the males are thus less susceptible.

Female rats are also more susceptible than the males to such organophosphorus insecticides as azinphosmethyl and parathion. Castration and hormone treatments may reverse this difference. However, unlike hexobarbital, parathion is metabolized more rapidly in the female rat than in the male. This faster metabolism of parathion results in a higher concentration of its metabolite, paraoxon, which is more toxic than the parent compound. This higher toxicity resulting form greater bioactivation in female rats, compared to males, is also true with aldrin and heptachlor, which undergo epoxidation. The female rat is also more susceptible to warfarin and strychnine. On the other hand, male rats are more susceptible than females to ergot and lead.

Differences in susceptibility between the sexes are also seen with other xenobiotics. For example, chloroform is acutely nephrotoxic in the male mouse but not in the females. Castration or the administration of estrogens reduces this effect in the males, and treatment with androgens enhances susceptibility to chloroform in the females. The greater susceptibility of male mice was explained on the basis of a much higher concentration of cytochrome P-450 (Smith et al, 1983). Nicotine is also more toxic to the male mouse, and digoxin is more toxic to the male dog. However, the female cat is more susceptible to dinitrophenol and the female rabbit is more to benzene.

Imbalances of non-sex hormones may also alter the susceptibility of animals to toxicants. Hyperthyroidism, hyperinsulinism, adrenalectomy, and stimulation of the pituitary-adrenal axis have all been shown to be capable of modifying the effects of certain toxicants (Dauterman, 1980).

There is some evidence that the pregnant rat (Ivankovic, 1969) is more susceptible to the carcinogenic activity of ethylnitrosourea. Highly malignant tumours, apparently of trophoblastic origin, developed in these animals and were rapidly fatal.

(iii) Age

It has been considered that neonates and very young animals, in general, are more susceptible to toxicants such as morphine. For a great majority of toxicants, the young are 1.5 -10 times more susceptible than adults (Goldenthal, 1971).

The available information reveals that the greater susceptibility of the juvenile or young animals to many toxicants may be attributed to deficiencies of various detoxication enzyme systems (Dauterman, 1980). Both phase I and phase II reactions of biotransformation may be responsible. For example, hexobarbital at a dose of 10 mg/kg induced a sleeping time of longer than 360 min in 1-day-old mice compared to 27 min in the 21-day-old. The proportion of hexobarbital metabolized by oxidation in 3 hours in these animals was 0% and 21-33%, respectively (Jondorf et al, 1959). On the other hand, chloramphenicol is excreted mainly as a glucuronide conjugate. When a dose of 50 mg/kg was given to 1-or 2-day-old infants, the blood levels were 15 μg/ml or higher over a period of 48 hours. In contrast, children aged 1-11 years maintained such blood levels for only 12 hours (Weiss et al, 1966).

However, not all xenobiotics are more toxic to the young. Certain chemicals, notably CNS stimulants, are much less toxic to neonates. Lu et al (1965) reported that the LD_{50} of DDT was more than 20 times greater in newborn rats than in adults, in sharp contrast to the effect of age on malathion (Table 19.1). This insensitivity to the toxicity of DDT may be significant in assessing the potential risk of this pesticide, because of one very much larger intake in young babies via breast feeding and cow's milk, especially on the unit body weight basis.

The effect of age on the susceptibility to other CNS stimulants, including other organochlorine insecticides, appears less marked (generally in the range of 2-10 times). Most organophosphorus insecticides are more toxic to the young; Schradan (octamethyl pyrophosphoramide) and phenylthiourea are notable exceptions (Brodeur and DuBois, 1963). Pandey and Shukla (1982) reported that fries and fingerlings of fishes are most sensitive stages for toxicants.

Table 19.1 : Effect of Age on Acute Toxicity of Malathion, DDT, and Dieldrin in Rats

Pesticide	Age	LD_{50} (mg/kg) with 95% confidence limits
Malathion	Newborn	135.4 (94.0 - 190.8)
	Pre-weaning	925.5 (679.0 - 1261.0)
	Adult	3697.0 (3179.0 - 4251.0)
DDT	Newborn	> 4000.0
	Pre-weaning	437.8 (346.3 - 535.9)
	Adult	194.5 (158.7 - 237.3)
Dieldrin	Newborn	167.8 (140.8 - 200.0)
	Pre-weaning	24.9 (19.7 - 31.5)
	Adult	37.0 (27.4 - 50.1)

Old animals as well as humans are also more susceptible to certain xenobiotics. This problem has not been studied as extensively as in the young. However, the available evidence reveals that the aged patients are generally more sensitive to many drugs. The possible mechanisms include reduced biotransformations and an impaired renal excretion. In addition, the distribution of xenobiotics in the body may also be altered because of increased body fat and reduced body water (Jarvik et al, 1981).

(iv) Nutritional Status

The toxicity of chemicals to organisms is affected by health and nutritional status. The biotransformation of toxicants is catalyzed by the microsomal mixed-function oxidase system (MMFO). A deficiency of essential fatty acids generally depresses MMFO activities. This is also true with protein deficiency. The decreased MMFO has different effects on the toxicity of xenobiotic chemicals. For example, hexobarbital and aminopyrine are detoxicated by these enzymes and are thus more toxic to male rats and mice with these nutrient deficiencies. On the other hand, the toxicities of alfatoxin, carbon tetrachloride, and heptachlor are lower in such animals because of depressed bioactivation of these toxicants. Rats fed low-protein diets were 2-26 times more sensitive to a variety of pesticides (Boyd, 1972). MMFO activities are decreased in animals fed with high levels of sugar.

Many carcinogenesis studies have revealed that restriction of food intake decreases tumour yield. Deficiency of protein generally lowers tumourigenicity of carcinogens, such as aflatoxin B1 and dimethylnitrosamine. The importance of diet on carcinogenesis is further demonstrated by the fact that rats and mice fed diets rich in fats have higher tumour incidences compared to those that are given a restricted diet.

Vitamin A deficiency depressed the MMFO. In general, this is also true with deficiencies of ascorbic acid i.e., vitamin C and also E. But thiamine deficiency has the opposite effect. Vitamin A deficiency, in addition, boosts the susceptibility of the respiratory tract to carcinogens (Nettesheim et al, 1979).

Some foods contain appreciable amounts of xenobiotic chemicals that are strong inducers of the MMFO, e.g., safrole, flavones, xanthines, and indoles. In addition, potent inducers such as DDT and polychlorinated biphenyls (PCBs) are present as contaminants in various foods.

(v) Diseases

The liver is the principal organ wherein biotransformation of xenobiotic chemicals takes place. Such diseases as acute and chronic hepatitis, cirrhosis of the liver, and hepatic necrosis often decrease the biotransformation of xenobiotics. The microsomal and nonmicrosomal enzyme systems as well as the phase II reactions may be involved.

Renal diseases may also affect the toxic manifestations of xenobiotics. This effect stems from disturbances of the excretory and metabolic functions of the kidney. Heart diseases, when severe, can increase the toxicity of chemicals by impairing the hepatic and renal circulation, thus affecting the metabolic and excretory functions of these organs.

(vi) Size

The toxicity of xenobiotics is also affected by the size of the animal. Often larger size animals are more resistant to toxicants and this has been found true in case of certain tropical fresh water fishes (Pandey and Shukla, 1982, Shukla and Pandey, 1984).

(vii) Acclimation

The animals acclimated to sublethal levels of toxicants may become more tolerant or more weakened, depending upon the mode of action of toxicants and potential of the biotransformational mechanisms of the animals.

Revision Questions

A. Long-Answer Questions

1. Categorise the factors affecting toxicity of xenobiotic chemicals and discuss the mechanism of chemical interactions.
2. Give an account of the factors pertaining to exposure that modify the toxicity of xenobiotics.
3. Describe various environmental factors that affect the toxicity of xenobiotics.
4. Give an account of the factors pertaining to organisms that modify the toxicity of xenobiotics.

B. Short-Answer Questions

1. Describe in brief in light of affecting the toxicity of xenobiotics :
   (i) Physico-chemical properties of xenobiotics
   (ii) Dose of toxicant
   (iii) Translocation of toxicant
   (iv) Biotransformation
   (v) Bioactivation.
2. Write short notes on :
   (i) Chemical interaction
   (ii) Environmental factors affecting toxicity
   (iii) Nutritional status affecting toxicity
   (iv) Age factors affecting toxicity.

20

Selective Toxicity

SYNOPSIS

Introduction, Definition, Need of Selective toxicity, Convincing demonstration of selective toxicity in man, Scientific basis of selective toxicity (Selective through accumulation, Selectivity through comparative biochemistry, Selectivity through comparative cytology), Mechanism of selective toxicity, Advantages of selective toxicity, Conclusion.

Introduction

Albert (1965) coined the term selective toxicity. Selective toxicity means that a chemical produces injury to one kind of living matter without harming another form of life even though the two may exist in intimate contact. The living matter that is injured is termed the uneconomic form (or undesirable), and the matter protected is called the *economic* form (or desirable). They may be related to each other as parasite and host or may be two tissues in one organism.

In agriculture, for example, there are fungi, insects, and even competitive plant life that injure the crop, and thus selective pesticides are needed. Similarly, animal husbandry and human medicine require agent such as antibiotics, that are selectively toxic to the undesirable form but do not produce damage to the desirable form.

Definition

From the above discussion, it becomes evident that selective toxicity is the injury of one species of living matter without harming another species with which the first is in intimate contact. The species which is to be eliminated is the uneconomic species and the species which is to be preserved is the economic species. Often these are related to one another as parasite and host, but in other cases they are symbiotic.

Need of Selective Toxicity

The need of selective toxicity is multi-disciplinary in life sciences. In agriculture, the need for selective toxicity arose in ancient period. Man learnt to grow crops of plants in order to obtain food and clothing. The seeds of these plants themselves became infected with all sorts of fungi, weeds and insect pests which threatened actually the economic success of the crops. Obviously, it is an achievement of the prime importance to have those selectively toxic agents which are applied in agriculture to remove many of the uneconomic species without deleterious effect or injury to the economic ones.

In animal husbandry a similar situation is clearly recognizable. Man's domestic as well as economic animals are afflicted by both ectoparasites as well as endoparasites. Application of potent selective toxic agent to get rid of the parasites clearly gives an explanation of selective toxicity needed for these animals.

Man himself is no less subject to various parasites than his domestic and economical animals. Chemotherapy is the name assigned to that branch of selective toxicity which is concerned with the removal of parasites from man and his tended animals. So far more has been investigated of the scientific principles underlying the practice of chemotherapy than of any other branch of selective toxicity.

Convincing Demonstration of Selective Toxicity in Man

General anesthetics admirably illustrate the selective use of toxicity. The more toxic the anesthetic, the more valuable it is but only if the toxicity is selective for the central nervous system and completely reversible with time. Morton's success with ether as a general anesthetics in 1846 was an early and convincing demonstration of selective toxicity. The accepted general anesthetics combine a high toxicity for the central nervous system with negligible toxicity to other tissues, all toxicity rapidly and completely disappears when administration is halted. The same mechanism is followed with local anaesthetics, muscle relaxants, and (but less rapidly the antagonists of histamine and neurotransmitters). Antiparasitic agents, on the other hand, although they must be selective against the parasite and sparing to the host, are preferably irreversible.

Scientific Basis of Selective Toxicity

There are three main principles by which an agent can exert a favourable selective effect. Either it can be equitoxic to all species, but accumulated principally by uneconomic species, or it may react with a cyto- or biochemical feature which plays a more important part in the uneconomic than in the economic species.

The scientific basis of selective toxicity :

1. Selectivity through accumulation
2. Selectivity through comparative bio-chemistry
3. Selectivity through comparative cytology.

1. Selectivity Through Accumulation

In fact, selectivity through accumulation is sometimes only a matter of gross morphology. Thus the comparatively large surface area (per unit weight) of an insect resting on a mammal, brings about a greater retention of sprayed material by the uneconomic species. In other cases, selective accumulation is achieved in a more positive way, for example, orally administered phenothiazine is much toxic to the intestinal worms in a sheep than to the sheep itself.

Accumulation implies some or all of the following :

(a) Efficient transport to the outside of the cell,

(b) A favourable permeability mechanism, and

(c) A satisfactory storage mechanism.

2. Selectivity Through Comparative Biochemistry

It is thought that all living matter — whether plant, or animal or microbes — had a common biochemistry, which, if universal, would offer no biochemical basis for selective toxicity. This thought is not true. It is a well-known fact that one species functions differently from another, which clearly indicates that there is actually marked biochemical difference. Strong species-differences are shown by homologous proteins, that is, those serving similar functions in related species. Thus, the N-terminal amino acids in the fibrinogen of seven mammals (Blombacs and Yamashina, 1958) was found to be :

Man	: Alanine and tyrosine
Pig	: Alanine (twice) and tyrosine
Ox	: Glutamic acid and tyrosine
Dog and Horse	: Threonine and tyrosine
Sheep and Goat	: Glycine, alanine, and tyrosine

The most striking biochemical difference between species are present not in degradative processes, *but in the choice and biosynthesis of enzymes* and smaller substances used in growth and division.

3. Selectivity Through Comparative Cytology

It is an established fact that plants and animals have outstanding cytological differences. For example, the cell wall and chloroplast are found in plants, but not in animals, likewise muscle cells and nerve cells are found only in animals but not in plants. In the last two decades, with the help of the electron microscope, it has been proved that the cell itself is full of organelles and that each kind of these components display strong species difference; and so much so that there are differences between cells from different tissues in the same species. It may, therefore, be conjectured that these differences certainly may assist selectivity in toxic agents.

Mechanism of Selective Toxicity

The selective toxicity of a chemical may be on account of any of the following mechanisms :

1. Due to differences in translocation factors
2. Due to differences in biotransformation reactions, and
3. Due to presence or absence of receptors.

1. Selective Toxicity due to Differences in Translocation of Chemical

Actually gross morphological, anatomical and cytological differences among various groups of organisms account for differential absorption, distribution and accumulation of chemicals which, in turn, may cause selective toxicity. For instance hexapods, i.e., insects, possess large surface area in comparison to mammals with respect to per unit weight. The large surface area of hexapods causes greater absorption and accumulation of any toxicant applied to kill them. It may lead to deleterious effect on hexapods without any or similar effect to other forms of life. As an example, DDT, an organochlorine insecticide, has been known to be readily absorbed through the chitinous exoskeleton of insects whereas the same insecticide is poorly absorbed through mammalian skin (Hayes, 1955). Consequently, DDT poses greater deleterious effects on insects only, but not on mammals. Another example is the application of 2, 4-D, a modern synthetic herbicide. It is generally used to destroy unwanted dicotyledonous weeds in the field of cereals. Actually the leaves of cereal crops are waxy and upright which prevent the penetration of 2, 4-D. On the other hand, it is accumulated in quantity sufficient to kill the dicotyledonous weeds from the fields which are rough in texture and wax-free, without affecting the cereals.

Mode of action of antibiotics in human is also related example of the aforementioned mechanism. Tetracycline, a potent antibiotic, is used by humans. It inhibits ribosomal protein synthesis of bacteria but not in the host i.e., human. This selectivity in action occurs as a result of favourable distribution of the drug. Bracteria may concentrate the antibiotics on account of selective permeability of their cell walls whereas the human cells are incapable of concentrating the antibiotics. Consequently, the concentrated drugs adversely affect the process of protein synthesis in bacteria without any otherwise effect or such effects to host cell.

Barriers to translocation of a chemical may be present in economical species whereas the same may be absent in uneconomical species.

2. Selective Toxicity due to Differences in the Biotransformation Mechanisms

Selective toxicity of chemicals may also be achieved owing to differences in the biotransformation processes. Biotransformation is a biocatalytic conversion of one form of toxicant into another form in the body of organisms. The toxicants may also be converted into more active forms othewise known as bioactivation which may, in turn, increase the toxicity of chemical in one group of organisms, whereas, in other group of organisms the same chemical may be converted into inactive forms and the toxicity of parent compound is diminished. Thus, the chemical may become nontoxic to other group of organisms. The biotransformation of toxicant may also alter the translocation of the chemical, hence the toxicity.

Actually the biotransformation mechanisms may vary in relation to genetic variations. One chemical may be toxic to certain strains and, at the same time, may be nontoxic to some other strains of a species. For instance, an antibiotic penicillin is not effective against certain strains of bacteria, particularly the gram-negative, because of the presence of an enzyme, *penicillinase*, which converts the antibiotic into inactive form. Similarly, malathion is ineffective against certain strains of houseflies and mosquitoes because they possess a specific enzyme (*esterase*) which biotransforms malathion into inactive form, which causes least toxic effects to these insects.

The various biotransformation reactions take place at different rates in different species which may cause selective toxicity of a chemical to one species without having such effect to the other. Heath (1961) reported the significance of rates of biocatalytic activation and inactivation of organophosphate pesticides as determinants of their selective toxicity to insect pests. These pesticides cause toxicity by bringing the inhibition of AChE activity. The organophosphorus pesticides can be grouped into two types :

1. Those having P=S group, and
2. Those having P=O group.

The pesticides having P=O group possess the ability to directly inhibit AChE activity, whereas those having P=S group are first oxidised into active fom (i.e., P=O from) in order to cause inhibition of enzyme, by microsomal mixed function oxidases (MMFO). The activated derivatives are then hydrolysed into inactive form by some other enzymes (i.e., *esterases*). In the rates of oxidation and hydrolysis of indirect type of AChE inhibitors differ in different species. The species in which the rate of oxidation of OP compounds is rapid and the rate of hydrolysis is slow, the active form of compound may accumulate which may induce the toxicity of the chemical. On the other hand, in other species, if the rate hydrolysis is rapid in comparison to the first species, the activated form of chemicals is rapidly converted into inactive form, hence the toxicity is diminished.

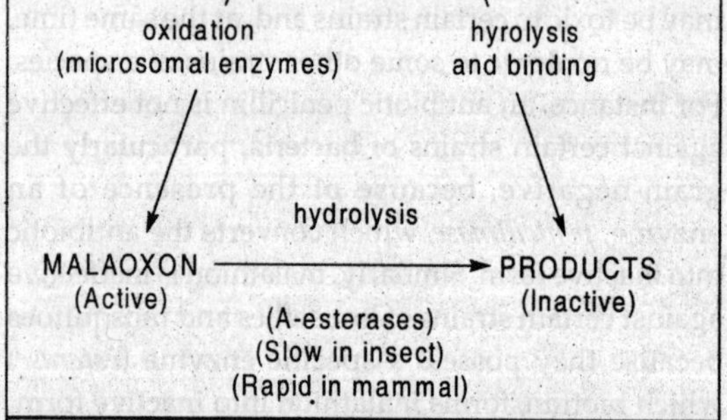

Fig. 20.1 : Relative rates of biotransformation of malathion in insects and mammals

In Fig. 20.1 biotransformation of malathion i.e., O, O-dimethyl, S (1,2-dicarbethoxyethyl) phosphordithioate and relative rates of various steps of biotransformation in insects and mammals has been represented. First and foremost malathion is oxidised into an active form, malaoxon, which is then hydrolysed into inactive products. Malathion may also be directly converted into products by hydrolysis and binding. The rate of oxidation in insects is rapid whereas the rate of hydrolysis and binding is very slow. Consequently, the active form (i.e., malaoxon) is accumulated in insects which kills them. On the other hand, the rates of oxidation, hydrolysis and binding are rapid in mammals, hence active form is rapidly converted into inactive products which cause no harm to mammals. Thus, malathion is selectively toxic to insects, i.e. uneconomical species, while causing no harm to mammals which are economical species.

3. Selective Toxicity due to Presence or Absence of Receptors

In case of preceding two mechanisms the selective toxicity is the function of active concentration of the chemical at specific sites; may be either different cells of an organism or different organisms. But in this mechanism all the cell or organisms are exposed to the same concentration and the toxicants are selective in their action due to existence or absence of receptors. The toxicants induce their effects on account of their interaction with certain receptors. It is also known that owing to cytological and biochemical differences in different groups of organisms, the specific receptors may be present in some whereas the same may be absent in others. Thus a chemical may be toxic to one form of life possessing appropriate receptors, at the same it may be nontoxic to other organisms which are devoid of such receptors.

The organo-phosphorus and carbamate insecticides owe their toxicity by interacting with AChE. These toxicants are similar in structure with that of natural substrate, ACh. Hence, they have close affinity for enzyme and bind to its active site. This binding actually involves strong covalent bonding forces leading to almost irreversible AChE-toxicants complex. Consequently, enzyme is not available for reaction with natural substrate, thereby normal neurophysiological function is impaired. Thus, these toxicants affect the nervous physiology without affecting other systems. In this way these substances reveal selective effects.

Some other toxicants are selectively toxic because they are sufficiently similar in geometry to normal enzyme substrate like those of OP and carbamate insecticides and compete with normal substrates for the active sites of enzymes in the body of animals.

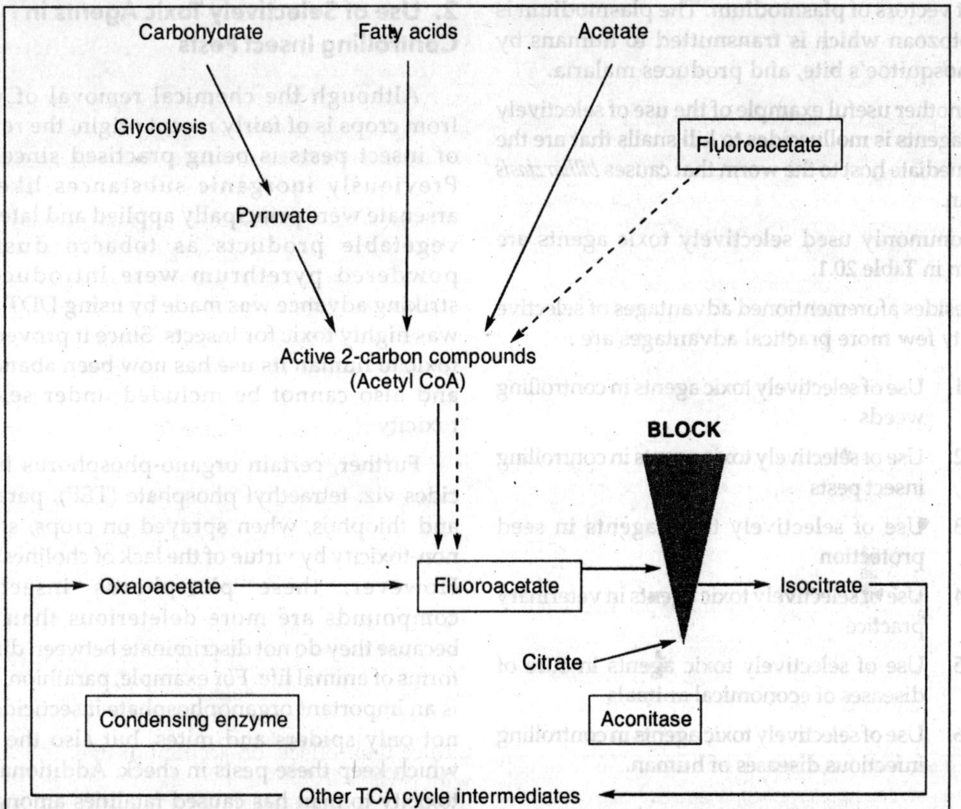

Fig. 20.2 : Diagrammatic representation of the mechanism of fluoroacetate-induced blockade of the TCA cycle

These toxicants also occupy the active sites of enzymes much in the same manner that normal substrate occupies and thus the enzymes do not remain available for natural substrate(s). One of the example of such selectively toxic chemical is sodium fluroacetate. In the body of organisms, it forms fluorocitrate by condensation with oxaloacetate (Fig. 20.2). The fluorocitrate then competes with citrate and binds at the active site of an enzyme, *aconitase,* leading to its inhibition. Thus, the sodium fluoroacetate blocks the TCA cycle in tissues and alters the physiology of organisms, ultimately leading to their death.

Advantages of Selective Toxicity

It is a great achievement of human beings that they have found selectively toxic agents for ills and also for use in animal husbandry, fields and forests. The continuance and even expansion of these advantages require uninterrupted investigations to improve upon the already known selectively toxic agents. One can feel it most unfortunate that the search for biological controls proves very expensive and has not often yielded practical results. Selective toxicity, on the other hand, may solve a high proportion of the problems of disease in plants, domestic animals and human beings. At present the most successful examples of biological control are those effected with selectively toxic agents. Thus trypanosomiasis, a protozoan disease of man and cattle, is controlled by chemical defoliation of those areas of African forests where tse-tse flies breed and then spraying organophosphorus insecticides on the exposed breeding sites. In this way, by attacking the insect vectors, which transfer trypanosomes to their mammalian hosts with every bite, the biological life cycle of these parasites is broken. For the same reason, houses in malarial areas and swampy grounds, which harbour anopheles mosquitoes, are regularly sprayed with insecticides to kill these insect vectors of plasmodium. The plasmodium is a protozoan which is transmitted to humans by the mosquitoe's bite, and produces malaria.

Another useful example of the use of selectively toxic agents is molluscides to kill snails that are the intermediate host to the worm that causes *bilharziasis* in man.

Commonly used selectively toxic agents are shown in Table 20.1.

Besides aforementioned advantages of selective toxicity few more practical advantages are :

1. Use of selectively toxic agents in controlling weeds
2. Use of selectively toxic agents in controlling insect pests
3. Use of selectively toxic agents in seed protection
4. Use of selectively toxic agents in veterinary practice
5. Use of selectively toxic agents in cure of diseases of economical animals
6. Use of selectively toxic agents in controlling infectious diseases of human.

1. Use of Selectively Toxic Agents in Controlling Weeds

Bonnet for the first time showed, in France, that yellow charlock could be killed in a field of oats, without injury to the crop, by spraying a solution of copper sulphate over the entire field. Further, in 1911, another Frenchman, Rabate, showed that solution of H_2SO_4 could be safely used on crops to destroy the weeds. Nowadays, the most versatile of the weed-killers are dinitro-o-cresol and 2, 4-D.

The introduction of organic weed killers has brought about tremendous economic gain. They are non-toxic to man and animals, and non-corrosive to equipment. They have resulted in an overall increase of 20% in grain crops. These substances have been successfully used on crops of flax and grasses, and also in removing weeds from lawns.

Somewhat akin to weed killing is defoliation (leaf-removal). In East Africa, trees are defoliated with the n-butyl ester of 2,4,5- trichlorophonxyacetic acid, which is sprayed in the air. The purpose is to eliminate tse-tse flies which carry trypanosomes.

2. Use of Selectively Toxic Agents in Controlling Insect Pests

Although the chemical removal of weeds from crops is of fairly recent origin, the removal of insect pests is being practised since long. Previously inorganic substances like lead arsenate were principally applied and later such vegetable products as tobacco dust and powdered pyrethrum were introduced. A striking advance was made by using DDT which was highly toxic for insects. Since it proved to be toxic to human its use has now been abandoned and also cannot be included under selective toxicity.

Further, certain organo-phosphorus insecticides viz. tetraethyl phosphate (TEP), parathion, and thiophos, when sprayed on crops, showed non-toxicity by virtue of the lack of cholinesterase. However, these phosphorus insecticidal compounds are more deleterious than DDT, because they do not discriminate between different forms of animal life. For example, parathion, which is an important organophosphate insecticide, kills not only spiders and mites, but also the birds, which keep these pests in check. Additionally, its toxicity to man has caused fatalities among farm workers who had not adopted requisite protective measures.

Keeping in view the principle of selective toxicity, organo-phosphorus insecticides require certain modifications :

(a) So that the chemical changes needed for activation can be accomplished by pests but not by birds, mammals, or else.

(b) So that total hydrolysis occurs slowly in pests but rapidly in birds and mammals. A promising substance of this kind is *tetram*. Actually it is systemic (i.e., rises with the plants sap) and protects the plant from infection by red spiders, mites etc. Apparently birds are not affected. Thus it can be regarded as an important selectively toxic agent.

3. Use of Selectively Toxic Agents in Seed Protection

There is another section of agriculture in which selectively toxic agents have been put to work. Seed — particularly grain seeds — are commonly dusted with phenol mercuric nitrate during storage to prevent infection by fungi. This treatment does not affect germination of the seeds. The phenol group increases the lipophilic properties of the mercury, and, hence, aids penetration into the fungus.

4. Use of Selectively Toxic Agents in Veterinary Practices

There have been many economically valuable uses of selectively toxic agents in veterinary practices in spite of the fact that the remedies devised for human illness cannot always be used in veterinary because of their high cost.

Bovine mastitis, which greatly reduces the yield of milk, is now cured with sulphanilamide.

5. Use of Selectively Toxic Agents in Cure of Diseases of Economical Animals

Worm infections in sheep, which lead to severe economic loss of wool and meat, are controlled with CCl_4 (carbon tetrachloride) and phenothiazine. The cheap and simple remedy piperazine is much used as selectively toxic agent in pigs, horses, dogs, cats and poultry.

6. Use of Selectively Toxic Agents in Controlling Infectious Diseases of Human

Fungal diseases of man, even when superficial, are being treated more successfully by certain antifungal compounds as selectively toxic agent than before by internal medication. There is, however, still great scope for improved remedies.

Also selectively toxic drugs viz., *suramin* (Bayer 205) or pentamidine for treatment of schistosomiasis are now available although better prophylactic drugs are needed.

Amebic dysentery is treated successfully with few selectively toxic drugs like emetine and synthetic drugs such as camoquine. Sulphonamides are used to control tropical bacillary dysentery.

Virus diseases like poliomyelitis, small pox, yellow fever, rabies and influenza have been controlled through chemotherapy. Vaccines exist to prevent these five viral diseases, but actually there is no perfect curative treatment.

Table 20.1 : Commonly used selectively toxic agents and their salient actions

Sl. No.	Name of Selective Toxicant	Salient actions
1.	Emetin	Selectively toxic to ameba
2.	Sulphonamides	Selectively toxic to bacillary dysentery
3.	Suramin	Selectively toxic to Schistosoma
4.	Mepacrine hydrochloride	Selectively toxic to Plasmodium
5.	p-amino salicyclic acid	Selectively toxic to bacteria
6.	Phenol mercuric nitrate	Selectively toxic to fungi attacking on seeds
7.	CCl_4	Selectively toxic to worm infections in pigs, dogs, cats and poultry
8.	2, 4-D	Selectively toxic to unwanted herbs
9.	2, 4, 5-T	Selectively toxic to unwanted weeds
10.	Methoxy DDT (methoxychlor)	Selectively toxic to insect pests of vegetables
11.	Dicofol	Selectively toxic to acarids, ticks and mites
12	Fenitrothion	Selectively toxic to insect pests
13.	Malathion	Selectively toxic to insect pests
14.	Abate	Selectively toxic to mosquito larvae
15.	Alpha-napthyl thiourea (ANTU)	Selectively toxic to rodents
16.	Norbormide	Selectively toxic to rodents
17.	Karathene	Selectively toxic to fungi
18.	Carboxin	Selectively toxic to fungi

Conclusion : Selective toxicity of any drug/chemical means the injury of one kind of living matter without harming another kind with which the first is in intimate contact. The living matter which is injured is conveniently referred to as the *uneconomic species*, and the unaffected is the *economic species*.

It is an achievement of prime importance for man that selectively toxic agents have been found not only for many of his ills, but also for use in his animal husbandry, fields and forests.

Drugs and other chemical agents used for selective toxic purposes are selective mainly for one of two reasons. Either (1) the chemical is equitoxic to both economic and uneconomic cells but is accumulated mainly by uneconomic cells, or

(2) it reacts fairly specifically with a cytological or a biochemical feature that is absent from or does not play an important role in the economic form (Albert, 1965).

Selectivity resulting from differences in distribution usually is caused by differences in the absorption, biotransformation, or excretion of the toxicant. The selective toxicity of an insecticide spray may be partly due to a larger surface area per unit weight that causes the insect to absorb a proportionally larger dose than does the mammal being sprayed. The effectiveness of radioactive iodine in the treatment of hyperthyroidism (as well as its thyroid carcinogenicity) is due to the selective ability of the thyroid gland to accumulate iodine.

A major reason why chemicals are toxic to one but not to another type of tissue is that there are differences in accumulation of the ultimate toxic compound in various tissues. This, in turn, may be due to differences in the ability of various tissues to biotransform the chemical into the ultimate toxic product.

Selective toxicity caused by differences in comparative cytology is exemplified by a comparison of plant and animal cells. Plants differ from animals in many ways; for example, absence of a nervous system, an efficient circulatory system, and muscles and the presence of a photosynthetic mechanism and cell walls. The fact that bacteria contain cell walls and humans do not has been utilised in developing selective toxic chemotherapeutic agents — such as penicillin and cephalosporins — that kill bacteria but are relatively nontoxic to mammalian cells.

Selective toxicity also can be a result of a difference in biochemistry in the two types of cells.

Revision Questions

A. Long-Answer Questions

1. Write an essay on selective toxicity and its application.
2. Define selective toxicity. Give its need and convincing demonstration in man.
3. Give an account of the scientific basis of selective toxicity.
4. Define selective toxicity and describe its mechanism.

B. Short-Answer Questions

1. *Briefly describe :*
   (i) Selective toxicity
   (ii) Commonly used selective toxic agents
   (iii) Application of selective toxicity.
2. *Write short notes on :*
   (i) Use of selectively toxic agent in controlling weeds
   (ii) Use of selectively toxic agent in control of insect pests
   (iii) Use of selectively toxic agent in seed protection
   (iv) Use of selectively toxic agent in veterinary practice
   (v) Use of selectively toxic agent in controlling infection
   (vi) Use of selectively toxic agent in cure of diseases of human.
3. *Briefly describe :*
   (i) Selective toxicity due to difference in translocation of chemical
   (ii) Selective toxicity due to presence or absence of receptors
   (iii) Selective toxicity due to difference in biotransformation mechanism.
4. *Write short notes on :*
   (i) Selectivity through accumulation
   (ii) Selectivity through comparative biochemistry
   (iii) Selectivity through comparative cytology.

21

Biotransformation of Toxicants

SYNOPSIS

Definition, Biotransformation sites, Objectives of biotransformation, Nature of biotransformation enzymes, Mechanism of biotransformation, Complex nature of biotransformation, Factors affecting biotransformation, Bioactivation.

As mentioned earlier, a xenobiotic (toxicant) may be absorbed into an animal via different routes. After absorption it is distributed to different parts of the body and, finally, is available for excretion. Many xenobiotics are known to undergo biotransformation while in the organs and tissues. Biotransformation is also termed metabolic transformation.

Definition of biotransformation : Biotransformation may be defined as "the biologically catalysed conversions of chemicals, other than the normal body constituents into other chemicals".

Or

Biotransformation is a process that, in general, converts the parent xenobiotic compounds or toxicants into their metabolites and then form conjugates and facilitate their release from the body.

Or

The biochemical modification of the xenobiotic (toxicant) molecules through the living cell is termed biotransformation or metabolic transformation.

Or

The act of reduction in the potentiality of the toxicant by the internal system of animal may be termed biotransformation.

During *biotransformation*, generally the parent toxic compounds undergo detoxification i.e., the formed metabolites and, finally, conjugates prove to be lesser toxic than the original or parent toxicants. But in certain cases the metabolites have been proved more toxic than the parent xenobiotics (toxicants) and such reaction is termed *bioactivation*.

Biotransformation Sites : The organ/tissue wherein the biotransformation takes place is called biotransformation site. The biotransformation of xenobiotics are often catalyzed by the enzymes which chiefly occur in the liver of vertebrates. In the vertebrates these enzymes also occur in the skin, kidney, lungs, intestine, placenta, gonads, embryonic liver, aorta, lymphocytes, blood platelets, adrenal cortex and medulla but not in nervous system, though their activity have been anticipated. In insects, such enzymes have been reported in mid-gut, fat body and Malpighian tubules. The most important biotransformation sites in the vertebrates including human are : liver, kidney, intestine, skin and lungs where the biotransformation takes place in the following order :

Liver > Kidney > Intestine > Lungs > Skin

Principal Objectives of Biotransformation : The principal objectives of the biotransformation processes are to detoxify the toxicants and their elimination from the body.

It includes :

1. Conversion of water-insoluble substances to water-soluble form,
2. Emulsification of substances with bile, so as to facilitate elimination in colloidal form with feces,
3. Incorporation of toxic molecules within inactive proteins, for example, metal molecules in metallothionien protein and formation of metalloprotein for elimination.

Nature of Biotransformation Enzymes : Actually the enzymes involved in biotransformation of xenobiotic compounds (toxicants) have relatively low degree of substrate specificity in comparison to those involved in the metabolism of constitutive chemicals.

The responsible enzymes in biotransformation of xenobiotics are monooxygenases or cytochrome P-450 species.

The reaction catalyzed by a mono-oxygenase is :

$$RH + O_2 + NADP + H^+ \rightarrow R\text{-}OH + HOH + NADP^+$$

RH represents a very wide variety of xenobiotics including endogenous compounds such as steroids.

The reaction catalyzed by cytochrome P-450 may also be represented as :

Reduced cyt. P-450 $\rightarrow$ Oxidized cyt. P-450.

$$RH + O_2 \rightarrow R\text{-}OH + HOH$$

The major monooxygenases in the ER are cytochrome P-450 species. Cytochrome P-450 is notable because of the fact that approximately 50% of the drugs ingested are metabolized by species of cytochrome P-450. The same enzyme also acts on various carcinogens and pollutants.

Salient Features of Cytochrome P-450 Species

1. These are hemoproteins.
2. These are present in high concentration in the membranes of the endoplasmic reticulum (ER) or liver.
3. These are also present in the mito-chrondria as well as in the ER of the adrenal gland.
4. Six closely related species of cytochrome P-450 present in the liver ER that act on a wide variety of drugs, carcinogens, and other xenobiotics. In recent years, the genes for these species of cytochrome P-450 have been isolated and studied.
5. The enzyme that uses NADPH in the reaction mechanism of cytochrome P-450 is called the *NADPH-Cytochrome P-450 reductase.*
6. The *phosphatidyl choline,* the major lipid found in the membranes of ER, is the principal component of the cytochrome P-450 system.
7. Most species of cytochrome P-450 are inducible e.g., the administration of phenobarbital (PB) causes a hypertrophy of the smooth ER and a 3 to 4 fold increase of the amount of cytochrome P-450 within 4-5 days. The mechanism of induction involves increased transcription of mRNA for cytochrome P-450.
8. Another species called cytochrome P-448 is specific for the metabolism polycyclic aromatic hydrocarbons (PAHs) and related molecules; hence it is said to be aromatic hydrocarbon hydroxylase (AHH). This enzyme is very important in the metabolism of PAHs and in carcinogenesis produced by these agents. Researches reveal that smokers have higher levels of this enzyme in some of their cells and tissues than do non-smokers. It is noticed that the activity of this enzyme may be increased in the placentae of women who smoke, thus altering the quantities of metabolites of PAHs to which the fetus is exposed.

The biological significance for the presence of these enzymes may be to evolve a protective device against xenobiotic compounds.

Mechanism of Biotransformation : R. T. Williams, in 1959, firstly studied the mechanism of biotransformation of xenobiotics and divided the entire steps taking part in enzymatic biotrans-formation into two phases :

1. Phase I reactions or nonsynthetic reactions involving oxidation, reduction and hydrolysis.
2. Phase II reactions or synthetic reactions involving the formation of a conjugate, i.e., metabolites of parent toxicant + endo-genous polar or ionic moiety.

To sum up, biotransformation is such a process in which the parent xenobiotic compound is converted into metabolites and then conjugates are formed. For example, benzene undergoes oxidation, a phase I reaction, to form phenol, which conjugates with sulphate, a phase II reaction. However, some chemicals form conjugates without proceeding in phase I reaction. For example phenol may conjugate with sulphate without proceeding in phase I reactions. The metabolites and conjugates are usually more water-soluble and polar, hence more readily excretable. Biotransformation can, therefore, be considered in general as a mechanism of detoxification by the host animal.

The type and rate of biotransformation of any xenobiotic compound differs from one species of animal to another and even from one strain to another, which often accounts for the difference of toxicity in animals. Apart from age and sex of the animal, exposures to other chemicals may also alter the biotransformation. Knowledge of such factors is important in designing the toxicological studies and in also the interpretation of possible health hazards to humans.

Phase I Reactions

There are three types of Phase I reactions :

1. Oxidation.
2. Reduction.
3. Hydrolysis.

1. Oxidation

The biotransformation of great variety of xenobiotic compounds involves oxidation process. The most important enzyme system catalysing the involved processes are cytochrome P-450 and NADPH cytochrome P-450 reductase. In these reactions, one atom of molecular oxygen is reduced to water and the other is incorporated into the substrate.

The cytochrome linked monooxygenases (oxidases) are located in the smooth endoplasmic reticulum. When a cell is homogenized, the endoplasmic reticulum splits into small vesicles known as microsomes. Because of the location of these enzymes and the great variety of chemicals that they may catalyze, these are also known as microsomal mixed-function oxidases (MMFO). In addition, oxidation of various toxicants is catalyzed by non-microsomal oxidoreductases that are located in the mitochondrial fraction.

Oxidation may take place in a variety of reaction forms and very often more than one metabolite is formed. Some examples are :

A. Microsomal Oxidations

1. **Aliphatic Oxidation** : Aliphatic oxidation involves oxidation of the aliphatic side-chains of aromatic chemicals :

Example :

n, propylbenzene → 3-phenylpropan-1, -ol, 3-Phenyl propan-2-ol, and 3-phenypropan-3-ol

2. **Aromatic hydroxylation** : Aromatic hydroxylation usually proceeds through an epoxide intermediate :

Example :

Naphthalene Naphthalene-1, 2-epoxide

↓

1-naphthol + 2-naphthol

3. **Expoxidation**

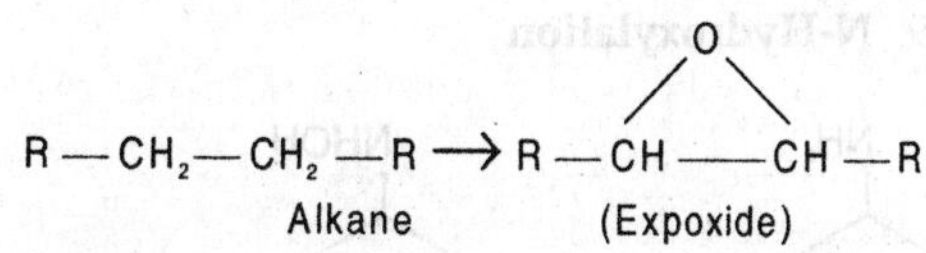

Example :

Aldrin → Dieldrin

4. **Oxidative deamination**

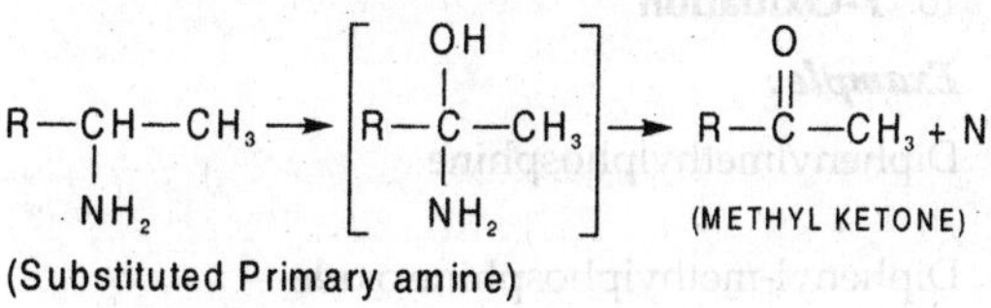

Example :

Amphetamine → Phenyl acetone

5. **N-dealkylation**

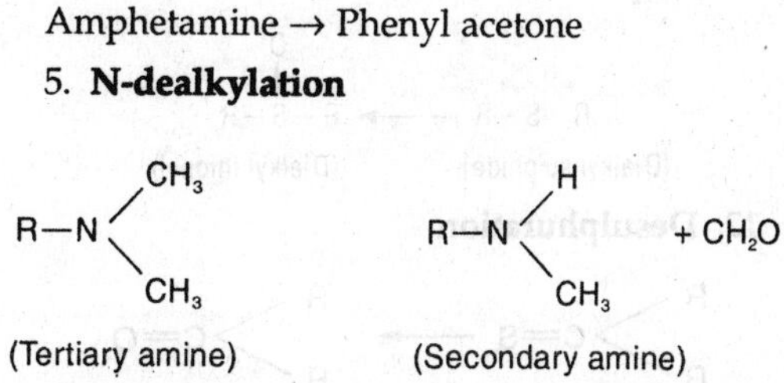

Example :

N, N-dimethyl-p-nitrophenyl carbamate

↓

N-methyl-p-nitrophenyl carbamate

6. **O-dealkylation**

$$R{-}O{-}CH_3 \longrightarrow R{-}OH + CH_2O$$

(Ether) (Primary alcohol)

Example :

p-nitroanisole → p-nitrophenol

7. **S-dealkylation**

$$R{-}S{-}CH_3 \longrightarrow RSH + CH_2O$$

(Thio-derivative) (Thioalcohol)

Example :

6-Methyl thiopurine → 6-Mercaptothiopurine

8. **N-Oxidation**

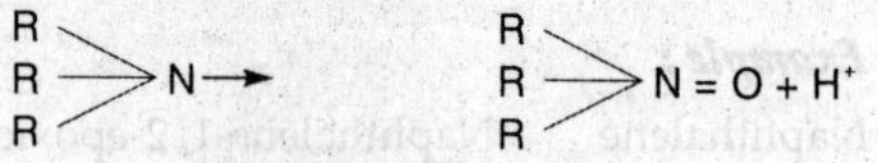

(Tertiary amine) (Nitroso compound)

Example :

Trimethylamine → Trimethylamine oxide

9. **N-Hydroxylation**

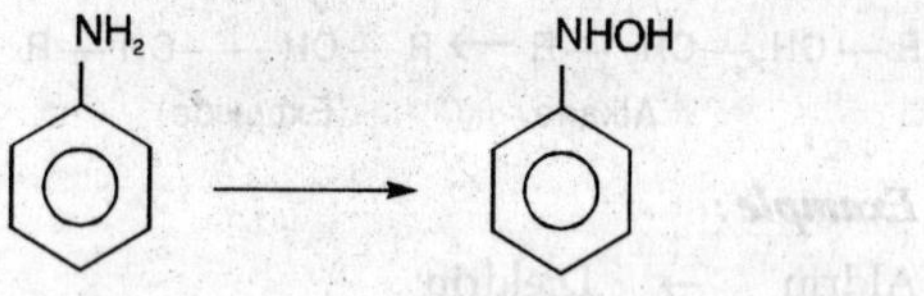

Aniline Phenyl hydroxyl amine

10. **P-Oxidation**

Example :

Diphenylmethylphosphine

↓

Diphenyl-methylphosphine oxide

11. **Sulphoxidation**

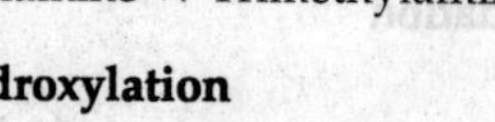

(Dialkyl sulphide) (Dialkyl thionyl)

12. **Desulphuration**

Thioketone Ketone

Example :

Parathion → Paraoxon

B. Nonmicrosomal Oxidations

1. **Amine Oxidation** : Monoamine oxidase located in mitochondria. Diamine oxidase is a soluble enzyme. Both participate in the oxidation of the primary, secondary and tertiary amines, such as 5-hydroxytryptamine and putrescine, into corresponding aldehydes :

$$RCH_2NH_2 \xrightarrow{O_2} RCH{=}NH \xrightarrow{HOH} RCHO + NH_3$$

Primary amine Imine Aldehyde

2. **Alcohol and aldehyde dehydrogenation** : It is catalysed by alcohol dehydrogenase and aldehyde dehydrogenase, respectively :

Example :

Ethanol → Acetaldehyde

Acetaldehyde → Acetic acid

Reaction showing Aldehyde dehydrogenation :

$$R{-}CHO + NAD^+ \longrightarrow R{-}COOH + NAD + H^+$$

2. Reduction

Xenobiotic chemicals may undergo reduction through the function of reductases. These reactions are actually less active in mammalian tissues but frequent in intestinal and intracellular bacteria. An important example is the reduction of prontosil to sulphanilamide.

Like oxidation, reduction may also be of two types, namely microsomal reduction and non-microsomal reduction :

A. Microsomal Reductions

1. **Nitro Reduction**

$ArNO_2$ → ArNO

(Nitrobenzene) (Nitrosobenzene)

ArNO → ArNHOH

(Nitrosobenzene) (Phenylhydroxylamine)

ArNHOH → $ArNH_2$

(Phenylhydroxylamine) (aniline)

$Ar = C_6H_5$

2. **Azo Reduction**

ArN = NAr → ArNHNHAr → $ArNH_2$

(Azobenzene) (Hydroazo Benzene) (Aniline)

B. Nonmicrosomal Reduction

Nonmicrosomal reduction occurs via the reverse reaction of alcohol dehydrogenases :

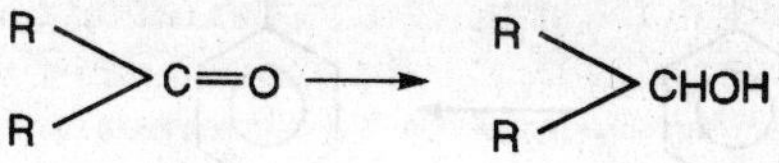

ketone Secondary alcohol

3. Hydrolysis

Various toxicants are having ester-type bonds and are subjected to hydrolysis. These are essentially esters, amides, and compounds of phosphate. Mammalian tissues, including the plasma, contain a large number of non-specific esterases and amidases, which take part in hydrolysis.

Testa and Jenner (1976) reported that the esterases, usually located in the soluble fraction of the cell, may be broadly grouped into four classes :

1. **Arylesterases** : These hydrolyse aromatic esters :

$$R{-}CO{-}O{-}R_1 \longrightarrow R{-}COOH + R_1{-}OH$$

(Ester) (Carboxylic acid)

2. **Cholinesterases** : These hydrolyse those esters in which the alcohol moiety is choline.

3. **Carboxylesterases** : These hydrolyse aliphatic esters.

4. **Acetylesterases** : These hydrolyse those esters in which the acid moiety is acetic acid.

In contrast to esterases, amidases cannot be grouped according to the substrate specificity. Furthermore, enzymatic hydrolysis of amides proceeds much slowly than that of esters, probably because of the lack of substrate specificity.

Phase II Reactions (CONJUGATION REACTIONS)

As described earlier in this chapter, Phase II reactions involve several types of endogenous metabolites that may form conjugates with the xenobiotics or their metabolites. In general, these conjugates are more soluble in water and more readily excretable. Examples of each type of conjugation have been illustrated next :

A. Glucuronide Formation

This is very common and the most important type of conjugation. The enzyme catalyzing this reaction is UDPGT (Uridine Diphosphate Glucuronyl Transferase) and the coenzyme UDPGA (Uridine 5'-diphospho- α -D-glucuronic acid). This enzyme is also located in the endoplasmic reticulum.

There are four classes of chemical compounds capable of forming conjugates with glucuronic acid :

1. aliphatic or aromatic alcohols,
2. carboxylic acid,
3. sulphydryl compounds, and
4. amines.

O-glucuronide formation

1. **Ether type**

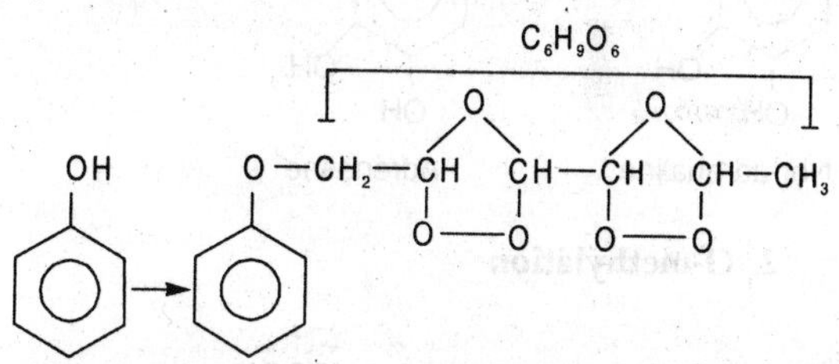

Phenol Phenoxyozonide (ether type)

2. **Ester type**

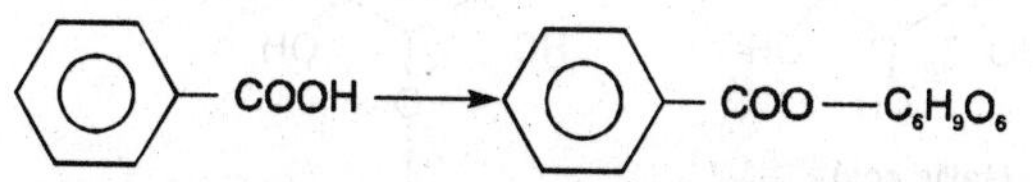

Benzoic acid Benzoate (Substituted ester)

3. **S-glucuronide formation**

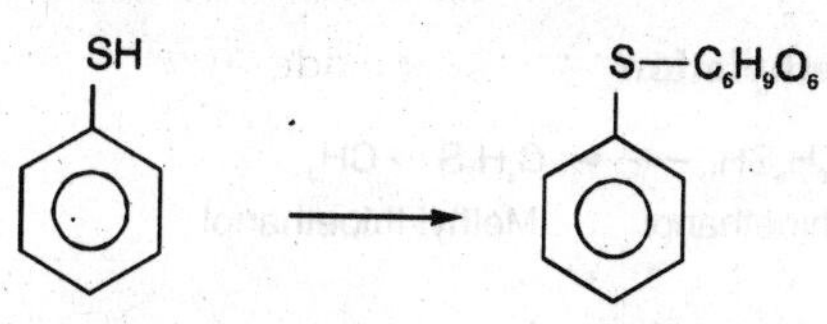

Thiophenol or Phenyl thioalcohol Substituted thioderivative

4. **N-glucuronide formation**

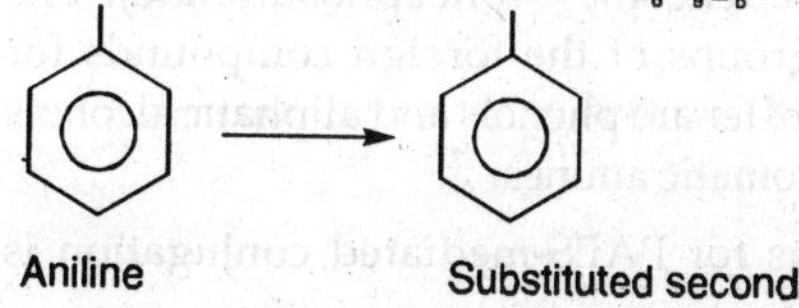

Aniline Substituted secondary amine

B. Methylation

This reaction is catalysed by methyl transferases. The coenzyme is the SAM (S-adenosylmethionine). However, methylation is not a major route of biotransformation of toxicants because of the broader availability of UDPGA, which leads to the formation of glucuronides. Futhermore, it does not always increase the water solubility of the methylated products.

Few reactions of methylations are :

1. **N-methylation**

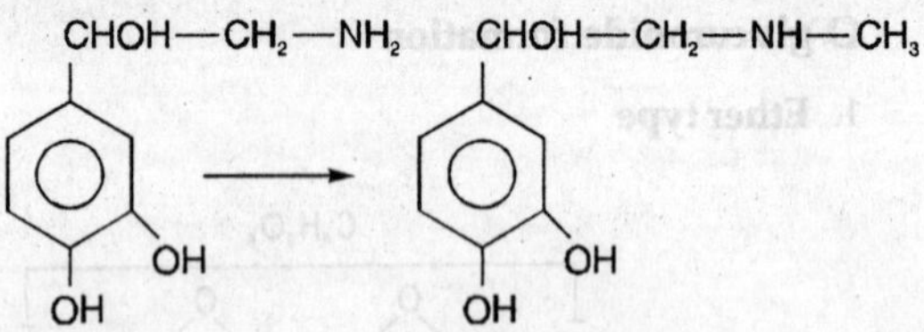

Nor-adrenaline Adrenaline

2. **O-methylation**

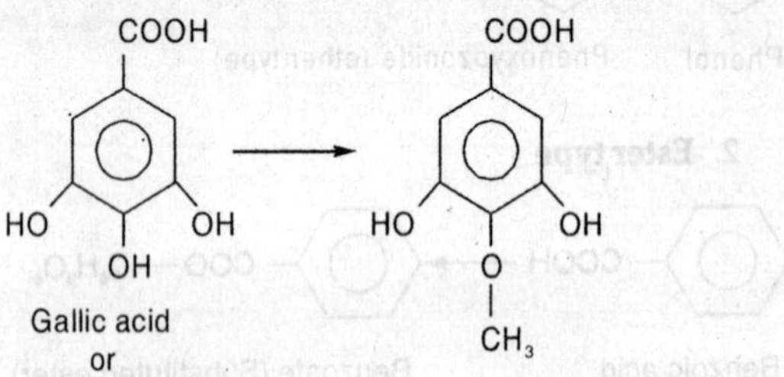

Gallic acid or 3, 4, 5 - Trihydroxy benzoic acid

3, 5 - Dihydroxy, 4 - Methoxy 1- benzoic acid

3. **S-methylation**

$C_2H_5SH \longrightarrow C_2H_5S-CH_3$

Thioethanol Methyl thioethanol

C. Sulphate conjugation

This reaction is catalysed by sulphotransferases. These enzymes are located in the cystosolic fraction of liver, kidney and intestine. The coenzyme is PAPS (3'-phosphoadenosine-5'-phosphosulphate). The functional groups of the foreign compounds for sulphate transfer are phenols and aliphatic alcohols as well as aromatic amines.

Reactions for PAPS-mediated conjugation is shown next :

Sulphate ester formation

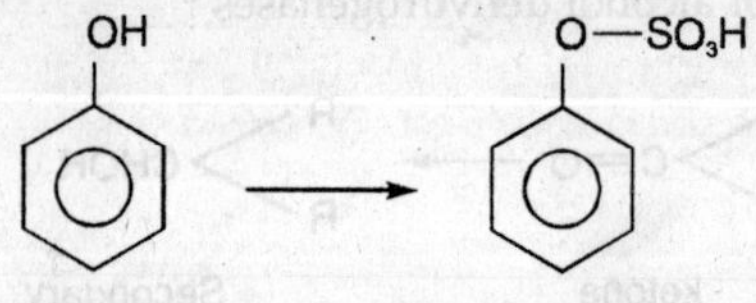

Phenol

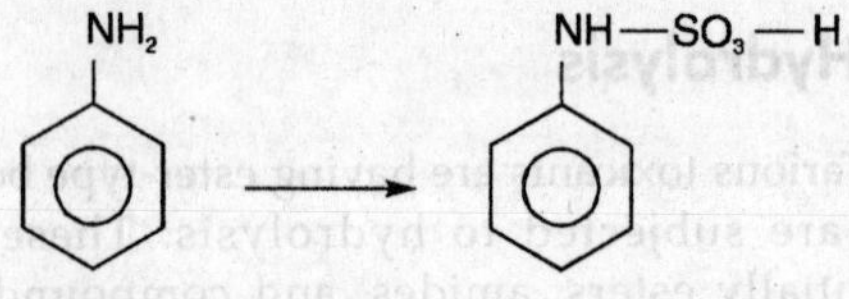

Aniline

D. Acetylation

Acetylation involves transfer of acetyl groups to primary aromatic amines, hydrazines, hydrazides, sulphonamides and certain primary aliphatic amines. The enzyme and coenzyme involved, respectively, are N-acetyl transferases and acetyl coenzyme A :

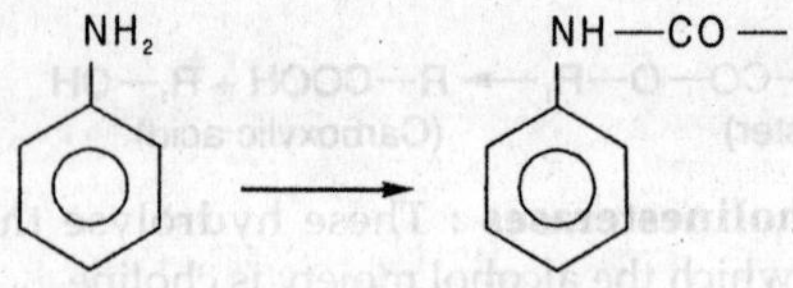

Aniline Actanilide

E. Amino acid conjugation

This conjugation is catalysed by amino conjugates and coenzyme A. Aromatic carboxylic acids, arylacetic acid, arylsubstituted acrylic acids can form conjugates with α -amino acids, mainly glycine, but also glutamine in humans and certain monkeys and ornithine in birds :

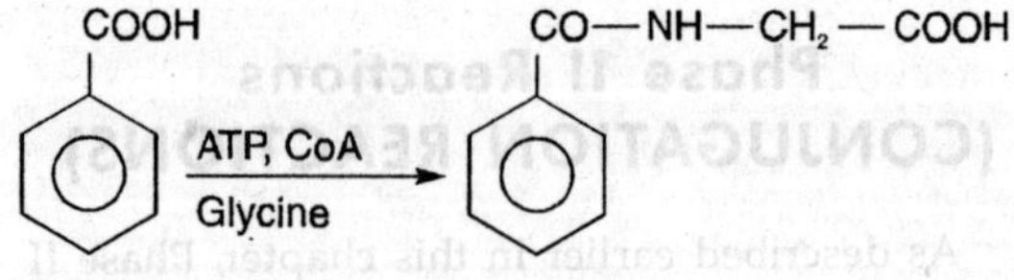

Benzoic acid N - Benzoyl amino acetic acid

F. Glutathione conjugation

This important reaction is effected by glutathione S-transferases and the cofactor glutathione. Glutathione conjugates subsequently undergo enzymatic cleavage and acetylation forming N-acetylcystenine (mercapturic acid) derivatives of the toxicants, which are readily excreted. Glutathione can also conjugate with unsaturated aliphatic compounds and displace nitro groups in chemicals.

It is rather important to mention that in the process of biotransformation of xenobiotics, a number or highly reactive electrophilic compounds are formed. Some of these compounds can react with cellular constituents and cause cellular death or induce tumour formation. The role of glutathione is to react with electrophilic compounds and thus prevent their harmful effects on the cells. However, exposure to very large amounts of such reactive substances can deplete the glutathione, thereby resulting in marked toxic effects.

Complex Nature of Biotransformation

Certain toxicants in their mode of action during biotransformation show a great degree of complexity because they generally undergo various types of biotransformation, forming a variety of metabolites and conjugates. Some of the metabolites and conjugates of bromobenzene and carbaryl are shown in Figs. 21.1 and 21.2, respectively.

WHO (1971) reported that organophosphorus insecticides, viz. fenitrothion, omethoate, etc., can be metabolised through dealkylation, oxidation, hydrolysis or desulphuration, yielding ten or more different metabolites. Parathion, which is an organophosphorus pesticide, is bioactivated in the liver to paraoxon, which is much more potent cholinesterase inhibitor.

Factors Affecting Biotransformation

As mentioned above, the liver is the main organ wherein biotransformation of xenobiotics takes place. However, diseases such as acute and chronic hepatitis, cirrhosis of the liver, and hepatic necrosis often decrease the biotransformation.

The biotransformation capability of infants of premature births is extremely low in comparison to the adults.

Certain other factors that affect the efficiency of liver to biotransform the toxicant/drug are : nutritional status, sex, age, procedure of administration of the drug and duration of drug administered.

Starvation of organisms results in decrease in the levels of biotransformation enzymes. Therefore, starved animals are often more sensitive to toxicants than the normally fed individuals.

The biotransformation of toxicants is catalyzed by the MMFO. A deficiency of essential fatty acids generally depresses MMFO activities. This is also true with protein deficiency. Deficiency of certain vitamins like vits. A, C and E depresses the MMFO. However, thiamine deficiency has the opposite effect.

Some foods contain appreciable amount of chemicals that are actually strong inducers of the MMFO, e.g., safrole, xanthines, and indoles. In addition, potent inducers such as DDT and PCB are present as contaminants in many foods.

The rate of biotransformation of xenobiotics varies according to the sex of the organism. For example, adult male rats biotransform xenobiotics at high rates than those of adult females.

Environmental factors like temperature and ionizing radiations also affect the biotransformation process. For instance, exposure of rodents to ionizing radiations reduce the rate of biotransformation of xenobiotics. It is because ionizing radiations have been reported to reduce the hydroxylation of steroids, desulphuration activity and glucuronide formation.

Bioactivation

The conversion of certain chemically stable compounds to highly chemically reactive metabolites is termed bioactivation. In other words, bioactivation is the biotransformation in which the formed metabolite proves to be highly toxic than the parent compound, i.e., toxicant or drug. These reactive compounds become covalently bound to tissue macromolecules and cause injury. In addition, other types of metabolites produce deleterious effects via other mechanism also.

The biotransformation of bromobenzene to its epoxide and subsequent reactions serve as an interesting example of bioactivation and its consequences. Although bromobenzene epoxide may become covalently bound to tissue macromolecules and cause injury, the alternative routes of metabolism may prevent or reduce the injury.

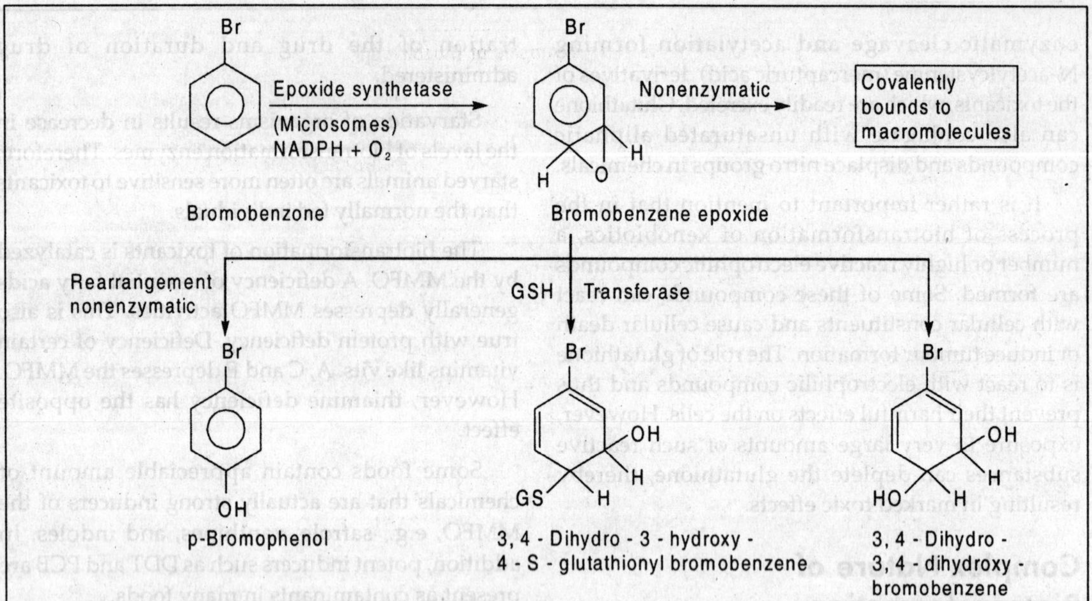

Fig. 21.1 : The bromobenzene epoxide formed from the parent chemical may covalently bind to macromolecules, but such binding may be minimal when the other metabolic routes are predominant. This reveals complex nature of biotransformation.

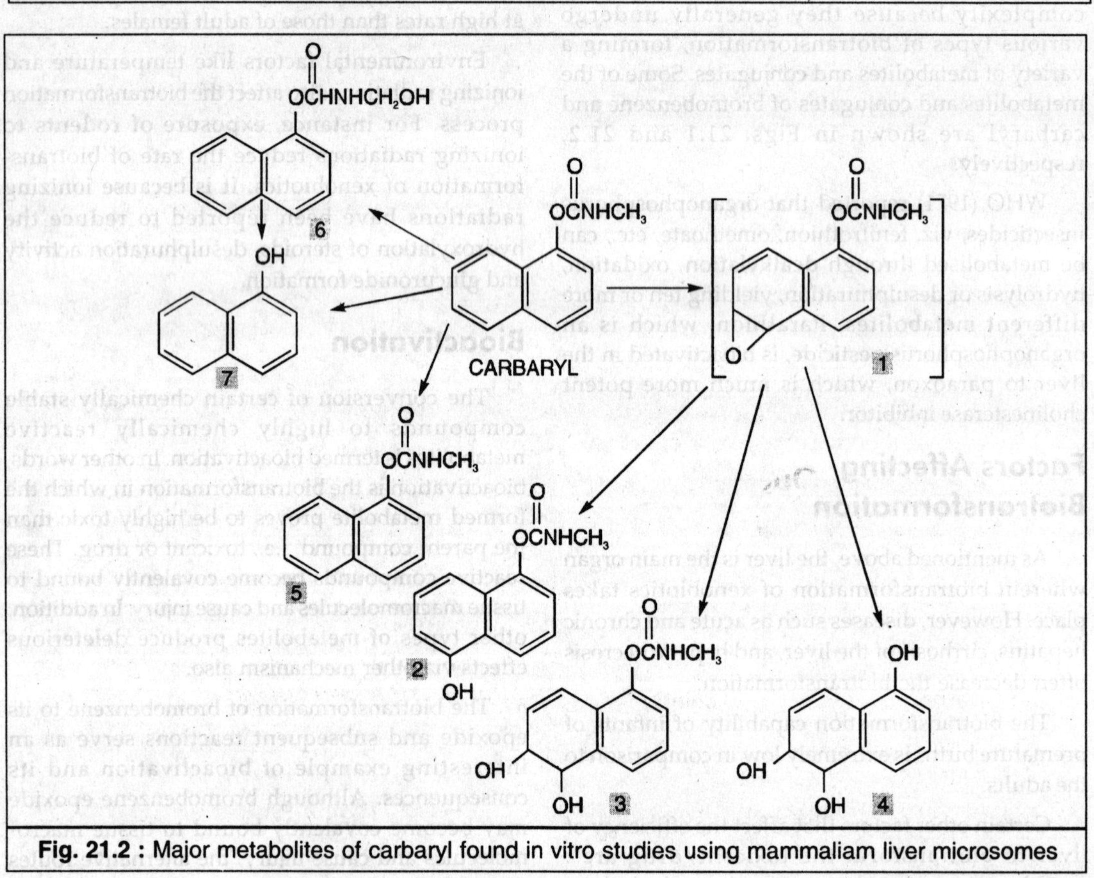

Fig. 21.2 : Major metabolites of carbaryl found in vitro studies using mammaliam liver microsomes

Table 21.1 : Examples of Bioactivation

Parent Compound	Toxic metabolite	Site of formation	Mechanism of toxicity	Toxicity
2-Acetylaminofluorene	N-Acetoxy-AAF, AAF N-sulphate	Liver	Covalent binding	Cancers
Allyl formate	Acrolein	Liver	Covalent binding	Hepatic necrosis
Aflatoxin B_1	Aflatoxin-2,3-epoxide	Liver	Covalent binding	Hepatic cancer
Benzene	Benzene epoxide	Liver	Covalent binding	Bone marrow depression
Bromobenzene	Bromobenzene epoxide	Liver and Lung	Covalent binding	Hepatic, renal bronchiolar necrosis
CCl_4	Trichloromethane free radical	Liver	Covalent binding	Hepatic necrosis, Hepatic cancer
Methanol	Formaldehyde	Liver and Retina	Complex	Retinal and general toxicity
Naphthylamine	N-Hydroxy naphthylamine	Liver	Covalent binding	Bladder cancer
Nitrites plus Secondary or Tertiary amines	Nitrosoamines	Stomach	Alkylation	Hepatic, and pulmonary cancer
Parathion	Paraoxon	Liver	Covalent binding to cholinesterase	Neuromuscular paralysis
Urethane	N-Hydroxy-Urethane	Liver	Alkylation	Cancers and Cytotoxicity

Likewise parathion, an organophosphorous pesticide, is bioactivated in the liver to paraoxon, which is much more potent cholinesterase inhibitor.

Few important examples of *bioactivation* have been illustrated in Table 21.1. Important site of bioactivation of many toxicants is liver; hence it is a common target organ. Liver is regarded as the largest metabolic gland. However, if the metabolites are sufficiently stable, they may affect other organs after being transported there (e.g., bromobenzene on the kidney). In rare cases, other sites may be lungs, stomach, adrenal, retina and bone marrow.

Revision Questions

A. Long-Answer Questions

1. Define biotransformation and describe the mechanism of biotransformation.
2. Point out the principal objectives of biotransformation and show biotransformation of any two xenobiotics studied by you.
3. Describe :
   (i) Phase I reactions of biotransformation
   (ii) Phase II reactions of biotransformation.
4. Describe the complex nature of biotransformation with suitable examples.
5. Define bioactivation. Illustrate it with suitable examples.

B. Short-Answer Questions

1. Write short notes on :
   (i) Biotransformation
   (ii) Biotransformation sites
   (iii) Objectives of Biotransformation
   (iv) MMFO.
2. *Briefly describe :*
   (i) Microsomal oxidation
   (ii) Non-microsomal oxidation
   (iii) Microsomal reduction
   (iv) Non-microsomal reduction
   (v) Cytochrome P-450 species.
3. *Write short notes on :*
   (i) Factors affecting biotransformation
   (ii) Bioactivation
   (iii) Hydrolysis of xenobiotics
   (iv) Amino acid conjugation
   (v) Sulphate conjugation.

22

Bioaccumulation of Xenobiotics

SYNOPSIS

Introduction, Process of bioaccumulation, Illustration of biomagnification of important xenobiotics, Factor affecting biotransformation, Elimination of xenobiotics, Important cases of biomagnification of xenobiotics.

Introduction

Bioaccumulation of xenobiotic compounds first gained public attention in the 1960s with the invention of DDT, DDD and methyl mercury residues in fishes and wildlife.

Mortalities and reproductive failures in fishes and fish-eating birds were linked to unusually high concentration of DDT or its metabolites in the fatty tissues of these animals. Since the top-level carnivores, especially birds, had higher residue concentration of these chemicals than the food they consumed, it was logical to postulate that accumulation occurred primarily by transfer through the food chain. Actually, any xenobiotic may accumulate in the biological system only when its rate of uptake exceeds the rate of elimination. The environmental fate of xenobiotic compounds is shown in Fig. 22.1 :

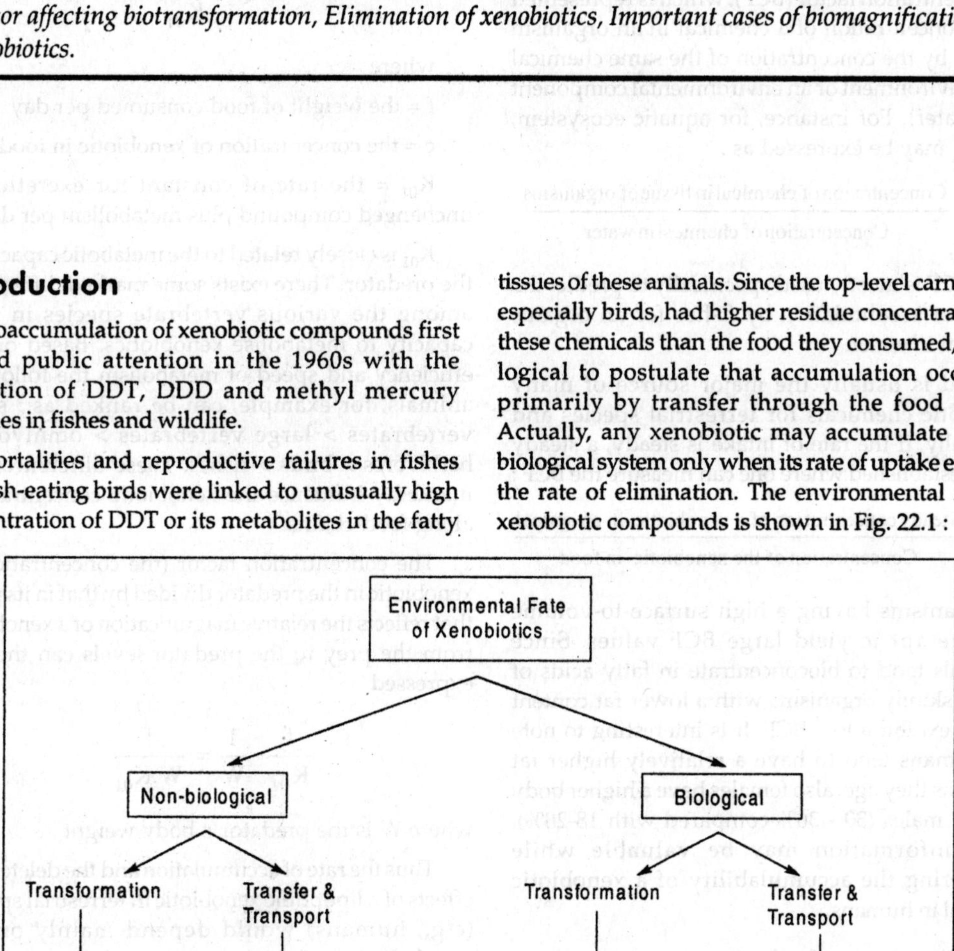

Fig. 22.1 : Environmental fate of xenobiotics

There are three related terms with the process of accumulation of xenobiotic in the organisms :

(i) Bioconcentration

(ii) Bioaccumulation, and

(iii) Bioamplification or Biomagnification.

(i) Bioconcentration

It refers to accumulation of xenobiotic chemicals in certain tissues of the organisms at concentrations above those of the immediate environment. The degree of bioconcentration is expressed as bioconcentration factor (BCF), which is represented as the concentration of a chemical in an organism divided by the concentration of the same chemical in the environment or an environmental component (e.g., water). For instance, for aquatic ecosystem, the BCF may be expressed as :

$$\text{BCF} = \frac{\text{Concentration of chemical in tissue of organisms}}{\text{Concentration of chemical in water}}$$

The BCF is mostly used specifically to predict the degree of accumulation by fishes for an organic pollutant in water.

Food is usually the major source of many xenobiotic chemicals for terrestrial species and eventually, if the rate of intake is steady, a steady state is established where one can measure the BCF :

$$\text{BCF} = \frac{\text{Steady concentration of a xenobiotic in an animal}}{\text{Concentration of the xenobiotic in food}}$$

Organisms having a high surface-to-volume ratio are apt to yield large BCF values. Since chemicals tend to bioconcentrate in fatty acids of tissues, skinny organisms with a lower fat content usually exhibit a low BCF. It is interesting to note that humans tend to have a relatively higher fat content as they age; also females have a higher body fat than males (30 - 36% compared with 18-28%). These information may be valuable while considering the accumulability of a xenobiotic chemical in humans.

(ii) Bioaccumulation

It is a wide term and refers to the accumulation of xenobiotic chemicals both from the medium and through the intake of food. Thus it is the uptake of xenobiotic chemicals dissolved/suspended in the medium and from the ingested food and sediment residues.

The major proportion of the intake of xenobiotic chemicals in case of a predator is by consumption of the preys that have already concentrated a xenobiotic. Walker and Moriarity (1987) suggested the following means for calculating the transfer levels of xenobiotics from a prey to a predator.

Supposing that a predator feeds on only one prey species and that all individuals of that prey species contain the same concentration of a xenobiotic (Q), then

$$Q = \frac{f.c}{K_{01}}$$

where

f = the weight of food consumed per day

c = the concentration of xenobiotic in food

K_{01} = the rate of constant for excretion of unchanged compound plus metabolism per day.

K_{01} is closely related to the metabolic capacity of the predator. There exists some marked differences among the various vertebrate species in their capacity to metabolise xenobiotics. Based on the efficiency and speed of metabolism the following animals, for example, can be ranked as : small vertebrates > large vertebrates > omnivores/herbivores > birds > fishes. These differences can markedly influence the xenobiotic concentrations along food-chains.

The concentration factor (the concentration of xenobiotic in the predator divided by that in its prey) that reflects the relative magnification of a xenobiotic from the prey to the predator levels can then be expressed :

$$\frac{f}{K_{01}} \cdot \frac{1}{W} = \frac{f}{W.K_{01}}$$

where W is the predator's body weight.

Thus the rate of accumulation and the deleterious effects of a lipophilic xenobiotic in terrestrial species (e.g., humans) would depend mainly on the metabolic fate of the former in the latter.

(iii) Biomagnification

It is also termed biological amplification. The xenobiotics such as mercurial salts (mercuric chloride), beryllium and lead compounds, aluminium cans, phenolic compounds, pesticides like

DDT etc., after getting their way into the environment, either do not degrade or degrade very slowly and are referred as non-biodegradable pollutants. These not only accumulate but are often biologically magnified or amplified : While energy decreases and becomes more dispersed at each trophic level in the food chain, some non-degradable pollutants become more concentrated with each link in a process that has now come to be known as "biological magnification or biological amplification". In simple words, the process of increase in the concentration of xenobiotic along the food chain may be termed as biological magnification or biomagnification. Actually biomagnification is a broader term which refers to the entire process of bioconcentration and bioaccumulation.

Many heavy metals and most organic insecticides concentrate biologically in the foodchain. As mentioned in the preceding section, the tendency of a xenobiotic to accumulate in the tissue of the organism is given by bioconcentration factor (BCF) which is represented in L/kg.

BCF may also be expressed as :

$$\text{BCF} = \frac{\text{(concentration of the xenobiotic in the organism, mass/vol)}}{\text{(concentration of the xenobiotic in the environment, mass/vol)}}$$

The BCF varies from species to species, and is affected by the metabolism and elimination system in the organisms. It is important to mention that the BCFs of most organic pesticides are much higher than those of inorganic pesticides. Further, it may be noted that BCFs are higher in aquatic ecosystems than in terrestrial ones. The BCFs of certain organic and inorganic xenobiotics in fishes are :

Chromium = 16 l/kg
Arsenic = 14 l/kg
Lead = 49 l/kg
Chlordane = 14,000 l/kg
DDT = 54,000 l/kg

Process of Bioaccumulation

In principle form, the bioaccumulation takes place when the rate of uptake of xenobiotic exceeds the rate of elimination. The process of uptake of xenobiotics from water, sediment and food have been worked out more in the aquatic systems, hence the general process of bioaccumulation applicable for the aquatic systems is being described below :

1. Uptake of chemicals from water

Many workers have reported direct uptake of xenobiotics from aquatic media by various organisms inhabiting in water (such as algae, annelids, insects, molluscs and fishes). The description of mechanism of uptake process of xenobiotics from water is concentration-based on following three transport processes :

(i) Diffusion

(ii) Special transport, and

(iii) Adsorption.

(i) Diffusion

Many xenobiotics enter the body of organisms by the process of diffusion. It is a physical process and occurs across permeable or semipremeable barriers against a concentration gradient (Δc). It does not require expenditure of energy.

Gills of the fishes are much vulnerable against xenobiotic insults. The lipid bilayers of semipermeable membranes of gills and lining of mouth and G.I. tract of fishes permit the rapid passage of lipophilic xenobiotics.

Weakly acidic and basic substances pass across biological membranes in unionised form. The water and other tiny ions up to moleculer weight of 100 pass through proteinous pores located in the biological membrane. Considerable proportion of xenobiotics have been found to passively diffuse across general body surface of various hexapods. The plasma proteins are known to facilitate the transport of xenobiotics into the blood.

(ii) Special Transport

Special transport of xenobiotics across biological membranes include both active and facilitated transport. The active transport takes place against concentration gradient and requires energy to move the xenobiotics. On the other hand, facilitated transport, does not require energy input and it does not concentrate xenobiotics against any concentration gradient. Metals have been found to be accumulated by both these transport processes. Low water salinity enhances the rate of uptake of many chemical substances.

(iii) **Adsorption**

It is also a physical process — a surface phenomenon. The adsorption of xenobiotics to the various adsorbents is a result of van der Walls' dipole-dipole interactions or coulombic forces between them. Xenobiotics are often absorbed either on the body surface of aquatic insects or directly to the gill surface of other aquatic organisms by covalent, electrostatic and molecular forces. It is particularly important as initial step in the accumulation process.

2. Bioavailability of xenobiotic in water

The uptake of the xenobiotic chemicals by the living forms largely depends upon their availability in the dissolved form in the aquatic media. It becomes obvious that the factors affecting the concentration of xenobiotics may also affect their uptake by the animals. The notable processes impairing bioavailability of xenobiotics in aquatic environment are adsorption to suspended solids, sediments, humic acids and other macromolecules, formation of colloidal suspensions, chelation, complexation and ionisation. Suspended particulates and adsorbents such as humic acids reduce the uptake of lipophilic xenobiotics by reducing their concentration in water.

3. Uptake of xenobiotics from food

Various xenobiotic chemicals absorbed through the gill surface and the integument are also readily absorbed by the G.I. tract by similar mechanisms of diffusion and transport. The lipophilic chemical compounds present in food are efficiently absorbed owing to long-term contact between the food and the membranes. Weak acids and bases are generally absorbed in unionised form. The pH of the stomach favours the diffusion of weak acids whereas the pH of the intestine favours the absorption of neutral or weakly basic xenobiotics. Uptake of metals from food depends on their forms viz., free or bound.

Illustration of Biomagnification of Important Xenobiotics

1. Pesticides (DDT as an example)

There are some established instances revealing the phenomenon of biomagnification of pesticides in developed countries, and Paul Miller won the Nobel Prize in 1948 for its (DDT) synthesis. However though its use is still widespread in developing countries, its use in the U.S.A. is banned since 1973. Salient example of biological magnification of DDT in an aquatic system is outlined below, where concentration of DDT continuously increased from water to birds :

Component	Concentration of DDT in ppm
Water	0.00005
Zooplanktons	0.04
Minnows	0.94
Predatory fishes	2.07
Fish-eating birds	22.8

Wordwell et al (1967) studied DDT accumulation in a typical food chain of aquatic ecosystem. The results are shown in Table 22.1 :

Table 22.1 : Concentration of DDT in a food chain of the aquatic ecosystem

Sl. No.	Component	DDT Residues (ppm)
1.	Water	0.00005
2.	Plankton	0.04
3.	Silverside minnow	0.23
4.	Sheephead minnow	0.94
5.	Pickerel (predatory fish)	1.33
6.	Needlefish (predatory fish)	2.07
7.	Heron (feeds on small aquatic animals)	3.57
8.	Tern (feeds on small aquatic animals)	3.91
9.	Herring gull (scavenger)	6.00
10.	Fish hawk (osprey) egg	13.8
11.	Merganser (fish-eating duck)	22.80
12.	Cormorant (feeds on larger fishes)	26.40

The end result of DDT use is that entire population of predatory birds such as the fish hawk (osprey) and of detritus feeders as fiddler crabs are wiped out. Birds are more vulnerable as DDT interferes with their egg shell formation by causing a breakdown in steroid hormones which results in fragile eggs that break before the young can hatch.

2. Metals (e.g. Mercury)

Besides pesticides, there are also some metals like mercury, lead, copper etc. that show a similar behaviour of biomagnification in food chain. Both inorganic and organic forms of mercury are found in industrial effluents, polluting chiefly the water bodies. Mercury pollution due to methyl mercury is a global problem.

Methyl mercury was responsible for the Minamata epidemic in Japan and Sweden.

Mercury is a very stable heavy metal having a half life of 18.2 years — a very long period of time that provides ample opportunity of its getting picked up by living organisms.

Inorganic mercury released into the environment is converted into more toxic methyl mercury compounds by the action of certain anaerobic bacteria; this transformation occurs in the sediments and bottom muds of waterways.

Under anaerobic conditions, *Clostridium cochlearium*, and, under aerobic conditions, *Pseudomonas sp.* and a fungus *neurospora crassa* convert Hg to more toxic monomethyl mercurials. Alkaline conditions favour the conversion of monomethyl mercury to volatile dimethyl mercury which could subsequently contaminate the media; under mild acidic conditions, dimethyl mercury is converted back to non-volatile methyl mercury, which may then be concentrated by aquatic biota and magnified through the food-chain.

Because of its affinity to tissue components of living organims, Hg rapidly binds with tissues and stays there for a long time. It accumulates in tissues like liver and muscles. It is soluble in both polar and apolar fractions.

The biomagnifications of Hg takes place along the following food-chain :

HUMAN

↑

Fishes

↑

Benthic diatoms and Dipteran larvae,

copepods, crustaceans

↑

Hg in aquatic media

The amount of Hg accumulated by fishes does not harm it. But man eating such contaminated fishes regularly accumulates more Hg in his tissues which, when it reaches the threshold level, starts producing toxicity. Only methyl mercury form of mercury is persistent and accumulates in food-chain.

Mercury concentration in blood and brain of the affected fetus has been reported to be about 20% higher than mother. Methyl mercury penetrates through placenta. Hg poisoning is caused due to inactivation of several sulfhydral enzymes by replacement of hydrogen atoms in sulfhydral groups.

3. Radioactive substances

Similar to pesticides and metals, the radioactive substances such as P-32, Strontium-90, Iodine-131, Caesium-137 etc. tend to accumulate in body tissues and lead to biomagnification. As an example, P-32 was found to be concentrated as much as 5,000 times above levels in the water in white fish downstream from the Hanford Atomic Power Plant on the Columbia river. Blue gills and crappies in the same river water were found to contain 20,000-30,000 times the concentration of P-32, while filamentous algae were found to have up to 100,000 times the water level of P-32. The concentration of P-32 increases from river water to egg yolk of ducks and geese (about two million times more than in water!).

Fe-59, Zn-65 and I-131 are the other radioactive isotopes that can be concentrated by the living organisms. These are the sources of radiation exposure to the general public when certain foods are eaten. Fishes may contain P-32, oysters and clams may contain Zn-65 and milk may contain I-131.

Concentration factors for Cs-137 is 250 in muscle and for Sr-90, it is 500 in bones of water birds as compared to concentration of these radionuclei in water. Concentration factors for Sr-90 in various components of an aquatic food web near an atomic power plant are shown in Fig. 22.2 :

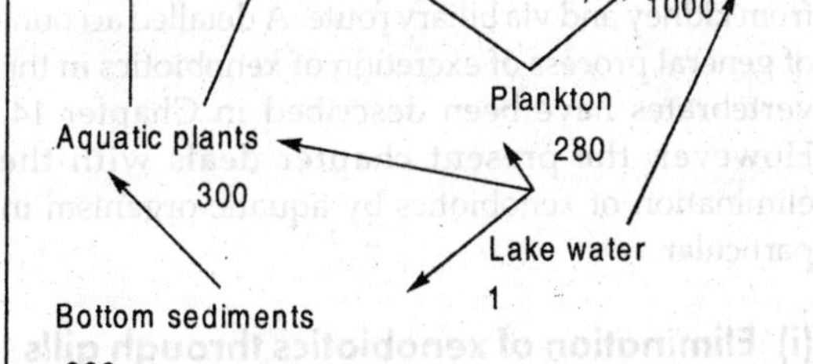

Fig. 22.2 : Concentration of strontium-90 in various parts of a food web of a lake receiving atomic wastes. Average concentration factors are shown in term of lake water = 1

Factors Affecting Biomagnification

Biomagnification or bioamplification of an environmental xenobiotic chemical depends on the following factors :

1. Physical properties of xenobiotics.
2. Duration of the persistence of the xenobiotic molecules in the environment.
3. Stability of the xenobiotic chemicals in the environment.
4. Availability of the primary pickers in the environment.
5. Affinity of the toxicants to various tissue components of the living organisms.
6. Availability of the next trophic level organism.
7. Number of the living organisms picking the xenobiotic molecules at different trophic levels in any ecosystem.
8. Amount of food intake by organisms.
9. Texture of the aquatic environment.
10. Body size and lipid content of the tissues of the organisms. In principal form — more the lipid content and the size of the organism, more is the accumulation of xenobiotics — and, consequently, more is the chance of biomagnification or bio-amplification.
11. Rate of elimination of xenobiotics by the living organism.

Elimination of Xenobiotics

Elimination of xenobiotics in the vertebrates may occur through various routes viz., transport across integument or respiratory surfaces, excretion from kidney and via biliary route. A detailed account of general process of excretion of xenobiotics in the vertebrates have been described in Chapter 14. However, the present chapter deals with the elimination of xenobiotics by aquatic organism in particular.

(i) Elimination of xenobiotics through gills

Elimination of nonpolar xenobiotics in fishes takes place through gills. Various xenobiotics such as weak electrolytes, weak bases, insecticides like DDT and pentachlorophenol are principally eliminated through gills of fishes.

(ii) Elimination of xenobiotics by liver and gall blader

Liver is the principal organ for biocatalytic conversion of lipophilic xenobiotics into their hydrophilic forms. The biotransformed toxicants are then transported from liver to the gall bladder. The gall bladder discharges these biotransformation products of xenobiotics with the bile into the small intestine and, ultimately, these are eliminated through feces. Various xenobiotics like Hg, Pb, As etc. are generally transported into the blood and excreted in the bile.

(iii) Elimination of xenobiotics through kidney

Kidney usually eliminates hydrophilic xenobiotics of low molecular weight — either by glomerular filtration or by diffusion or by secretory processes of tubules. Xenobiotics bound to plasma proteins are retained by the glomeruli and, hence, are not eliminated unless they become free.

Important Cases of Biomagnification of Xenobiotics

1. Minamata Disease

Mercury pollution, especially due to methyl mercury, is a global problem. Mercury, a byproduct of the production of vinyl chloride, is used in many chemical industries. It is also a byproduct of some incinerators, power plants and laboratories. In Japan, illness and even death occurred in the 1950s among fishermen who consumed fishes, crabs and shell-fish contaminated with methyl mercury from the industrial city of Minamata in Japan. The main unit of this city, Shin-Bihon Chisso Hiryo Co., began to produce vinyl chloride and acetaldehyde by the catalytic conversion of acetylene, since 1949. During this process, some of the $HgCl_2$ catalyst was unknowingly converted to methyl mercury. This factory kept ejecting its effluent wastes into the Minamata Bay of Kyushu in Japan during 1953-1961, and at Nigata, also in Japan, in 1965. The people who took the fishes, crabs, and shell-fish suffered from Minamata disease. In Japan, in 1952, due to this disease 17 people died and 23 were disabled permanently.

Further, in Sweden also, many rivers and lakes became contaminated due to widespread use of mercury compounds as fungicides and algecides in paper and pulp industries, and in agriculture. Chloral alkali plants were the chief source of mercury containing effluents. Methyl mercury is stable, persistent and accumulates in food-chain. It is soluble in lipids and, thus, after being taken by animals, it accumulates in fatty tissues.

Fish may accumulate the methyl mercury ions directly. There may be nearly 3,000 times more mercury in fish than in water. Swedish fish-eaters have been reported to have high mercury content in their blood.

Symptoms of Minamata Disease

The symptoms of Minamata include numbness, visual disturbances, dysphasia, ataxia, mental deterioration, convulsions and, finally, death. Mercury readily penetrated the CNS of children born in Minamata causing teratogenic effects. In drosophilla, methyl mercury (0.25 ppm) treatment brought chromosomal disjunction in gametes.

2. Itai-Itai Disease

It is a case of Cd (cadmium) biomagnification. It was first reported in Japan. People at Toyana, Prefecture, Japan, consumed Cd-contaminated fishes during 1940-1960 and fell victim to this disease. Cadmium binds with tissue protein in living organisms and accumulates. For detail see Chapter 5.

3. Yusho Disease

It is a case of biomagnification of PCBs. PCBs are commonly used in electrical capacitors, paints, transformers, heat-transfer fluids and varnishes. Polychlorinated biphenyl (PCBs) are lipophilic and highly stable xenobiotics. These accumulate in the fatty tissues of the living organisms.

PCB magnification has been reported in bald eagles and human beings in the following manners :

PCBs → Crustaceans → Fishes → Human

Or

PCBs → Crustaceans → Small birds → Eagle

Or

PCBs → Crustaceans → Eagle

It reveals that both man and eagle — being on the top of the food-chain — accumulate maximum amount of PCBs. PCBs are excreted with breast milk. High concentration of PCBs impairs reproduction and proves fatal for bald eagles.

Symptoms of Yusho-Disease

Nausea, vomiting, weakness, anorexia, fatigue, numbness of extremities and abnormalities in pigmentation are the principal symptoms due to PCB intoxication. In children, PCB produces abnormal tooth formation.

Revision Questions

A. Long-Answer Questions

1. Define bioaccumulation and describe the processes involved in bioaccumulation.
2. Define biomagnification. Illustrate the biomagnification of :
   (i) Pesticides
   (ii) Metals
   (iii) Radioactive substances.
3. Describe two important cases of biomagnification studied by you.
4. Describe various processes for the elimination of xenobiotics.

B. Short-answer Questions

1. *Describe in brief :*
   (i) Bioconcentration
   (ii) Bioaccumulation
   (iii) Biomagnification.
2. *Write short notes on :*
   (i) Factors affecting biomagnification
   (ii) Minamata disease
   (iii) Itai-Itai disease
   (iv) Yusho-disease.

23

Antidotal Procedures

SYNOPSIS

Introduction, Classification of antagonists, Pharmacological assessments of certain antidote action, Principles of antidotal treatment, Antidotal procedures (pathways and measures).

Introduction

The chemical compounds which produce deleterious effects on living tissue of animals are termed 'poisons'. The effect of most poisons depends on the quantity consumed and the age and physical condition of animals. Different kinds of poisons are listed in Table 23.1 :

Table 23.1 : Different kinds of poisons

1. Corrosives	Like strong acids and alkalies which cause local destruction of tissue
2. Irritants	Produce congestion of the organ when they come in contact with them
3. Neurotoxins	Adversely affect the nervous system. It comprises of narcotics, barbiturates, alcohol and anaesthetics
4. Hemotoxins	These combine with blood and prevent oxygen from forming oxyhemoglobin viz., CO and hydrogen cyanide.
5. Hepatotoxins	Affect the liver cells
6. Nephrotoxins	Cause injurious effects on nephrons
7. Cardiotoxins	Adversely affect the functioning of heart

An antidote is strictly defined as 'a remedy for counteracting a poison' or substances which prevent action of poisons are called '*antidotes*', and procedures applied to reduce toxicity in acute poisoning are termed *antidotal procedures*.

Toxicity of a chemical in an animal may also be either increased or decreased by simultaneous or consecutive exposure to another chemical. If this combined effect is equal to the sum of the effect of each substance given alone, the interaction is considered to be additive, e.g., combinations of most organphosphorus pesticides on cholinesterase activity. If the combined effect is greater than the sum, the interaction is considered to be synergistic, e.g., CCl_4 and ethanol on the liver. The term 'potentiation' is used to describe the situation in which the toxicity of substances on an organ is markedly increased by another substance which itself has no effect on that organ. For example, Isopropanol has no effect on the liver, but it can considerably increase the hepatotoxicity in presence of CCl_4.

The exposure of a chemical to an organism may thus reduce the toxicity of another and the chemical reaction thus produces a less toxic product. This phenomenon is known as **chemical antagonism**, e.g., Chelation of heavy metals by dimercaprol.

In toxicology, the poison or toxicant may also be referred to as agonist and its antidote as 'antagonist'. In practice, an antidote may be considered a physiological agent which reverses or prevents the development of symptoms and signs of poisoning. Such antidote may either be chemical or pharmacological or physiological in nature.

Classification of Antagonists

Based on the mechanism of action, the antagonists are classified as :

1. Competitive Antagonist
2. Non-Competitive Antagonist
3. Chemical Neutralizers
4. Metabolic Inhibitors
5. Oxidation-reduction Reactions
6. Chelating agents
7. Mechanical agents
8. Universal antidote
9. Other examples.

1. Competitive Antagonist

These agents compete directly with agonist for receptor sites and if sufficient concentration is reached, the agonist is displaced from the receptor site and its actions are terminated, e.g. :

(i) Naloxone for opiate (sleep drug) poisoning
(ii) O_2 for CO_2 intoxications
(iii) Naloxone for morphine and other related narcotics
(iv) Edrophonium for neuromuscular exciting agents, viz., curare.

Therefore, competitive antagonism exists only when the agonist and antagonist act on the same receptor.

2. Non-Competitive Antagonist

These antidotes act by producing a physiologic effect which competes with that produced by the agonist. An example of this mechanism is use of atropine for treating organophosphate or carbamate insecticide poisoning. The organophosphate or carbamate insecticide inhibit the enzyme acetylcholinesterase at the nerve ending and the result is an elevated level of acetylcholine producing excessive stimulations. Atropine competes with the effects of the elevated level of acetylcholine, but not directly with insecticides. Other example of non-competitive antagonist is the use of haloalkylamine for cyclopropane, a serious toxic agent. Non-competitive antidotes are also termed physiological antidotes.

3. Chemical Neutralizers

These antidotes, in fact, neutralise the reactions of agonists or poisons or toxicants, e.g., administration of sodium thiosulphate for cyanide poisoning results in neutralization of the cyanide ion (CN^-) by forming a non-toxic compound, sodium thiocyantae. Another example of this type is $AgNO_3$ (silver nitrate) as agonist and NaCl (Sodium chloride) as antagonist or antidote. The reaction is :

$$\underset{\text{(Silver nitrate)}}{AgNO_3} + \underset{\text{(Sod. chloride)}}{NaCl} \rightarrow \underset{\text{(Sod. nitrate)}}{NaNO_3} + \underset{\text{Sil.chlo. (ppt)}}{AgCl}$$

The silver chloride so formed in the form of precipitate is removed.

The third example of this class is Trivalent organic arsenicals as agonist and Ferric (Fe^{3+}) oxide (hydrate) as antagonist. Arsenic is found in two forms — trivalent and pentavalent. Former is more toxic than the latter.

4. Metabolic Inhibitors

When sufficient amount of ethanol (ethyl alcohol) is administered to the person poisoned with methanol (methyl alcohol) or ethylene glycol, the ethanol blocks the metabolism of the methanol or ethylene glycol. Thus serious toxicity — which results from the formation of toxic metabolites — is prevented. In this way ethyl alcohol acts as an antidote (antagonist) for methyl alcohol (the agonist or poison).

5. Oxidation-Reduction Reactions

The person with methemoglobinemia have a significant portion of iron (Fe^{3+}) in their circulating hemoglobin. In this state, the oxygen-carrying capacity of the red blood cells is decreased. Methylene blue reduces the methemoglobin in normal hemoglobin and, therefore, reverses the toxic effects.

6. Chelating Agents

The chelating-agents with multiple free valancies bind metal molecules, forming a stable complex which is excreted from the body and thus prevent the toxicity. The most commonly used chelating agents are :

(i) Deferoxamine for iron poisoning
(ii) Calcium EDTA (Ethylene-diamine-tetra-acetic acid) for lead and zinc poisoning, and
(iii) Dimercaprol (BAL) (British-anti-Lewistic) for arsenic poisoning. Important chelating agents are described ahead in this chapter.

7. Mechanical Agents

These antidotes check the further translocation of poisons from the portal entry.

Example 1 : In case of poisoning by strychnine (an alkaloid poison), finely powdered activated charcol in 150 mg/g dosage, absorbs the poison molecules within its pores. Powdered charcoal is very effective as its one gm absorbs 5,000 mg of strychnine.

Example 2 : Intake of fats, egg-albumen etc. forms a thick coating over gastrointestinal mucosa and prevents the translocation of the toxicants or poisons into the blood. Thus, by using mechanical antidotes, the systemic as well as the local toxicity of poisons is prevented. These antidotes are chiefly used in the cases of irritants poisoning.

8. Universal Antidote

The chemical substances which are applied over unknown toxicities or poisoning are known as universal antidotes. For example — two parts of burned powdered toast, one part of milk of magnesia and one part of strong tea may be applied in any unknown poisoning.

The burned powdered toast is a source of carbon, the milk of magnesia is a source of magnesium oxide and the strong tea is a source of tannic acid. The carbon absorbs poisons, magnesium oxide has a soothing effect on mucus membrane of the stomach and a laxative action that tends to neutralise poisons, and tannic acid tends to neutralise caustic alkaline substances, if present in poison.

9. Others

N-Acetylcysteine is used as an antidote for acetaminophen acting as agonist and diphenylhydramine for treating phenothiazine-induced adverse reactions.

Certain other practices should also be adopted to minimise the toxic effects in the body, based on the principles of antidotes. First, the poison must be diluted. This may be accomplished by letting the sufferer drink as much water as possible. A large volume of water also promotes vomiting. Next, the stomach should be emptied. This is usually done by stimulating the vomiting reflex. Warm salt water, soapy water of substances such as tartar or mustard often induce vomiting. If this fails, the sufferer may gag himself into vomiting by holding his finger to the base of the tongue. After the sufferer has been induced to vomit, it is useful to give milk or universal antidote.

Another way in the treatment of poisons is to eliminate it from the systemic circulation. However, as with ethyl alcohol, methyl alcohol, ether, benzene and acetone, the elimination may be aided by breathing deeply in order to expel the substance with the exhaled air. Drinking as much fluid as possible will aid in excretion through kidneys. The use of sodium sulphate and sodium phosphate may aid in the elimination of some poisons from blood by way of the liver, bile and digestive system.

If sleep-producing drugs — such as opium or morphine — have been taken, it is best to keep the patient awake by giving strong coffee. In instance of strychnine poisoning, any stimulant should be avoided and the sufferer should be kept under doctor's supervision as quiet as possible.

As mentioned above in this chapter, antidotes to poison may be physiological, chemical or pharmacological in nature.

To date, very few direct antidotes are known. Their number is increasing with increasing knowledge of the pharmacology and biochemistry of poisons. In Table 23.2 important well-known specific antidotes (antagonists) for poisons (agonists) are mentioned briefly.

Table 23.2 : Poisons (Agonists) and their specific antidotes (Antagonists)

Sl. No.	Toxicants or Poisons	Antidotes
1.	Arsenic	Dimercaprol (BAL)
2.	Mercury	Dimercaprol (BAL)
3.	Copper	Pencillamine; EDTA
4.	Lead	Dimercaprol; EDTA
5.	Pesticide (Cholinesterase inhibitors) except carbamate group of insecticides	Paralidoxime
6.	Heparine (drug)	Protamine
7.	Formaldehyde	Ammonia
8.	Cyanide	1. Methemoglobin (It is formed by nitrite administration) 2. Thiosulphate
9.	CO_2	O_2
10.	Fluroacetate	Monoaceitin
11.	Bromide (Br^-)	Chloride(Cl^-)
12.	Strontium	Calcium salts
13.	Radium	Calcium salts
14.	Curare (Neuromuscular blocking agent)	Edrophonium
15.	Morphine and other related narcotics	Naloxone
16.	CO	O_2
17.	Thallium	Potassium salts
18.	Cyclopropane	Haloalkylamine
19.	Histamine	Antihistamines
20.	Morphine	Atropine

Pharmacological Assessments of Certain Antidote Actions

1. Antidote for Halogens, Cyanides and Hyrocyanic Acid

The sodium thiosulphate ($Na_2S_2O_3$) is employed as an antidote in poisoning by halogens, cyanide compounds and hydrocyanic acid. The use of sodium thiosulphate is based on its property of evolving sulphur.

The reaction of thiosulphate against potassium cyanide poisoning may be expressed as :

$$KCN + Na_2S_2O_3 \xrightarrow{H+} \underset{\text{(Pot. Thiocyanate)}}{KSCN} + Na_2SO_3$$

The formed potassium thiocyanate is much less poisonous than potassium cyanide. Therefore, sodium thiosulphate should be taken as first aid in poisoning by hydrocyanic acid or its salts.

Sodium nitrite is also used as an antidote in cyanide poisoning. In this case it is introduced intravenously in form of 1% solution up to 5 ml.

Sodium thiosulphate with HCl acid has been found to heal itching. The formed finely dispersed sulphur and the SO_2 — that has a high parasiticidal effect — stops the itching :

$$Na_2S_2o_3 + 2HCl \rightarrow 2NaCl + SO_2\uparrow + S\downarrow + HOH$$

Sodium thiosulphate can also be used as an antidote in poisoning by arsenic, mercury and lead. It is administered intravenously in allergic diseases, in the form of a 30% aqueous solution.

2. Arsenic Antidote

All the arsenic compounds are highly poisonous. Arsenic poisoning results in sharp collapse, muscular weakness and, in the long run, paralysis and death. In acute arsenic poisoning apomorphine, an alkaoid, is given subcutaneously as an emetic and gastric lavage is used. A preparation called arsenic antidote (Antidotum Arsenic, a mixture of burnt magnesia and iron (III) sulphate) can be employed. These substances are kept separately in the pharmacy and are poured together when needed. The reaction of these two substances form iron (III) hydroxide ($Fe(OH)_3$) and magnesium sulphate ($MgSO_4$) :

$$3MgO + Fe_2(SO_4)_3 + 3HOH \rightarrow 2Fe(OH)_3\downarrow + 3MgSO_4$$

The freshly precipitated iron hydroxide is, in fact, the antidote, because with arsenic it forms iron (III) arsenite ($FeAsO_3$), a compound insoluble in the gastric juice. Moreover, the freshly precipitated iron hydroxide adsorbs arsenic compounds, while the magnesium sulphate formed in the reaction acts as a purgative and facilitates the excretion of the poison from the body. This is why it is recommended to prepare this antitodes extempore because only freshly precipitated iron hydroxide has adsorbing properties.

A more effective method against arsenic poisoning is the use of a supersaturated alkaline solution of hydrogen sulphide in water with the addition of magnesium sulphate. The preparation is dispensed in 100 ml bottles. Its composition (in grams) is H_2S 0.5 - 0.7, NaOH 0.1, $NaHCO_3$ 1.25, and $MgSO_4$ 0.38. The mechanism of action of this antidote consists in that the reagent forms insoluble compounds with arsenic and, therefore, precipitates are not absorbed.

In acute arsenic poisoning, the administration of this antidote is preceded by giving 3 gm of citric acid or 4 gm of tartaric acid or 1.5 tablespoons of vinegar essence diluted in 300 ml of water. In this case, it is recommended to drink albumin water.

3. Mercury Antidote

Soluble mercury salts are very poisonous. The first sign of mercury poisoning is a sharp metallic taste in mouth, pain in the stomach, vomiting, loose bowels (diarrhea) and cardiac depression. Persons suffering from kidney ailments are especially sensitive to mercury preparations.

First aid in mercury poisoning consists of intaking of a metal antidote (Antidotum metallosum) or a suspension of activated carbon. One must simultaneously take albumin water, milk, or burnt magnesia with water. Sodium chloride and food product containing it must not be used because this can enhance the poisoning owing to the formation of another soluble and poisonous mercury dichloride.

Another antidote used in mercury salt poisoning is sodium calcium edetate, also known as tetacin calcium, a white crystalline powder. In its chemical structure tetacin calcium is a cyclic compound in which the calcium can be substituted by ions of heavy metals with the formation of water-soluble complexes having a low toxicity. The complexes are rapidly excreted with the urine, which underlies the use of this preparation as an antidote in poisoning by heavy metals salts :

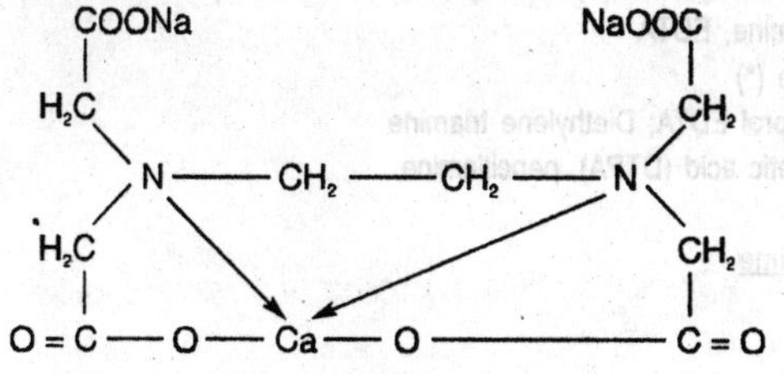

Tetacin Calcium

A 10% solution of tetacin calcium for injection is a pharmacopial preparation

Principles of Antidotal Treatment

Antidote to poisons may be chemical, physiological or pharmacological in nature. It counteracts the principal effects of a given poison. Yet a very few direct antidotes are known. The number is increasing with increasing knowledge of the pharmacology and biochemistry of poisons. A few of the antidotes along with their mechanisms of action are summarised in Table 23.3.

Mechanism I

BAL, or dimercaprol, is often useful as an antidote for poisoning by lead, arsenic, and mercury. It is most effective against arsenic. BAL exhibits its greatest efficacy when it is administered soon after the exposure to the toxicant. Once poisoning is well-established, BAL has little value in any poisoning.

Calcium ethylene diamine tetraacetate (calcium EDTA) has usefulness as a chelating agent in lead poisoning.

Mechanism II

Oximes, such as protopam chloride, 2-PAM, and DAM are useful in the treatment of organophosphorus compounds. Protopam and other members of this group act to reverse cholinesterase inhibition and are, therefore, specific antidotes. They do require some time to act; therefore, they should be used in conjunction with atropine.

Mechanism III

Chemicals like cystine and monoacetin provide protection against poisoning because they inhibit the metabolic oxidation of the compounds to more toxic ones from less toxic precursors.

Mechanism IV

There is still another group of antidotes which specifically accelerate the excretion of poisons, for example calcium salts promote excretion of strontium and radium.

Mechanism V

Another group of antidotes include chemicals which compete with poisons for essential receptors, for example, nalorphin competes with morphine antidote for the receptor sites.

Vitamin K is useful in any poisoning involving a prolongation of clotting time due to a prothrombin deficiency, such as that due to warfarin.

Mechanism VI

Atropine and antihistamines act by blocking receptors for the agents which are responsible for the toxic effects e.g., organophosphorus insecticides and histamines.

Mechanism VII

Methylene blue can be used to produce methemoglobinemia in cyanide poisoning, and, oddly enough, it can be used to reduce methemoglobinemia in nitrite and chlorate poisoning. The activity of methylene blue may be enhanced by the simultaneous use of ascorbic acid to reduce methemoglobinemia.

It may be pertinent to mention that some of the antidotes have toxic properties of their own and can complicate the clinical picture as to make it impossible to save the patient. Thus, toxicology has assumed a great practical importance in search of superior antidotes.

Table 23.3 : TOXIC AGENTS, SPECIFIC ANTIDOTES, AND MECHANISM OF ANTIDOTAL ACTION

Toxic Agent	Specific Antidote
Mechanism I : Complexes with poison	
Metals	
Arsenic, mercury, iron	Dimercaprol (BAL)
	Sodium ferrocyanide (by mouth): deferroxime
Copper	Pencillamine, EDTA
Thallium	Dithizone (*)
Lead, and Radioactive materials	Dimercaprol EDTA; Diethylene triamine pentaacetic acid (DTPA), pencillamine
Pesticides	
Cholinestease inhibitors (except carbamate group of insecticides)	Pralidoxime
Drugs and other chemicals	
Heparin	Protamine
Digitalis	Steroid-binding resins
Formaldehyde	Ammonia (by mouth)
Cyanide	Methemoglobin (forced by nitrite administration)
Bacterial Toxins	
Botulinus toxins	Botulinus antitoxins
Mechanism 2 : Metabolic conversion of poison to non-toxic product	
Cyanide	Thiosulphate
Mechanism 3 : Antidote blocks metabolic formation of poison from less toxic precursor	
Selenocystathionine	Cystine
Fluoroacetate	Acetate; monoacetin
Mechanism 4 : Antidote specifically accelerates excretion of poison	
Bromide	Chloride
Strontium, Radium	Calcium salts
Mechanism 5 : Antidote competes with poison for essential receptors	
Neuromuscular blocking agents e.g., curare	Neostigmine; edrophonium
Morphine and other related narcotics	Naloxine; other related antagonists
Coumarin Anticoagulants	Vitamin K
Thallium	Potassium salts
Carbon monoxide	Oxygen
Mechanism 6 : Antidote blocks receptors that are responsible for toxic effects	
Cholinesterase inhibitors	Atropine
Cyclopropane	α-adrenergic blocking agents i.e., haloalkylamine
Histamine	Antihistamines
Mechanism 7 : Antidote restores normal function by repairing or bypassing effect of poison	
Agents that produce methemoglobinemia	Methylene blue
Digitalis	Potassium salts, poison-adrenergic blocking agents; procaineamide
Botulinus toxin	Guanidine
5-Fluorouracil	Thymidine
6-Mercaptopurine	Purines

***indicates practical efficacy of antidotes not well-established or doubtful.**

Antidotal Procedures (Pathways and Measures)

Antidotal therapy, in fact, is chiefly based upon the following three aspects :

1. Procedures which diminish absorption/translocation of poisons
2. Procedures which enhance termination of action of poisons, and
3. Procedures which elevate the threshold level of toxicity.

1. Procedures applied to diminish absorption/translocation of poisons

By this procedure efforts are made to prevent absorption/translocation of poisons into the blood and to eliminate them from their sites of contact especially in the case of contact or ingested poisons.

In case of contact poisons, viz., acids and pesticides, the affected part is washed thoroughly with warm water to eradicate the poison molecules. The wounds are then subjected to healing agents.

In case of ingested poisons, elimination of poison from the sufferer's (victim) gut is done by the following methods :

(i) By the use of emetics which induce vomiting
(ii) By the use of gastric lavage
(iii) By the use of cathartics
(iv) By the use of chemicals.

(i) By the use of emetics (vomiting-inducing agents)

Vomiting may be induced mechanically also, viz., by tickling the back of the pharynx in case of conscious patients.

Chemically, vomiting may be induced by using the following :

(a) **Ipecac syrup**

Ipecac grain or syrup, if available, may be given to induce emesis. The conventional emetic dose for children is 15 ml and for adolesecent and adults is 30 ml — followed by 200-250 ml of water.

(b) **Dry mustard**

One teaspoonful of dry mustard is dissolved in one glass of lukewarm water. One-fourth of this mixture is given to the victim followed by a glass of warm water. The procedure is repeated after every 1-2 minute which induces vomiting.

(c) **Warm salt water**

Excess table salt dissolved in lukewarm water in large quantities induces vomiting.

(d) **Salts of heavy metals**

A mixture of $ZnSO_4$ and $CuSO_4$ in 1% solution may induce vomiting.

(e) **Apomorphine**

Apomorphine hydrochloride has been proved a potent vomiting-inducing agent. The major disadvantage of apomorphine hydrochloride is its inactivity by oral intake. It is, therefore, generally given through parenteral routes. Subcutaneous injection of this drug in 0.006 g concentration produces vomiting in few minutes.

(ii) By the use of gastric lavage

When vomiting fails after emetic treatment and, sometimes, even after vomiting, the unabsorbed poison is physically removed from the stomach by gastric lavage. This method is applied by physicians in hospitals. In this method a wide tube is applied to wash the stomach and suck out the stomach contents with fluid.

(iii) By the use of cathartics

The use of cathartics hasten the passage of poisons through the G.I. tract. After about one hour of the ingestion of the poison — and in case of decreased bowel motility which slows down the passage of ingested poisons through the G.I. tract — certain oil or saline cathartics enhance the passage of bowel, thereby the poisons. Thus, the deleterious effects of poisons is diminished. In general, saline cathartics are more effective than the oil cathartics.

(iv) By the use of chemicals

In addition to physical removal of poisons from the stomach, their virtual removal may be accomplished by the use of chemical substances that effectively react with, firmly bind with, or adsorb, those poisons.

Such chemical substances may be grouped into three :

(a) Neutralisers
(b) Chelating agents, and
(c) Adsorbents.

(a) Neutralisers

When the nature of the metal poison is unknown, universal antidote (described earlier) may be applied to the victim and, if the chemical nature of the poison is known, specific chemical substances may be used to neutralise the poisonous effect of the toxicant as given below :

Chemical substances	Applied for
1. Acetic acid	Alkali
2. Ferrous oxide	Arsenicals
3. MgO	Acids
4. CaO	Acids
5. Glucosides, $KmnO_4$	Phosphorus
6. $NaHCO_3$ 70% solution	CH_3OH
7. Activated Charcoal	Cyanides
8. Barbiturates	$AgNO_3$
9. NaCl	$AgNO_3$
10. Glucosides	Amidopyrin, Antipyrin
11. Antivenom	Snake venom

(b) Chelating agents

The chemical substances which form more stable complex with metallic toxicants — thereby affecting their absorption and ultimately reduce their toxicity — are called chelating agents. In brief, chelating agents are used to reduce the absorption and translocation of metallic poisons/toxicants.

Some common chelating agents are :

1. **British anti-Lewistic (BAL) or Dimercaprol**

It is applied in inactivating the compounds of arsenic, mercury, antimony, bismuth, chromium, cobalt, copper, nickel, zinc and thallium. Dimercaprol-metal complexes are relatively non-toxic and are subsequently excreted in the urine or feces. Dimercaprol contains thio (-SH) group.

2. **Calcium-Sodium BAL**

It chelates with lead.

3. **Ethyl diamine tetra acetate (EDTA)**

It is a commonly used chelating agent and forms stable and relatively non-toxic complexes with metallic compounds of As, Pb, Hg, Fe, Ca and Co. The complexes are easily excreted by the kidney along with urine.

4. **Sodium-Calcium EDTA**

It is one of the widely used chelating agents which especially binds with lead compounds. It is usually given in the form of intravenous infusion. It may also be given intramuscularly, especially to children and orally in case of mild lead poisoning.

5. **Hexa-fluro-monothioacetyl aetone (HFTA)**

It binds with copper, nickel, cadmium, zinc and lead. It prevents these metals from converting to ionic forms and thus prevents their deleterious effects.

6. **Desferrioxamine**

It is used in iron poisoning by reducing its absorption from gut. It converts iron into inactive, non-toxic water soluble form and is eliminated in urine.

7. **Desferrioxamine mesylate**

This chelating agent is especially used in lead poisoning.

8. **Pencillamine**

It forms stable complexes with Cu, Pb and Hg and thus facilitates the excretion of these metals through the kidney. It is usually recommended to be ingested orally and thus has advantage over Na-Ca EDTA.

(c) Adsorbents

Activated charcoal is the commonly used adsorbent to diminish absorption and translocation of toxicants. If activated charcoal is not available, several other adsorbents may be used. For instance, a highly adsorptive clay known as *montmorillonite* is more palatable but inferior than powdered charcoal.

2. Procedures applied to enhance termination of action of poisons

The procedures employed to enhance the termination of action of poisons may broadly be divided into two groups :

(a) Physical procedures, and
(b) Chemical procedures.

(a) Physical procedures

Enhancing termination of action of poisons, the commonly applied physical procedures include (i) diuresis, (ii) dialysis, and (iii) hemoperfusion.

(i) **Diuresis**

Ion-trapping and increased urine flow are the two basic aspects involved in diuresis. Once the poison is ionized, then reabsorption from the renal tubules is impaired, ultimately resulting in more elimination in the urine. For certain poisons, forced diuresis by the use of fluids and diuretics may enhance the rate of their elimination. Commonly used diuretic drugs are urea, mannitol and furosomatidedide. These are actually osmotic diuretics and prove more effective than water diuretics. These drugs inhibit ADH release, reduce water reabsorption and raise urine output.

(ii) **Dialysis**

This technique is based on the passage of poisons through a semipermeable dialysis membrane and ultimate removal of poisons in dialysate against the concentration gradient. Since the dialysis involves movement of poison molecules, it is partly dependent on the molecular weight of the poisonous compounds which are highly serum-protein bound and, having large volume of distribution, are either not dialyzable or poorly removed by this method. The lipid dialysis is recommended for the removal of lipophilic poisons but, owing to their large volume of distribution, only a small percentage is removed even after 4-6 hours of dialysis.

(iii) **Haemoperfusion**

It is a recent technique. The blood is passed through a column of charcoal or adsorbent resin for the removal of extracorporeal poisons. The poisons occupying large volume of distribution are poorly removed by this technique. However, some poisons are better removed because of absorptive capacity of the column.

(b) Chemical procedures

The poisons (toxicants) exert deleterious effects on account of their interaction with suitable receptors at the specific sites. The removal of poison from the effector sites before their interaction with receptors terminates their action. The chelates — as described in the preceding section — have been proved effective for establishing inactive complexes which detoxify especially the metallic poisons. Not only this, the chelates also promote the elimination of poisons in urine. Chelates also raise the water-solubility of the poisons.

3. Procedures applied to elevate threshold level of toxicity

A certain level of concentration of a poison exerts deleterious effect. This concentration of the poison is termed the *threshold concentration* and toxicity exerted is termed *threshold level of toxicity*. Actually, antidotes increase threshold of toxicity without affecting the concentration of poisons at the effector sites by the following mechanisms :

(i) They directly antagonize the system affected by the poisons and enhance a physiologically opposing system, and

(ii) They block the response exerted by the poisons on some physiological mechanism.

Antidotal procedures — where the toxicity levels of the poisons are elevated by specific antidotes — are termed *elevating threshold level of toxicity of poisons*.

Various physiological antidotes are used to elevate threshold level of toxicity. Some important physiological antidotes are described below which elevate threshold level of toxicity or produce opposite physiological function to that of poison molecule :

(I) Atropine for Morphine Poisoning

Morphine depresses respiratory center, initiates unconciousness and death. As a physiological antidote, atropine stimulates respiratory center and thus prevents unconciousness.

(II) Atropine for Pilocarpine Poisoning

Pilocarpine produces miosis, i.e., contraction of the eye. As a physiological antidote, atropine produces mydriasis, i.e., dilation and thus inhibits miosis.

(III) Coramine, Caffine, Cardiazole and Strychnine for Morphine Poisoning

As mentioned earlier, morphine depresses respiratory centre, initiates unconciosusness and death.

The mentioned three physiological antidotes raise the sensitivity of the respiratory center and bring back normal blood pH.

(IV) Naloxone for Morphine Poisoning

Naloxone exerts same effects as atropine.

(V) Nalorphine for Morphine Poisoning

As morphine exerts Cheyne-Stokes respiration, the nalorphine given intravenously prevents deleterious effects of morphine by stimulating respiratory center.

(VI) Thiamine hydrochloride for Ethyl alcohol Poisoning

Thiamine hydrochloride diminishes the toxicity of ethyl alcohol and helps its excretion in urine.

(VII) Calcium for Magnesium Poisoning

Magnesium has been known to cause depression of CNS and paralysis of motor neurons. Calcium as physiological antidote exerts opposite effect and reduces magnesium poisoning.

(VIII) Calcium for Lead Poisoning

Lead, after deposition in bones, causes damage to bone. Calcium intake in equal doses helps in repairing bones.

(IX) Sodium Thiosulphate and Dicobalt EDTA for Cyanide Poisoning

Cyanide exerts toxicity to cellular respiration. The physiological antidotes named above produce stimulation of cellular respiratory system.

(X) Niacinamide for Vacor (Rodenticide) Poisoning

Vacor, a popular rodenticide, acts on NAD and obstructs their functions. Niacinamide counters the destructed NAD and helps in formation of normal NAD.

(XI) Atropine and Phenobarbital for BHC Poisoning

BHC, a widespread organochlorine insecticide, poisons the nervous system and raises the B.P. Atropine prevents nerve poisoning and phenobarbital lowers the B.P. These two thus act as physiological antidotes.

(XII) Pyridoxine Aldosine Methiodide (PAM) for Organophosphorus Insecticide Poisoning

Organophosphorus insecticides actually block action of cholinesterase enzyme and thus raise action of acetylcholine that produces excessive muscular activity. PAM as a physiological antidote inhibits release of acetylcholine and prevents muscular convulsions.

(XIII) Inhalation of O_2 with 7% CO_2 for respiratory poisons like CO and HCN

CO and HCN convert hemoglobin into methemoglobin and inhibit normal gaseous transport. O_2 with 7% CO_2, when inhaled, stimulates respiratory center of brain and raises the rate of respiration. In this way it lowers the toxicity of CO and HCN.

(XIV) Picrotoxin for Barbiturate Poisoning

Barbiturate acts as a muscle poison by depressing muscular activity. Picrotoxin produces strong muscular convulsions.

Revision Questions

A. Long-Answer Questions

1. Define antidotes, classify them on the basis of mechanism of action.
2. Describe the principles of antidotal treatment.
3. Define antidotes and describe antidotal procedures.

B. Short-Answer Questions

1. Describe in brief :
   (i) Different kinds of poisons
   (ii) Chelating agents
   (iii) Chemical neutralizers
   (iv) Universal antidote.
2. Give the pharmacological assessments for antidote actions for :
   (i) Antidote for halogens, cyanogens and HCN acid
   (ii) Antidote for arsenical poisoning
   (iii) Antidote for mercury poisoning.

3. Write short notes on :
   (i) Procedures applied to diminish absorption/translocation of poisons
   (ii) Procedures applied to enhance termination of action of poisons
   (iii) Procedures applied to elevate threshold level of toxicity.
4. Give a brief account of the antidote applied for :
   (i) Morphine poisoning
   (ii) Pilocarpine poisoning
   (iii) Ethyl alcohol poisoning
   (iv) Magnesium poisoning
   (v) Lead poisoning
   (vi) Cyanide poisoning
   (vii) Rodenticide poisoning
   (viii) BHC poisoning
   (ix) Organophosphorus insecticide poisoning
   (x) CO poisoning.

24

Environmental Impact and Risk Assessment

SYNOPSIS

Introduction, EIA : Definition and Scope, Characteristics, Objectives, Components, Methodology, Procedure for obtaining clearance, Document.
Risk Assessment : Hazard identification, Toxicity assessment, Exposure assessment, Risk characterisation.

Introduction

With the advent of improved scientific and technological innovations there has been a spurt in industrial and developmental activities. This has caused enormous strain on environmental resources and dynamism resulting in gross ecological imbalance. In fact, every major developmental activity is bound to have positive or adverse impacts on environment. We have witnessed multifarious environmental consequences on account of developmental activities; be it in agriculture, power generation, mega dams, and nuclear or mining sector. Global environmental consequences in terms of global warming, rising sea levels and depleting trend of genetic resources etc. have already rang alarm-bells for threatened future of mankind and should be treated as an alarm for the impending dangers to mankind.

It has thus become imperative to maintain a balance between capacity of the environment and developmental activities. This can be achieved by various strategies of Environmental Management Plan (EMP). Environmental Impact Assessment (EIA) and Environmental Risk Assessment (ERA) are two most important approaches of EMP.

A : ENVIRONMENTAL IMPACT ASSESSMENT (EIA)

The term 'EIA' originated from the U.S. National Environmental Policy Act (NEPA) of 1969, which emphasized on the inclusion of environmental impact considerations in every project planning or developmental activity along with technical and socio-economic evaluations. Thus, it can be said that EIA is the prior assessment of the future impact of consequences of any major developmental activity on the quality and dynamism of total human environment. In fact, EIA is an effective management tool, which aims at integrating environmental factors into project planning and decision making so as to achieve ecologically sustainable development.

More than 100 countries of the world have adopted EIA legislation. Canada in 1973, Australia and New Zealand in 1974, Japan in 1981, the European Community in 1984, The Netherlands in 1986, and the UK in 1988, have enacted similar legislations at different times to conduct EIA. Government of India also brought "The Environmental Impact Assessment Notification", 1994, under the provisions of Environment (Protection) Act, 1986, and the Environmental (Protection) Rules, 1986. This notification was further amended (S.O. No. 60E) on 10th April 1997 making public hearing mandatory for environmental clearance. The public hearing will be conducted by the State Pollution Control Boards (SCPBs) before the project proposals are sent to the Ministry of Environment and Forests (MoEF) for obtaining environmental clearance.

In fact, EIA is a systemic approach for planning, operation and monitoring of developmental projects. Keeping this in mind the RIO Declaration of the Earth Summit, 1992, has clearly laid emphasis on significance of EIA and suggested : "Environmental Impact Assessment, as a national instrument, shall be undertaken for proposed activities that are likely to have a significant adverse impact on environment and are subject to a decision of a competent national authority."

Definition and Scope of EIA

The EIA can be designated as an approach that seeks to improve development by a priority assessment. Munn, 1979, suggests that EIA aims at predicting and identifying the impacts on bio-geographical environment and on man's health and well-being of legislature proposals, policies, programmes, projects and operational procedures and to interpret and communicate information about the impacts. Hagerty & Heer, 1997, defined EIA as "an activity that aims at establishing quantitative values for selected parameters, which indicate the quality of environment before, during, and after the proposed activities". UNEP (1987) defined EIA as an examination, analysis and assessment of planned activities with a view to ensure environmentally sound and sustainable development. In fact, EIA provides wider perspectives and consequences of any developmental project/operation, which include administrative programmes and policies, legislative actions, and physical, chemical, biological, cultural and socio-economic compartments of the total human environment.

Although the scope of EIA is limited to major developmental projects having wider consequences on human environment but, theoretically speaking, any action, project, operation, administrative policies, plans and programmes and legislative actions etc. that have the potential to bring about significant environmental changes should be subjected to EIA exercise. As per "The Environmental Impact Assessment Notification", 1994, of Ministry of Environment and Forest, Government of India, expansion of modernization of any activity (if pollution load is to exceed the existing one) or a new project listed in Schedule 1 of the notification shall not be undertaken in any part of India unless it has been accorded environmental clearance by the Central Government in accordance with the procedure hereinafter specified in this notification. In India, the following projects are listed in Schedule I of the aforesaid notification :

(i) Nuclear Power and related projects such as heavy water plants, nuclear fuel complex, and rare earths.

(ii) River valley projects including hydel power, major irrigation and their combination including flood control.

(iii) Ports, harbours, airports (except minor ports and harbours).

(iv) Petroleum Refineries including crude and product pipelines.

(v) Chemical fertilizers (Nitrogenous and Phosphatic, other than single super-phosphate).

(vi) Pesticides (Technical).

(vii) Petrochemical complexes (Both Olefinic and Aromatic) and Petrochemical intermediate, such as DMT, Caprolactam, LAB etc. and production of basic plastics, such as LDPE, HDPE, PP, PVC.

(viii) Bulk drugs and pharmaceuticals.

(ix) Exploration for oil and gas and their production, transportation and storage.

(x) Synthetic rubber.

(xi) Asbestos and asbestos products.

(xii) Hydrocyanic acid and its derivatives.

(xiii) (a) Primary metallurgical industries (such as production of Iron and Steel, Aluminium, Copper, Zinc, Lead and Ferro-alloys).

(b) Electric arc furnaces (Mini Steel Plants)

(xiv) Chlor-alkali industry.

(xv) Integrated paint complex including manufacture of resins and basic raw materials required in the manufacture of paints.

(xvi) Viscose Staple fibre and filament yarn.

(xvii) Storage batteries integrated with manufacture of oxides of lead and lead antimony alloy.

(xviii) All tourism projects between 200 m - 500 meters of High Tide line or at locations with an elevation of more than 1,000 meters with investment of more than Rs. 5 crores.

(xix) Thermal power plants.

(xx) Mining projects (major minerals) with leases more than 5 hectares.

(xxi) Highway Projects.

(xxii) Tarred Roads in Himalayas and/or Forest areas.

(xxiii) Distilleries.

(xxiv) Raw Skins and Hides.

(xxv) Pulp, Paper and Newsprint.

(xxvi) Dyes.

(xxvii) Cement.

(xxviii) Foundries (individual).

(xxix) Electroplating.

Characteristics of EIA

The EIA gives a clear-cut picture of important environmental considerations which are to be incorporated in the decision-making process while setting up of the major projects.

It has many important characteristics :

(i) It provides a systemic evaluation of almost all significant environmental consequences of a developmental project

(ii) EIA is a structured, systemic and comprehensive approach

(iii) All EIA processes draw driving force by legislation

(iv) EIA is one of the most valuable and interdisciplinary decision-making tools

(v) EIA can force the policy-makers to reconsider the project proposals

(vi) It can be used to ensure regional planning for sustainable development

(vii) It ensures the accountability of decision-makers to the public

(viii) It provides possible alternate development options against identified environmental impacts

(ix) EIA document should be circulated for objective review of its results by others

(x) EIA cannot be taken in isolation, rather, it is related with other factors, viz., Environmental Statement, Environmental Audit and New System of National Accounts etc.

(xi) It seeks public participation in decision-making.

Objectives of EIA

The principal objective of EIA is to ensure that the development is sustained with minimal environmental degradation. Adverse effects on environment should be minimized by incorporating suitable measures in the project planning itself on the basis of predictions regarding environmental degradation on account of project operation. The Ministry of Environment and Forest, Government of India, has been assigned the responsibility for appraisal of projects with regards to their future environmental implications.

Three primary objectives of EIA can be summarized :

(i) EIA predicts the likely environmental impacts of projects.

(ii) It presents these predictions and options to decision-makers.

(iii) It finds ways to reduce unacceptable impacts to shape the project so that it suits the local environment.

Components of EIA

The key steps in EIA process are :

(i) *Project definition and identification :* During project identification and definition, the project proponent conducts feasibility studies, defines the usefulness of the study, considers alternatives, files a notice of intention to seek EIA clearance and, ideally, initiates an inter-agency and public consultation process.

(ii) *Screening :* At this stage, the EIA agency determines whether the project may proceed as planned or it should be modified partially or completely. For this, the EIA agency consults the proponent and other agencies and public participants to determine the requirements of further studies.

(iii) *Scoping :* This is also an early planning stage and deals with a more detailed plan of study for the project to identify major concerns and key impacts, and to decide assessment methods and models to be used.

(iv) *Agencies :* Agencies and public representatives concerned with the project or the project area are consulted.

(v) *Baseline data collection :* Baseline data collection and analyses are very important in project planning. They play an important role in primary monitoring of the environment in the project area.

(vi) *Identification of impact :* Impact prediction and assessment are the most important parts of the technical process. This can be achieved by employing suitable models and a careful evaluation of inputs and outputs of environmental impacts.

(vii) *Alternative evaluation criteria :* They include legally-mandated criteria, technical/scientific criteria, and social acceptability criteria. Alternative sites and design process should be critically examined to maximize the positive environmental impacts, socio-economic benefits, profitability, and minimize the temporary adverse impacts.

(viii) *Management plan :* After the identification of environmental impacts, the mitigation measures are now needed to be defined. A good management plan should have flexible project planning so that it can adopt the modified or entirely new project alternatives. It should aim to minimize adverse environmental impacts.

(ix) *Publication of EIA report :* Circulation/Publication of EIA report is an important step. It brings public comments, which would be definitely helpful for primary stages of EIA like screening and scooping.

(x) *Formal approval (with or without conditions) :* The decision on the EIA report is put forward in a written record with conditions that the project proponent must comply with the provisions mentioned in the the document.

(xi) *Monitoring and compliance :* A careful monitoring ensures the compliance of provisions mentioned in the management plan. It not only provides confidence to workers, public agencies, and communities involved in dealing with negative impacts, but also gives useful feedback on the accuracy of the EIA's impact predictions.

Environmental Impact Assessment Methodology

After the inception of the NEPA, 1969, a number of methodologies and models are now available for evaluation and assessment of environmental impacts. Computer-based models are more improved than conventional subjective models.

Some of the important methods for EIA are :

1. *Ad hoc approach :* This is the oldest and perhaps the crudest approach to EIA. This method provides very little guidance for assessing impacts on specific parameters beyond suggesting impacts on broad areas like impact on land, forests, population, water, wildlife, etc. This approach does not count secondary impacts.

2. *Checklist :* It is the updated version of the ad hoc approach in which specific areas of potential impacts are listed. Checklists contain environmental parameters. Each environmental parameter for adverse, beneficial, or no effects due to proposed project activities can be simply ticked in the checklist and overall environmental impact can be evaluated. This method is usually very large, subjective, and provides little guidance that can aid in the decision-making process.

3. *Matrix method :* The method is attributed to Leopold et al (1971). The method uses a matrix format to relate project actions and environmental components.

Procedure for obtaining EIA Clearance

In India, Environmental clearance for mega-developmental projects can be obtained from Ministry of Environment and Forest by submitting prescribed application form (Schedule II, Sub-para I (a) of Para II of The Environmental Impact Assessment Notification, 1994). For this project authorities have to submit detailed information regarding the future environmental implications of the project concerned.

Duties of the Project Authorities

The project authorities have to furnish the following documents for environmental appraisal of a developmental project :

1. Detailed project report
2. Filled-in questionnaire
3. Environmental Impact Statement (EIS) along with Environmental Management Plan.
4. Positive or negative impact of the project.

Some of the issues to be included are :

(i) Impact on soil, water and air quality.
(ii) Impact on land use, forests, agriculture, fisheries, tourism, recreation etc.
(iii) Socio-economic impacts including short and long-term impact on population.
(iv) Impact on health.
(v) Impact on fauna and flora — particularly endemic and endangered species.
(vi) Cost-benefit analysis including the measures for environmental protection.

There is a separate EIA Wing for this purpose.

EIA Wing

The Ministry of Environment and Forests (MoEF), Government of India, has an environmental impact assessment wing, comprising three divisions, each dealing with projects in specific areas :

(a) Impact Assessment Division I
River Valley project, major irrigation projects and hydel power projects.
(b) Impact Assessment Division II
Industrial projects, thermal power projects and mining projects.
(c) Impact Assessment Division III
Ports and harbour projects, tourism projects, human settlement projects, projects in ecologically fragile areas and communication projects.

For the purpose of seeking environmental clearance of mega-development projects, MoEF has a multi-disciplinary **Environmental Appraisal Committee** (EAC).

The EAC has following specific sectors :

1. River valley, multipurpose, irrigation and hydel power.
2. Industrial projects.
3. Mining projects.
4. Thermal power projects.

There are provisions for constituting expert committees for EIA as per guidelines of Schedule III of The Environmental Impact Assessment Notification, 1994.

1. The committees will consist of experts in the following disciplines :
   (i) Eco-System Management
   (ii) Air/Water Pollution Control
   (iii) Water Resource Management
   (iv) Flora/Fauna Conservation and Management
   (v) Land Use Planning
   (vi) Social Sciences/Rehabilitation
   (vii) Project Appraisal
   (viii) Ecology
   (ix) Environmental Health
   (x) Subject Area Specialists
   (xi) Representatives of NGOs/Persons Concerned with Environmental Issues.
2. The Chairman will be an outstanding and experienced ecologist or environmentalist or technical professional with wide managerial experience.
3. The representative of IAA will act as Member-Secretary.
4. Chairman and members will serve in their individual capacities except those specifically nominated as representatives.
5. The membership of a Committee shall not exceed 15.

Preparation of EIA Document

The general format for submitting EIA document should have following heads and their details :

(i) Cover Page
(ii) Abstract
(iii) Table of Content
(iv) Introduction
(v) Purpose and Need (Objective)
(vi) Available Alternatives
(vii) Comparison of Alternatives
(viii) Affected Environment and Consequence
(ix) Mitigation of Adverse Effects and Mitigation Action Plan
(x) Comments and Responses to the Comments
(xi) References

(xii) Index

(xiii) Appendices

(a) Minutes of the meeting held

(b) List of Persons/Institutions, to whom the copies of EIA Reports are being sent.

The EIA document should be circulated for comments in respect of Completeness, Adequacy and Merit Revision.

B. RISK ASSESSMENT

Risk Assessment is the process of determining the probability and magnitude of harm to human life, welfare and environment, potentially caused by the release of hazardous chemical, physical or biological pollutants.

Assessment of risks is performed in four steps :

1. Hazard Identification
2. Toxicity Assessment
3. Exposure Assessment
4. Risk Characterisation.

Hazard Identification

The first step in the risk assessment process is 'hazard identification'. Hazard Identification is designed to screen the hazardous contaminants/pollutants and select those of greatest public health and of environmental concern. Essentially, this process assists to identify those environmental pollutants which can cause an increase in the incidence of adverse health.

Hazard Identification conditions can be studied in three major steps :

(a) Structure-Activity Relationships

(b) In vitro and short-term tests

(c) Animal Bioassays.

The hazard potential of a chemical is chiefly associated with its intrinsic toxic properties; but some characteristics of its source and the environment receiving inputs also determine the totality of the adverse effects.

Major properties of a substance signifying its hazardous potential are its acute toxic potential, water solubility/lipophilicity, persistence, pH, stability, sensitivity, electrophilicity, chemical reactivity and carcinogenicity etc. Chief properties of the pollutant source to be considered are the quantities released, geographical distribution, distribution time of release, and the medium into which pollutants are released. The properties of the environment receiving inputs include its physical and chemical characteristics, water removal rates, ventilation pattern, species present and their trophic status. All these factors together determine the hazardous potential of a substance.

The properties of a pollutant primarily determine the degree of harmfulness resulting from its exposure when the properties of the pollutant source and of the environment receiving inputs do not have much influence. A useful classification of hazardous substances, adopted by the EEC nations, was first developed by Canton and Sloof (1975). With the help of this, it is easy to identify safe and harmful substance and their relative potential of either property. Some examples of each category are given in Table 24.1. Of course, along with these guidelines, such properties like carcinogenicity, teratogenicity and mutagenic potentials of the substance need be given serious consideration.

Classification of Hazardous Substances (Canton and Sloof, 1975)

Substances that are positive carcinogens in animals and/or humans are placed in "black" and negative carcinogens in "white" lists, however, wherever the evidence is unequivocal they may be placed in "grey"-list, depending upon the degree of conclusive evidence obtained in various test species. Biocides are not included in the grey-list. Thus, at the end of this stage, the utility/justification of making risk estimation for exposure to a chemical can be made.

Table 24.1 : Classification of Hazardous Substances

Black List	Grey List
Zinc	Organophosphorus
Copper	Organotines
Nickel	Mercury and its Compounds
Chromium	Tin and its Compounds
Lead	Cadmium and its Compounds
Selenium	Persistent Organochlorines
Arsenic	Hydrocarbons of Petroleum Origin
Antimony	Substance with Carcinogenic Potentials
Molybdenum	
Titanium	
Tin	
Barium	
Boron	
Uranium	
Vanadium	
Tellurium	
Cobalt	

Toxicity Assessment

Hazard is defined as the potential of a chemical to cause an adverse health effect as a result of sufficient exposure. Hazard assessment involves the determination of the nature and extent of adverse health or environmental effects affiliated with exposure to the pollutant in question.

This is essentially a two-step process involving

(i) Characterisation of the dose-response relationship, and

(ii) Extrapolation to human exposure conditions.

Toxicological evaluation is done by exposing laboratory animals to varying doses of a pollutant via inhalation; ingestion (oral) or injection and the observed toxic effects are then extrapolated to human exposure conditions, this warranting the inclusion and uncertainty factor.

The dose response assessment actually determines the relationship between the dose of a chemical and the incidence of an adverse effect. Toxic effects (excluding carcinogenicity) are evaluated by estimating and acceptable daily intake. This assumes that there is a dose below which no adverse effects occur; this dose level is called the threshold dose (for non-carcinogens).

Further extrapolations of the low dose in animals to low dose in humans are done to complete the process. When we try to convert dose data from laboratory experiments to humans, there are several issues to consider, viz., scaling for the difference in size, life-span, metabolism, difference in routes of exposure, difference in biochemical and pharmacokinetic behaviour and the differential susceptibility between species to the deleterious effects of a pollutant. The process of extrapolation is a complicated and a controversial one, which involves the inclusion of a safety factor, to provide a range of safetiness to humans from the toxic effects of xenobiotics that have been predicted with other animal systems experimentally.

For extrapolating non-carcinogenic effects from animal studies to humans a safety factor of 100 is used (a factor of 10 for compensating for the species variation e.g., rats and human) and another 10 to account for the inter-individual human variation. Thus, it is believed that there exists a threshold for such compounds and that no observable effect levels (NOEL) can be established for them. An example is the case of nitrites. Nitrites induce methaemoglobinemia by oxidizing the iron moiety in the hemoglobin. But at doses at or below NOEL, the production of methemoglobinemia is non-functional or physiologically insignificant. Only when the amount of methemoglobinemia produced by nitrites is so high as to interrupt the normal O_2 supply to cell, it is of concern.

On the other hand, dose-response relations for carcinogens do not assume a threshold. This is because the cells affected by these chemicals are thought to be incapable of repairing completely or compensating for the damage done. Even though carcinogens have so far been considered to exert such damages, recent observations suggest that the central nervous system damage caused by lead in children may assume no threshold as well. Even exposure to small levels of carcinogens is assumed to cause some damage. Thus, toxic effects resulting from the exposure to carcinogens at lower levels have always been a matter of scientific interest.

Exposure Assessment

Exposure assessment is the process of measuring or estimating the intensity, frequency, and duration of human exposures to an environmental pollutant. This exposure to pollutants can be calculated as a daily dose rate averaged for a population.

The major factors in the process of assessing the exposure of human populations to pollutants present in the environment are (I) intensity (II) frequency, and (III) duration of exposure.

When the pollutant in question is in the fluid medium, its concentration and the contact time with the respiratory tract and skin are very important in assessing the exposure. But when it is in the solid materials, the amount of solid material ingested and the concentration of the pollutant in those tissues/organs are of prime importance.

One simple approach of determining exposure of a toxicant/pollutant is to consider its physiochemical characteristics, for example, vapour pressure of a chemical. The higher the vapour pressure of a chemical, the greater is the possibility that it would be found in the air. Similarly, the higher the solubility of a chemical in gaseous solutions, the more would be the chances of detecting it in water. The more insoluble is an inorganic compound in air and water, the greater are the probabilities of soil/sediments being a major pathway of exposure.

Finally, the higher the lipophilicity, the brighter are the possibilities of the food chain being the major pathway. A set of parameters that determines the fate and distribution of a pollutant in various environmental compartments is listed in Table 24.2 :

Table 24.2 : Physical and chemical characteristics of a xenobiotic (environmental toxicant) that may determine its environmental fate

Sl. No.	Physical and Chemical Characteristics
1.	Adsorption coefficient
2.	Biodegradation rate
3.	Boiling point
4.	Henry's Law Constant
5.	Hydrolysis rate
6.	Melting point
7.	Molecular weight
8.	Octanol/water partition coefficient
9.	Photolysis rate
10.	Vapour pressure
11.	Volatilisation rate
12.	Water solubility

RISK CHARACTERISATION

Risk characterisation is the process of estimating the incidence and severity of a health effect resulting from the exposure to a xenobiotic. For the characterisation of risk, exposure to a xenobiotic through different pathways should be considered.

Risk is most often determined as the uncertainty concerning an undesired event, where uncertainty is expressed as the probability of occurrence. The probability of a detrimental effect for an individual or a population, though difficult to quantify, can be estimated. This estimate is called **population risk (PR)**. It is the number of cases of effects per lifetime for existing exposure level (Cothern and Marcus, 1984) and can be expressed as

$$PR = \sum C_i R_i P_i$$

where

C_i is the concentration of xenobiotic (in water/air) to which the population is exposed.

R_i is the number of cases or effects per lifetime per unit concentration per individual.

P_i is the number of people exposed to concentration C of xenobiotic.

Though it is not possible to accurately predict the levels of ill-effects caused by human actions, the concept of risk has led to substantial improvements in the environmental assessment and protection process. Risk analysis provides a more rational basis for decision that may otherwise be highly subjective by

(i) emphasizing probabilities and frequencies of events, and

(ii) explicitly quantifying uncertainty.

Sampocolo and Binetti (1986) suggested the models for quantifying risk arising from personal, domestic and professional exposures, based on prioritisation concepts. With the use of the approach the risk resulting from exposure to environmental pollutants can be calculated by the following expression :

$$\text{Risk (environmental exposure)} = (PCP + TP + ETP) \times R \times Q \times ED \times P \times BC \times RP$$

where

PCP Sum of points ascribed to physiochemical properties of the substance.

TP Sum of points ascribed to toxicological properties of the substance.

ETP Sum of points ascribed to ecotoxicological properties of the substance.

R Coefficient of priority relevant to PCP + TP + ETP.

Q Quantity present.

ED Environmental diffusion.

P Persistence.

BC Bio-concentration.

RP Size of population at risk.

In general, three major components determine the risk (the probability associated with the occurrence of a hazard due to a pollutant). These are :

(a) Intrinsic toxic properties of the pollutants (hazard).

(b) Quantities of the pollutant released into the environment (exposure), and

(c) The size of population at risk.

Thus, a chemical may be extremely toxic but if it has not been released into the environment, apparently there is no risk associated with it. On the other hand, if the toxicity of a pollutant is high and also the quantities released into environment are sufficiently large to pose a potential risk to the exposed population, a decision is needed to be taken to minimize its use in the future or force for an immediate ban (depending upon the outcome of the risk-assessment process).

Risk assessment forms the basis of risk management, which is the policy-making process.

Revision Questions

A. Long-answer questions

1. Define EIA. Discuss its objectives and scope in relation with developmental plans of our country.
2. Describe EIA in following terms :
   (i) Components of EIA
   (ii) Objectives of EIA
   (iii) EIA wing
   (iv) Environmental Appraisal Committee
   (v) Duties of project authorities
   (vi) EIA Methodology.
3. Define risk-assessment and describe various steps in performing risk-assessments.
4. What is EIA? Describe the procedure for obtaining EIA clearance for mega development projects.

B. Short-answer questions

1. Write briefly about :
   (i) Objectives of EIA
   (ii) Characteristics of EIA
   (iii) Risk assessment
   (iv) EIA Document.
2. Write short notes on :
   (i) Risk characterisation
   (ii) Hazard identification
   (iii) Toxicity assessment
   (iv) Exposure assessment.

Glossary

Acaricides : Chemicals used to kill small phytophagous mites.

Acute : Having a sudden onset, lasting a short time. Can be used to define either the exposure or the response.

Acute toxicity : Toxic effects produced quickly within a short period of time by large doses of toxicant.

Accumulation : Increase in quantity of toxicant.

Additive : Any substance not usually present in the food but is added, either deliberately or incidentally, in order to improve some quality of the food.

Additive toxicity : The toxicity of a mixture of chemicals which is approximately equivalent to that expected from a simple summation of known toxicities of the individual chemicals present in the mixture (i.e. algebraic summation of effects).

Affinity : The readiness of a toxicant to combine chemically with a substance and remain in combination.

Agonist : A chemical (drug or toxicant) that initiates a pharmacological response after binding with a receptor.

Alkaloid : Any of various nitrogenous organic bases found in plants having specific physiological action.

Anabasine : An alkaloid closely related to nicotine and occurs in small woody perennial *Anabasis aphylla*.

Analogue : Something regarded in terms of its parallelism to something else.

Anaphylaxis : Repeat administration of a drug or toxicant producing violent effect.

Anorexia : Loss of appetite.

Antagonist : A chemical which opposes or resists the action of other chemical (toxicant).

Antidote : An agent that neutralizes the poison or counteracts its effect.

Antidotal procedure : The procedure through which the effect of procedure a poison is removed or procedures employed to counteract the effect of toxicant or poison.

Application factor (AF) : A numerical, unitless value, calculated as the threshold chronically toxic concentration of a chemical divided by its acutely toxic concentration. An AF is generally calculated by dividing the limits (no observed adverse effect concentration (NOAEC) and lowest observed effect concentration (LOEC) on the maximum acceptable toxicant concentration (MATC) by the time-independent LC_{50}, if available, or the 96-h LC_{50} (or 48-h EC_{50} or 48-h LC_{50} for daphnids) from a flow through acute toxicity test. The AF is usually reported as a range and is multiplied by the median lethal concentration of a chemical as determined in a short-term (acute) toxicity test to estimate an expected no-effect concentration under chronic exposure.

Arsenicals : Various compounds of the element arsenic are collectively called arsenicals.

Arrhythmia : Loss of heart rhythm.

Ataxia : Loss of the power of muscle contraction.

Asymptotic threshold : The concentration of a chemical at which some **concentration**

percentage of a population of test organisms is in a state of approximate homeostasis for some prolonged period of time (not necessarily absolute). It may be demonstrated as the concentration at which the toxicity curve is approximately asymptomatic (parallel) to the time axis. The asymptotic LC_{50} is the concentration at which acute mortality has essentially ceased and will not change substantially with exposure time. That is, there is no evidence of significantly increasing effects resulting from increasing duration of exposure.

Autochthonous : From within. Indigenous organisms or organic material.

Bioaccumulation : Increase in concentration of xenobiotics through food and water.

Biological indicators : Plants or animals or their activities which give an indication of the presence of toxicants.

Bioassay : Test used to evaluate the relative potency of a chemical by comparing its effect on a living organism with the effect of standard preparation on the same type of organism. Bioassays are frequently used in the pharmaceutical industry to evaluate the potency of vitamins and drugs, "bioassay" and "toxicity test" are not synonymous — refer "toxicity tests".

Bioavailability : Usually defined as the amount, or fraction of a chemical product that reaches the systemic circulation as unchanged (not metabolized) compound.

Bioconcentration : A process by which there is net accumulation of a chemical directly from ambient medium into organisms resulting from simultaneous uptake and elimination.

Bioconcentration factor (BCF) : A unitless value describing the degree to which a chemical can be concentrated in the tissue of an organism in the environment. At apparent equilibrium during the uptake phase of a bioconcentration test, the BCF is the concentration of chemical in one or more tissues of the organisms divided by the average exposure concentration in the test.

Biodegradation : The breakdown of complex organic compounds into simpler ones by microorganisms.

Biological Control : Use of natural enemies to control pests.

Biomagnification : Result of the processes of bioconcentration and bioaccumulation by which tissue cocentrations of bioaccumulated chemicals increase as the chemical passes up through the successive trophic levels. The term implies an efficient transfer of chemical from food to consumer, so that residue concentrations increase systematically from one trophic level to the next.

Biomonitoring : Use of living organisms as "sensors" in environmental quality surveillance to detect changes in an effluent or environment and to indicate whether life may be endangered.

Biotransformation : Metabolic transformation of an element or compound within the body of living organisms.

Botanical Insecticides : Plant products having insecticidal properties.

Bradycardia : Slowness of the heart beat.

Cachexia : A general lack of nutrition and wasting occurring in the course of a chronic disease.

Carcinogen : Cancer producing agent.

Carcinogenicity : Uncontrolled somatic division including formation of malignant tumors.

Chemical Additive : It is a substance for which the exact chemical nature is known.

Chelation : Bond formation between a metal ion and two or more polar groupings of a single molecule.

Chelate : A complex formed through chelation.

Cheilitis : Inflammation of lips.

Chronic : Involving a stimulus that is lingering or continues for a long time; often signifies periods from several weeks to years, depending on the reproductive life cycle of the species. Can be used to define either the exposure or the response to an exposure (effect). Chronic exposure typically induces a biological response of relatively slow progress and long continuance. The chronic toxicity test is used to study the effects of continuous, long-term exposure of a chemical or other potentially toxic material on organisms.

Chronic disease : Denoting a disease of slow progress and long continuance.

Chronic toxicity : It is long term toxicity.

Contact Insecticides : These insecticides kill insects by contacting and entering their bodies either directly through the integument into the blood or by penetrating the respiratory system, through the spiracles into the tracheae.

Concentration-response curve : A curve describing the relationship between exposure concentration and percentage response of the test population.

Control : A treatment in toxicity test that duplicates all the conditions of the exposure treatments but contains no test materials. The control is used to determine the absence of toxicity of basic test conditions (e.g., health of test organisms, quality of dilution water).

Cumulative : Brought about, or increased in strength, by successive additions at different times or in different ways.

Cumulative effect : Effects produced due to accumulation of toxicant above critical level.

Deactivation effect : The process of recovering from the toxic effect.

Delayed effects : Effects or responses that occur some time after exposure. Carcinogenic effects of chemicals typically have a long latency period; the occurrence of a tumor may take years after the initial exposure.

Delayed action : Postponement of onset of symptoms, due to adaptation of organisms.

Deleterious effect : Toxic effect.

Detoxification : Reduction in toxicity by metabolic activity of the animal.

Distal effects : Effects appearing long after administration of toxicant.

Dipping method : Insects are simply picked up by a pair of forceps and dipped into the insecticide preparation.

Depuration : A process that results in elimination of a material from an organism.

Dose : The quantifiable amount of material introduced into the animal by injection or ingestion.

Dose-response curve : Similar to concentration-response curve except that the dose (i.e., the quantity) of the chemical administered (e.g., by injection) to the organisms is known. The curve is plotted as dose versus response.

Dose response : Relation between the degree of reaction and relationship between the concentration of the toxicant.

Dynamic Air Flow System : The animal is maintained in a chamber which is constantly supplied with an air current containing a fixed amount of pesticide.

Dyspnea : Rapid breathing.

Economic Toxicology : Toxicological study for evaluating economic gain or loss incurred in terms of money by the use of toxic substances.

Ecotoxicology : Study of toxicant that produces adverse effects on environment.

Effluent : Liquid industrial waste or sewage released into river or sea.

Environment Toxicology : It is one of the most important sub-fields of Toxicology concerned with the effects of all environmental pollutants on all forms of biological system.

Estimated (or expected) environmental concentration (EEC) : The concentration of material estimated as being likely to occur in environment to which organisms are exposed as a result of planned manufacture, use and disposal.

Eutrophication : Enrichment of inland water and coastal sea, particularly with sewage causing retrogressive succession.

Exposure : The contact reaction between a chemical or physical agent and a biological system.

Fate : Disposition of a material in various environmental compartments (e.g., soil or sediment, water, air, biota) as a result of transport, transformation, and degradation.

Fatal dose : The smallest quantity of a poison which has been authentically recorded as having caused the death of an average adult.

Flow through system : A toxicity test where the contaminated test solution is changed continuously by following it through one end of the test vessel and taking it out from the other end.

Fog : Suspension of tiny water droplets forming a cloud close to the ground surface.

Forensic Toxicology : Toxicological knowledge applied to investigate crime.

Frequency response : Percent population showing a particular response against a particular dose of the toxicant.

Fumigants : These are substances which when volatilized are capable of destroying insects.

Fungicides : Chemicals used to control fungus growth.

Gastric lavage : Washing of stomach with luke-warm water.

Gnotobiotic : Composition completely known. Often referring to mixed population of organisms.

Graded-dose-response : Study of response in a single test animal against progressively increasing doses of toxicant.

Greenhouse Effect : The warming of earth's surface as a result of the trapping of long wave radiations by carbon dioxide and some other gases.

Haemoglobinemia : Presence of free haemoglobin in blood.

Haemoglobinuria : Presence of haemoglobin in urine.

Half-life : Time required to reduce by one-half the concentration of a material in a medium (e.g., soil or water).

Hazard : Likelihood that a chemical will cause injury or adverse effect under the conditions of its production, use, or disposal.

Hazard evaluation : Identification and assessment of the potential adverse effect that could result from manufacture, use, and disposal of a material in a specified quantity and manner.

Heavy Metals : A group of metals with high specific gravity, usually above 5.

Hepatotoxic : Toxic to liver.

Herbicides : Chemical used to control unwanted plants (weeds) growing where man does not want them.

Incipient LC_{50} : The concentration of a chemical which is lethal to 50% of the test organisms as a result of exposure for periods sufficiently long that acute lethal action has essentially ceased. The asymptote (part of the toxicity curve parallel to the time axis) of the toxicity curve indicates the value of the incipient LC_{50}, approximately.

Intake : The entry of a substance into lungs, gastrointestinal tract or subcutaneous tissues of animals. Its fate will be governed by absorption.

Insecticides : Substances which kill insects by their chemical action.

Integrated Pest Management : It is a new approach to pest control which combines many different methods such as biological, chemical, physical, and economic methods.

Itai-Itai Byo : A disease caused by eating cadmium contaminated fish.

Joint action : Two or more chemicals exerting their effects simultaneously.

Lethal : Causing death by direct action. Death of organisms is the cessation of all visible signs of biological activity.

LT_{50} : The time required to kill 50% of the population at a certain dose or concentration is often represented as LT_{50}.

Lowest observed effect concentration (LOEC) : The lowest concentration of a material used in a toxicity test that has statistically significant adverse effect on the exposed population of test organisms as compared with the controls. When derived from life cycle or partial life cycle test, it is numerically the same as the upper limit of the MATC.

Malignant : Cancerous.

Maximum acceptable toxicant concentration (MATC) : The hypothetical toxic threshold concentration lying in a range bounded at the lower end by the highest tested concentration having no observed effect (NOEC) and at the higher end by the lowest tested concentration having a significant toxic effect (LOEC) in a life cycle (full chronic) or partial life cycle (partial chronic) test. This may be represented as NOEC < MATC < LOEC. Calculation of a MATC requires quantitative life cycle toxicity data on the effects of a material on survival, growth and reproduction.

Median effective concentration (EC_{50}) : The concentration of material in medium to which test organisms are exposed that is estimated to be effective in producing some sublethal response in 50% of the test organisms. The EC_{50} is usually expressed as a time-dependent value (e.g., 24-h of 96-h EC_{50}). The sublethal response elicited from the test organisms as a result of exposure to the test material must be clearly defined. For example, test organisms may be immobilized, lose equilibrium, or undergo physiological or behavioural changes.

Median effective dose (ED_{50}) : The dose of material estimated to be effective in producing some sublethal response in 50% of the test organisms. It is appropriately used with test animals such as rats, mice, and dogs, but it is rarely applicable to aquatic organisms because it indicates the quantity of a material introduced directly into the body by the injection or ingestion rather than the concentration of the material in water in which aquatic organisms are exposed during toxicity test.

Median lethal concentration (LC_{50}) : The concentration of material in medium to which test organisms are exposed that is estimated to be lethal to 50% of the test organisms. The LC_{50} is usually expressed as a time-dependent value (e.g., 24-h or 96-h LC_{50}; the concentration estimated to be lethal to 50% of the test organisms after 24 of 96h of exposure). The LC_{50} may be derived by observation (i.e., 50% of the test organisms may be observed to be dead in one test material concentration), by interpolation (i.e., mortality of more than 50% of the test organisms occurred at one test concentration and mortality of fewer than 50% of the test organisms occurred at lower test concentration, and the LC_{50} is estimated by interpolation between these two data points), or by calculation (i.e., the LC_{50} is statistically derived by analysis of mortality data from all test concentrations).

Median lethal dose (LD_{50}) : The dose of material that is estimated to be lethal to 50% of the test organisms. It is appropriately used with test animals such as rats, mice, and dogs, but is rarely applicable to aquatic organisms because it indicates the quantity of material introduced directly into the body by injection or ingestion rather than the concentration of the material in water in which aquatic organisms are exposed during toxicity test.

Medial tolerance limit (TL_m or TL_{50}) : The concentration of material in medium at which 50% of the test organisms survive after a specified time of exposure. That TL_{50} (equivalent to the TL_m) is usually expressed as a time-dependent value (e.g., 24-h or 96-h TL_{50}, the estimated concentration and lethal dose, the terms tolerance limit is applicable in designating a level of any measurable lethal condition (e.g., extremes in pH, temperature, dissolved oxygen) TL_m or TL_{50} have been replaced by median lethal concentration (LC_{50}) and median effective concentration (EC_{50}).

Minamata disease : A disease caused by eating Hg contaminated fish.

Monitoring test : A test designed to be applied on a routine basis, with some degree of control, to ensure that the quality of environment or effluent has not exceeded some prescribed criteria range. In a biomonitoring test, organisms are used as "sensors" to detect changes in the quality of environment of effluent. A monitoring test implies generation or information, on a continuous or other regular basis.

Mixed Function Oxidase System (MFO) : It is the term used for NADPH requiring general oxidation system. It is located in the microsomal portions of various tissues, particularly the liver.

Molluscides : Chemicals used to kill molluscs, like slugs and snails, which cause serious damage to crops, especially potatoes.

Mono-oxygenase : Each system containing cytochrome P_{450} is called Mono-oxygenase and many Mono-oxygenases together constitute MFO system.

Mutation : All abnormal genetic inheritance processes including abnormal somatic cell division.

Mutagenicity : Mutangenic actions occurring at cellular level, i.e., in chromosonal DNA and other cellular and nuclear constituents related with cell division.

Mutagen : A substance the produces mutation.

Natural Organic Insecticides : A number of plant products possess insecticidal properties. These collectively constitute natural organic insecticides.

Necrosis : Death of tissue.

Nematicides : Chemical compounds employed for the eradication of nematodes.

Neoplasia : Formation of tumours.

Nephrotoxic : Toxic to kidney.

No observed adverse effect concentration (NOAEC) : The highest concentration of a material in a toxicity test that has no statistically significant adverse effect on the exposed population of test organisms as compared with the controls. When derived from a life cycle or partial life cycle test, it is numerically the same as the lower limit of the MATC.

Oncogen : Agent that exerts cancerous tumors.

Osteomalacia : Graded softening and bending of bones.

Parts per billion (ppb) : One unit of chemical (usually expressed as mass) per 1,000,000,000 (10^9) units of the medium (e.g., water) or organism (e.g., tissue) in which it is contained.

Parts per million (ppm) : One unit chemical (usually expressed as mass) per 1,000,000 (10^6) units of the medium (e.g., water) or organism (e.g., tissues) in which it is contained.

Parts per trillion (pptr) : One unit chemical (usually expressed as mass) per 1,000,000,000,000 (10^{12}) units of the medium (e.g., water) or organism (e.g., tissue) in which it is contained.

Pest : Any living organism that interferes with human activity in a negative way.

Pesticides : Chemicals and biological materials that are used by man to reduce pest organisms.

Pharmaco-dynamics : Study of drug action.

Pharmacology of insecticides : Study of action of insecticides.

Poison : Any substance which, taken into or formed in the body, destroys life or impairs health.

Potentiation : Increase in toxicity.

Proximal effect : Effects that appear earlier but faster.

Preliminary test : See Screening test.

Quality assurance (QA) : A program organized and designed to provide accurate and precise results. Included are selection of proper technical methods, tests, or laboratory procedures; sample collection and preservation; selection of limits; evaluation of data; quality control and qualifications and training of personnel.

Quality control (QC) : Specific actions required to provide information for the quality assurance program. Included are standardizations, calibration, replicates, and control and check samples suitable for statistical estimates of confidence of the data.

Quantal dose response : Study of a said response in a population against a known dose of toxicant.

Radioactive substances : The fission byproducts formed by spontaneous disintegration of certain heavy metals like radium, uranium, thorium, actinium etc.

Receptors : Molecules in or on cell to which a specific antigen may bind.

Resistance time : The period of time for which an organism can live beyond the incipient lethal level.

Ring-test (or Round-robin test) : (1) A conjoint test conducted under strictly standardized and uniformly applied conditions to assess the precision and accuracy with which different laboratories can determine the toxicity of a chemical or effluent. (2) A test designed to measure statistically the reproducibility of a test method, or to compare the results obtained from the use of different test methods.

Risk : A statistical concept defined as the expected frequency or probability of undesirable effect resulting from a specified exposure to know or potential environmental concentrations of a material. A material is considered safe if the risks associated with its exposure are judged to be acceptable. Estimates of risk may be expressed in absolute or relative terms. Absolute risk is the excess risk due to exposure. Relative risk is the ratio of the risk in exposed population to the risk in the unexposed population.

Rodenticide : Chemicals employed for the control of rodents like rat, mice, rabbit, moles, ground squirrels, etc.

Safe concentration (SC) : Concentration of material to which prolonged exposure will cause no adverse effect.

Safety : The practical certainty that adverse effects or injury will not result from exposure to a material when used in the quantity and the manner proposed for its use. "Practical certainty" must be defined, for example, in terms of a numerically specified low risk. Another view is that safety is a value judgment of the acceptability of risk.

Screening test (Preliminary test or Range-finding test) : (1) A test conducted to estimate the concentrations to be used for a definitive test. (2) A short-term test used early in a testing program to evaluate the potential of a chemical (or other material) to produce some selected adverse effect (e.g., mortality).

Solvent or Carrier : An agent (other than water) in which the test chemical is mixed to make it miscible with dilution water before distribution to test chambers. Solvents or carriers are used in tests where the concentrations of the test chemical are extremely low and a very small amount of test material must be added to the test chambers.

Static system : An exposure system for aquatic toxicity tests in which the test chambers contain still solutions of the test material or control water.

Static Air Flow System : In this system the test animal is kept in a closed chamber, usually an enclosed glass container with small opening for outlet and inlet, and exposed to one or several spray or aerosol applications.

Stomach Poisons : Chemicals lethal to insects on ingestion and are generally applied against chewing insects.

Sublethal : Below the concentration that directly causes death. Exposure to sublethal concentrations of a material may produce less obvious effects on behaviour, biochemical and/or physiological function, and histology of organisms.

Survival time : The time interval between initial exposure of an organism to a harmful chemical and death.

Synergism : A phenomenon in which the combined effect of two compounds is more than the sum of the individual effects.

Systemic Poisons : Chemicals which, when applied to seeds, roots, stems, or leaves of plants, are absorbed and translocated to the various plant parts in amounts lethal to insects feeding thereon.

Tachyphylaxis : Development of tolerance to a toxicant after repeated administration.

Teratogenicity : Somatic mutations occurring at embryonic stage are teratogenic effects.

Teratogen : Substance causing teratogenicity.

Teratology : The study of teratogens and their actions.

Test material : A chemical, formulation, effluent, sludge, or other agent or substance that is under investigation in toxicity test.

Test solution (or Test treatment) : Medium containing the material to be tested to which the test organisms are exposed. Different test solutions contain different concentrations of the test material.

Threshold : The minimum dose that evokes response.

Threshold concentration : A concentration above which some effect (or response) will be produced and below which it will not.

Time-independency (TI) test : An acute toxicity test with no predetermined temporal end point. This type of test, sometimes referred to as a "threshold" or "incipient" lethality test, is allowed to continue until acute toxicity (mortality or a defined sublethal effect) has ceased or nearly ceased and the toxicity curve (plot of effect versus time of exposure) indicates a threshold or incipient concentration. With most test materials, this point is reached within 7-10d, but it may not be reached within 21d. Practical or economic reasons may dictate that the test be stopped at this point and a test be designed for longer duration.

Toxic : Harmful (deleterious).

Toxic unit : The strength of a chemical (measured in some unit) expressed as a fraction or proportion of its lethal threshold concentration (measured in the same unit). The strength may be calculated as :

$$\text{Toxic unit} = \frac{\text{actual concentration of chemical in solution}}{\text{lethal threshold concentration}}$$

If this number is greater than 1.0, more than half of a group of organisms will be killed by the chemical. If it is less than 1.0, half the organisms will not be killed. 1.0 toxic unit = the incipient LC_{50}.

Toxicant : An agent or material capable of producing an adverse response (effect) in a biological system, seriously injuring structure or function or producing death.

Toxicity : The inherent potential or capacity of a material to cause adverse effects in a living organism.

Toxicity curve : The curve obtained by plotting the median survival times of a population of test organisms against concentration on a logarithmic scale.

Toxicity test : The means by which the toxicity of a chemical or other test material is determined. A toxicity test is used to measure the degree of response produced by exposure to a specific level of stimulus (or concentration of chemical).

Toxicology : It is that branch of medical science that deals with the nature, properties, effect and the detection of poisons.

Tumors : A new growth of cells in the body without inflammation, a morbid swelling or enlargement.

Tumorigenicity : Uncontrolled somatic cell division including formation of benign tumors is Tumorigenesis.

Ultimate median tolerance limit : The concentration of a chemical at which acute toxicity ceases. Also called incipient lethal level, lethal threshold concentration, and a asymptotic LC_{50}.

Uptake : The absorption of a chemical into extracellular fluid for systemic circulation. Its fate will be governed by metabolic processes.

Xenobiotics : Organic compounds other than fat, carbohydrate, protein, vitamin, steroid, or mineral, are known as foreign organic compounds or Xenobiotics.

Selected Readings

1. Albert, A. 1960. Selective Toxicity, Wiley, New York.
2. Ariens, E. J., Simonis, A. M. and Offermerier, J. 1976 Introduction to General Toxicology.
3. Boyland, E. and Goulding, R. 1968. Modern Trends in Toxicology, Butterworths, London.
4. Butter, G. C. (Ed) 1978. Principles of Ecotoxicology. SCOPE 12, ICSU-SCOPE, John Wiley & Sons, New York.
5. Carsons, R. 1962 Silent Spring. Houghtan, Mifflin, Boston.
6. Casarett, L. J. and Doull, J. 1980. Toxicology, A Basic Science of Poisons, 2nd ed. The Macmillian Co., New York.
7. Duffs, J. H. and Worth Howard, G. J. (1996). Fundamental Toxicology for Chemicals. Royal Society of Chemistry, Cambridge (U. K.).
8. Fairhall, L. T. 1969. Industrial Toxicology, Hafner Publishing Co., New York.
9. Frank C. Lu (1985). Basic Toxicology (Fundamentals, Target Organs and Risk Assessment). Hemisphere Publishing Corporation, Washington.
10. GESAMP (Joint Group of Experts on the Scientific Aspects of Marine Pollution). 1980. Report, 11th Session, Dubrovnik, Yugoslavia.
11. Gupta, P. K. and Salunkhe, D. K. 1985. Modern Toxicology. Vols I, II, III, Metropolitan Book Co. Pvt. Ltd., New Delhi.
12. Kamrin, M. A. (1997). Pesticide Profiles. Lewis Publishers.
13. Loomis, T. A. 1968 Essentials of Toxicology, Lea & Febiger, Philadelphia.
14. Matsumuram F. (1976). Toxicology of Insecticides. Plenum Press. N. Y. and London.
15. Moure, R. (1975). Man in the Environment. Bronzoi Book N. Y.
16. Mehlman, M. A. (Ed) 1976. Advances in Modern Toxicology. (series) USA. Hemisphere Publishing Crop.
17. Moriarity, F. 1983. Ecotoxicology. The Study of Pollutants in Ecosystem, Academic Press, London.
18. Omkar. Concept of Toxicology. S. L. Nagin Chand & Co. Jalandhar (India).
19. Paget, G. E. 1970. Methods in Toxicology. Blackwell Scientific Publ., Oxford.
20. Satoskar, R. S. and Bhandarkar, S. D. : Pharmacology and Pharmacotherapeutics. Popular Prakashan, Mumbai (India).
21. Pandey K. & J. P. Shukla. 1982. Deleterious effects of Arsenic on the growth of the fingerlings of a tropical fresh-water fish Colisa fasciatus. Actapharmacol et. Toxicol. 50, 398-400.
22. Pandey K. & J. P. Shukla. 1982. Hydrobiological parameters as an indicator of water pollution in Chilwa lake, Gorakhpur. Actahydrochim Hydrobiol. 10, 113-116.
23. Pandey K. & J. P. Shukla. 1983. Acute toxicity of carbamitde and its long term effects on the growth of fingerlings of Colisa fasciatus. Actahydrochim Hydrobiol. 11, 145-149.
24. Shukla J. P. & Pandey K. 1984. Impaired ovarian function in Arsenic treated fish, Colisa fasciatus. Toxicol. Lett. 21, 1-3.
25. Shukla J. P. & Pandey K. 1984. Altered Nucleic Acids Metabolism in the ovary exposed to Arsenic stress in Colisa Fasciatus. Actahydrochim Hydrobiol. 12, 277-279.
26. Shukla J. P. & Pandey K. 1984. Arsenic induced structural changes during the ovarian cycle of Colisa fasciatus. Bull. Inst. Zool. Acad. Sinica. 24, 89-94.
27. Shukla J. P. & Pandey K. 1985. Impaired spermatogenesis in Arsenic treated fresh-water fish, Colisa fasciatus. Toxicol. Left. 21, 191-195.

28. Shukla J. P. & Pandey K. 1984. Arsenic Induced cellular & biochemical alteration in testicular cycle of fresh-water fish, Colisa fasciatus. Cellular & Mol Biology. 30, 227-231.

29. Shukla J. P. & Pandey K 1984. Carbamide induced histological alterations during different phases of the testicular cycle of Colisa fasciatus. Bull. Inst. Zool. Acad Sinica. 23, 181-186.

30. Shukla J. P. & Pandey K. 1985. Toxicity and long term effects of Arsenic on the gonadal protein metabolism in a tropical fresh-water fish, Colisa fasciatus. Actahydrochim Hydrobiol. 13, 127-131.

31. Shukla J. P. 1985. Deleterious effects of Arsenic on the Nucleic Acids and Protein Metabolism in the liver of H. fossilis. Actaphydrochim Hydrobiol. 13, 611-615.

32. Shukla J. P. & Pandey K. 1986. Zinc induced changes in the Nucleic Acids and Protein Metabolism in the fingerlings of a fresh-water murrel, Channa Punctatus. Acata Hydrochim Hydrobiol. 14(2), 195-197.

33. Shukla J. P. & Pandey K. 1987. Deleterious effects of Malathion on survivality and growth of the fingerlings of Channa Punctatus, a freshwater murrel. Actahydrochim Hydrobiol. 15(6), 653-657.

34. Shukla J. P. & Pandey K. 1987. Survivility and impaired growth in Arsenic treated fingerlings of Channa Punctatus, a fresh-water murrel. Actahydrochim Hydrobiol. 15(3), 307-311.

35. Shukla J. P. 1988. Malathion induced biochemical alterations in the liver of fingerlings of Channa Punctatus. Actahydrochim Hydrobiol. 6, 621-624.

36. Shukla J. P. & Pandey K. 1987. Survivility and impaired growth in Arsenic treated fingerlings of Channa Punctatus, a fresh-water murrel. Actahydrochim Hydrobiol. 15(3), 307-311.

37. Shukla J. P. & Pandey K. 1988. Toxicity and long term effects of a sublethal concentration of Cadmium on the growth of the fingerlings of Channa Punctatus, a fresh-water murrel. Actahydrochim Hydrobiol. 16, 537-540.

38. Pandey K. & J. P. Shukla 1979. Arsenic toxicity in a tropical fresh-water fish, Puntius sophore. Nat. Acad. Sci. Lett. 2, 425-426.

39. Pandey K. & J. P. Shukla 1979. Studies on the toxicity of Ammonia to a fresh-water fish, Channa Punctatus. J. Env. Res. 3, 11-12.

40. Pandey K. & J. P. Shukla 1982. Toxicity of Arsenic to a fresh-water fish Channa Punctatus. Env. Ind. 2, 63-64.

41. Pandey K. & J. P. Shukla 1980. Studies on the toxicity of urea to a fresh-water fish, Colisa fasciatus. Env. Ind. 3, 73-74.

42. Rombke, J. and Moltmann, J. F. (1995). Applied Ecotoxicology. Lewish Publishers.

43. Sharma, P. D. (2000). Toxicology. Rastogi Publications, Meerut, U. P.

44. Trainter, D. C. 1966. A Handbook of Industrial Toxicology. Angus & Robertson Ltd., Sydney.

45. Truhaut, R. 1974. Ecotoxicology — A New Branch of Toxicology. In Ecological Toxicology Research (Eds. A. D. McIntyre and C. F. Mills) pp. 3-24, Plenum Press. New York.